PHYSICOCHEMICAL ASPECTS OF FOOD ENGINEERING AND PROCESSING

Contemporary Food Engineering

Series Editor

Professor Da-Wen Sun, Director

Food Refrigeration & Computerized Food Technology
National University of Ireland, Dublin
(University College Dublin)
Dublin, Ireland
http://www.ucd.ie/sun/

Physicochemical Aspects of Food Engineering and Processing, *edited by Sakamon Devahastin* (2010)

Infrared Heating for Food and Agricultural Processing, *edited by Zhongli Pan and Griffiths Gregory Atungulu* (2010)

Mathematical Modeling of Food Processing, *edited by Mohammed M. Farid* (2009)

Engineering Aspects of Milk and Dairy Products, *edited by Jane Sélia dos Reis Coimbra and José A. Teixeira* (2009)

Innovation in Food Engineering: New Techniques and Products, *edited by Maria Laura Passos and Claudio P. Ribeiro* (2009)

Processing Effects on Safety and Quality of Foods, *edited by Enrique Ortega-Rivas* (2009)

Engineering Aspects of Thermal Food Processing, *edited by Ricardo Simpson* (2009)

Ultraviolet Light in Food Technology: Principles and Applications, *Tatiana N. Koutchma, Larry J. Forney, and Carmen I. Moraru* (2009)

Advances in Deep-Fat Frying of Foods, *edited by Serpil Sahin and Servet Gülüm Sumnu* (2009)

Extracting Bioactive Compounds for Food Products: Theory and Applications, *edited by M. Angela A. Meireles* (2009)

Advances in Food Dehydration, *edited by Cristina Ratti* (2009)

Optimization in Food Engineering, *edited by Ferruh Erdoğdu* (2009)

Optical Monitoring of Fresh and Processed Agricultural Crops, *edited by Manuela Zude* (2009)

Food Engineering Aspects of Baking Sweet Goods, *edited by Servet Gülüm Sumnu and Serpil Sahin* (2008)

Computational Fluid Dynamics in Food Processing, *edited by Da-Wen Sun* (2007)

Contemporary Food Engineering Series

Da-Wen Sun, Series Editor

PHYSICOCHEMICAL ASPECTS OF FOOD ENGINEERING AND PROCESSING

Edited by

SAKAMON DEVAHASTIN

CRC Press
Taylor & Francis Group
Boca Raton London New York

CRC Press is an imprint of the
Taylor & Francis Group, an **informa** business

CRC Press
Taylor & Francis Group
6000 Broken Sound Parkway NW, Suite 300
Boca Raton, FL 33487-2742

First issued in paperback 2019

ISBN-13: 978-1-4200-8241-8 (hbk)
ISBN-13: 978-0-367-38373-2 (pbk)

Library of Congress Cataloging-in-Publication Data

Physicochemical aspects of food engineering and processing / edited by Sakamon Devahastin.
p. cm. -- (Contemporary food engineering)
Includes bibliographical references and index.
ISBN 978-1-4200-8241-8
1. Food--Analysis. 2. Food--Composition. 3. Food--Microbiology. I. Devahastin, Sakamon, 1974- II. Title. III. Series.

TX541.P487 2010
664'.07--dc22 2010003954

Visit the Taylor & Francis Web site at
http://www.taylorandfrancis.com

and the CRC Press Web site at
http://www.crcpress.com

To my parents for their endless love

To Peung for her extraordinary patience

Contents

Series Preface

Food engineering is the multidisciplinary field of applied physical sciences combined with the knowledge of product properties. Food engineers provide the technological knowledge transfer essential to the cost-effective production and commercialization of food products and services. In particular, food engineers develop and design processes and equipment in order to convert raw agricultural materials and ingredients into safe, convenient, and nutritious consumer food products. However, food engineering topics are continuously undergoing changes to meet diverse consumer demands, and the subject is being rapidly developed to reflect market needs.

In the development of food engineering, one of the many challenges is to employ modern tools and knowledge, such as computational materials science and nanotechnology, to develop new products and processes. Simultaneously, improving quality, safety, and security remain critical issues in the study of food engineering. New packaging materials and techniques are being developed to provide more protection to foods, and novel preservation technologies are emerging to enhance food security and defense. Additionally, process control and automation regularly appear among the top priorities identified in food engineering. Advanced monitoring and control systems are developed to facilitate automation and flexible food manufacturing. Furthermore, energy saving and minimization of environmental problems continue to be important issues in food engineering, and significant progress is being made in waste management, efficient utilization of energy, and reduction of effluents and emissions in food production.

The *Contemporary Food Engineering* book series, which consists of edited books, attempts to address some of the recent developments in food engineering. Advances in classical unit operations in engineering related to food manufacturing are covered as well as such topics as progress in the transport and storage of liquid and solid foods; heating, chilling, and freezing of foods; mass transfer in foods; chemical and biochemical aspects of food engineering and the use of kinetic analysis; dehydration, thermal processing, nonthermal processing, extrusion, liquid food concentration, membrane processes, and applications of membranes in food processing; shelf life, electronic indicators in inventory management, and sustainable technologies in food processing; and packaging, cleaning, and sanitation. These books are aimed at professional food scientists, academics researching food engineering problems, and graduate-level students.

The editors of these books are leading engineers and scientists from all parts of the world. All of them were asked to present their books in such a manner as to address the market needs and pinpoint the cutting-edge technologies in food engineering. Furthermore, all contributions are written by internationally renowned experts who have both academic and professional credentials. All authors have attempted to provide critical, comprehensive, and readily accessible information on the art and science of a relevant topic in each chapter, with reference lists for further information. Therefore, each book can serve as an essential reference source to students and researchers in universities and research institutions.

Da-Wen Sun
Series Editor

Series Editor

Born in southern China, Professor Da-Wen Sun is a world authority on food engineering research and education, he is a member of the Royal Irish Academy which is the highest academic honor in Ireland. His main research activities include cooling, drying, and refrigeration processes and systems; quality and safety of food products; bioprocess simulation and optimization; and computer vision technology. Especially, his innovative studies on vacuum cooling of cooked meats, pizza quality inspection by computer vision, and edible films for shelf-life extension of fruits and vegetables have been widely reported in national and international media. The results of his work have been published in over 200 peer-reviewed journal papers and more than 200 conference papers.

Sun received his BSc honors (first class), his MSc in mechanical engineering, and his PhD in chemical engineering in China before working in various universities in Europe. He became the first Chinese national to be permanently employed in an Irish university when he was appointed as college lecturer at the National University of Ireland, Dublin (University College Dublin), in 1995, and was then continously promoted in the shortest possible time to senior lecturer, associate professor, and full professor. He is currently the professor of Food and Biosystems Engineering and the director of the Food Refrigeration and Computerized Food Technology Research Group at University College Dublin.

Sun has contributed significantly to the field of food engineering as a leading educator in this field. He has trained many PhD students who have made their own contributions to the industry and academia. He has also regularly given lectures on advances in food engineering in international academic institutions and delivered keynote speeches at international conferences. As a recognized authority in food engineering, he has been conferred adjunct/visiting/consulting professorships from 10 top universities in China including Zhejiang University, Shanghai Jiaotong University, Harbin Institute of Technology, China Agricultural University, South China University of Technology, and Jiangnan University. In recognition of his significant contribution to food engineering worldwide and for his outstanding leadership in this field, the International Commission of Agricultural Engineering (CIGR) awarded him the CIGR Merit Award in 2000 and again in 2006. The Institution of Mechanical Engineers (IMechE) based in the United Kingdom named him Food Engineer of the Year 2004. In 2008, he was awarded the CIGR Recognition Award in honor of his distinguished achievements in the top 1% of agricultural engineering scientists in the world.

Sun is a fellow of the Institution of Agricultural Engineers and a fellow of Engineers Ireland (the Institution of Engineers of Ireland). He has received numerous awards for teaching and research excellence, including the President's Research

Fellowship and the President's Research Award of University College Dublin on two occasions. He is a member of the CIGR Executive Board and an honorary vice-president of CIGR; the editor-in-chief of *Food and Bioprocess Technology—An International Journal* (Springer); former editor of the *Journal of Food Engineering* (Elsevier); and an editorial board member for the *Journal of Food Engineering* (Elsevier), the *Journal of Food Process Engineering* (Blackwell), *Sensing and Instrumentation for Food Quality and Safety* (Springer), and *Czech Journal of Food Sciences*. He is also a chartered engineer.

Preface

Although a number of books have been written on food engineering and processing, there is still a need for up-to-date information and critical evaluation of advances in this rapidly growing field. A comprehensive source of information on the effects of various processes on the physical, chemical, as well as biochemical properties of diverse food products is still lacking. A gap does exist between books focusing on the engineering aspect and those focusing on the scientific aspect of food processing. This book is designed to bridge this gap by providing an in-depth analysis of the essential physicochemical aspects of food engineering and processing.

The book is divided into two major parts that cover the effects of selected food processing technologies on the physicochemical properties of selected food products. In Part I, some important processing technologies, i.e., microencapsulation, frying, microwave-assisted thermal processing, high-pressure processing, pulsed electric field processing, and freezing, and their effects on the physicochemical properties of various food products are reviewed critically. A chapter is also included in this part (Chapter 4) on the effects of various processing technologies on microbial growth and inactivation. In Part II, the focus is on products. Multiphase food systems made of proteins, seafoods, red meats, and pet foods, and the physicochemical changes upon processing are reviewed. Links are provided to connect various chapters since, in many cases, the information is necessarily presented in different chapters. With this scope in mind, it is hoped that the reader will have a complete overview of the physicochemical aspects of food engineering and processing from equally important angles.

The chapters in this book have been contributed by 20 authors from 6 countries who are experts in the physicochemical aspects of food engineering and processing. Whenever possible, the most recent advances on the topics have been included and discussed. The book should thus be useful for both graduate students and academic researchers engaged in food processing research, as well as for industrial researchers seeking up-to-date and new information on the increasingly important combination of the two aspects of food research and development.

I would like to express my sincere gratitude to all the contributing authors for answering my calls and for recognizing the need for the book leading to their contributions. I would also like to thank Taylor & Francis and its staff for their great support and patience during the long production process of the book. Last but not the least, I thank my parents and Peung for their support and encouragement during the entire process.

Although we have tried to make the book as error free as possible, some flaws might still exist, but they all remain mine.

Sakamon Devahastin
Bangkok, Thailand

Editor

Dr. Sakamon Devahastin is currently an associate professor in the Department of Food Engineering at King Mongkut's University of Technology Thonburi in Bangkok, Thailand. His main research interests are in drying of foods and biomaterials, from the development of novel drying technologies for heat- and oxygen-sensitive materials to the study of their physicochemical changes during drying, and also in particulate systems and computational fluid dynamics and heat transfer. He has so far published more than 75 papers in refereed international journals and given some 50 presentations at various international conferences. He is an author/coauthor of seven book chapters and the editor of a book on drying. He has also served as an associate editor of an archival journal, *Drying Technology* (published by Taylor & Francis), and been on the editorial boards of various other journals in chemical and food engineering. Among many awards bestowed upon him, Dr. Devahastin was awarded the Young Technologist Award by the Foundation for the Promotion of Science and Technology under the Patronage of H.M. the King of Thailand in 2004; the TRF-CHE-Scopus Researcher Award in Engineering and Multidisciplinary Category by the Thailand Research Fund, Commission on Higher Education, and Scopus; as well as the Taylor & Francis Award for Sustained Exemplary Service to *Drying Technology* and Excellence in Drying Research Contributions by Taylor & Francis, both in 2009. Dr. Devahastin received his PhD in chemical engineering from McGill University, Montreal, Canada, in 2001.

Contributors

Akinbode A. Adedeji
Department of Bioresource Engineering
McGill University
Ste-Anne-de-Bellevue, Québec, Canada

Malek Amiali
Ecole Nationale Supérieur
Agronomique d'Alger
Algiers, Algeria

Soottawat Benjakul
Department of Food Technology
Prince of Songkla University
Songkhla, Thailand

Naphaporn Chiewchan
Department of Food Engineering
King Mongkut's University of
Technology Thonburi
Bangkok, Thailand

Adriana E. Delgado
Agriculture and Food Science Centre
University College Dublin
Dublin, Ireland

Sakamon Devahastin
Department of Food Engineering
King Mongkut's University of
Technology Thonburi
Bangkok, Thailand

Takeshi Furuta
Department of Biotechnology
Tottori University
Tottori, Japan

Parichat Hongsprabhas
Department of Food Science and
Technology
Faculty of Agro-Industry
Kasetsart University
Bangkok, Thailand

Arun Muthukumaran
Department of Bioresource Engineering
McGill University
Ste-Anne-de-Bellevue, Québec, Canada

Tze Loon Neoh
Department of Applied Biological Science
Kagawa University
Kagawa, Japan

Michael O. Ngadi
Department of Bioresource Engineering
McGill University
Ste-Anne-de-Bellevue, Québec, Canada

Chalida Niamnuy
Faculty of Engineering
Mahasarakham University
Mahasarakham, Thailand

Valérie Orsat
Department of Bioresource Engineering
McGill University
Ste-Anne-de-Bellevue, Québec, Canada

G. S. Vijaya Raghavan
Department of Bioresource Engineering
McGill University
Ste-Anne-de-Bellevue, Québec, Canada

Navin K. Rastogi
Department of Food Engineering
Central Food Technological Research Institute
Council of Scientific and Industrial Research
Mysore, Karnataka, India

Apinan Soottitantawat
Department of Chemical Engineering
Faculty of Engineering
Center of Excellence in Particle Technology
Chulalongkorn University
Bangkok, Thailand

Da-Wen Sun
Agriculture and Food Science Centre
University College Dublin
Dublin, Ireland

Bethany Uttaro
Agriculture and Agri-Food Canada
Lacombe, Alberta, Canada

Wonnop Visessanguan
National Center for Genetic Engineering and Biotechnology
National Science and Technology Development Agency
Pathum Thani, Thailand

Hidefumi Yoshii
Department of Applied Biological Science
Kagawa University
Kagawa, Japan

Part I

Physicochemical Changes of Foods: A Focus on Processes

1 Effect of Microencapsulation on Food Flavors and Their Releases

Takeshi Furuta, Apinan Soottitantawat, Tze Loon Neoh, and Hidefumi Yoshii

CONTENTS

1.1 INTRODUCTION

Microencapsulation technology has undergone remarkable development during the past three decades. As the applications of microencapsulation diversify into various fields, a number of products that rely on microencapsulation to provide their respective unique attributes have emerged. Within the food industry, microencapsulation generally refers to an action of capturing a core material in a shell or coating to provide protection against degradative reactions and/or to prevent losses of flavors. In addition, microencapsulation is also used to control the release of flavors during food processing and storage. Microencapsulated products can be divided into five main categories: flavorings, vitamins and minerals, oils and fats (such as ω-3s and ω-6s), herbs and bioactives, and other food ingredients (Barobosa-Canovas et al., 2005; Reineccius, 2005).

Microencapsulation of flavors is a technology of enclosing flavor compounds (core materials) in a carrier matrix. An amorphous or metastable solid is normally used as a carrier matrix. Microencapsulation is useful for improving the chemical stability of flavor compounds, providing controlled release of flavor compounds from microencapsulated flavor products, providing a free-flowing powder with improved handling properties and physical protection of volatile properties of flavor.

Various encapsulation methods of flavors have been proposed. Among them, spray drying is the most common technique to produce flavor powders from food flavor emulsion (Reineccius, 1988). After the spray drying process, food powder at low moisture content may exist in a glassy state at room temperature. In this glassy state, flavor release and stability of flavors depend on a kinetic rather than a thermodynamic mechanism; the kinetics can be predicted through the knowledge of the glass transition temperature of carrier materials. However, plasticization by water adsorption under high humidity condition may cause a reduction in the glass transition temperature to below room temperature, leading to a structural change in wall materials, resulting in flavor release from the rubbery carrier matrices (Ubbink and Schoonman, 2003).

Cyclodextrins (CDs) are commonly used to improve the stability of flavor via the encapsulation of certain specific ingredients that naturally exist in food materials. The method is often called "molecular encapsulation" because the flavor ingredients are encapsulated in the molecular cavity of CDs. CDs form inclusion complexes with a variety of molecules including flavors, fats, and colors. Most natural and artificial flavors are volatile oils or liquids, and the complexation with CDs provides a promising alternative to the conventional encapsulation technologies for flavor protection.

1.2 ENCAPSULATION TECHNOLOGIES FOR FOOD FLAVORS

A significant objective of flavor encapsulation is the improvement of stability during storage of flavors and the modification of the release characteristics of encapsulated flavors during processing as well as during consumption. The encapsulation of flavors in a powdery form decreases the water activity of the powder, resulting in lowered bioavailability of water inside the matrix of the powder. In addition, the reactivity and mobility of the flavor decrease, so adverse effects such as evaporation and oxidation of flavors could be avoided. From the viewpoint of the powder properties, lowering of the moisture content implies an increase in the glass transition temperature of the powder matrix. It is noted that the mobility and chemical reactivity of the flavors inside the matrix increase exponentially with the moisture content and temperature, thus the need to control the glass transition temperature of the matrix (Ubbink and Schoonman, 2003).

Various techniques have been used to transform liquid food flavors into the powdery form (Shahidi and Han, 1993; King AH, 1995; Gibbs et al., 1999). Traditional, but commercially available processes include spray drying, freeze drying, fluidized bed drying, extrusion, molecular encapsulation, and coacervation. The former four methods are sometimes called "glass encapsulation," since all these techniques involve the use of water-soluble carbohydrates and the encapsulation principle is based on encapsulation in the amorphous carbohydrate glasses (Ubbink and Schoonman, 2003). Molecular encapsulation, on the other hand, primarily uses CDs as an encapsulant; flavors are entrapped in the molecular cavity as an inclusion complex. Recently, more sophisticated techniques have been developed, such as the use of crystal transformation and adsorption in porous carbohydrates. Among these techniques, spray drying is the most widely used technique to produce particulate dried powder, because of its relatively simple and continuous operating system.

1.3 ENCAPSULATION OF FLAVORS BY SPRAY DRYING

1.3.1 Principle and Outline of Spray Drying System

Spray drying is by definition a transformation of a feed of liquid or paste material (solution, dispersion, or paste) into a dried particulate powder. The liquid feed is atomized into tiny droplets of several dozens to a few hundred microns in size, which are then dried into the powder by contacting with a hot drying medium

TABLE 1.1
Typical Spray-Dried Products

Type of Material	Ingredients
Flavoring agents	Oil, spices, seasonings, sweeteners
Vitamins	Vitamin E, β-carotene, ascorbic acid
Minerals	Calcium, magnesium, and phosphorus
Oils and fats	Fish oils (DHA, EPA), sea buckthorn oil, lycopene
Herbs and bioactives	Creatine, probiotic bacteria
Others	Enzymes, leavening agents, psyllium, yeast

(Masters, 1991). The main advantages of spray drying over other drying methods are rapid drying; drying can be completed within a very short time, between 5 and 30 s. Spray drying can produce a powdery spherical product directly from a solution or dispersion. Therefore, spray drying is widely applied to the manufacturing of powdery products; typical spray-dried food products are listed in Table 1.1. Particularly, more than 80% of the production of food flavor powders is by spray drying.

A typical spray dryer commonly consists of a feed pump, an atomizer, an air heater, an air dispenser, a drying chamber, and systems for exhaust air cleaning and powder recovery, as shown in Figure 1.1. A pressure spray nozzle or a centrifugal-disc-type atomizer is most commonly employed for the formation of fine sprayed droplets. Atomization is the key technology in spray drying, which determines the droplet size distribution and powder sizes. A feed liquid is atomized to the greatest possible degree in order to increase its surface area. For instance, to process 1 cm^3 of a solution into fine droplets of 100 μm or 1 μm diameter, the surface area increases 100- or 10,000-fold, respectively. Usually, the atomized spray has a droplet diameter of 10–200 μm, and the typical drying time is 5–30 s (Furuta et al., 1994a). A centrifugal rotary disc atomizer is favorably applied to a highly viscous liquid and slurry feed, and has an advantage of easy control of the droplet diameter by varying the rotational speed of the disk. A two-fluid nozzle is an alternative atomizer for a highly viscous fluid and is highlighted as a candidate for the production of nanosized particles.

1.3.2 Theory and Mechanism of Flavor Encapsulation by Spray Drying

In the case of spray drying of liquid food, the content of flavors in a product powder is often an important quality factor of the product powder. Since high-temperature air is generally used as a drying medium, one would expect that a large extent of flavors could be lost during the drying process. In practice, although over 90% of water is evaporated, most volatile flavors are expectedly retained when optimum drying conditions are followed. The following question may arise: "How can we get flavor compounds in the powder although most of these compounds are highly volatile with respect to water and could be easily lost during drying?" Reasons for good retention of volatiles have been the subject of substantial research in the 1970s.

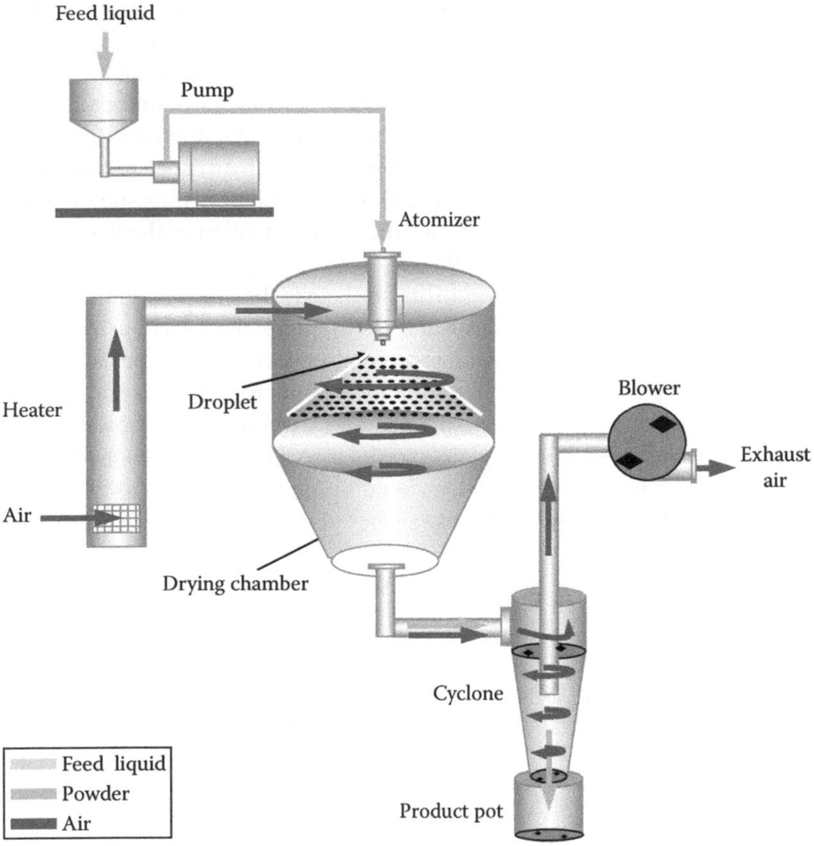

FIGURE 1.1 Typical spray drying system.

1.3.2.1 Microencapsulation of Hydrophilic Flavors by Spray Drying

There are two kinds of food liquid flavors, hydrophilic and hydrophobic flavors. The mechanisms of flavor encapsulation, which are at work during drying processes, depend on the type of flavors. For hydrophilic flavors like alcohol and acetone, the accepted explanation for flavor retention is given by the "selective diffusion theory" (Thijssen and Rulkens, 1968; Coumans et al., 1994; King CJ, 1995). The actual retention of the volatile compounds in spray-dried foods is, however, considerably higher than what would be predicted from equilibrium considerations alone (Bomben et al., 1973). In food materials containing sugars and/or polymers, the reduction of water content increases the glass transition temperature and the resulting amorphous matrix is impermeable to organic compounds. Permeability to water, however, remains finite. As noted by Thjiissen and Rulkens (1968), the water concentration at the droplet surface decreases as drying progresses; the diffusivities of the flavor components decrease by several orders of magnitude, more sharply than that of water. While water continues to permeate at a significant rate through the amorphous film, the flavors diffuse into the film at a negligible rate. Therefore, this dry surface acts as a semipermeable membrane permitting the continued loss of water (evaporation), but efficiently retaining (or trapping) organic flavor molecules. The result of this phenomenon is that the diffusion coefficients of other substances become much less than that of water above some dissolved solids content. The selective diffusion theory has been well formulated mathematically and can be solved numerically. On the basis of the numerical results, flavor retention during spray drying increases with an increase in the solid content of the feed, inlet gas temperature, and inlet gas flow rate, and a decrease in the inlet gas humidity, as shown in Figure 1.2. All of these conditions would enhance the early formation of the dry skin on the surface of the droplet (Furuta et al., 1984).

As a counterpart of the selective diffusion theory, the "microregion entrapment theory" was proposed by Flink and Karel (1970). This theory is basically similar to the selective diffusion theory, but has more physicochemical aspects. According to this theory, dehydration results in a formation of microregions containing highly concentrated solutions of flavor compounds and carbohydrates. Lowering of the moisture content in these microregions creates associations of carbohydrate molecules with flavor compounds by hydrogen bonding, decreasing the loss of the volatile compounds. A similar entrapment theory as the "glass state theory" was proposed by

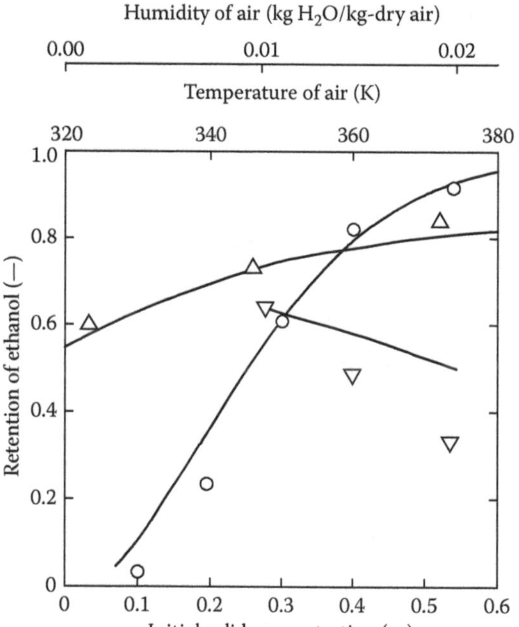

FIGURE 1.2 Retention of ethanol in a single suspended droplet dried in hot air. ○ Retention of ethanol versus initial solid (MD) concentration (air temp. = 323 K, velocity of air = 1.15 m/s, humidity of air = 0.008 kg H_2O/kg-dry air). △ Retention of ethanol versus air temperature (initial solid concentration = 0.3, velocity of air = 1.2 m/s, humidity of air = 0.008 kg H_2O/kg-dry air). ▽ Retention of ethanol versus air humidity (initial solid concentration = 0.3, velocity of air = 1.15 m/s, air temp. = 323 K). Solid curves are the calculated results on the basis of the selective diffusion theory. (From Furuta, T. et al., *Dry. Technol.*, 2, 311, 1984. With permission.)

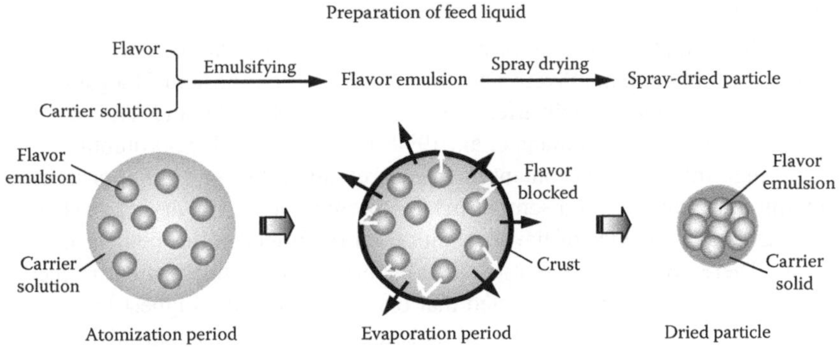

FIGURE 1.3 Spray drying scheme of emulsified flavor solution. Hydrophobic flavor has to be emulsified into an emulsion (O/W [oil-in-water] emulsion).

Franks et al. (1991), in which the stabilization of flavor results from entrapment in a particular conformation region in highly viscous glass carbohydrates.

1.3.2.2 Spray Drying of Emulsified Hydrophobic Flavor

In spray drying of hydrophobic flavors, the flavors are emulsified into an aqueous solution of carrier material, usually a carbohydrate. A mechanism to retain hydrophobic flavors could be explained as in the case of hydrophilic flavors. The emulsion liquid is pumped through a spraying atomizer and sprayed into a high-temperature chamber (Figure 1.3). The selection of a suitable carrier material (carbohydrate) is an important initial step of successful spray drying, since the wall material would influence the emulsion stability before drying as well as the flowability of spray-dried particles and shelf life of the powder. An ideal carrier material should have a good emulsifying property and film-forming characteristics; have low viscosity, even for a highly concentrated solution; exhibit low hygroscopicity; and be able to release the encapsulated flavors at the reconstitution of the product (Ubbink and Schooman, 2003). Some main types of capsule wall materials are carbohydrates (starch, maltodextrins [MDs], corn syrup solids, and CDs), cellulose esters and ethers (carboxymethylcellulose, methylcellulose, and ethylcellulose), gums (gum Arabic [GA, or gum acacia], agar, and sodium alginate), lipids (wax, paraffin, fats, and oils), and proteins (gelatin, soy protein, and whey protein).

1.3.2.3 Factors Affecting Retention of Spray-Dried Emulsified Hydrophobic Flavors

1.3.2.3.1 Type of Flavor (Solubility of Flavor)

The solubility of flavor ingredients influences the retention of flavors. Retentions of several esters in GA encapsulation were studied (Rosenberg et al., 1990). Retentions of partially water-soluble esters, such as ethyl propionate and ethyl butyrate (EB), were less than those of esters having lower solubility (ethyl caproate). A similar trend was also reported by Liu et al. (2001) and Soottitantawat et al. (2003). A higher retention of D-limonene (insoluble flavor) was observed compared to that of EB (moderate solubility).

1.3.2.3.2 Solid Content of Infeed Solution

The most important factor determining the retention of volatiles during drying is the infeed solids content (Sivetz and Foote, 1963; Menting and Hoogstad, 1967; Reineccius and Coulter, 1969; Menting et al., 1970; Rulkens and Thijssen, 1972; Rosenberg et al., 1990; Coumans et al., 1994; Liu et al., 2001; Soottitantawat et al., 2005). A high infeed solids content increases retention during drying, primarily by enhancing the formation of a semipermeable membrane on the drying particle surface. A strong dependence of flavor retention on the infeed solids content is readily apparent. There are two different results concerning the solids content in feed liquid. Some researchers have suggested that one should use the highest infeed solids content possible; an increase of the solids content resulted in a continuous increase in volatiles retention (Leahy et al., 1983; Sankarikutty et al., 1988; Rosenberg et al., 1990; McNamee et al., 2001). On the other hand, other investigators have reported that each encapsulating material has a characteristic optimum concentration for maximum retention (Reineccius and Bangs, 1982; Bhandari et al., 1992; Ré, 1998).

1.3.2.3.3 Emulsion Size

Flavor retention is dependent not only on the solubility of the flavor but also on the emulsion size. For example, the retention of D-limonene was found to decrease with an increase in the emulsion size, as illustrated in Figure 1.4 (Soottitantawat et al., 2003). The increased loss of esters at small emulsion sizes was due to the larger surface areas of the fine emulsions; the increase in surface areas of emulsion droplets would result in an accelerated dissolution of the flavor in the carrier solution, and loss

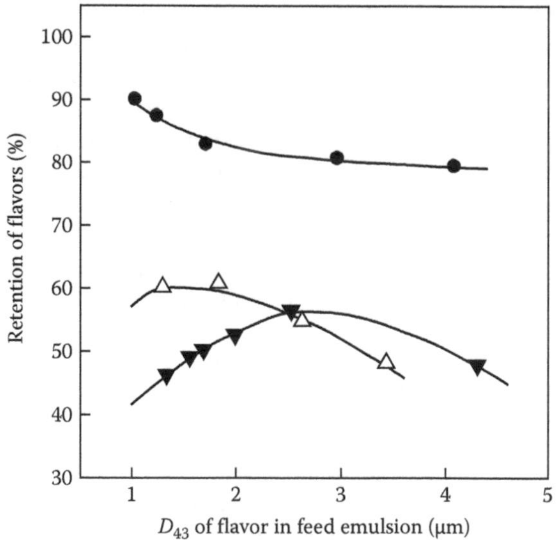

FIGURE 1.4 Effect of emulsion size in feed liquid on retention of D-limonene. D_{43} is the volume-averaged mean diameter of emulsions. Wall materials: blend of GA-MD. ● D-limonene, △ EB, ▼ ethyl propionate. (From Soottitantawat, A. et al., *J. Food Sci.*, 68, 2256, 2003. With permission.)

from its surface during drying (Soottitantawat et al., 2003). More water-soluble volatiles (e.g., EB and ethyl propionate), on the other hand, have an optimum emulsion size for flavor retention.

1.3.2.3.4 Flavor Load

Higher flavor loading generally results in poorer flavor retention (Reineccius and Coulter, 1969; Rulkens and Thijssen, 1972; Rosenberg et al., 1990). This is anticipated since higher loading results in greater proportions of volatiles close to the drying surface, thereby shortening the diffusion length to the particle interface. Since higher volatile loadings result in greater losses, most volatile flavorings are generally dried at carrier solids to a flavor mass ratio of 4 to 1 (Bhandari et al., 1992; Risch, 1995; McNamee et al., 1998; Ré, 1998; Gibbs et al., 1999). This ratio has been reported optimal for encapsulating materials like GA and other carbohydrate derivatives (Reineccius, 1988).

1.3.3 Stickiness of Spray-Dried Powder

The stickiness and subsequent adhesion of particles on the dryer surface during spray drying is one of the serious problems encountered in the spray drying of sugar-rich foods (Bhandari and Howes, 1999; Adhikari et al., 2005). The deposition of powder inside a drying chamber causes unfavorable agglomeration and leads to lower product yields and performance of the dryer. The stickiness of the powder has a significant impact on whether a liquid food can be successfully spray-dried to produce a free-flowing powder. In recent years, it has been well recognized that stickiness is closely related to state changes of carrier materials (Levine and Slade, 1986; Roos, 2003). Such state changes are the secondary ones and are observed at a particular temperature, namely, the glass transition temperature. The glass transition temperature exhibits a strong dependence on the moisture content (Ubbink and Schoonman, 2003); the lower the moisture content, the higher the glass transition temperature.

As mentioned earlier, atomized fine droplets are in contact with hot and dry air, resulting in the sudden formation of a low-moisture carbohydrate membrane, a phenomenon commonly called "casehardening." Since the surface water of the droplets is evaporated so rapidly, low-moisture films form in an amorphous state, which has a higher glass transition temperature than the original liquid. Therefore, the spray drying process can be recognized as the state change from liquid to amorphous solid, which has a higher glass transition temperature. The formation of the glassy particle surface is the basic requirement for successful spray drying, since such a solidified surface does not allow agglomeration between contacting particles or adhesion on the inner surface of the dryer (Roos, 1995). Therefore, the temperature–water content relationship of the particle should be appropriately controlled and adjusted according to the glass transition–water content relationship (Roos, 2003). Figure 1.5 schematically illustrates the glass transition temperature–water content relationship of lactose and also its sticky region. Commonly, the stickiness line is estimated to be 10°C–20°C higher than the T_g line (Bhandari and Howes, 1999). During spray drying, the particle temperature–water content relationship would change from point A to B or C (or C′) along the hypothetical line. To avoid particle agglomeration and

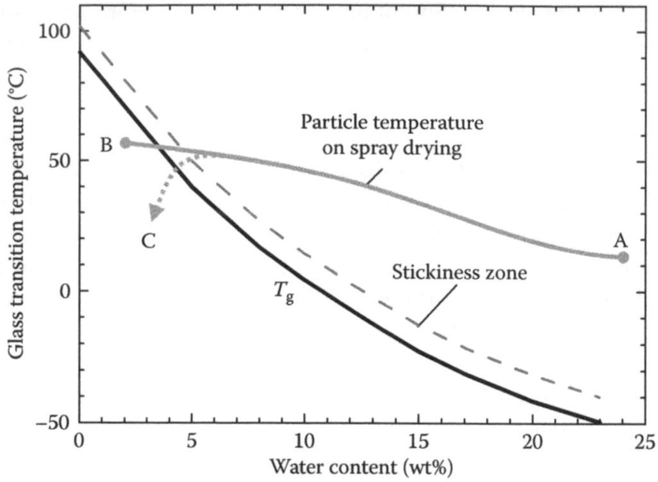

FIGURE 1.5 Hypothetical glass transition temperature–water content relationship of lactose and its sticky region.

adhesion to the dryer wall, the particle moisture content should be lowered so as to increase the particle glass transition temperature. Suppressing the final particle temperature might be an alternative technique to avoid undesirable results (Roos, 2003; Adhikari et al., 2005).

1.4 RELEASE AND OXIDATION OF ENCAPSULATED FLAVORS DURING STORAGE

1.4.1 Implication of Glass Temperature on Storage Stability of Encapsulated Flavors

The primary objectives of the microencapsulation of food liquid flavors or oils are to provide dry and free-flowing powders, provide protection against degradative reactions, and prevent losses of flavors during subsequent processing and storage. Besides, the encapsulation gives the benefits of controlled release of flavors and retardation of oxidation of oils. These functional properties of encapsulated flavors and oils are closely related to the structure of carrier matrix materials. Several physicochemical properties of the carrier matrix (commonly, carbohydrates) are directly or indirectly functions of the glass transition temperature of the amorphous glass. In an amorphous state, the encapsulation carrier matrix is physically stable and not undergoing any significant structural changes. If the temperature and the water content increase, the rheological property of the carrier matrix changes from a glassy, brittle state to a rubbery, viscoelastic state, causing some undesirable structural variations. The glass transition temperature is drastically altered with the water content; higher water content corresponds to lower glass transition temperature. The glass transition temperature also depends on the molecular weight of the carrier materials. A decrease in the average molecular weight of the carrier materials leads to a reduction of the glass transition temperature.

In general, the rotational and vibrational motions are limited in the amorphous glassy state. In the rubbery state, on the other hand, large-scale molecular motion, such as translational motion, is possible (Ubbink and Schoonman, 2003). Therefore, the encapsulated flavors or oils exist stably in the amorphous glassy state, but in the rubbery state some deterioration may take place. Since an amorphous state is not an equilibrium state, a thermodynamic driving force tends to shift the amorphous state to a more stable crystal state, resulting in a time-dependent crystallization, solidification of powders, and caking.

1.4.2 Release of Flavors from Spray-Dried Powder during Storage

1.4.2.1 Mathematical Modeling of Flavor Release

The analysis of flavor release from spray-dried powder is complex and difficult, since the phenomena of flavor release take place with intrinsically overlapping mechanisms, mainly the diffusion of flavors and water as well as the oxidation of flavors. Since the diffusion of molecules in a low-moisture carrier matrix is strongly dependent on the glass transition temperature of the matrix (Roos, 1995), the release and the oxidation of the encapsulated flavors are related to the structural changes of the matrix (collapse and crystallization) (Labrousse et al., 1992). A better understanding of the effect of storage relative humidity (RH) on the properties of capsule matrices would be useful in the quality control and applications of these powders (Whorton and Reineccius, 1995; Gunning et al., 1999; Beristain et al., 2002).

A simple equation has been proposed for the correlation of the release time-course of a spray-dried ethyl-n-butyrate powder during storage (Yoshii et al., 2001):

$$R = \exp[-(k_{\mathrm{R}}t)^n] \quad (1.1)$$

where

R is the retention of flavor in the powder
t is the storage time
k_{R} is the release rate constant
n is a parameter representing the release mechanism

Equation 1.1 is also called the Weibull distribution function, which has been successfully applied to describe the shelf-life failure (Gacular and Kubala, 1975). The release mechanism of flavor can simply be characterized by the value of n in Equation 1.1. A zero-order release occurs when the core ingredient is a pure material (flavor) and releases through the wall of a reservoir microcapsule (carbohydrate, for example) (Karel and Langer, 1988). A half-order release ($n \approx 0.5$) generally occurs when the flavor diffusion inside the carrier matrix primarily governs the release phenomena. A first-order release occurs when the core material is actually a solution. In the case of $n > 1$, the induction period of flavor release can be observed and the burst of flavor emulsion might occur. Note that the classification of the release mechanism is valid only for the release occurring from a given single microcapsule; a mixture of microcapsules would include a distribution of capsules varying in size and wall thickness. Since a spray-dried powder of an emulsion is essentially a mixture of microcapsules with varying properties, the parameter n in Equation 1.1 varies depending on the property of the powder. Equation 1.1 is essentially analogous to the equation of Kohlrausch–Williams–Watt (KWW) as follows (Williams and Watts, 1970):

$$G(t) = G_0 \exp\left[-\left(\frac{t}{\lambda}\right)^{\beta}\right] \tag{1.2}$$

where

- $G(t)$ corresponds to the residual amount of flavor in the powder
- G_0 is the initial content of flavor
- λ is the relaxation time, which corresponds to the inverse of the release rate of flavor
- β is the relaxation constant
- t is time

Equation 1.2 has been originally proposed to express the relaxation phenomena in a polymer. The relaxation constant, β, represents the breadth of energy distribution in the polymer relaxation phenomena. $\beta = 1$ means a simple relaxation, and smaller values of β mean larger breadth of the energy distribution. Since a spray-dried powder consists of various particles having different release characteristics, its total (or overall) release behavior might be the sum of the KWW relaxation equation of an individual particle i:

$$G(t) = \sum_i G_i \exp\left[-\left(\frac{t}{\lambda_i}\right)^{\beta_i}\right] \tag{1.3}$$

The release of flavor from a spray-dried powder during storage is recognized as a kind of relaxation phenomena in an amorphous glass, inside which numerous emulsion droplets of different sizes are distributed. Therefore, it might be necessary to develop an alternative correlation equation of flavor release from a statistical perspective. Considering the distribution of activation energy for the rate constants, the following equation is suggested for the correlation of the complicated time-dependent phenomena (Kawamura et al., 1981):

$$\phi = \frac{C}{C_0} = \frac{RT}{\sqrt{2\pi}\sigma} \int_{-\infty}^{+\infty} \exp\left[-\frac{R^2T^2(\ln k_1 - \ln k_{10})^2}{2\sigma^2}\right] \cdot \exp(-k_1 t)d(\ln k_1),$$

$$\left[k_1 = \frac{\kappa T}{h} \exp\left(-\frac{\Delta G}{RT}\right)\right] \tag{1.4}$$

where

- C is the flavor concentration in spray-dried powder
- C_0 is the initial flavor concentration
- R is the gas constant
- T is the absolute temperature
- t is the time

k_{10} is the rate constant corresponding to the mean activation energy
ΔG is the activation energy of flavor release
σ is the standard deviation if the activation energy, ΔG, is assumed to be represented by the Gaussian distribution function of ln k_1

This equation was originally developed to express the inactivation kinetics of α-chymotrypsin and glucoamylase covalently bound to a water-insoluble support in an aqueous system (Kawamura et al., 1981). Equation 1.4 was successfully applied to express the oxidation kinetics of fish oil (EPA, eicosapentaenoic acid; Yoshii et al., 2003) and linoleic acid powder (Ishidoh et al., 2002, 2003). The above-mentioned three equations are equivalent from the perspective of simulation of flavor release from spray-dried powder. All of the parameters, n in Equation 1.1, β in KWW's equation, and a Gaussian distribution with the standard deviation σ in Equation 1.4, can be viewed as a consequence of the activation energy distribution of the release rate.

1.4.2.2 Release Rate of Flavors under Various Relative Humidities and Temperatures

The release behavior of flavors varies not only with the RH but also with the compositions of the carrier materials. A combination of moisture and temperature strongly influences the degree of plasticization of the carrier materials during storage, thereby governing the rates of flavor losses.

A typical example of the release time-courses of encapsulated flavor in a spray-dried powder is illustrated in Figure 1.6, which reports the residual fraction of D-limonene in the powder stored at 50°C and at 23%, 51%, 75%, and 96% RH. Four kinds of capsule matrices, that is, a blend of GA and MD, soybean soluble polysaccharide (SSPS) and MD, HI-CAP 100® and MD, and HI-CAP 100®, were tested (Soottitantawat et al., 2004a). The RH greatly affected the release rate of D-limonene, but the relationship was not simple. Considering only 23% and 96% RH, the release of D-limonene increased with increasing RH. However, when comparing the results at 51% and 75% RH, the release of D-limonene at 51% RH was higher than that

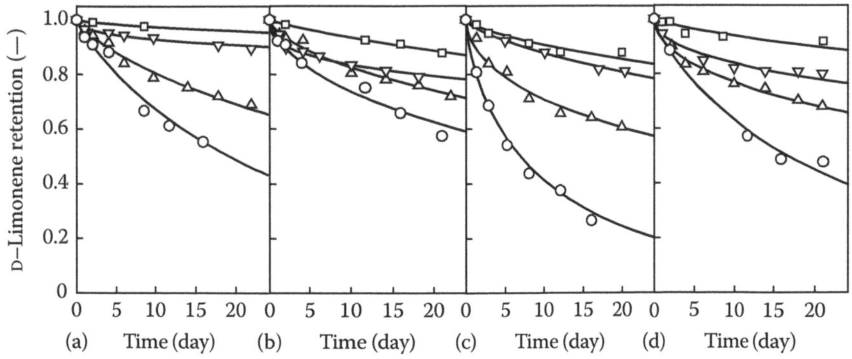

FIGURE 1.6 Release time-courses of encapsulated D-limonene in spray-dried powders stored at various relative humidities at 50°C. (a) GA-MD blend; (b) SSPS-MD blend; (c) HI-CAP 100®-MD blend; (d) HI-CAP 100®. □ 23% RH, □ 51% RH, ▽ 75% RH, ○ 96% RH. (From Soottitantawat, A. et al., *J. Agric. Food Chem.*, 52, 1269, 2004a. With permission.)

observed at 75% RH. These results showed that the release of D-limonene was closely related at least to the water activity of the powder (Levi and Karel, 1995; Yoshii et al., 2001). The loss of D-limonene during storage might be caused by two mechanisms, the diffusion of D-limonene through the matrices of the carrier materials and the oxidation of D-limonene. However, the loss by oxidation was at most 5%–6% of the initial D-limonene content (Soottitantawat et al., 2004a). Therefore, the release of D-limonene might result mainly from diffusion (Whorton and Reineccius, 1995). In case of other flavors, an analogous release time-courses behavior was also noted (Yoshii et al., 2001). To evaluate the release rate constant of D-limonene, Equation 1.1 was applied to the release time-courses of D-limonene; the results are illustrated by the solid curves in Figure 1.6. Equation 1.1 could correlate successfully the release curves of D-limonene; the release rate constant k could be estimated as a function of RH for different carrier solid matrices (Soottitantawat et al., 2004a).

Figure 1.7 illustrates the relationship between the release rate constants and water activity (a_w) of the carrier matrices. In brief, the release rate constants k_R first increase with an increase in a_w; this is followed by a decrease in a_w of around 0.7. At a higher a_w, the release rate tends to increase again because the powder matrices are destroyed. A similar phenomenon was observed in the study of the equilibrium headspace volatile concentration of roasted coffee as a function of water activity. The volatile concentrations were very low at low and high water activity values, and were highest at intermediate water activity, in the range of 0.25–0.35 for light- and medium-roasted coffee and about 0.7 for dark-roasted coffee (Anese et al., 2005). The variation pattern of k_R against a_w is very similar, irrespective of the types of carrier materials. The reverse relationships of k_R at 51% and 75% RH could be expected

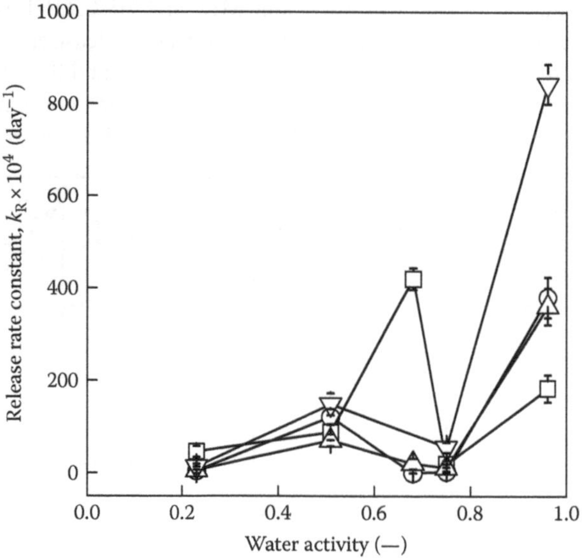

FIGURE 1.7 Relationship between release rate constant (k_R) and water activity of carrier matrices (a_w). Symbols are the same as in Figure 1.6. (From Soottitantawat, A. et al., *J. Agric. Food Chem.*, 52, 1269, 2004a. With permission.)

from the release time-courses of D-limonene in Figure 1.6 and might be closely related to the particle-matrix structural changes from the amorphous glassy state to the rubbery state by the plasticizing effect of moisture adsorption.

In order to explain the structural changes of the powder at different a_w, scanning electron microscopy (SEM) was used to observe the outer structure of the powder during storage, as shown in Figure 1.8 (Soottitantawat et al., 2004a). At a low a_w (≈0.23), the spray-dried particles remained in the original shape and the carrier solids seemed to be still in the glassy state (Figure 1.8a). As compared with the release rate at $a_w \approx 0.23$, the rate at $a_w \approx 0.51$ was higher, although the structural changes could not be clearly observed (Figure 1.8b). This suggested a higher mobility of D-limonene due to the plasticization of the capsule matrices. When the water activity increased further to around 0.7, the powder began to be rehydrated (Figure 1.8c). At this stage, it might be assumed that the effective surface area decreased, resulting in a decrease of D-limonene evaporation from the surface of the powdery particles. Most particles were observed to be clumped and adhered together into a paste-like mass, which represented the rubbery state of the carrier matrices. The regime corresponded to the minimum value of k_R. At a very high a_w, the release rate increased again (Figure 1.7). The most likely explanation is that the emulsion droplets of D-limonene in the powder crushed due to the destruction of the capsule matrices (Whorton and Reineccius, 1995).

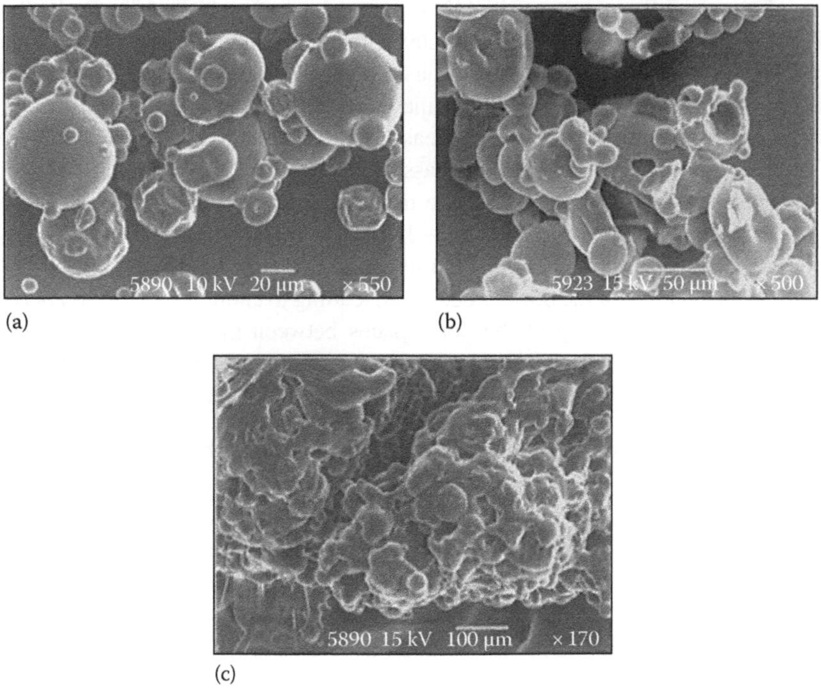

FIGURE 1.8 Outer structural changes of spray-dried powder of GA-MD blend as wall capsules stored at 50°C. (a) Storage at 23% RH for 1 week; (b) storage at 51% RH for 1 week; (c) storage at 75% RH for 1 day. (From Soottitantawat, A. et al., *J. Agric. Food Chem.*, 52, 1269, 2004a. With permission.)

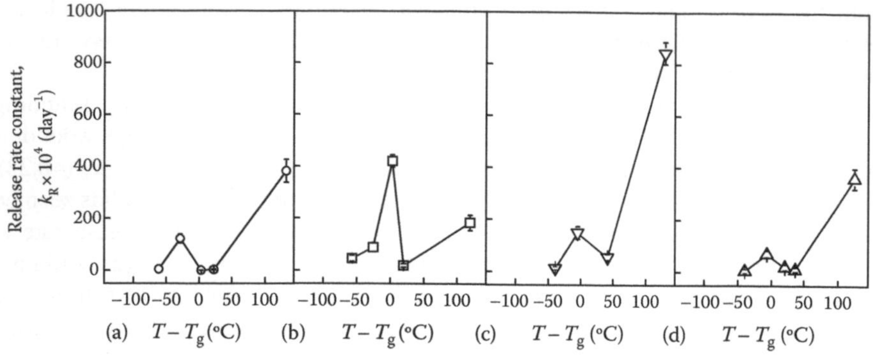

FIGURE 1.9 Release rate constants, k_R, versus $T - T_g$. (a) GA-MD blend; (b) SSPS-MD blend; (c) HI-CAP 100®-MD blend; (d) HI-CAP 100®. The error bars indicate 95% confidence level. (From Soottitantawat, A. et al., *J. Agric. Food Chem.*, 52, 1269, 2004a. With permission.)

Some physicochemical changes caused by the phase transition of carbohydrates from the amorphous glassy state to the rubbery state are commonly expressed with the temperature difference between the storage temperature (T) and the glass transition temperature (T_g) of the carrier matrices, $T - T_g$. The idea is based on the fact that the viscosity (or the relaxation time) of the carrier matrices follows the Williams–Landel–Ferry (WLF) equation, which is expressed as a function of $T - T_g$ (Williams et al., 1955). Therefore, in this case, the release rate constants, k_R, were also correlated with $T - T_g$, as shown in Figure 1.9 (Labrousse et al., 1992; Levi and Karel, 1995). The release rate constants increased with an increase in $T - T_g$ up to the point of roughly equal to zero; this was followed by a decrease and then an increase again with increasing $T - T_g$. Higher release rates near the glass transition temperature could be explained by the increasing mobility of D-limonene molecules. On the other hand, low release rates were observed in the range of $0 < T - T_g < 50$, when the powders changed into the rubbery state. In this region, higher mobility of D-limonene should occur; but at the same time, the collapse of the powder occurred, leading to clump and adhesion between particles, resulting in closing of the pore spaces between the particles and decreasing the surface area for D-limonene evaporation. Whorton and Reineccius (1995) also reported similar results and defined the state as the re-encapsulation of flavor.

1.4.2.3 PTR-MS as a New Methodology for Analyzing Flavor Release

The traditional equilibrium method of flavor release study mentioned above is extremely time consuming, and several weeks are commonly needed to obtain full release profiles of flavors from powders. Recently, thanks to the pioneering work of Dronen and Reineccius (2003), proton transfer reaction mass spectrometry (PTR-MS) has been used as a rapid analysis to measure the release time-courses of flavors from spray-dried powders. The PTR-MS method has been applied extensively to analyze the release kinetics of volatile organic compounds from roasted and ground coffee beans. The release profiles could then be mathematically analyzed by means of Equation 1.1 to obtain the release kinetic parameters, k_R and n (Mateus et al., 2007).

The PTR-MS method is quite sophisticated but very fast in obtaining the results. However, as suggested by Dronen and Reineccius (2003), comparing the release profiles obtained using PTR-MS with those obtained using the traditional method, the PTR-MS method has a disadvantage that it cannot obtain the time-dependent phenomena (relaxation phenomena) such as collapse and glass transition of the carrier matrices. As mentioned earlier, the relaxation phenomena of state changes of carrier matrices have significant effects on the flavor release characteristics, which the PTR-MS method cannot detect and properly account for.

1.4.3 Oxidation of Encapsulated Flavors during Storage

In addition to the release of flavors through the walls of spray-dried particles, the oxidation of encapsulated flavors is also an important index of stability. Many oxidation products are known to be formed during the oxidation reactions of flavors. For the encapsulation of the model flavor D-limonene, limonene oxide (limonene-1, 2-epoxide) and carvone are commonly chosen as indicators of the oxidation of D-limonene, which are generated side by side during the oxidation reaction of D-limonene (Anandaraman and Reineccius, 1986). The formation of the oxides depends significantly on the RH and increases initially during storage. During the initial period, the formation of oxides increases linearly with time, so the apparent oxidation rate constant can be calculated on the basis of the zero-order kinetic reaction schemes (Anandaraman and Reineccius, 1986). However, over a longer storage time, the formation of oxides tends to decrease, particularly at higher RH. This might be explained by an accelerated degradation to form other oxide compounds and the release of the oxides into the surroundings (Soottitantawat et al., 2004a). The initial oxidation rate constants (zero-order reaction rate), k_X, are plotted against a_w in Figure 1.10 for the carrier matrices of GA-MD during storage at 50°C. The changes in k_X with respect to a_w are analogous to those of the release rate constants, k_R, shown in Figure 1.7. The oxidation rate constant, k_X, reached a maximum at $a_w \approx 0.5$

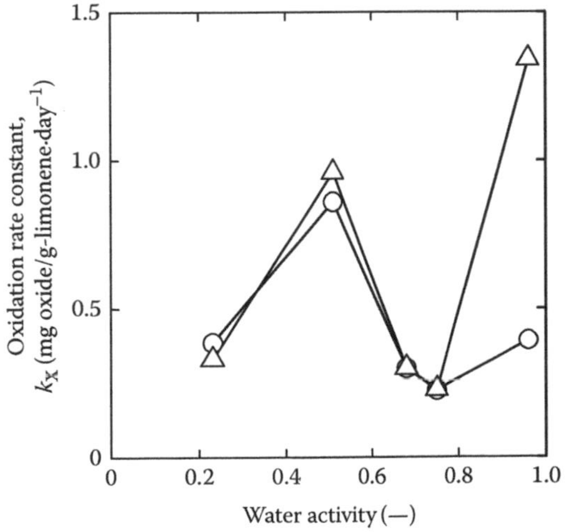

FIGURE 1.10 Effect of water activity on the oxidation rate constant of D-limonene for carrier matrices of GA-MD. ○ Limonene oxide, △ carvone. (From Soottitantawat, A. et al., *J. Agric. Food Chem.*, 52, 1269, 2004a. With permission.)

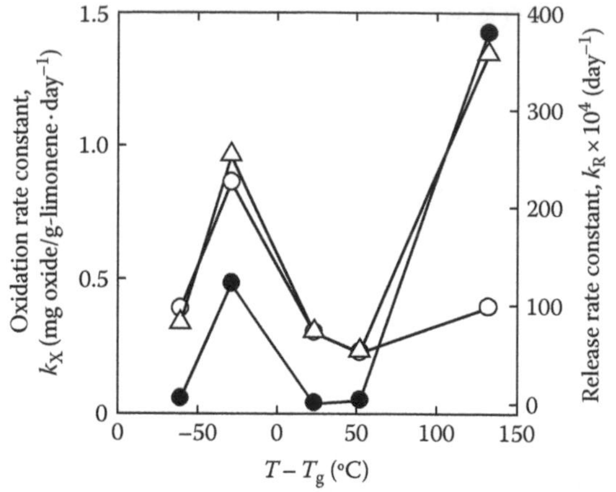

FIGURE 1.11 Correlation of k_R and k_X with $T - T_g$. ● D-Limonene (released), ○ limonene oxide, △ carvone.

where the powder matrices' structure began to change from a glassy to a rubbery state, and became minimum at $a_w \approx 0.7$.

Since the effect of water activity on the oxidation process of D-limonene may be analogous to that of its release, the oxidation rate constants, k_X, are also plotted against $T - T_g$, as shown in Figure 1.11, together with the relation between the release rate constant (k_R) and $T - T_g$. The two curves have very similar functional relationships with $T - T_g$. The oxidation rate constants, k_X, increased with an increase in $T - T_g$ and reached a maximum at roughly $T - T_g$ equals to zero, which is analogous to the behavior of k_R. This implies that the release rate and the oxidation reaction rate have close relations with T_g. The rubbery state and collapse of carrier matrices are associated with greater stability as compared to the glassy state, since the decrease of the

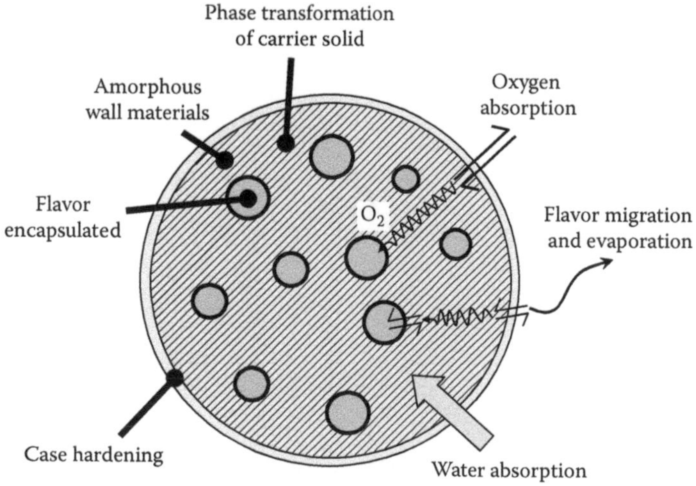

FIGURE 1.12 Schematic illustration of a spray-dried particle in humid air environment.

surface area occurred with the aggregation of carrier matrices; the oxygen uptake from the surface to the interior might also be restricted (Nelson and Labuza, 1992).

Figure 1.12 is a schematic illustration of a spray-dried particle in a humid air environment in which the particle would adsorb water vapor; this is then followed by state changes of carrier matrices from the amorphous state to a rubbery state. The encapsulated flavors can easily move in the matrix of the carrier matrices. At the same time, the oxygen uptake into the wall matrix becomes higher and the oxidation of the encapsulated flavors progresses. The most interesting point is that around the glass transition temperature, both release and oxidation rate constants change nearly in the same trends with $T - T_g$, as shown in Figure 1.11. This implies that the flavor diffusion and the oxygen uptake can be treated as a similar behavior.

1.5 MORPHOLOGY OF SPRAY-DRIED PARTICLES

1.5.1 Structural Analysis of Spray-Dried Particles by SEM

The morphology of spray-dried particles is very important since it influences the powder flowability, redispersibility, density as well as stability of the encapsulated active compounds. Typically, a study of the morphology (porosity and surface integrity) of encapsulated flavor powder is performed using SEM (Rosenberg et al., 1985; Kim and Morr 1996; Ré and Liu, 1996; Soottitantawat et al., 2003, 2004, 2007). Several techniques have also been developed to provide information on the inner microstructure of the microcapsules (Moreau and Rosenberg, 1993). Figure 1.13 shows typical SEM photographs of outside and inside structures of spray-dried particles (Soottitantawat et al., 2003). The inner structure of a spray-dried particle obtained from emulsion would be that the active core material (e.g., D-limonene emulsion) is in the form of small droplets embedded in the shell region of the carrier matrix. In the center of the capsule, a large void can be observed, which occupies most of the capsule volume. The formation of the central void is related to the expansion of the capsule (Chang et al., 1988; El-Sayed et al., 1990; Maa et al., 1997; Walton and Mumford, 1999a,b; Hecht and King, 2000a,b; Walton, 2000). These quality factors are a direct result of the design and operating

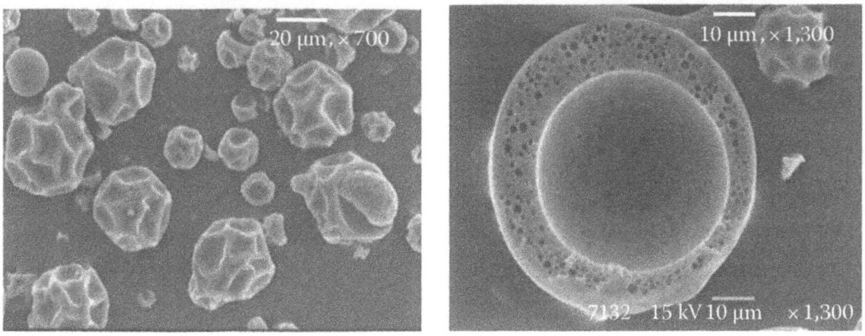

(a) (b)

FIGURE 1.13 SEM photographs of (a) outside and (b) inside structures of spray-dried particles.

conditions of a spray dryer, such as type of atomizer, air drying temperature, feed rate, and feed concentration as well as viscosity or nature of the carrier materials (El-Sayed et al., 1990; Onwulata et al., 1996; Finney et al., 2002). Hollow particles are produced at higher air temperatures due to the expansion of air bubbles in the droplets during drying (Verhey, 1972a,b, 1973; Greenwald and King, 1981; Alamilla-Beltran et al., 2005).

1.5.2 Inner Structural Analysis of Spray-Dried Particle by Confocal Laser Scanning Microscopy

Recently, confocal laser scanning microscopy (CLSM) has shown to be a helpful tool for microparticle characterization (Lamprecht et al., 2000a–c). CLSM minimizes scattered light from out-of-focus structures and permits the identification of several compounds through the use of different fluorescence labels. Therefore, CLSM can be applied as a nondestructive visualization technique for microparticles. Moreover, CLSM allows visualization and characterization of structures, not only on the surface but also inside the particles, provided that the carrier matrices can be fluorescently labeled. Besides, by collecting several coplanar cross sections, a three-dimensional reconstruction of the inspected objects is possible. Figure 1.14 shows the application of CLSM to investigate the cross-sectional structures of spray-dried powders of MD (DE 3 and 20). Fluorescein sodium salt was dissolved in the feed solution as a fluorescent probe of carrier solid. The laser supplied a 485 nm excitation wavelength, which allowed an emission of the fluorescein sodium with excitation/emission

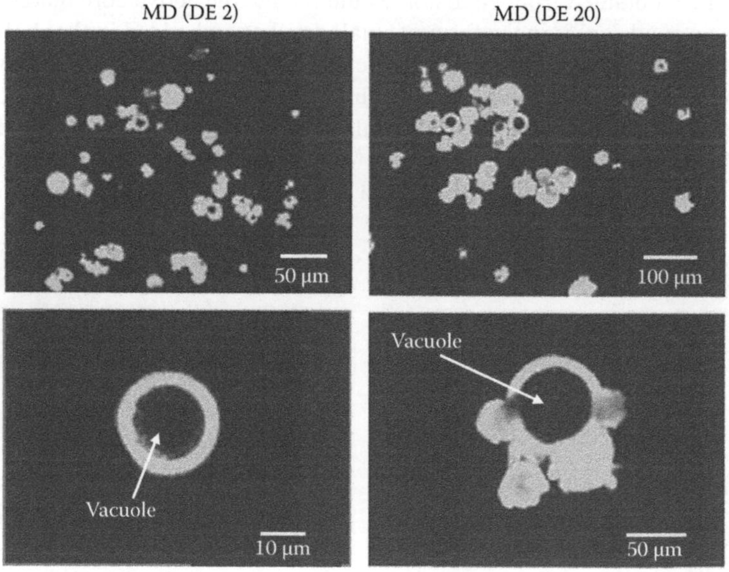

FIGURE 1.14 Cross-sectional structures of spray-dried powders by CLSM. Wall materials: MD (DE 2 and 20).

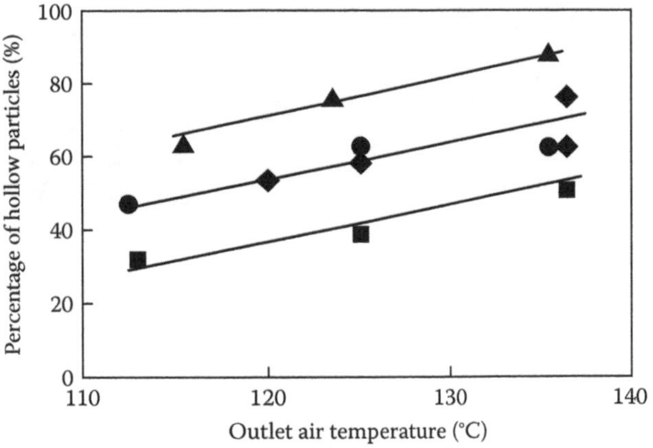

FIGURE 1.15 Effect of outlet temperature and additives on the formation of hollow particles in spray-dried powders. Additives: ■ control (blend of GA/MD), ● gelatin (1%), ▲ decaglycerin monolaurate (0.14 wt%), ◆ ethanol (5%). (From Paramita, V. et al. *Dry. Technol.*, 28, 323, 2010. With permission.)

wavelengths of 485/530 nm, respectively. As only a fluorescent material could be detected by CLSM, spray-dried particles containing a vacuole could be observed as a white fluorescent ring (Soottitantawat et al., 2007).

The effect of the outlet air temperature and additives on the formation of hollow particles in spray-dried powders is illustrated in Figure 1.15. The percentage of hollow particles increases linearly with an increase in the outlet air temperature. When 0.14 wt% of surfactant (HLB 10) was added to the GA-MD solution, the percentage of hollow particles was the highest; particularly at 135°C, the percentage reached about 87%. This phenomenon might be related to the bubble formation in the sprayed droplets. Higher percentages of hollow particles were also obtained with the addition of a small amount of ethanol or gelatin. Gelatin is an enhancer of film formation; the film-forming property might promote the expansion of the sprayed droplets (Paramita et al., 2010).

1.5.3 Observation of Morphology and Flavor Droplets in Spray-Dried Particles by CLSM

In order to observe encapsulated flavor droplets (emulsion) inside a spray-dried particle, ethyl-*n*-butyrate was used as a model flavor. Nile red was dissolved in ethyl-*n*-butyrate and used as a fluorescein probe of the oil phase (ethyl-*n*-butyrate emulsion) of the solution. The labeled flavor was added to the carrier solution and was emulsified.

Figure 1.16 shows the application of CLSM to investigate the morphology and arrangement of encapsulated flavor droplets in the spray-dried powder. Figure 1.16 shows a merged image of confocal from an argon laser (for sodium fluorescence in gray) and from an He-Ne laser (for Nile red in white) and a transmitted image with

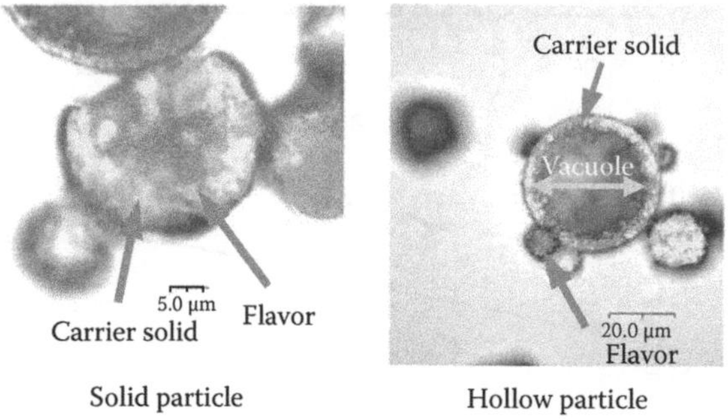

FIGURE 1.16 CLSM photographs of the morphology and arrangement of encapsulated flavor droplets in spray-dried powder. The merged image of confocal (for sodium fluorescence in gray and for Nile red in white) and transmitted image with and without vacuole of powder. (From Soottitantawat, A. et al., Structural analysis of spray-dried powders by confocal laser scanning microscopy, in *Proceedings of the Annual Meeting of IFT 2004*, Las Vegas, NV, 17-G-22. 2004b.)

and without a vacuole of powder (Soottitantawat et al., 2004b). These photographs show the cross section of the spray-dried powder containing flavor droplets (in white color) with a vacuole inside the particle. This illustrates that CLSM represents a new challenging tool for studying and observing encapsulated flavors in a nondestructive fashion. CLSM can be applied to study the release characteristics of flavors from powder in real time (Yoshii et al., 2007).

1.6 CYCLODEXTRINS AND THEIR INCLUSION ABILITY

CDs are doughnut-shaped cyclic oligosaccharides with an interior cavity capable of forming specific inclusion complexes with many organic compounds. CDs are made from starch using cyclodextrin glycosyltransferase (CGTase) to hydrolyze and cyclize the starch to form closed circular molecules known as CDs. CDs are a family of three well-known, industrially produced major cyclic oligosaccharides and several minor, rare ones. The three major CDs are crystalline, homogeneous, nonhygroscopic substances with truncated molecular structures, built up from glucopyranose units. They are α-, β-, and γ-CD, which comprise six, seven, and eight D-glucopyranosyl units, respectively, linked by α-(1,4)-glycosidic bonds, and often called native CDs. The hydrogen and glycosidic oxygen atoms face toward the center of the molecule and form an electron-dense or apolar lining in the cavity, which can interact with a hydrophobic compound that matches the size of the CD cavity to form an association or complex. On the other hand, the polar hydroxyl groups of the glucose monomers face toward the outside of the CD molecule and are responsible for the aqueous solubility of the molecule itself and of its complex. Table 1.2 summarizes some of the characteristic values of the native CDs (Valle, 2004). Notable among these figures are the aqueous solubilities of 14.5, 1.85, and 23.2 g/100 g-H_2O at 25°C for α-, β-, and γ-CD, respectively.

After a century of continuous research and development, CDs have gained certain recognition in various fields including foods (Kllengode and Hanna, 1997; Reineccius

TABLE 1.2
Characteristic Values of Native CDs

	α-CD	β-CD	γ-CD
Number of glucopyranose units (—)	6	7	8
Molecular weight (g/mol)	972	1135	1297
Cavity diameter (nm)	0.57	0.78	0.95
Outer diameter (nm)	13.7	15.3	16.9
Height of torus (nm)	0.78	0.78	0.78
Cavity volume (nm^3)	0.174	0.262	0.427
Solubility in water at 25°C (g/100 g-H_2O)	14.5	1.85	23.2

et al., 2003; Hădărugă et al., 2006), pharmaceuticals (Loftsson et al., 2005; Sheehy and Ramstad, 2005; Bakkour et al., 2006), agriculture (Lezcano et al., 2002; Biebel et al., 2003), analytics (Juvancz and Szejtli, 2002; Tang et al., 2006; Ward, 2006), cosmetics and personal care (Scalia et al., 2002, 2004; Lantz et al., 2006), etc. Their applications are mainly intended for the entrapment of smaller molecules, for the stabilization of reactive intermediates, for catalysis through encapsulation, and as potential molecular transport and drug delivery devices (Rudkevich and Leontiev, 2004). In food-related applications, flavor compounds are being encapsulated into CDs for better retention and protection from various possible means of deterioration as well as for controlled delivery (Szente and Szejtli, 1986, 2004; Hedges et al., 1995; Hedges and McBride, 1999).

CDs are also being utilized as carriers for controlled release of particular compounds entrapped within the cavities. Particularly in the pharmaceutical industry, controlled release systems are desirable to give optimized efficacy, safety, and convenience because they can be designed to deliver a drug at a specified rate, for a specific period of time, and even at a desired location (Furuta et al., 1993). The performance of CDs as carriers in controlled release systems can be evaluated by the release characteristics of inclusion complexes. The application of a specific inclusion complex is to a great extent dependent on its release characteristic.

1.7 FUNDAMENTALS OF INCLUSION

The most notable feature of CDs is their ability to form inclusion complexes with a very wide range of solid, liquid, and gaseous compounds. Complex formation is a dimensional fit of appropriately sized guest molecules inside the host cavity. Van der Waals forces, hydrophobic interaction, and hydrogen bonds are those binding forces involved in the binding of the guest molecules to the cavity of CDs (Szejtli, 1988). The possibility of inclusion complex formation with selected CDs depends mostly on the properties of the guest to be included; its polarity and geometry are crucial factors. A medium in which encapsulation occurs also plays an important role.

1.7.1 Minimum Water Content Needed for Inclusion

Recently, considerable attention has been focused on the water content required to form an inclusion complex of a guest molecule with CDs in order to investigate the possibility of formation of inclusion complexes at low moisture content. It is well known that the presence of water is essential for the formation of inclusion complexes between CDs and hydrophobic substances. As schematically illustrated in Figure 1.17, some water molecules initially included in the cavity would be expelled outside by guest molecules upon the formation of inclusion complexes. This implies that the guest molecules are included in place of water molecules. However, few studies have been carried out to determine as to how many water molecules are needed for the inclusion of guest molecules in CDs. Yoshii et al. (1994) determined the minimum number of water molecules required to encapsulate D-limonene (as a guest molecule) in α-, β-, and γ-CD by means of a micro-aqueous method, which is a common method for studying enzymatic reactions in organic solvents. The inclusion ratio (molar ratio of D-limonene to CD) of an inclusion complex was significantly influenced by the initial water content of the mixture during the inclusion process, as shown in Figure 1.18. Inclusion complexes were hardly formed in all dry mixtures.

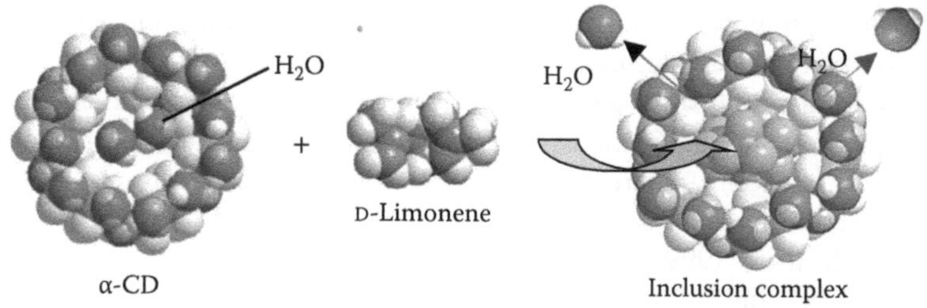

FIGURE 1.17 Schematic illustration of the inclusion of D-limonene in α-CD.

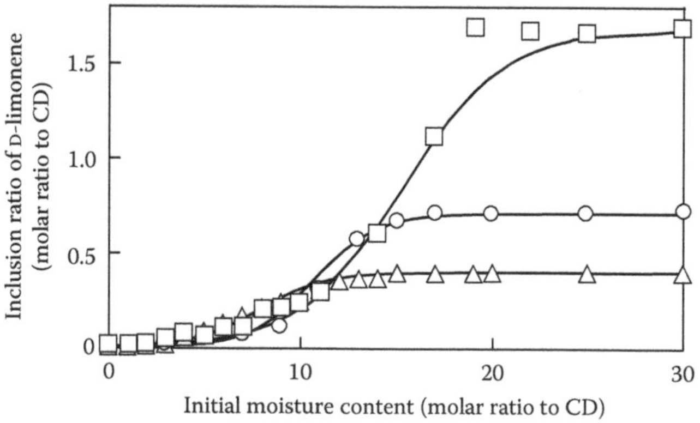

FIGURE 1.18 Formation of inclusion complexes between D-limonene and CDs at various water contents. △ α-CD, ○ β-CD, □ γ-CD. (From Yoshii, H. et al., *J. Biochem.*, 115, 1035, 1994. With permission.)

The amount of inclusion complexes gradually increased in the low-water-content region. However, over a specific water content for each CD, which was roughly 2, 4, and 10 for α-, β-, and γ-CD, respectively, the inclusion ratio increased exponentially and then reached a maximal plateau (R_{max}). The maximum inclusion ratios of α-, β-, and γ-CD were around 0.4, 0.7, and 1.68, respectively. These inclusion ratios suggested that the inclusion stoichiometries between CDs and D-limonene were roughly 2:1 ($R_{max} \approx 0.5$), 1:1 ($R_{max} \approx 1.0$), and 2:3 ($R_{max} \approx 1.5$) for α-, β-, and γ-CD, respectively.

1.7.2 Inclusion Complexes in Organic Solvents

The most popular method for the preparation of inclusion complexes is to form the complexes in an aqueous solution of CD. However, since various pharmaceutical drugs have low solubility in water, it is very important to develop a new method for the effective formation of their inclusion complexes. Very poorly soluble guests cannot form inclusion complexes in any acceptable concentration without using a solvent. It has been shown that the addition of a selected third component, such as an alcohol or a surfactant, can affect the formation of CD complexes (Yoshii et al., 1998). However, few studies have been carried out on the preparation of inclusion complexes in pure organic solvents.

Figure 1.19 shows the molar ratio of included D-limonene or ethanol to β-CD against the amount of ethanol content in the solution. With an increase in the amount of ethanol added, the inclusion molar ratio of D-limonene increased up to a maximum value at the molar ratio of ethanol to β-CD of about 20, beyond which the inclusion ratio started to decrease. On the other hand, the inclusion ratio of ethanol to β-CD decreased to a minimum at the molar ratio of about 30, and then increased with an increase in the ethanol content in the slurry. At higher ethanol contents, D-limonene might be prevented by ethanol from the formation of inclusion complexes with β-CD.

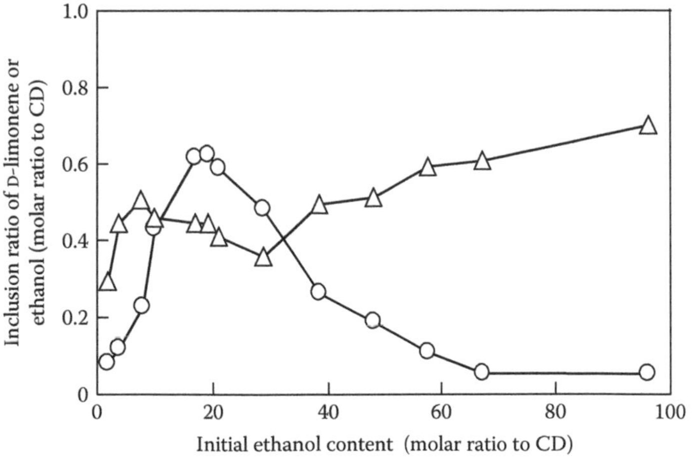

FIGURE 1.19 Formation of inclusion complexes of D-limonene and ethanol at various ethanol contents. ○ D-limonene, △ ethanol. Initial molar ratio of D-limonene to β-CD was 1.7. (From Yoshii, H. et al., *Biosci. Biotechnol. Biochem.*, 62, 2166, 1998. With permission.)

1.7.3 Competitive Inclusion of Flavors in Cyclodextrins

In Section 1.7.2, it is noted that there are some competitions for inclusion between two potential guest compounds. In order to elucidate the competitive inclusion of a target flavor molecule into CDs, a few binary model systems were tested, namely, phenyl ethanol/D-limonene and methyl *n*-hexanoate/D-limonene; phenyl ethanol and methyl *n*-hexanoate were used as target flavors, and D-limonene was used as an inhibitor (Shiga et al., 2002). For α-CD, the inclusion ratio of phenyl ethanol increased rapidly with an increase in the initial amount of phenyl ethanol, approaching the plateau values that were dependent on the respective amount of D-limonene. On the other hand, D-limonene included in α-CD decreased with an increase in the initial phenyl ethanol content. This implies that phenyl ethanol and D-limonene might be included competitively in the molecular cavity of CD. Similar to the enzyme inhibition kinetics, double reciprocal plots for phenyl ethanol at three different D-limonene concentrations are depicted in Figure 1.20a through c for α-, β-, and γ-CD, respectively. The competition between phenyl ethanol and D-limonene for inclusion into each CD could be successfully represented by the double reciprocal plot for the enzyme inhibition kinetics of Michaelis–Menten type (Shiga et al., 2002). At low concentration of phenyl ethanol, the reciprocals of the inclusion ratio of phenyl ethanol in α- and β-CDs were well correlated by linear lines of different slopes, intersecting the *x*-axis at definite points. The slopes of the correlation lines were dependent on the mixing ratio of D-limonene to phenyl ethanol. This means that a noncompetitive inhibition of inclusion applied between phenyl ethanol and D-limonene for α- and β-CD. For γ-CD, on the other hand, the uncompetitive type of inhibition seemed to take place, since the correlation lines of the same slope were obtained. The difference of the inhibition type might be attributed to the cavity diameter of a given CD. A larger cavity of γ-CD could include both phenyl ethanol and D-limonene together,

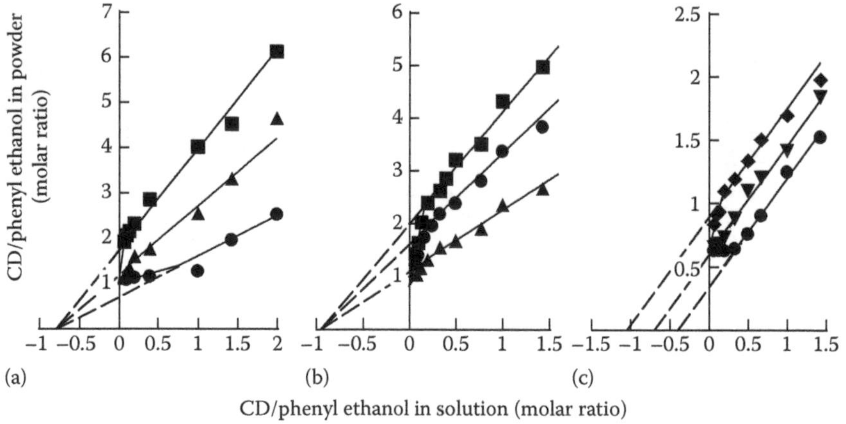

FIGURE 1.20 Correlation of the inclusion fraction of phenyl ethanol in (a) α-CD, (b) β-CD, and (c) γ-CD by the inhibited enzyme kinetics of the Michaelis–Menten type (● D-limonene/CD [molar ratio] = 0, ▲ 2, ■ 5, ▼ 10, ◆ 20). (From Shiga, H. et al., *J. Chem. Eng. Jpn.*, 35, 468, 2002. With permission.)

resulting in the uncompetitive inhibition of inclusion. At higher concentrations of phenyl ethanol, the correlation lines converged to a point on the *y*-axis, independent of the concentration of D-limonene and the type of CD. This implies that in the higher-concentration region, a competitive inhibition of inclusion was true between phenyl ethanol and D-limonene.

1.7.4 Complexation Techniques

No universal method exists for the preparation of CD complexes. The method has to be tailor-made for a particular guest compound. Nevertheless, co-precipitation is the most widely used method in the laboratory. CD is dissolved in water, and the guest is added while agitation of the CD solution is performed. In many cases, the mixture of the CD solution and guest must be cooled while stirring for the inclusion complexes to precipitate. The precipitate may be washed with a small amount of water or any other water-miscible solvent, such as ethyl alcohol, methanol, or acetone. The main disadvantage of this method lies in the scaling up of the process. Because of the limited solubility of CD, large volumes of water have to be used. CD need not be completely dissolved, but can be in a slurry state. A complete complexation can be achieved by stirring the guest into an aqueous slurry suspension of CD. The amount of time required to complete the complexation varies with the guest. Additives such as ethanol have been reported to promote complex formation in a solid or semisolid state (Furuta et al., 1993).

Paste complexation and extrusion are variations of the slurry method, and are known as the kneading method (Furuta et al., 1994b; Kllengode and Hanna, 1997; Bhandari et al., 1999). Only a small amount of water is added to CD to form a paste using a mortar and pestle, or on a large scale using a kneader. The resultant complex can be dried directly or washed with a small amount of water and collected by filtration or centrifugation. Extrusion has the advantage of being a continuous process that uses very little water. However, the major drawback of this method is that some heat-labile guests may decompose during the process because of the generated heat.

1.8 INCLUSION AND OXIDATION OF POLYUNSATURATED FATTY ACIDS IN CYCLODEXTRINS

ω-3 polyunsaturated fatty acids (PUFAs), such as EPA and docosahexaenoic acid (DHA), have important physiological functions. These PUFAs are chemically quite reactive, requiring proper encapsulation in a powder form to protect against autoxidation (Szente and Szejtli, 2004). A molecular inclusion by CD has indeed been applied for the encapsulation of PUFAs into a powder form. For example, inclusion complex powders of fatty acid methyl/ethyl esters (FAME/FAEE) with CD were prepared with a self-cleaning twin-screw kneader in a nitrogen atmosphere. Figure 1.21 shows a plot of the amount of FAME included in α-CD against the initial amount of FAME added to α-CD (Yoshii et al., 1995). The dotted line indicates the theoretical value of the included FAME if it would be completely included into α-CD. The included FAME in α-CD increased linearly with the amount of added FAME and reached a plateau, indicating the maximum inclusion value. This maximum value decreased with the length of the FAME. From

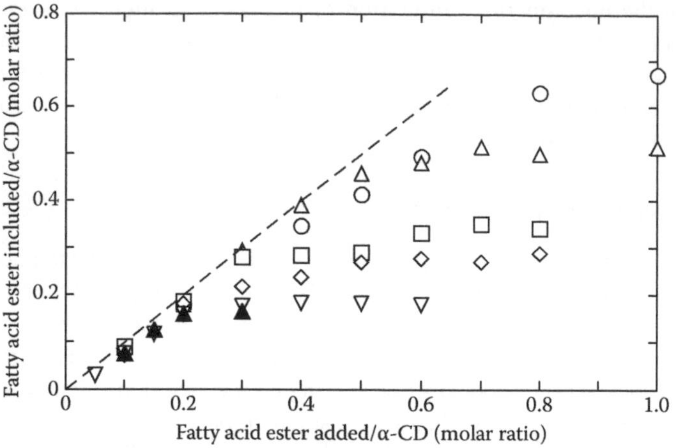

FIGURE 1.21 Effect of initial molar ratio of various fatty acid esters to α-CD on the inclusion fraction of fatty acid esters in α-CD. ○ CAPME, △ CPRME, □ MYRME, ◇ LINME, ▽ EPAEE, ● DHAEE. Abbreviations: CAPME, *n*-caproic acid methyl ester; CPRME, *n*-caprylic acid methyl ester; LINME, linoleic acid methyl ester; EPAEE, ethyl eicosapentaenoate; DHAEE, ethyl docosahexaenoate. (From Yoshii, H. et al., *Oyo Toshitsu Kagaku*, 42, 243, 1995. With permission.)

the plateau value, it was estimated that the average number of α-CD molecules required for one mole of ethyl eicosapentaenoate (EPAEE) or ethyl docosahexaenoate (DHAEE) is about 6.

The time courses of oxidation of EPAEE included in CDs are illustrated in Figure 1.22 at 50°C, together with the oxidation profile of liquid EPAEE (Yoshii

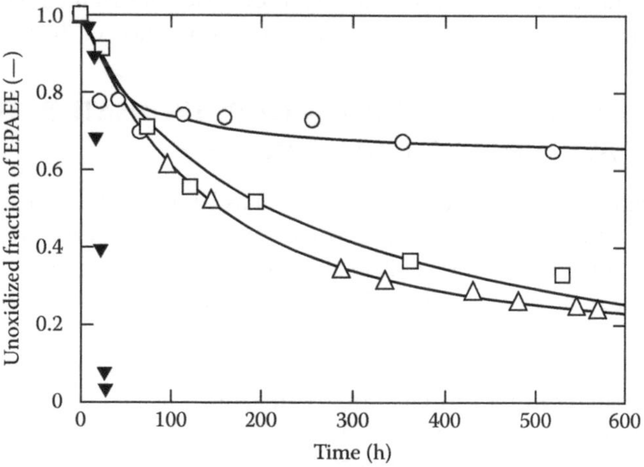

FIGURE 1.22 Correlation of the oxidation process of EPAEE by a model of statistically distributed activation energy. ○ α-CD ($\sigma = 1.20$, $k_{10} = 1.795$), △ β-CD ($\sigma = 0.56$, $k_{10} = 0.000713$), □ γ-CD ($\sigma = 1.15$, $k_{10} = 0.0058$), ▼ EPAEE liquid. (From Yoshii, H. et al., *Jpn. J. Food. Eng.*, 4, 25, 2003. With permission.)

et al., 2003). The inclusion complexes of EPAEE with all the CDs were quite stable compared to those of liquid EPAEE. Oxidation time courses were correlated by Equation 1.4, as shown by the solid lines in which the activation energy of the oxidative reaction could be represented by the Gaussian distribution curve with good correlations. Since the powders were not washed, EPAEE of various forms might exist in the powder, such as in the unincluded form adsorbed on the surface of the CD powders and the various included forms in the cavity of CDs. Therefore, it would be reasonable to assume that the activation energy of the oxidative reaction of EPAEE has the Gaussian distribution.

1.9 INCLUSION OF FOOD FLAVORS IN CYCLODEXTRINS

1.9.1 Formation of Inclusion Complexes of Flavors in Native Cyclodextrins and Their Releases

For food applications of inclusion complexes, a controlled release property is of importance. This is true for many food products, such as microwave entrees, snacks, and desserts. The pharmaceutical industry has also utilized the technique as a drug delivery system. However, the release kinetics of guest compounds from the inclusion complex powders has not yet been fully understood.

Inclusion complex powders can be prepared by kneading with a twin-screw kneader. Among many tested flavors are D-limonene, allyl isothiocyanate (AITC), and hinokitiol, which is an antibacterial compound. The release time-courses of these flavors complexed with β-CD are shown in Figure 1.23 under various RH values at 70°C. The retention of flavors was calculated from the molar ratio of the residual

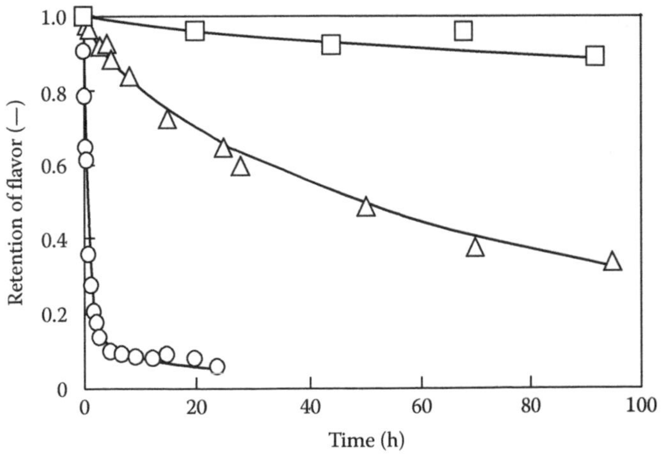

FIGURE 1.23 Release time-courses of various flavors included in β-CD under various RH values at 70°C. ○ AITC (75% RH), △ D-limonene (75% RH), □ hinokitiol (90% RH). (From Shiga, H. et al., Release characteristics of allyl isothiocyanate encapsulated in cyclodextrins, in *Proceedings of the 10th International Cyclodextrin Symposium*, Ann Arbor, MI, 2000, pp. 553–558.)

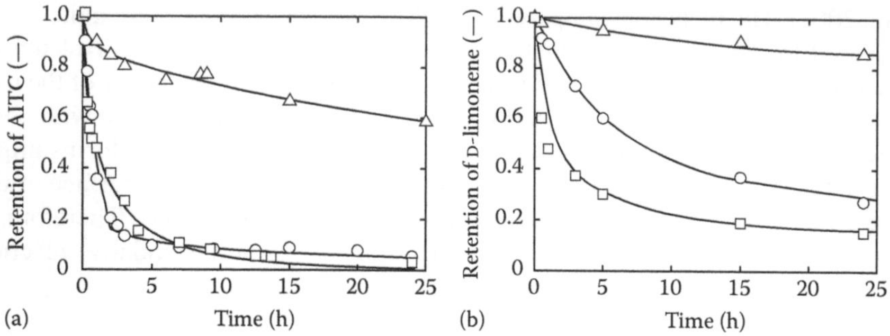

FIGURE 1.24 Release time-courses of (a) AITC and (b) D-limonene included in different CDs. △ α-CD, ○ β-CD, □ γ-CD. (From Shiga, H. et al., Release characteristics of allyl isothiocyanate encapsulated in cyclodextrins, in *Proceedings of the 10th International Cyclodextrin Symposium*, Ann Arbor, MI, 2000, pp. 553–558.)

flavors in the powder to their initial amounts. AITC showed a rapid release, while D-limonene had a moderate release rate. Hinokitiol exhibited an extended release profile and good controlled released properties even at higher RH.

Figure 1.24a shows the retention of AITC included in α-, β-, and γ-CD against the release time at 50°C and 75% RH. The AITC included in α-CD exhibited an extended release. After 25 h, 60% of the initial amount of AITC still remained in the powder at such a high temperature and humidity. On the other hand, the AITC included both in β- and γ-CD released significantly during the initial period of release, showing a quantitatively similar release behavior in both cases. This implies that the AITC included in α-CD had good controlled released properties. Qualitatively similar results are observed for D-limonene included in the native CDs, as shown in Figure 1.24b. These results imply that the release characteristics of included flavors in CDs depend on the combination of the types of flavors and CDs.

The release rate constants of AITC and D-limonene included in α-, β-, and γ-CD can be estimated by means of Equation 1.2. Since the release rate of the flavor is very sensitive to the environmental humidity, it can be presumed that the included flavor was replaced by the water molecules. In Figure 1.25a and b, the release rate constant, k_R, is plotted against moisture concentration in humid air flowing through the sample bottle. The k_R values are well correlated with the moisture concentration in the humid air, C_w. In both AITC and D-limonene cases, k_R of α-CD are much lower than those of β- and γ-CD, indicating the controlled release characteristics. Concerning the release of AITC, the values of k_R of β- and γ-CD are quantitatively the same.

1.9.2 Inclusion of Flavors in Modified Cyclodextrins and Their Releases

In this section, an example case on the use of spray drying to prepare inclusion complex powders with randomly methylated β-CD (RM-β-CD) is given. The inclusion complexes were subjected to release experiments under constant RH and temperature, to investigate the effects of the two parameters on the release rates.

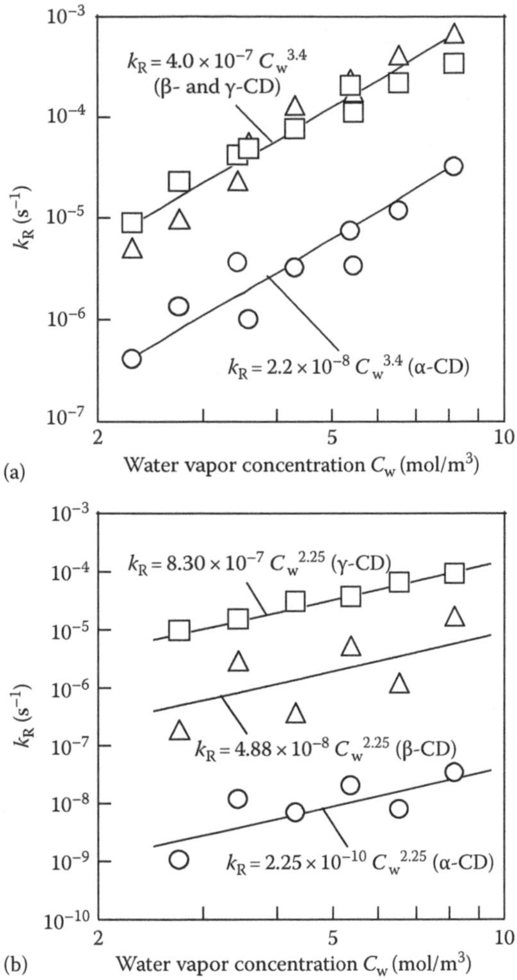

FIGURE 1.25 Plot of release rate constant, k_R, against moisture concentration, C_w, in humid air flowing through the sample bottle for (a) AITC and (b) D-limonene. ○ α-CD, △ β-CD, □ γ-CD. (From Shiga, H. et al., Release characteristics of allyl isothiocyanate encapsulated in cyclodextrins, in *Proceedings of the 10th International Cyclodextrin Symposium*, Ann Arbor, MI, 2000, pp. 553–558.)

Figure 1.26 shows the release time-courses of four flavors included in RM-β-CD at 50°C and 31%–75% RH (Furuta et al., 2008). D-limonene and L-menthol were more stable than AITC and EB under various RH conditions. At 75% RH, except for L-menthol, the flavors initially released markedly, nonetheless the release slackened off noticeably after 24 h. This might be caused by the occurrence of collapse of the powders, in which case the powder surface was covered with a hydrated layer, resulting in a decrease in the flavor evaporation flux, as mentioned in Section 1.4.2.2.

To estimate the activation energy of release, which is an index of release during storage, the inclusion complex powder was subjected to a thermo-gravimetrical

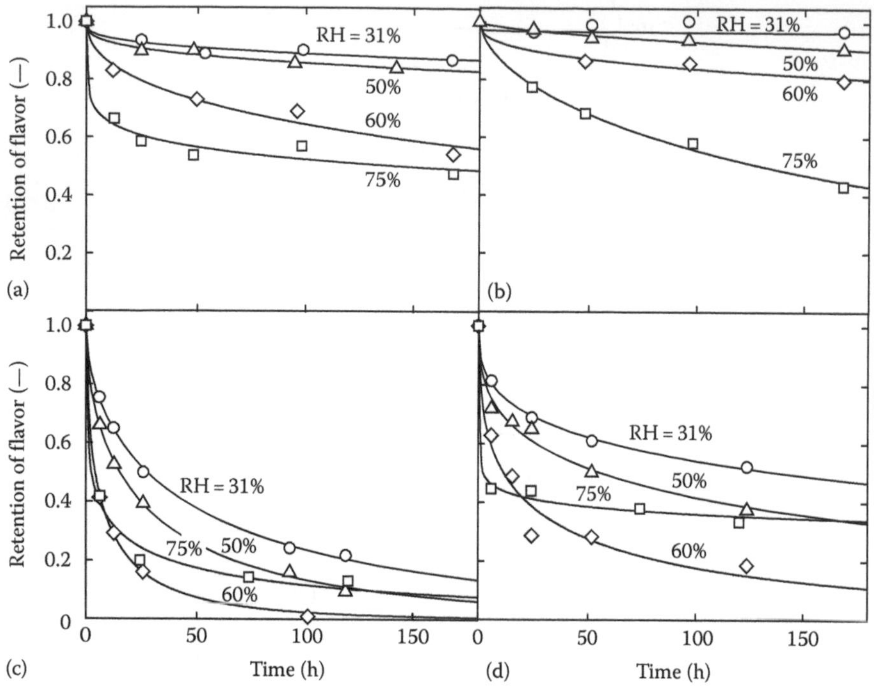

FIGURE 1.26 Release time-courses of four flavors included in RM-β-CD at 50°C and 31%–75% RH. (a) ○ RH = 31%, (b) △ RH = 50%, (c) ◇ RH = 60%, (d) □ RH = 75%. (From Furuta, T. et al., Release kinetics of various flavors encapsulated in modified cyclodextrins, in *Proceedings of the 14th International Cyclodextrin Symposium*, Kyoto, Japan, 2008.)

analysis (Neoh et al., 2008). Table 1.3 shows the activation energies of release for the four flavors together with the release rate constants of the flavors at 50°C and 50% RH. It can be concluded that the activation energies of hard-to-release flavors (D-limonene and L-menthol) are less than those of the easy-to-release flavors (AITC and EB).

TABLE 1.3
Comparison of Release Activation Energy, E_R, and Release Rate Constant, k_R, of Various Flavors

Flavor	Activation Energy, E_R (kJ/mol)	Release Rate, k_R, at 50°C, 50% RH ($\times 10^{-3}$ h^{-1})
D-Limonene	123	0.027
L-Menthol	111	0.14
AITC	54	120
Ethyl *n*-butyrate	66	36

1.10 CONCLUDING REMARKS

The microencapsulation of food flavors, particularly by spray drying and molecular inclusion with CDs, is reviewed mainly on the basis of our recent works. Microencapsulation can simplify food-manufacturing processes by converting liquid flavors into solid powders, and hence improve the convenience of foods. The controlling factor for efficient encapsulation of hydrophobic flavors by spray drying is the size and stability of emulsion of the flavors in the feed liquids. Molecular encapsulation by CDs, on the other hand, is affected by size compatibility between the guest molecules and the cavity of CDs. For powders obtained from both encapsulation techniques, the stability of the encapsulated core materials or guest molecules (e.g., flavors) is very sensitive to the humidity of the storage environment. The higher the humidity, the higher will be the release rate of flavors from the encapsulated powders.

REFERENCES

Adhikari B, Howes T, Leocomte D, Bhandari BR. 2005. A glass transition temperature approach for the prediction of the surface stickiness of a drying droplet during spray drying. *Powder Technology* 149:168–179.

Alamilla-Beltrán L, Chanona-Pérez JJ, Jiménez-Aparicio AR, Gutiérrez-López GF. 2005. Description of morphological changes of particles along spray drying. *Journal of Food Engineering* 67:179–184.

Anandaraman S, Reineccius GA. 1986. Stability of encapsulated orange peel oil. *Food Technology* 40(11):88–94.

Anese M, Manzocco L, Maltini E. 2005. Effect of coffee physical structure on volatile release. *European Food Research and Technology* 221:434–438.

Bakkour Y, Vermeersch G, Morcellet M, Boschin F, Martel B, Azaroual N. 2006. Formation of cyclodextrin inclusion complexes with doxycyclin-hyclate: NMR investigation of their characterisation and stability. *Journal of Inclusion Phenomena and Macrocyclic Chemistry* 54(1–2):109–114.

Barbosa-Cánovas GV, Ortega-Rivas E, Juliano P, Yan H. 2005. *Food Powders: Physical Properties, Processing, and Functionality*. New York: Springer-Verlag.

Beristain CI, Azuara E, Vernon-Carter EJ. 2002. Effect of water activity on the stability to oxidation of spray-dried encapsulated orange peel oil using mesquite gum (*Prosopis juliflora*) as wall material. *Journal of Food Science* 67:206–211.

Bhandari BR, Howes T. 1999. Implication of glass transition for the drying and stability of dried foods. *Journal of Food Engineering* 40:71–79.

Bhandari BR, Dumoulin ED, Richard HMJ, Noleau I, Lebert AM. 1992. Flavor encapsulation by spray drying: Application to citral and linalyl acetate. *Journal of Food Science* 57:217–221.

Bhandari BR, D'Arey BR, Padukka I. 1999. Encapsulation of lemon oil by paste method using b-cyclodextrin: Encapsulation efficiency and profile of oil volatile. *Journal of Agricultural and Food Chemistry* 47:5194–5197.

Biebel R, Rametzhofer E, Klapal H, Polheim D, Viernstein H. 2003. Action of pyrethrum-based formulations against grain weevils. *International Journal of Pharmaceutics* 256:175–181.

Bomben JL, Bruin S, Thijssen HAC, Merson RL. 1973. Aroma recovery and retention in concentration and drying of foods. In: Stewart GS, Mrak E, Chichester CO (Eds.), *Advances in Food Research*, Vol. 20, pp. 1–111. New York: Academic Press.

Chang YI, Scire J, Jacobs B. 1988. Effect of particle size and microstructure properties on encapsulated orange oil. In: Risch SJ, Reineccius GA (Eds.), *Flavor Encapsulation*, pp. 87–102. Washington, DC: American Chemical Society.

Coumans WJ, Kerkhof P, Bruin S. 1994. Theoretical and practical aspects of aroma retention in spray drying and freeze drying. *Drying Technology* 12:99–149.

Dronen DM, Reineccius GA. 2003. Rapid analysis of volatile release from powders using dynamic vapour sorption atmospheric pressure chemical ionization mass spectrometry. *Journal of Food Science* 68:2158–2162.

El-Sayed TM, Wallack DA, King CJ. 1990. Changes in particle morphology during drying of drops. I. Effects of composition and drying conditions. *Industrial and Engineering Chemistry Research* 29:2346–2354.

Finney J, Buffo R, Reineccius GA. 2002. Effect of type of atomization and processing temperatures on the physical properties and stability of spray-dried flavors. *Journal of Food Science* 67(3):1108–1114.

Flink JM, Karel M. 1970. Effects of process variables on retention of volatiles in freeze-drying. *Journal of Food Science* 35:444–447.

Franks F, Hately RHM, Mathias SF. 1991. Material science and the production of shelf-stable biologicals. *BioPharm* 4:38–42.

Furuta T, Tsujimoto S, Okazaki M, Toei R. 1984. Effect of drying on retention of ethanol in maltodextrin solution during drying of a single droplet. *Drying Technology* 2:311–327.

Furuta T, Yoshii H, Miyamoto A, Yasunishi A, Hirano H. 1993. Effects of water of inclusion complexes of D-limonene and cyclodextrins. *Supramolecular Chemistry* 1:321–325.

Furuta T, Hayashi H, Ohashi T. 1994a. Some criteria of sprat dryer design for food liquid. *Drying Technology* 12:151–177.

Furuta T, Yoshii H, Kobayashi T, Nishitarumi T, Yasunishi A. 1994b. Powdery encapsulation of D-limonene by kneading with mixed powders of β-cyclodextrin and maltodextrin at low water content. *Bioscience, Biotechnology and Biochemistry* 58:847–850.

Furuta T, Konaka T, Neoh T-L, Yoshii H. 2008. Release kinetics of various flavors encapsulated in modified cyclodextrins. In *Proceedings of the 14th International Cyclodextrin Symposium*, Kyoto, Japan.

Gacular MC Jr., Kubala JJ. 1975. Statistical models for shelf life failures. *Journal of Food Science* 40:404–409.

Gibbs BF, Kermasha S, Alli I, Mulligan CN. 1999. Encapsulation in the food industry: A review. *International Journal of Food Sciences and Nutrition* 50:213–224.

Greenwald GC, King CJ. 1981. The effects of design and operating conditions on particle morphology for spray-dried foods. *Journal of Food Process Engineering* 4:171–187.

Gunning YM, Gunning PA, Kemsley EK, Parker R, Ring SG, Wilson RH, Blake A. 1999. Factors affecting the release of flavor encapsulated in carbohydrate matrixes. *Journal of Agricultural and Food Chemistry* 47:5198–5205.

Hădărugă NG, Hădărugă DI, Păunescu V, Tatu C, Ordodi VL, Bandur G, Lupea AX. 2006. Thermal stability of the linoleic acid/α- and β-cyclodextrin complexes. *Food Chemistry* 99(3):500–508.

Hecht JP, King CJ. 2000a. Spray drying: Influence of developing drop morphology on drying rates and retention of volatile substances. 1. Single-drop. *Industrial and Engineering Chemistry Research* 39:1756–1765.

Hecht JP, King CJ. 2000b. Spray drying: Influence of developing drop morphology on drying rates and retention of volatile substances. 2. Modeling. *Industrial and Engineering Chemistry Research* 39:1766–1774.

Hedges A, McBride C. 1999. Utilization of β-cyclodextrin in food. *Cereal Foods World* 44:700–704.

Hedges A, Shieh WJ, Sikorski CT. 1995. Use of cyclodextrins for encapsulation in the use and treatment of food products. In: Risch SJ, Reineccius GA (Eds.), *Encapsulation and Controlled Release of Food Ingredients*, pp. 60–71. Washington, DC: ACS.

Ishido E, Hakamata K, Minemoto Y, Adachi S, Matsuno R. 2002. Oxidation process of linoleic acid encapsulated with a polysaccharide by spray-drying. *Food Science and Technology Research* 8(1):85–88.

Ishido E, Minemoto Y, Adachi S, Matsuno R. 2003. Heterogeneity during autoxidation of linoleic acid encapsulated with a polysaccharide. *Journal of Food Engineering* 59:237–243.

Juvancz Z, Szejtli J. 2002. The role of cyclodextrins in chiral selective chromatography. *Trends in Analytical Chemistry* 21(5):379–388.

Karel M, Langer R. 1988. Control release of food additives. In: Risch SJ, Reineccius GA (Eds.), *Flavor Encapsulation*, pp. 67–77. Washington, DC: American Chemical Society.

Kawamura Y, Nakanishi K, Matsuno R, Kamikubo T. 1981. Stability of immobilized Greek small letter alpha-chymotrypsin. *Biotechnology and Bioengineering* 23:1219–1236.

Kim YD, Morr CV. 1996. Microencapsultaion properties of gum arabic and several food proteins: Spray-dried orange oil emulsion particles. *Journal of Agricultural and Food Chemistry* 44:1314–1320.

King AH. 1995. Encapsulation of food ingredients: A review of available technology, focusing on hydrocolloids. In: Risch SJ, Reineccius GA (Eds.), *Encapsulation and Controlled Release of Food Ingredients*, pp. 26–39. Washington, DC: ACS.

King CJ. 1995. Spray drying: Retention of volatile compounds revisited. *Drying Technology* 13:1221–1240.

Kllengode ANR, Hanna MA. 1997. Cyclodextrin complexed flavors retention in extruded starches. *Journal of Food Science* 62:1057–1060.

Labrousse S, Roos Y, Karel M. 1992. Collapse and crystallization in amorphous matrices with encapsulated compounds. *Sciences des Aliments* 12:757–769.

Lamprecht A, Schafer UF, Lehr CM. 2000a. Characterization of microcapsules by confocal laser scanning microscopy: Structure, capsule wall composition and encapsulation rate. *European Journal of Pharmaceutics and Biopharmaceutics* 49:1–9.

Lamprecht A, Schafer UF, Lehr CM. 2000b. Structural analysis of microparticles by confocal laser scanning microscopy. Structural analysis of microparticles by 246 confocal laser scanning microscopy. *AAPS Pharm Sci Tech* 1:E17.

Lamprecht A, Schafer UF, Lehr CM. 2000c. Visualization and quantification of polymer distribution in microcapsules by confocal laser scanning microscopy (CLSM). *International Journal of Pharmaceutics* 196:223–226.

Lantz AW, Rodriguez MA, Wetterer SM, Armstrong DW. 2006. Estimation of association constants between oral malodor components and various native and derivatized cyclodextrins. *Analytica Chimica Acta* 557(1–2):184–190.

Leahy MM, Anandaraman S, Bangs WE, Reineccius GA. 1983. Spray drying of food flavors, II. A comparison of encapsulating agents for the drying of artificial flavors. *Perfume Flavor* 8(5):49–52.

Levi G, Karel M. 1995. The effect of phase transitions on release of *n*-propanol entrapped in carbohydrate glasses. *Journal of Food Engineering* 24:1–13.

Levine H, Slade L. 1986. A polymer physico-chemical approach to the study of commercial starch hydrolysis products (SHPs). *Carbohydrate Polymers* 6:213–244.

Lezcano M, Al-Soufi W, Novo M, Rodríguez-Núñez E, Tato JV. 2002. Complexation of several benzimidazole-type fungicides with α- and β-cyclodextrins. *Journal of Agricultural and Food Chemistry* 50(1):108–112.

Liu XD, Atarashi T, Furuta T, Yoshii H, Ohkawara S. 2001. Microencapsulation of emulsified hydrophobic flavors by spray drying. *Drying Technology* 19:1361–1374.

Loftsson T, Össurardóttir IB, Thorsteinsson T, Duan M, Másson M. 2005. Cyclodextrin solubilization of the antibacterial agents triclosan and triclocarban: Effect of ionization and polymers. *Journal of Inclusion Phenomena and Macrocyclic Chemistry* 52(1–2):109–117.

Maa YF, Costantino HR, Nguyen PA, Hsu CC. 1997. The effect of operating and formulation variables on the morphology of spray-dried protein particles. *Pharmaceutical Development and Technology* 2:213–223.

Masters K. 1991. *Spray Drying Handbook*, 5th edn. Essex, U.K.: Longman.

Mateus M-L, Lindinger C, Gumy J-C, Liardon R. 2007. Release kinetics of volatile organic compounds from roasted and ground coffee: Online measurements by PTR-MS and mathematical modelling. *Journal of Agricultural and Food Chemistry* 55:10117–10128.

McNamee BF, O'Riordan ED, O'Sullivan M. 1998. Emulsification and microencapsulation properties of gum arabic. *Journal of Agricultural and Food Chemistry* 46:4551–4555.

McNamee BF, O'Riordan ED, O'Sullivan M. 2001. Effect of partial replacement of gum arabic with carbohydrates on its microencapsulation properties. *Journal of Agricultural and Food Chemistry* 49:3385–3388.

Menting LC, Hoogstad B. 1967. Volatiles retention during the drying of aqueous carbohydrate solutions. *Journal of Food Science* 32:87–90.

Menting LC, Hoogstad B, Thujssen HAC. 1970. Diffusion coefficient of water and organic volatiles in carbohydrate-water systems. *International Journal of Food Science and Technology* 5:111–126.

Moreau DL, Rosenberg M. 1993. Microstructure and fat extractability in microcapsules based on whey proteins or mixtures of whey proteins and lactose. *Food Structure* 12:457–468.

Nelson KA, Labuza TP. 1992. Relationship between water and lipid oxidation rates. In: Angelo AJ St (Ed.), *Water Activity and Glass Transition Theory*, pp. 93–103. Washington, DC: American Chemical Society.

Neoh T-L, Yamauchi K, Yoshii H, Furuta T. 2008. Kinetic study of thermally stimulated dissociation of inclusion complex of 1-methylcyclopropene with α-cyclodextrin by thermal analysis. *Journal of Physical Chemistry B* 49:15914–15920.

Onwulata CI, Smith PW, Cooke PH, Holsinger VH. 1996. Particle structures of encapsulated milk fat powders. *Lebensmittel-Wissenschaft und-Technologie* 29:163–172.

Paramita V, Iida K, Yoshii H, Furuta T. 2010. Effect of additives on the morphology of spray-dried powder. *Drying Technology* 28:323–329.

Ré MI. 1998. Microencapsulation by spray drying. *Drying Technology* 16:1195–1236.

Ré MI, Liu YJ. 1996. Microencapsulation by spray drying: Influence of wall systems on the retention of the volatiles compounds. In *Proceedings of the 10th International Drying Symposium (Drying'96)*, Kraków, Poland, pp. 541–549.

Reineccius GA. 1988. Spray-drying of food flavors. In: Risch SJ, Reineccius GA (Eds.), *Flavor Encapsulation*, pp. 55–66. Washington, DC: American Chemical Society.

Reineccius G. 2005. *Flavor Chemistry and Technology*, 2nd edn, pp. 351–384. Boca Raton: CRC Press.

Reineccius GA, Bangs WE. 1982. Spray drying of food flavors. III. Optimum infeed concentrations for the retention of artificial flavors. *Perfume Flavor* 9:27–29.

Reineccius GA, Coulter ST. 1969. Flavor retention during drying. *Journal of Dairy Science* 52(8):1219–1223.

Reineccius TA, Reineccius GA, Peppard TL. 2003. Flavor release from cyclodextrin complexes: Comparison of alpha, beta, and gamma types. *Journal of Food Science* 68(4):1234–1239.

Roos Y. 1995. *Phase Transitions in Foods*. Food Science and Technology International Series, p. 360. San Diego, CA: Academic Press.

Roos Y. 2003. Phase and state transitions in food dehydration. In *Proceedings of the Symposium EU Drying 03*, Crete, Greece, pp. 153–168.

Rosenberg M, Kopelman IJ, Talmon YY. 1985. A scanning electron microscopy study of microencapsulation. *Journal of Food Science* 50:139–144.

Rosenberg M, Kopelman IJ, Talmon Y. 1990. Factors affecting retention in spray-drying microencapsulation of volatile materials. *Journal of Agricultural and Food Chemistry* 38:1288–1294.

Risch SJ. 1995. Encapsulation: Overview of uses and techniques. In: Risch SJ, Reineccius GA (Eds.), *Encapsulation and Controlled Release of Food Ingredients*, pp. 2–7. Washington, DC: American Chemical Society.

Rudkevich DM, Leontiev AV. 2004. Molecular encapsulation of gases. *Australian Journal of Chemistry* 57:713.

Rulkens WH, Thijssen HAC. 1972. The retention of organic volatiles in spray-drying aqueous carbohydrate solutions. *Journal of Food Technology*. 7(1):95–105.

Sankarikutty B, Sreekumar MM, Narayanan CS, Mathew AG. 1988. Studies on microencapsulation of cardamon oil by spray drying technique. *Journal of Food Science and Technology* 25(6):352–356.

Scalia S, Casolari A, Iaconinoto A, Simeoni S. 2002. Comparative studies of the influence of cyclodextrins on the stability of the sunscreen agent, 2-ethylhexyl-*p*-methoxycinnamate. *Journal of Pharmaceutical and Biomedical Analysis* 30(4):1181–1189.

Scalia S, Molinari A, Casolari A, Maldotti A. 2004. Complexation of the sunscreen agent, phenylbenzimidazole sulphonic acid with cyclodextrins: Effect on stability and photo-induced free radical formation. *European Journal of Pharmaceutical Sciences* 22(4):241–249.

Shahidi F, Han XQ. 1993. Encapsulation of food ingredients. *Critical Reviews in Food Science and Nutrition* 33(6):501–547.

Sheehy PM, Ramstad T. 2005. Determination of the molecular complexation constant between alprostadil and alpha-cyclodextrin by conductometry implications for a freeze-dried formulation. *Journal of Pharmaceutical and Biomedical Analysis* 39(5):877–885.

Shiga H, Furuta T, Yoshii H, Hayashida K, Linko P. 2000. Release characteristics of allyl isothiocyanate encapsulated in cyclodextrins. In *Proceedings of the 10th International Cyclodextrin Symposium*, Ann Arbor, MI, pp. 553–558.

Shiga H, Nishimura H, Yoshii H, Furuta T, Linko P. 2002. Competitive inclusion of binary flavor mixtures in α-, β- and γ-cyclodextrin. *Journal of Chemical Engineering of Japan* 35:468–473.

Sivetz M, Foote HE. 1963. *Coffee Processing Technology*. Westport, CT: AVI Publishing Co.

Soottitantawat A, Yoshii H, Furuta T, Ohgawara M, Linko P. 2003. Microencapsulation by spray drying: Influence of emulsion size on the retention of volatile compounds. *Journal of Food Science* 68:2256–2262.

Soottitantawat A, Yoshii H, Furuta T, Ohgawara M, Forssell P, Partanen R, Poutanen K, Linko P. 2004a. Effect of water activity on the release characteristics and oxidative stability of D-limonene encapsulated by spray drying. *Journal of Agricultural and Food Chemistry* 52:1269–1276.

Soottitantawat A, Peigney J, Uekaji Y, Yoshii H, Furuta T, Ohkawara M, Linko P. 2004b. Structural analysis of spray-dried powders by confocal laser scanning microscopy. In *Proceedings of the Annual Meeting of IFT 2004*, Las Vegas, NV, 17-G-22.

Soottitantawat A, Takayama K, Okamura K, Muranaka D, Yoshii H, Furuta T, Ohkawara M, Linko P. 2005. Microencapsulation of *l*-menthol by spray drying and its release characteristics. *Innovative Food Science and Emerging Technologies* 69:163–170.

Soottitantawat A, Peigney J, Uekaji U, Yoshii H, Furuta T, Ohgawara M, Linko P. 2007. Structural analysis of spray-dried powders by confocal laser scanning microscopy. *Asia-Pacific Journal of Chemical Engineering* 2:41–46.

Szejtli J. 1988. *Cyclodextrin Technology*. Dordrecht, the Netherlands: Kluwer Academic Publishers.

Szente L, Szejtli J. 1986. Molecular encapsulation of natural and synthetic coffee flavour with β-cyclodextrin. *Journal of Food Science* 51(4):1024–1027.

Szente L, Szejtli J. 2004, Cyclodextrins as food ingredients. *Trends in Food Science and Technology* 15:137–142.

Tang W, Muderawan IW, Ng S-C, Chan HSO. 2006. Enantioselective separation in capillary electrophoresis using a novel mono-6^A-propylammonium-β-cyclodextrin as chiral selector. *Analytica Chimica Acta* 555(1):63–67.

Thijssen HAC, Rulkens WH. 1968. Retention of aromas in drying food liquids. *De Ingenieur* JRG 80 NR 47:45–56.

Ubbink J, Schoonman A. 2003. Flavor delivery systems. In: Seidel A (Ed), *Kirk-Othmer Encyclopedia of Chemical Technology*. New York: Wiley.

Valle EM. 2004. Cyclodextrind M.D. and their uses: A review. *Process Biochemistry* 39:1033–1046.

Verhey JGP. 1972a. Vacuole formation in spray-dried powder particles. 1. Air incorporation and bubble expansion. *Netherlands Milk and Dairy Journal* 26:186–202.

Verhey JGP. 1972b. Vacuole formation in spray-dried powder particles. 2. Location and prevention of air incorporation. *Netherlands Milk and Dairy Journal* 26:203–224.

Verhey JGP. 1973. Vacuole formation in spray-dried powder particles. 3. Atomization and droplet drying. *Netherlands Milk and Dairy Journal* 27:3–18.

Walton DE. 2000. The morphology of spray-dried particles, a qualitative view. *Drying Technology* 18:1943–1986.

Walton DE, Mumford CJ. 1999a. Spray-dried products characterization of particle morphology. *Transactions of IchemE* 77(A):21–38.

Walton DE, Mumford CJ. 1999b. The morphology of spray-dried particles, the effect of process variables upon the morphology of spray-dried particles. *Transactions of IchemE* 77(A):442–460.

Ward TJ. 2006. Chiral separations. *Analytical Chemistry* 78(12):3947–3956.

Whorton C, Reineccius GA. 1995. Evaluation of the mechanisms associated with the release of encapsulated flavor materials from maltodextrin matrices. In: Risch SJ, Reineccius GA (Eds.), *Encapsulation and Controlled Release of Food Ingredients*, pp. 143–160. Washington, DC: ACS.

Williams M, Landel RF, Ferry JD. 1955. The temperature dependence of relaxation mechanisms in amorphous polymers and other glass-forming liquids. *Journal of American Chemical Society* 77:3701–3707.

Williams G, Watts DC. 1970. Non-symmetrical dielectric relaxation behavior arising from a simple empirical decay function. *Transactions of the Faraday Society* 66:80–85.

Yoshii H, Furuta T, Yasunishi A, Hirano H. 1994. Minimum number of water molecules required for inclusion of D-limonene in the cyclodextrin cavity. *Journal of Biochemistry* 115:1035–1037.

Yoshii H, Furuta T, Kawasaki K, Hirano H, Morita Y, Shiina C, Nakayama S. 1995. Quantitative analysis of α-cyclodextrin inclusion complexes with fatty acid methyl/ethyl ester by x-ray diffractometry. *Oyo Toshitsu Kagaku* 42:243–249.

Yoshii H, Kometani T, Furuta T, Watanabe Y, Linko Yu-Y, Linko P. 1998. Formation of inclusion complex of cyclodextrin with ethanol under anhydrous condition. *Bioscience, Biotechnology and Biochemistry* 62:2166–2170.

Yoshii H, Soottitantawat A, Liu X-D, Atarashi T, Furuta T, Aishima S, Ohgawara M, Linko P. 2001. Flavor release from spray-dried maltodextrin/gum arabic or soy matrices as a function of storage relative humidity. *Innovative Food Science and Emerging Technologies* 2:55–61.

Yoshii H, Furuta T, Fujiwara M, Linko P. 2003. Oxidation stability of powdery ethyl eicosapentaenoate included in cyclodextrins and polysaccharide/cyclodextrin mixtures. *Japan Journal of Food Engineering* 4:25–30.

Yoshii H, Kawamura D, Neoh T-L, Furuta T. 2007. Visualization of flavor release in the spray-dried particle by confocal laser scanning microscopy. *Proceedings of the 5th Asia-Pacific Drying Conference*, Hong Kong, China, pp. 317–322.

2 Physicochemical Changes of Foods during Frying: Novel Evaluation Techniques and Effects of Process Parameters

Akinbode A. Adedeji and Michael O. Ngadi

CONTENTS

2.1 INTRODUCTION

Frying is one of the oldest methods of food preparation (Stier, 2004). It entails the application of heat from frying oil to achieve cooking and drying as well as flavor, crust, and color development in fried foods. In spite of the high calorific content and related health concerns, fried foods remain among the most consumed foods and their market demand has been increasing steadily (Piper, 2001). It is the unique flavor associated with fried foods and the satiety feeling, among other factors, that make deep-fried foods desirable.

The application of heat to food through immersion in hot oil causes several physical and chemical changes in both the frying oil and the food, resulting in changes that define the quality of the fried product (Moreira et al., 1999; Dobarganes et al., 2000). During frying, moisture migrates from the interior of food to its surface as a combination of vapor and water due to pressure and concentration gradients generated by heating, creating a porous system within the food (Sahin and Sumnu, 2009). Food either shrinks or swells during frying, depending on its chemical composition and other frying conditions. Products like potato and tortilla chips as well as chicken nuggets may undergo shrinkage, whereas leavened and aerated products such as donuts and *akara* (fried cowpea paste) may experience swelling during frying (Moreira and Barrufet, 1998).

The vapor escape paths eventually become an access for fat intrusion into a frying product. The type and quality of frying oil, food composition, and pretreatment prior to frying determine to a great extent the amount of oil that is absorbed and, subsequently, the quality of the fried product. A number of pretreatments, namely, coating, pre-drying, microwave precooking, and osmo-dehydration, have been applied to improve the quality of fried products (Huse et al., 1998; Garcia et al., 2004; Rimac-Brnčić et al., 2004; Adedeji et al., 2009). The frying process leads to the formation of a crispy hard crust and golden brown coloration in some foods due to high frying temperatures, significant moisture loss, and chemical changes including starch gelatinization, protein denaturation, caramelization, and Maillard reaction.

The market for fried foods is expanding, and the need for quick, efficient, and innovative technologies to measure and study quality changes during frying is essential. In light of this, a number of novel techniques are being developed to monitor and assess fried food quality. This chapter discusses some of the physicochemical changes that occur during frying, the effect of various pretreatments, methods of quality measurement, and advanced techniques that are being used to study and predict these changes in foods during frying.

2.2 PHYSICOCHEMICAL CHANGES DURING FRYING

Moisture loss, fat absorption, protein denaturation, starch gelatinization, browning due to caramelization of sugars and Maillard reaction, crust formation, pore development, and shrinkage are among the many changes that occur in foods during frying. Apparently, these changes ultimately define the color, flavor, texture, and quality of fried foods. Frying conditions including the frying temperature, the type of frying oil, and the composition of food contribute to the degree of change. The assessment of these changes can be performed either by a subjective or an objective method. The subjective method involves the use of human assessors and is often subject to a perception error as a result of variation in human judgment. However, this method remains one of the most widely used techniques in the sensory analysis of food properties. On the other hand, chemical analysis is used to determine various chemical constituents of fried foods. This type of analysis is useful in tracking how these constituents change during frying. Other instrumental methods are also being used to assess various physicochemical properties of fried foods. These include the use of a texture analyzer for mechanical properties' determination and a colorimeter to capture the color indices of products.

Most chemical measurement techniques are standard and are specified in the Association of Official Analytical Chemists (AOAC) standard methods (AOAC, 2006) or the ASABE (American Society of Agricultural and Biological Engineers) standards, such as ASAE D245.6 and ASAE S353 (ASABE, 2009). However, instrumental methods of quality measurement can be challenging and vary from product to product. In some cases, by contacting the product, measuring tools can come on the way of obtaining true representative data. There are newer, emerging, and advanced techniques that are being developed to address some of these challenges. In their optimized condition, these techniques use quick, nondestructive, and relatively less expensive systems for measuring or monitoring fried food quality changes.

2.2.1 Moisture Loss and Fat Uptake

Deep-fat frying is a simultaneous heat and mass transfer process. In simplistic terms, heat is transferred by convection from hot oil to the food surface and by conduction to the inner part of the food. In high-moisture foods, the interior temperature scarcely rises above the boiling temperature of water during frying at atmospheric pressure. However, in structurally intact, thin-layered products with medium-to-low moisture content, interior temperatures might indeed rise beyond the boiling point of water during frying (Farkas et al., 1996).

The heating process leads to water vapor formation, which finds its way to the surface and subsequently into the frying oil. On the other hand, oil intrusion into many fried products, especially starchy and relatively thin products, has been shown to occur during the cooling period after frying (Moreira et al., 1997; Saguy et al., 1997). However, some fried products, such as chicken products, namely, chicken nuggets and fried chicken meat, show a significant amount of oil being absorbed during the frying process (Kassama, 2003; Ngadi et al., 2006, 2007; Adedeji et al., 2009; Ngadi et al., 2009). These different phenomena might be associated with the surface characteristics of different products (Adedeji et al., 2009).

The mode of mass transfer in fried products is still a subject of debate, since several authors have described it differently. A general assumption that mass transfer in fried products is diffusion governed is often made, which implies that the concentration gradient is largely the main driving force for mass exchange between the product and the surrounding fluid (Ateba and Mittal, 1994; Chen and Moreira, 1997). Ngadi et al. (2006) used a modification of the solution of Fick's second law of diffusion (Equation 2.1) for an infinite slab to predict moisture transfer during deep-fat frying of chicken nuggets. The equation was further simplified to a linear expression as in Equation 2.2.

$$M - M_{\mathrm{e}} = \frac{8}{\pi^2}(M_{\mathrm{o}} - M_{\mathrm{e}})\exp\left(-\frac{\pi^2 Dt}{4L}\right) \tag{2.1}$$

$$M = A_1 \exp(-A_2 t) + A_3 \tag{2.2}$$

where
- M is the instantaneous moisture content
- M_e is the equilibrium moisture content
- M_o is the initial moisture content
- D is the effective diffusivity
- t is the frying time
- L is the effective product thickness
- A_1 and A_2 are parameters related to initial and equilibrium moisture contents, A_3 is the kinetic constant for moisture loss

Other studies have indicated that apart from diffusion, other driving mechanisms such as pressure gradient, resulting from condensation during cooling; capillary force; and surface wetting can also contribute to mass transfer in fried products. The pressure gradient is created as a result of underpressure that develops when vapor condenses during cooling, driving all surface oil into pores that were burrowed by the evaporating vapor during frying (Rice and Gamble, 1989; Saguy and Pinthus, 1995; Vitrac et al., 2000). This has been described as a condensation mechanism, which is completely independent of the food-surface-wetting characteristic or capillary force (Rice and Gamble, 1989; Bouchon et al., 2001; Mellema, 2003). The capillary pressure effect has been shown to be a function of pore diameter, d; interfacial tension between the surrounding air and the oil, γ; and the wetting angle, θ:

$$P = \frac{4\gamma_{\mathrm{lg}}\cos\theta}{d} \tag{2.3}$$

This equation shows that wider pores with θ greater than 90° would produce smaller capillary pressure. For tortilla chips, Moreira et al. (1997) showed that smaller pores led to higher fat uptake than bigger pores. Pores with an average radius of 1 μm are reported to have pressure equivalent to 1 atm. This implies that when pore radius is less than 1 μm, oil can be absorbed even if there is water vapor inside the pore (Mellema, 2003). The tortuosity and depth of pores are additional factors that determine the degree of oil uptake by capillary pressure, such that long and continuous pores would allow more oil absorption.

The moisture-loss profile during deep-fat frying follows a pattern typical of most high-temperature dehydration processes. There is a quick initial drying rate followed by a much reduced rate when most of the moisture in food has been removed. Farkas et al. (1996) identified four stages of frying, corresponding to an initial heating of the product surface to the medium temperature followed by surface boiling, falling rate, and bubble-end-point stages. Two distinct regions, namely, the core and crust regions, form in a product during frying. Minimal moisture loss and fat uptake occur in the core region, whereas the crust (or the coat region) experiences profound moisture and fat changes. Adedeji et al. (2009) and Ngadi et al. (2006) presented results of moisture loss during deep-fat frying of chicken nuggets. These investigators

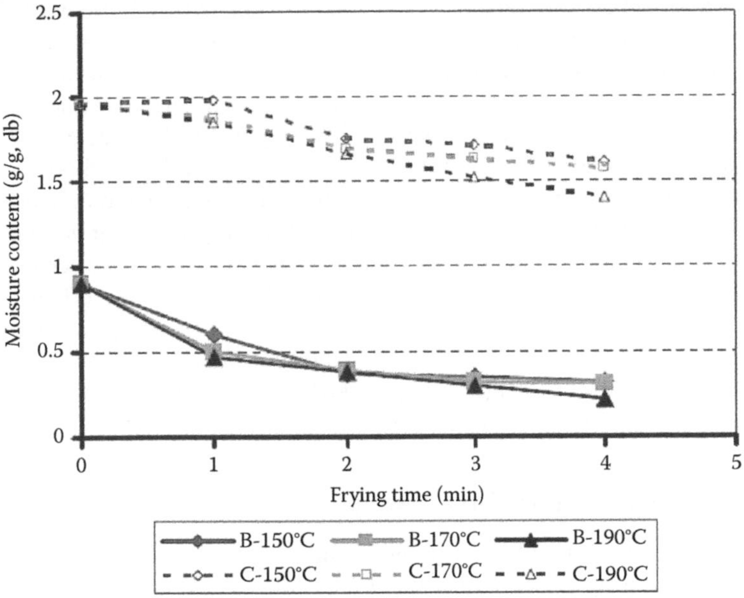

FIGURE 2.1 Moisture profiles in the breading (B) and core (C) regions of chicken nuggets during deep-fat frying at different temperatures. (From Ngadi, M. et al., *Int. J. Food Eng.*, 2(3), Art. 8, 1, 2006.)

observed a higher loss of moisture during the first few minutes of frying, and subsequently a reduced rate of moisture loss (Figure 2.1). It was also observed that the moisture profile in the core portion was different from that in the breading region. Although it was difficult to delineate, it seemed that the first stage of frying might have occurred at the surface of the breading coating within the first 2 min of frying. The core showed mostly constant moisture, especially at the initial stage. This was due to reduced heat transfer to the core from the coated surface of the product. A similar moisture-loss pattern has been reported by other authors (Indira et al., 1999; Budžaki and Šeruga, 2005).

The determination of moisture and oil contents of fried products is very important because they are strong indicators of quality. Poor-quality products have undesirable moisture and oil contents. There are several methods that can be used to measure moisture and oil contents of fried products. Over the years, some new innovative techniques have been developed. The conventional technique for moisture determination is the gravimetric method, in which masses of a sample before and after oven-drying at a specified elevated temperature are compared. Other methods include toluene distillation, vacuum desiccation, vacuum oven-drying, and the Karl Fischer method (Karmas, 1980). A rapid moisture determination technique involves measuring the dielectric properties of a material and correlating them to the amount of moisture in the material. This technique is nondestructive and gives accurate reading with a small merging of error (Kroff, 1984). Hyperspectral imaging (HSI) is another method that has been explored for moisture-content determination. El Masery et al. (2007) illustrated the potential of HSI for moisture-content determination by applying the technique to strawberry; correlation coefficient values of 0.90 and 0.96 were obtained for the moisture content of the training and validation sets, respectively.

Oil content determination is usually performed by Soxhlet solvent extraction; this usually takes up to 4 h, but it is the standard method recommended by the AOAC. On the other hand, a method called Foss-Let takes only 5–10 min to determine the fat content in meat products. The typical error associated with this method is reported to be less than 1% when compared to the standard AOAC method (Kroff, 1984). Several other methods of fat analysis include nuclear magnetic imaging (NMI) and electronic meat-measuring equipment (EMME), which can be used for measuring the fat content in lean meat at an accuracy of 0.1%–0.2% of the official method (Kroff, 1984). It functions on the basis of the difference in electrolyte components of lean and fat components. Another method is the Vico-fat measuring system, which uses a perchloroethylene agent (Pettinati, 1982) and the HSI technique. Some of these techniques are rapid and amenable to online, real-time control.

2.2.2 Porosity and Pore Development

The microstructure of fried foods is a significant factor that influences moisture loss and fat uptake during frying (Bouchon et al., 2001). Pores and their distributions are important components of the microstructure of fried products. The porosity of a product is normally a result of interactions between the product and its surrounding medium at elevated frying temperatures. Some food materials are naturally porous, especially such products as pastry, donut, and *akara*, which are whipped to incorporate air into them before frying (Huse et al., 1998). Preexisting pores are normally quick and open channels for mass transfer during frying and could significantly affect the amount of oil entrained in the final product. For example, *akara*, which is a traditional deep-fat fried food in West Africa and is produced from cowpea (*Vigna unguiculata*) paste through a process of whipping air into it to form a foam-like paste, is among the products that are of interest to many investigators. The formation of pores by whipping prior to frying had a significant effect on the amount of oil absorbed by the product (Prinyawiwatkul et al., 1994; Huse et al., 1998; Singh et al., 2004; Huse et al., 2006). No data are currently available and no study has been reported on the porosity of this product, however.

During frying, pores are developed due to moisture evaporation from within a food material (Moreira et al., 1999; Kassama and Ngadi, 2005). The formed voids are filled with oil through different mechanisms. In certain cases, quick formation of an impervious crust leads to the formation of large pores within fried foods (Shyu et al., 2005). A crust is formed when the surface moisture content is reduced, thereby raising the surface temperature close to that of the frying oil. This elevated temperature induces chemical changes such as starch gelatinization, protein denaturation, and gelation of polysaccharides such as hydrocolloids, which result in the formation of crust in fried foods. Another major factor influencing pore development during frying is the frying conditions, which include frying duration and temperature. Kassama and Ngadi (2005) reported an increase in the porosity of fried chicken meat with time. However, the change in the porosity leveled out after about 400 s of frying (Figure 2.2); this phenomenon was attributed to fat intrusion, which saturated

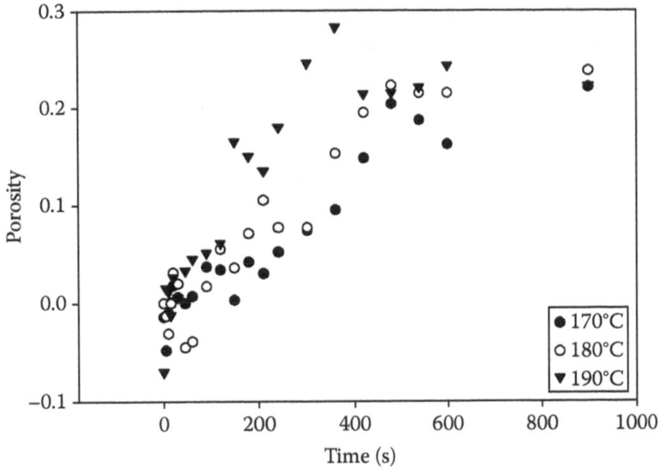

FIGURE 2.2 Effect of frying conditions on pore development in fried chicken meat. (From Kassama, L.S. and Ngadi, M.O., *Dry. Technol.*, 23, 907, 2005.)

pore paths at a longer frying time. The effect of temperature was also clearly demonstrated in this experiment, where the increase in the frying temperature led to a significant increase in the porosity. Adedeji and Ngadi (2008) later reported the effect of formulation on pore development in the batter coating of chicken nuggets. The addition of hydrocolloid at various levels of rice flour, which was used as a substitute for wheat flour, resulted in a significant change in the porosity of the fried coating.

Pore characteristics have been evaluated by various physical methods, such as pycnometry (Kassama and Ngadi, 2004) and mercury intrusion porosimetry (Farkas and Singh, 1991; Karathanos et al., 1996; Ngadi et al., 2001), and imaging methods (Miri et al., 2006; Adedeji and Ngadi, 2008). The latter technique (imaging) is novel and includes a number of different approaches, namely, microscopy, computer vision (CV), and x-ray micro-computed tomography (micro-CT). The principle of porosimetry is based on the non-wetting characteristics of mercury, which is used as the medium for pore properties' determination. This technique involves the introduction of mercury into a porous food under pressure in a stepwise manner, and then the determination of the pore characteristics based on the relation between the amount of mercury allowed into the sample of a specific diameter and the pressure required. This relationship is defined by the Washburn equation (Dullien, 1992; Giesche, 2006). For this expression to be valid, certain assumptions are made, e.g., that the pore shape is cylindrical. Pycnometry is a technique that utilizes density measurement to calculate the porosity of a material. An extensive documentation on how to use this technique is given by Webb (2001).

2.2.3 Shrinkage

Fried foods shrink when the moisture is lost and the food cells collapse as a consequence of heating and evaporation during frying. It is a phenomenon described as a decrease in the product dimension when heat-induced evaporation/drying occurs (Krokida et al., 2000). The occurrence of shrinkage affects food microstructural properties, such as porosity, pore-size distribution, and textural characteristics. Shrinkage can be expressed in terms of length or volume (Taiwo and Baik, 2007):

$$S = \left(\frac{D_0 - D_{(t)}}{D_0} \right) \tag{2.4}$$

$$S_v = 100\left(\frac{V_0 - V_{(t)}}{V_0} \right) \tag{2.5}$$

where

S and S_v are shrinkage expressed, respectively, in terms of dimension (thickness/diameter) and volume
D_0 is the initial dimension
V_0 is the initial volume
$D_{(t)}$ and $V_{(t)}$ are the dimension and volume after frying at time t

Shrinkage starts as a surface occurrence, since drying during frying initiates at the surface, and then progresses into the sample with the frying time. When the drying rate is high, the crust forms quickly, hence reducing the rate of moisture, which causes the shape of the product to form early with minimal shrinkage. Kawas and Moreira (2001), in their work on tortilla chips, reported that shrinkage was closely related to mass transfer changes during the frying operation. These investigators also showed that most of the diameter shrinkage experienced by tortilla chips occurred during the first 5 s of frying, which was attributed to quick crust formation.

Kassama and Ngadi (2003) reported an exponential decrease in the volumetric shrinkage of deep-fat fried deboned chicken breast meat. The rapid increase in shrinkage at an early stage of frying and the correlation with moisture loss were also observed in this study. Taiwo and Baik (2007) studied the effects of different pretreatments, such as blanching, freezing, air drying, and osmotic dehydration, on the shrinkage of fried sweet potato. They also reported quick crust formation, leading to early shrinkage setting in the samples. These investigators also established a relationship between shrinkage and moisture loss. However, they showed that fried sweet potato samples experienced some increase in volume after the initial decrease, and this was attributed partially to quick crust formation that led to pressure buildup within the sample, causing puffing. The effects of pretreatments were shown to be significant on the shrinkage of the samples. Shrinkage, apart from being an important quality definer, is also a very useful tool in transport phenomena modeling of the frying process.

2.2.4 Texture

Texture is a sensory attribute that influences the acceptability of fried foods and is associated with the mechanical, geometrical, and acoustic configuration of the foods (Szczesniak, 1987). Textural properties of fried foods are developed as a consequence of many physical, chemical, and structural changes taking place during frying, which include heat and mass transfer and chemical reactions (Pedreschi and Zúñiga, 2009; Troncoso et al., 2009). Starch gelatinization, protein denaturation, and pore formation are among the major factors that determine the textural properties of fried foods (Suderman, 1983; Loewe, 1993; Ngadi et al., 2007).

The degree of textural development in a fried food is closely related to the frying temperature and time as well as the type of frying oil. Food fried in oil with a high level of hydrogenation would result in a crisper texture, since hydrogenated oils are known to give higher heat transfer rates. Ngadi et al. (2007) reported that 100% hydrogenated oil gave chicken nuggets with a higher maximum load, when tested on a universal testing machine, than samples from less hydrogenated oils. A similar result was reported by Kita et al. (2005). When vegetables and meat are fried in oil, a crust develops and certain microstructural changes are induced by the heat and mass transfer processes. These microstructural changes also define the textural nature of the food (Aguilera and Stanley, 1999; Aguilera, 2005).

Food texture is measured by sensory analysis or by an instrumental method. Using a human inspector for a textural evaluation is subject to some errors because of variations in perception, even when trained panelists are used and a well-defined standard is referenced. However, Katz and Labuza (1981) compared sensory results and cohesiveness values from force–deformation curves for potato chips, popcorn, puffed corn curls as well as saltines, and obtained a good agreement between the two sets of data. A similar comparison was made by van Loon et al. (2007) for the crispness of French fries; comparable results were also noted.

An instrumental method is predicated on such parameters as force, power spectrum, and fractal dimensions. The technique is easy to use, quick, and gives room for flexibility of control. The force–deformation curve or puncture/compression tests are most commonly applied, especially in potato and battered meat products. The maximum force required to penetrate a sample at a certain displacement rate is interpreted as the textural quality of hardness, crispness, or crunchiness. Ateba and Mittal (1994) presented texture analysis results of fried meatballs, showing the effect of frying time. The break force peak was shown to increase with the frying time. Potato chips always show a two-stage hardness, with initial softening characteristics, indicating cooking, and the later-stage hardness due to crust formation as a result of starch gelatinization. These characteristics of potato chips were elucidated by an instrumental method of force–displacement test according to Pedreschi and Moyano (2005). For French fries, the formation of hard crust provides a covering for the inner soft core, which presents the unique texture characteristic synonymous with these types of fried potato products (Agblor and Scanlon, 2000). Kumar et al. (2006) studied the hardness of a traditional Indian fried sweet, *gulabjamun*, using a compression test and demonstrated the effects of frying time and temperature. Higher temperature and longer frying time produced *gulabjamun* with a higher maximum break force. The modern approach in texture measurement involves the use of an imaging technique and an analytical/statistical method to obtain texture data noninvasively (Qiao et al., 2007a).

2.2.5 Color

Color, because of its superficiality, is a major acceptability index of most foods by consumers. It is the first sensory attribute assessed, even before a food touches the mouth. Consumers are apt to relate color to other quality attributes of food, such as flavor, safety, and nutrition, because color indeed correlates well with these physical, chemical, and sensory qualities of food (Pedreschi et al., 2006). Color development during deep-fat frying is based on two main reactions. The first reaction is an interaction between the carbonyl group of sugars (reducing sugars) and the nucleophilic amino group of the amino acids. These two groups form a variety of molecules/compounds responsible for a range of flavors and colors, usually requiring thermal energy (heat). The second type of reaction is caramelization, which occurs as a result of pyrolysis of some sugars when intense heat is applied. Both processes produce similar results, although their paths of production differ. These reactions are responsible for the golden brown coloration in some fried foods. The extent of color development in fried foods is determined by the moisture content, water activity, frying oil quality, food composition, and heat intensity. The measurement of color in fried foods is performed in three ways: visual assessment, instrumentation evaluation, and use of CV. The latter is a new approach with added advantage over the conventional methods (Sun, 2000; Du and Sun, 2004; Pedreschi et al., 2006).

2.3 EFFECT OF OIL QUALITY ON PHYSICOCHEMICAL PROPERTIES OF FRIED FOODS

The quality of oil is usually a function of its origin, composition, and frying conditions such as frying time and temperature (Garcia et al., 2004). Vegetable oils are generally known to contain no cholesterol. However, certain vegetable oils, such as coconut and palm oil, contain much saturated fat, which is relatively stable under storage but of great concern in terms of its *trans* fat content. On the other hand, other oils such as rapeseed, soybean, and sunflower oils contain more of unsaturated fatty acids compositions of triglycerides, which are more healthful but unstable during frying and storage (Ngadi et al., 2007; Chemicalland21, 2008). The hydrogenation process is frequently used to reduce the unsaturated-fatty-acids content of frying oil. This process not only changes the chemical constituent of frying oil but also its physical properties such that the lipid produced is more viscous than its unsaturated counterpart (Fernández et al., 2005).

The oil used for frying is often produced from a blend of oils of different attributes to balance the qualities and to impart the desired value to the final product (Ngadi et al., 2007). Obviously, oil quality defines its frying characteristics such as thermal (heat transfer coefficient), nutritional, sensory, and mass transfer properties (viscosity and surface tension). Oil quality changes as it is used repeatedly, due to chemical changes such as hydrolysis, polymerization, and oxidation, leading to the formation of free fatty acids (FFA), volatiles, acylglycerols, and acrylamides (Lalas, 2009). Hydrolysis occurs when moisture released into the frying oil reacts with triglycerides to form FFA. Oxidation results from oil exposure to the atmosphere during frying at elevated temperatures. Multiple double-bond unsaturated fatty acids, such as linolenic acid, are more susceptible to oxidation than single-bond fatty acids (e.g., oleic acids) (Lalas, 2009). The rate of oxidation becomes higher at higher temperatures. It is indeed imperative to determine the "fry time" (point at which oil would no longer give acceptable product quality) of oil in order to prevent undesirable physicochemical changes of food.

Qualities of frying oil and of fried foods are very closely related (Blumenthal, 1991; Dunford, 2005), which is why adequate attention should be placed on the selection and condition of oil used in the frying operation. Indicators of oil quality such as foaming, smoke emission, color change, pH change, presence of sediments, and detection of free radicals through rapid means are important. A number of rapid means of determining these quality indicators have been developed (O'Brien, 1998). Kazemi et al. (2005) used the HSI technique to predict the quality of frying oil in terms of its acid value, total polar compounds, and viscosity. The investigators obtained comparable results with those obtained from conventional methods. Such a technique is discussed further in the chapter.

Correlations have been established between oil and fried food properties. Ngadi et al. (2007) studied the effect of oil hydrogenation on color change, texture, and oil content of chicken nuggets fried at various time intervals. Chicken nuggets fried in oil with a high level of hydrogenation were substantially different from samples fried in non-hydrogenated oil. Figures 2.3 and 2.4 show the effect of the level of oil hydrogenation on the moisture and oil contents during frying of chicken nuggets, respectively. There was also a significant increase in the maximum load (Figure 2.5) for penetration into the sample, which depicts crispness, as the level of hydrogenation in the frying oil increased. The color (chroma-value color saturation) of the fried

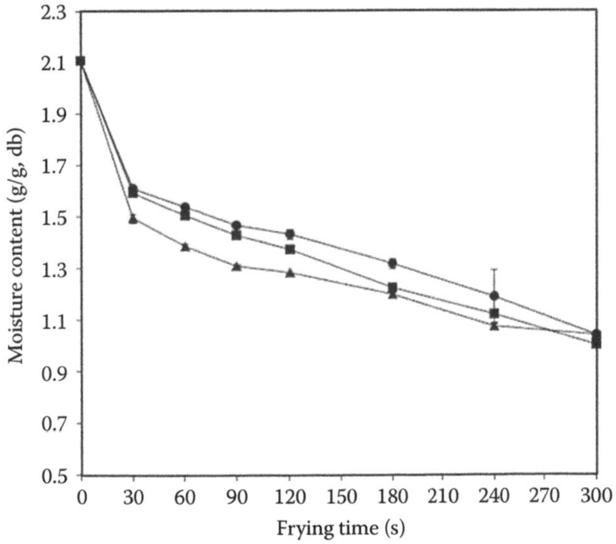

FIGURE 2.3 Average moisture content of chicken nuggets fried in oils with different degrees of hydrogenation (•: 0%; ■: 60%; ▲: 100%). The percentage refers to the w/w ratio of hydrogenated oil. Error bars show standard deviations. (From Ngadi, M.O. et al., *LWT Food Sci. Technol.*, 40, 1784, 2007.)

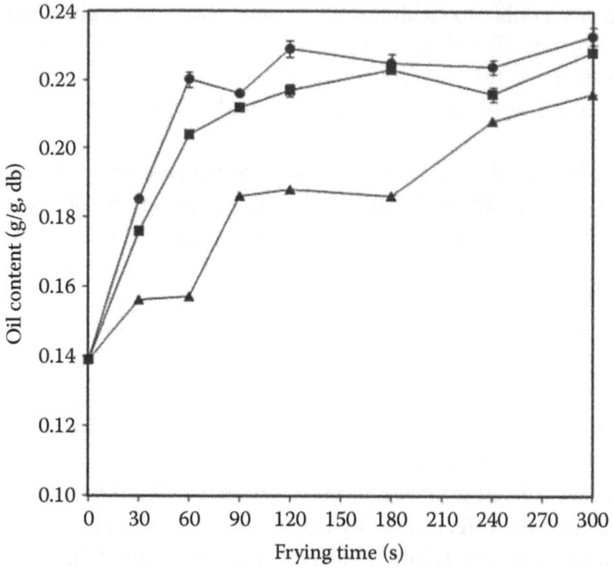

FIGURE 2.4 Average oil content of chicken nuggets fried in oils with different degrees of hydrogenation (•: 0%; ■: 60%; ▲: 100%). The percentage refers to the w/w ratio of hydrogenated oil. Error bars show standard deviations. (From Ngadi, M.O. et al., *LWT Food Sci. Technol.*, 40, 1784, 2007.)

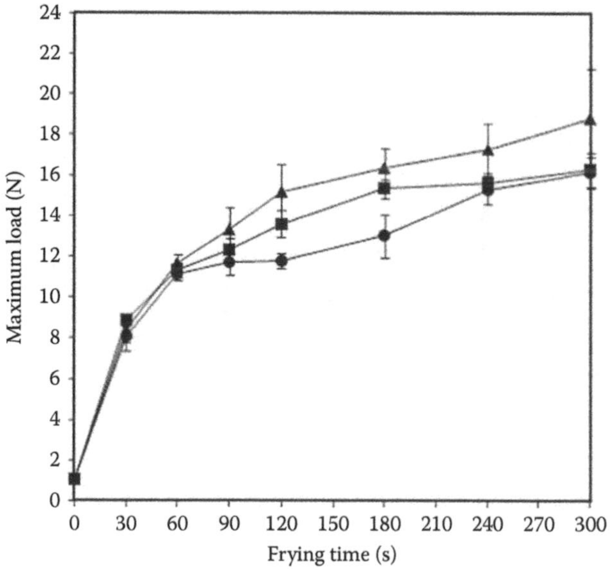

FIGURE 2.5 Average maximum load of chicken nuggets fried in oils with different degrees of hydrogenation (•: 0%; ■: 60%; ▲: 100%). The percentage refers to the w/w ratio of hydrogenated oil. Error bars show standard deviations. (From Ngadi, M.O. et al., *LWT Food Sci. Technol.*, 40, 1784, 2007.)

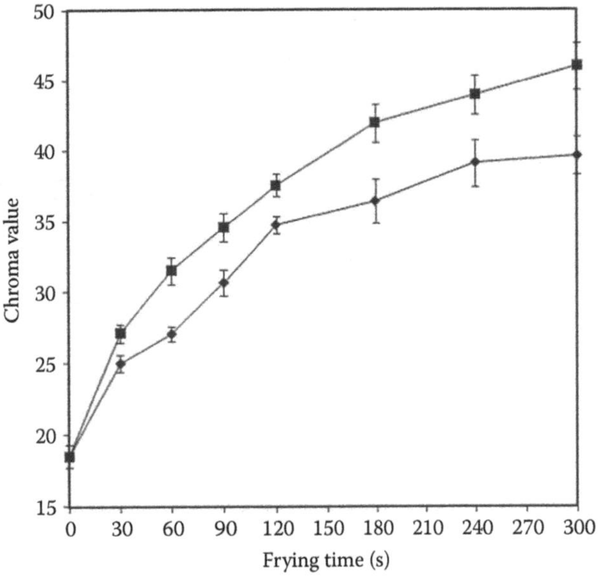

FIGURE 2.6 Average chroma values (C^*) of chicken nuggets fried in oils with different degrees of hydrogenation (•: 0%; ■: 60%; ▲: 100%). The percentage refers to the w/w ratio of hydrogenated oil. Error bars show standard deviations. (From Ngadi, M.O. et al., *LWT Food Sci. Technol.*, 40, 1784, 2007.)

chicken nuggets was equally affected by the degree of unsaturation of the frying oil, as shown in Figure 2.6. Moisture and oil contents decreased with the frying time as the degree of hydrogenation in the frying oil increased. Kita et al. (2005) concluded that French fries fried in hydrogenated oil were harder than those fried in less hydrogenated liquid rapeseed. The results further establish the significant influence of oil type and quality on the physicochemical properties of fried foods.

2.4 EFFECTS OF PRETREATMENTS ON PHYSICOCHEMICAL CHANGES OF FRIED FOODS

There are several pretreatments that can be applied to foods prior to frying, including preheating of oil, par-frying, microwave precooking, partial drying, pre-dusting, blanching, osmo-dehydration, and application of coatings (Huse et al., 1998; Krokida et al., 2001a,b; Rimac-Brncic et al., 2004; Ngadi et al., 2009). Pretreatments are applied for various reasons, such as value addition, reduced fat uptake, and improved product quality. The type of pretreatment a food is subjected to determines the degree of the physicochemical changes observed. Par-frying, partial drying, and blanching are known to modify the structural configuration of fried food surfaces to modulate mass transfer during frying. Besides blanching, par-frying, partial drying, and osmo-dehydration would reduce the product moisture content prior to frying, which in turn lowers the underpressure that propels oil uptake during the cooling process. Pre-dusting and coating are applied to modify the product surface to form a blanket

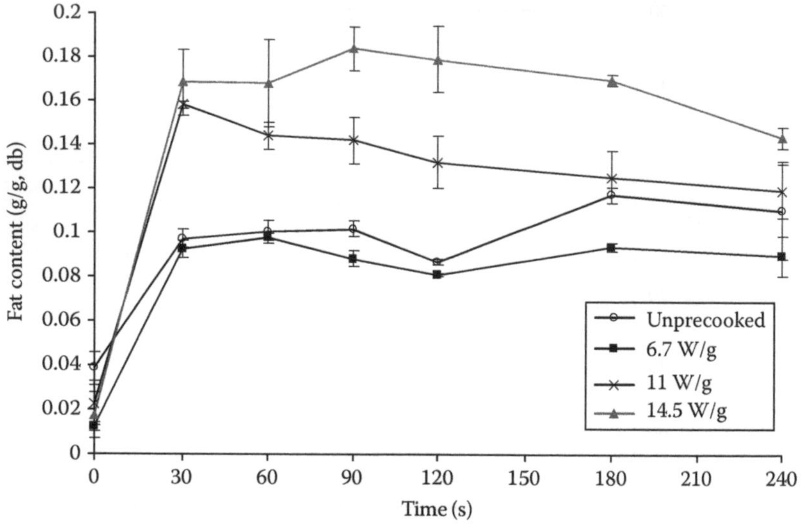

FIGURE 2.7 Effect of microwave precooking on fat content of breading part of chicken nuggets fried at 190°C. (From Adedeji, A.A. et al., *J. Food Eng.*, 91, 146, 2009.)

of screen that would reduce mass transfer during the frying operation. These surface modification processes also cause physical changes such as browning, crust formation, and flavor development.

Adedeji et al. (2009) precooked chicken nuggets in microwave at different power levels and reported a tremendous reduction in oil uptake and a significant change in moisture loss in the fried products when a lower level of microwave power density

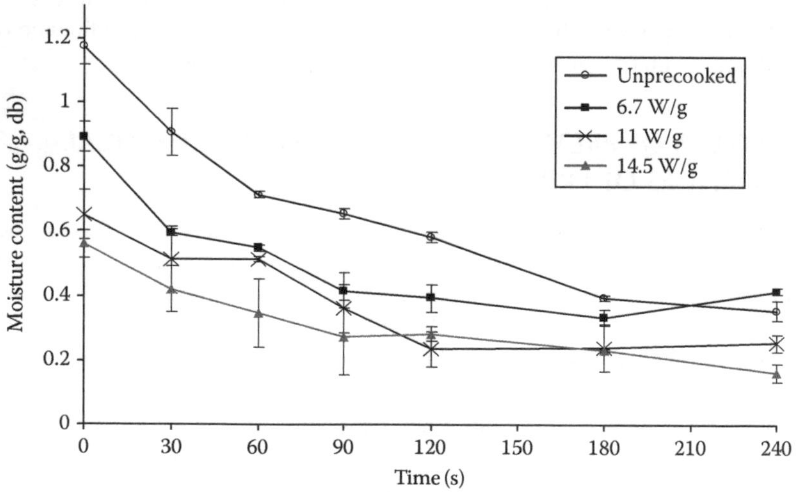

FIGURE 2.8 Effect of microwave precooking at different power densities on moisture loss in breading coating of chicken nuggets fried at 190°C. (From Adedeji, A.A. et al., *J. Food Eng.*, 91, 146, 2009).

was applied (Figures 2.7 and 2.8). Shyu et al. (2005) reported the combined use of blanching, freezing, and osmotic dehydration as pretreatments prior to vacuum frying of carrot chips. These investigators observed that osmotic immersion in a fructose solution for up to 60 min increased the total solids content of the fried carrots; increased sugar solution pretreatment reduced oil absorption. An examination of the fried carrot chips via scanning electron microscopy showed various effects of pretreatments on the physical properties, such as shrinkage and porosity. Krokida et al. (2001a) investigated the effect of pre-drying on moisture and fat contents, color, and porosity of fried potatoes (French fries) and reported that air drying significantly decreased oil and moisture contents of the sample, while the product porosity increased. There was a negative correlation between the color indices and the pre-drying time. In addition, the effects of different osmotic media (sugar, NaCl, maltodextrin-DE 12, and maltodextrin-DE 21) were studied on the quality attributes of French fries; similar results to the pre-drying treatment were observed (Krokida et al., 2001b).

2.5 ADVANCED TECHNIQUES FOR MEASUREMENT OF FRIED FOOD PROPERTIES

Fried foods quality is defined by their properties, such as moisture and oil contents, porosity, color, taste, and nutritional content (Dogan et al., 2005). Conventional methods of measuring these properties include physical (instrumental), chemical, and sensory methods (Qiao et al., 2007a). Some of these procedures are subjective, cumbersome, and time-consuming, and often require the destruction or modification of a product. During the past decade, a number of novel methods for determining food properties have been proposed, and some of them are presented below.

2.5.1 Hyperspectral Imaging for Fried Food Quality Analysis

Quality control in food processing requires quick and reliable feedback systems, especially at critical points. Conventional methods of analyses, such as high-performance liquid chromatography (HPLC), mass spectrometry, and titration, are quite cumbersome, time-consuming, and expensive, and require material destruction. HSI is an emerging technology that combines both traditional imaging and spectroscopy to noninvasively obtain spatial and spectral information from an object (Gowen et al., 2007). It was originally designed for application in remote sensing (Goetz et al., 1985), but later found other uses in fields like agriculture (Monteiro et al., 2007), pharmaceuticals (Roggo et al., 2005), medicine (Zheng et al., 2004), and food processing (Kazemi et al., 2005; Qiao et al., 2007b). The nondestructive, rugged, and flexible nature of HSI makes it a process analytical tool of the twenty-first century, especially in the food industry.

A hyperspectral image is made up of a series of wave bands for each spatial position of an area studied. Therefore, each pixel contains a spectrum, which is a representative of the characteristic of the position. This is like a *fingerprint* for the characteristic of the position, and this would match any other positions within the sample with a similar characteristic. Hyperspectral images are called *hypercubes*, three-dimensional (3D) blocks of data, comprising of two spatial dimensions and

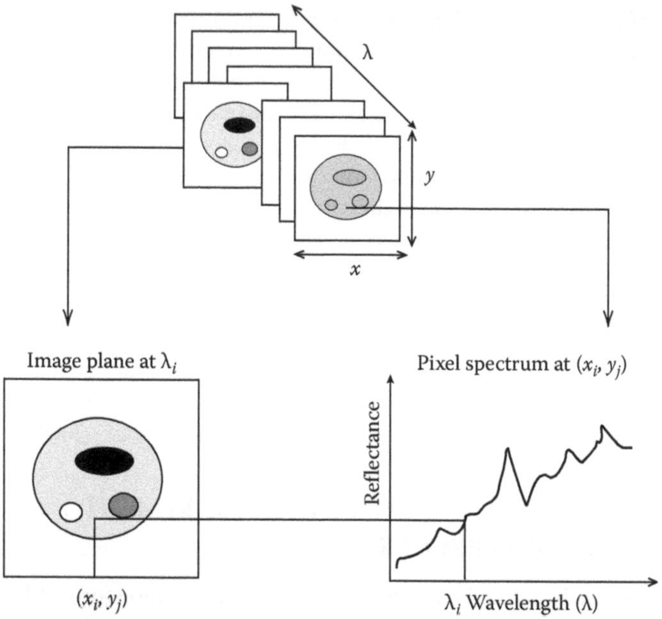

FIGURE 2.9 A schematic diagram of hypercube showing the relationship between spatial and spectral dimensions. (From Gowen, A.A. et al., *Trends Food Sci. Technol.*, 18, 590, 2007.)

one wavelength dimension. In HSI, a 2D spatial image of a product is stacked with the spectral data to generate a 3D hypercube (Figure 2.9). The hypercubes obtained are taken through a series of processes, such as reflectance calibration, preprocessing, classification, and image processing (Kim et al., 2001; Gowen et al., 2007). Reflectance calibration is carried out to account for the background spectral response of the instrument and the "dark" camera response. The corrected reflectance value is calculated as Equation 2.6 (Gowen et al., 2007):

$$R = \frac{\text{sample} - \text{dark}}{\text{background} - \text{dark}} \tag{2.6}$$

Preprocessing is performed to eliminate nonchemical biases from the spectral information and to further prepare the image for processing; classification enables the identification of regions with similar spectral characteristics through such analytical methods as principal components analysis (PCA) (Xing et al., 2007), multilinear regression (MLR) (El Masry et al., 2007), and artificial neural network (Qiao et al., 2007b); while image processing is carried out to translate the contrast developed by the classification procedure into a picture depicting component distribution (Gowen et al., 2007). Components of HSI include the following: objective lens, spectrograph, camera, acquisition system, translation stage, illuminator, and computer (Figure 2.10).

HSI can be carried out in three modes, namely, transmission (fluorescence), emission, and reflectance modes. The reflectance mode is the most common mode used in

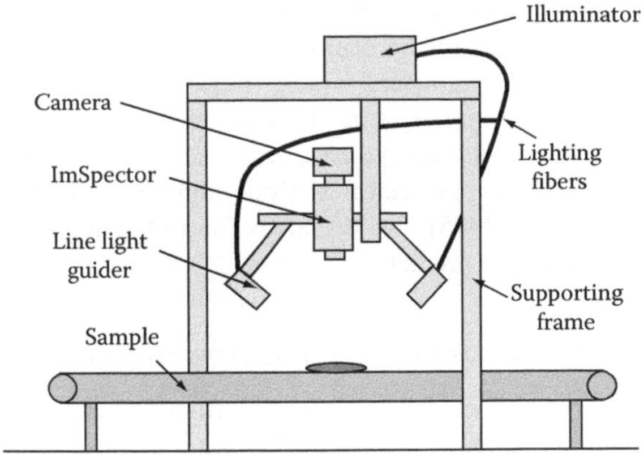

FIGURE 2.10 The HSI system. (From Qiao, J. et al., *J. Food Eng.*, 83, 10, 2007b.)

HSI, and it ranges within the wave band of 400–1000 nm for a visible-near-infrared (VIS-NIR) spectrum or a 1000–1700 nm range for NIR, and it has been used to detect defects, contaminants, and quality attributes of fruits, vegetables, and meat products (Gowen et al., 2007).

Qiao et al. (2007a) developed a hyperspectral-image-based method to predict three mechanical properties (maximum load, energy to break point, and toughness) of chicken nuggets fried for different time intervals (Figure 2.11). A co-occurrence matrix was used to extract five image texture indices, i.e., angular second moment

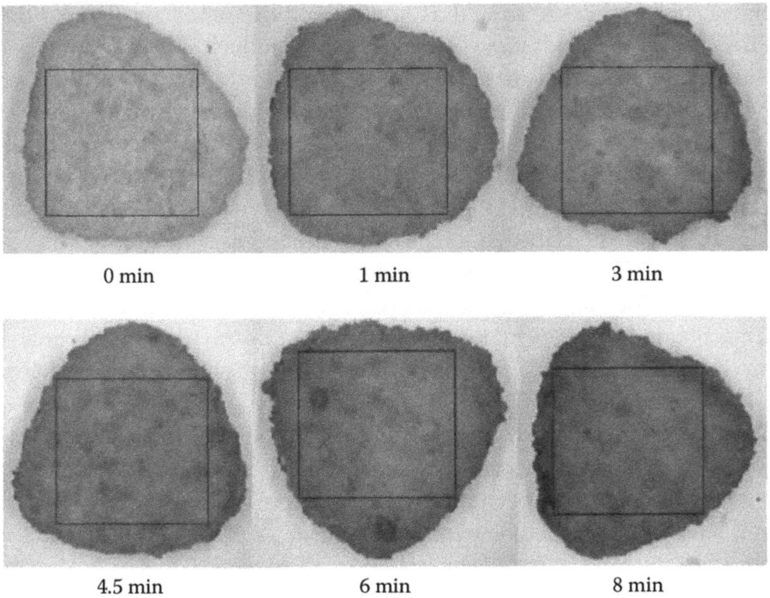

FIGURE 2.11 Images of fried chicken nuggets at different frying times (the rectangular areas were the regions of interest). (From Qiao, J. et al., *J. Food Eng.*, 79, 1065, 2007a.)

(ASM), contrast (CON), entropy (ENT), homogeneity (HOM), and correlation (COR); a multiple-layer feed-forward neural network model was established to predict the three mechanical parameters. These investigators obtained a correlation coefficient above 0.84 in comparison to a conventional method. Also, heated-oil qualities (acid value, total polar component, and viscosity) were quantitatively predicted using the VIS-NIR hyperspectral analysis and partial least squares calibration models (Kazemi et al., 2005). The R^2 values for all the quality parameters were above 0.92, indicating a good prediction.

2.5.2 Computer Vision for Fried Food Color Measurement

The color measurement of fried foods is traditionally done either instrumentally or visually through trained panelists. Human evaluation is quite subjective, tedious, and subject to error of perception. Hence, it is least recommended for the color determination of fried foods. The instrumentation procedure involves the use of color spaces and numerical values to create, visualize, and represent color in 2D and 3D spaces (Trusell et al., 2005). Food color is usually measured using the $L^*a^*b^*$ color indices. This is the international standard color measurement technique, adopted by the Commission Internationale d'Eclairage (CIE) in 1976. L^* is a measure of luminance or lightness on a scale of 0–100, a^* is a measure of greenness to redness, and b^* is a measure of blueness to yellowness. This technique provides measurement similar to how human eye perceive colors (Hunt, 1991).

A new approach to color measurement is the use of CV, which decreases some of the constraints of $L^*a^*b^*$ color space measurement, such as inability to measure a large surface area (>2 cm^2), need to measure several spots on a sample to get a true representative of color, as well as homogeneity requirement in color distribution of the surface to be measured (Mendoza and Aguilera, 2004). CV is a noninvasive, objective procedure of measuring the color pattern on a nonuniform surface; other physical properties such as texture and morphological elements could also be evaluated (Mendoza and Aguilera, 2004; Jinorose et al., 2009). The combination of computational technique, digital camera, and image processing software has made color measurement in foods less expensive and more versatile. The color model used in this technique is the RGB (red, green, blue) model, in which the sensor (charge-coupled device, CCD camera) captures color in the red, green, and blue spectra as images. These color images are then analyzed with the aid of an image processing software to obtain the interpretation of measurement taken. Figure 2.12 shows a flow diagram of the processing steps for color assessment of food using CV. León et al. (2006) and Pedreschi et al. (2006) developed a method to convert the color measured in RGB to $L^*a^*b^*$ that removes the limitation of the RGB measurement such as variation in color intensity, which results from the sensor's sensitivity variance and illumination source. CV has been used to study blooming of chocolate (Briones and Aguilera, 2005), color formation in microwaved pizza (Yam and Papadakis, 2004), and pore characteristics in pork (Du and Sun, 2006) as well as ripening of banana (Mendoza and Aguilera, 2004).

Several authors have used CV to study various color components of fried foods. Gökmen et al. (2006) used CV to study the brown color formation on fried potato

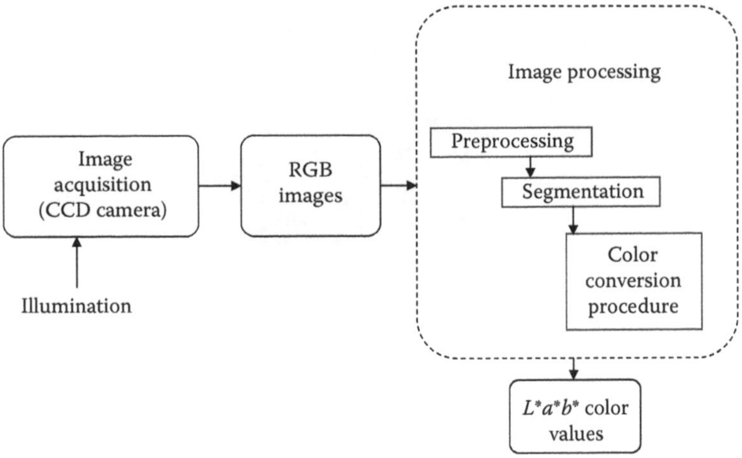

FIGURE 2.12 A flow diagram of CV system for color measurement. (Adapted from Pedreschi, F. et al., *Food Res. Int.*, 39(10), 1092, 2006.)

chips, and used this to correlate and predict the acrylamide content in the product. Pedreschi et al. (2006) also used CV to predict $L^*a^*b^*$ of fried potato chips from the RGB measurement. Further development and improvement of this technique holds so much for the future of food color measurement.

2.5.3 Use of X-Ray Micro-CT Imaging for Evaluating Microstructural Properties of Fried Foods

X-ray micro-CT is a new generation of the x-ray imaging system that has the capability of acquiring images from small objects and producing images with excellent resolution (in a few microns range). The major advantage of the x-ray micro-CT technique lies in its ability to noninvasively obtain and render high-resolution 3D images of an object with minimal or no sample preparation under the native environment of the system (Sasov and Van Dyck, 1998; Sanguansri and Augustin, 2006; van Dalen et al., 2007). The principle of the x-ray imaging technique stands on its capability to delineate between material components based on the differences in their mass densities, thereby making it possible to differentiate components in a sample based on their densities (van Dalen et al., 2003; Lim and Barigou, 2004). A good example of the use of x-ray micro-CT is in the quantitative analysis of pore characteristics, e.g., porosity determination, where air and solid components of a food material are differentiated based on the density difference between these two components.

Originally, x-ray systems were used in the medical field to study human anatomy, such as soft tissues, bones, and teeth (Ketcham and Carlson, 2001; Trater et al., 2005). The development of powerful computers, better detectors (CCD camera), image processing/analysis software, and improved focal spot with higher resolving power, reduced time of processing, and feedback, however, has enabled the extension of the technique to other applications, such as materials and biological sciences including

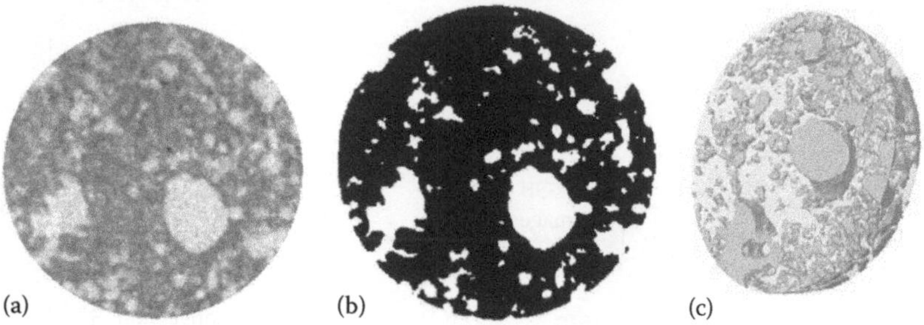

FIGURE 2.13 (a) Grayscale; (b) binarized; and (c) 3D images of formulated fried batter coating obtained by the x-ray micro-CT imaging technique. (From Adedeji, A.A. and Ngadi, M.O., Microstructural properties of deep-fat fried chicken nuggets coating with different batter formulation. *International Journal of Food Properties,* 5(4), Art. 11, 2009.)

food science and technology (Patel et al., 2003; Miri et al., 2006; Jinorose et al., 2009). Its application in studying food microstructural properties, such as porosity, pore-size distribution, pore shape, degree of pore interconnectivity as well as crust formation of fried foods has been reported in the literature. Adedeji and Ngadi (2008) evaluated various microstructural properties of formulated deep-fat fried chicken nuggets' coating using x-ray micro-CT. Grayscale images were obtained from an x-ray micro-CT scanning of the samples; these images were reconstructed and produced as 3D images by digitally stacking a series of 2D images, obtained at different angles, together (Figure 2.13).

Table 2.1 lists some of the pore parameters obtained from a 3D image analysis of a fried product. Miri et al. (2006) studied the effect of frying time and temperature on crust formation and pore characteristics of French fries using the x-ray micro-CT technique, and concluded that the method allowed not only a fundamental understanding of the frying process, but also a process design that would result in specific microstructures.

2.5.4 Fat Content Evaluation Using Confocal Laser Scanning Microscopy

The use of the confocal laser scanning microscopy (CLSM) technique in evaluating the amount of fat in fried foods is a relatively new approach. CLSM has been used previously to qualitatively assess food products in terms of changes in structure as well as chemical constituent, and to make correlations with other quality properties (Keogh and Auty, 1998). Image component differentiation in CLSM is achieved by the application of dyes/probes that stain different components of a biological material in different shades of color. The use of probes that are capable of differentiating fat within the matrix of food component and the ability to obtain 3D images through optical sectioning create a possibility of quantifying fat within a food using the partially invasive process. Some of the dyes used in food applications include Sudan Red, Congo Red, Nile Blue A, and Nile Red. Images so obtained can then be analyzed using an appropriate image processing software, such as the MATLAB® image

TABLE 2.1
Pore Properties of Deep-Fat Fried Chicken Nuggets Coating Formulated from Wheat and Rice Flour Obtained from X-Ray Micro-CT

Sample	Porosity	Structure Model Index	Fragmentation Index, 1/mm	OS/OV, 1/mm	Fractal Dimension	Number of Pores
W100R0 + CMC0%	24.33[ab]	2	9.82[a]	28.62[a]	2.16[a]	291[bc]
W100R0 + CMC1%	18.21[a]	3	13.72[c]	31.35[a]	2.04[a]	341[c]
W70R30 + CMC1%	32.18[b]	2	9.47[ab]	27.44[a]	2.29[a]	181[ab]
W50R50 + CMC1%	31.77[b]	2	10.02[b]	26.00[a]	2.24[a]	196[ab]
W30R70 + CMC1%	28.17[b]	2	9.94[b]	24.60[a]	2.17[a]	234[abc]
W0R100 + CMC1%	27.99[b]	2	7.51[a]	19.31[a]	2.11[a]	127[a]

Source: Adedeji, A.A. and Ngadi, M.O., Microstructural properties of deep-fat fried chicken nuggets coating with different batter formulation. *International Journal of Food Properties*, 5(4), Art. 11, 2009.

Note: Means with different letter superscripts in the same column are significantly different. W, wheat; R, rice; CMC, carboxymethyl cellulose; W100R0, 100% wheat and 0% rice; CMC0%, 0% CMC; same nomenclature applies to other sample IDs; OS, object surface; OV, object volume.

processing toolbox. Pore characteristics such as porosity and pore-size distribution can also be determined from CLSM images. Another advantage of CLSM is that images can be reconstructed for presentation as a 3D piece due to the optical sectioning capability of a CLSM system. Achir et al. (2009) presented a 3D reconstruction of oil distribution and cell walls in French fries obtained by UV-VIS (ultraviolet-visible) CLSM.

2.5.5 Use of Fourier-Transform Infrared Spectroscopy in Qualifying and Quantifying Oil Content in Fried Foods

Fourier-transform infrared (FTIR) is an offshoot of mid-infrared spectroscopy (IR), which is an instrumental analytical technique used for fat qualification. Its development started between 1945 and 1965, after which it was neglected due to the development of other less expensive methods. However, a renewed interest started in the 1970s when FTIR was developed, since FTIR could overcome some of the limitations associated with IR, making it possible to use the technique in quantitative analyses of fat components (van de Voort et al., 2008; Sherazi et al., 2009).

FTIR spectroscopy is a technique based on the molecular *fingerprint* produced when an electromagnetic energy (infrared light) source is impinged on a material of interest; collected data are then converted from an interference pattern into a spectrum. The wavelength of light absorbed is characteristic of the chemical bond. FTIR spectroscopy is perhaps the most powerful tool for identifying different types of chemical bonds (functional groups) found in fresh and used oil. It provides rapid means for monitoring oxidation, evaluating oxidative stability, and quantifying the amount of different oil constituents in fried products (Russin et al., 2004). Sherazi et al. (2009) used the FTIR technique to quantify the amount of *trans* fat in different types of edible oils and fats, and found comparable results with gas chromatography, but with slightly better sensitivity and accuracy for low *trans* fat values.

2.6 CONCLUDING REMARKS

The market for deep-fat fried foods keeps expanding and so are the ways of studying the frying process to obtain better quality fried products. Fried foods undergo several physicochemical changes, which are induced by various factors, such as heat and mass transfer processes during frying, product composition, pretreatment process, and type and quality of frying oil. Conventional methods of assessing and determining these quality attributes are rapidly being replaced by novel techniques that provide various merits, including time saving, noninvasiveness, and online applications. The benefits that would accrue from these innovative ways of quality assessment and control would profit both the manufacturers in maximizing profit from the production of high-quality and healthy items, and the consumers by the availability of convenient and nutritious foods.

REFERENCES

Achir, N., Vitrac, O., and Trystram, G. (2009). Heat and mass transfer during frying. In S. Sahin and S. G. Sumnu (Eds.), *Advances in Deep-Fat Frying of Foods* (pp. 21–22). Boca Raton, FL: CRC Press.

Adedeji, A. A. and Ngadi, M. O. (2008). Microstructural characterization of deep-fried breaded products using x-ray micro-computed tomography. Paper presented at the *10th ICEF: International Congress of Engineering and Food*, Viña del Mar, Chile. April 20–24, 2008.

Adedeji, A. A. and Ngadi, M. O. Microstructural properties of deep-fat fried chicken nuggets coating with different batter formulation. *International Journal of Food Properties*, 5(4), Art. 11, 2009.

Adedeji, A. A., Ngadi, M. O., and Raghavan, G. S. V. (2009). Kinetics of mass transfer in microwave precooked and deep-fat fried chicken nuggets. *Journal of Food Engineering*, 91(1), 146–153.

Agblor, A. and Scanlon, M. (2000). Processing conditions influencing the physical properties of French fried potatoes. *Potato Research*, 43(2), 163–177.

Aguilera, J. M. (2005). Why food microstructure? *Journal of Food Engineering*, 67(1–2), 3–11.

Aguilera, J. M. and Stanley, D. W. (1999). *Microstructural Principles of Food Processing and Engineering*. Gaithersburg, MD: Aspen Publishers.

AOAC. (2006). *Official Methods of Analysis* (18th edn.). Gaithersburg, MD: Association of Official Analytical Chemists International.

ASABE. (2009). American Society of Agricultural and Biological Engineers Standards, http://asae.frymulti.com/standards.asp (accessed February 19, 2009).

Ateba, P. and Mittal, G. S. (1994). Dynamics of crust formation and kinetics of quality changes during frying of meatballs. *Journal of Food Science*, 59(6), 1275–1278.

Blumenthal, M. M. (1991). A new look at the chemistry and physics of deep fat frying. *Food Technology*, 45(2), 68–72, 94.

Bouchon, P., Hollins, P., Pearson, M., Pyle, D. L., and Tobin, M. J. (2001). Oil distribution in fried potatoes monitored by infrared microspectroscopy. *Journal of Food Science*, 66(7), 918–923.

Briones, V. and Aguilera, J. M. (2005). Image analysis of changes in surface color of chocolate. *Food Research International*, 38(1), 87–94.

Budžaki, S. and Šeruga, B. (2005). Moisture loss and oil uptake during deep fat frying of "Kroštula" dough. *European Food Research and Technology*, 220(1), 90–95.

Chemicalland21. (2008). Palm oil, http://chemicalland21.com/lifescience/UH/PALMOIL.htm (accessed February 10, 2009).

Chen, Y. and Moreira, R. G. (1997). Modelling of a batch deep-fat frying process for tortilla chips. *Food and Bioproducts Processing*, 75(3), 181–190.

Dobarganes, C., Márquez-Ruiz, G., and Velasco, J. (2000). Interactions between fat and food during deep-frying. *European Journal of Lipid Science and Technology*, 102(8–9), 521–528.

Dogan, S. F., Sahin, S., and Sumnu, G. (2005). Effects of soy and rice flour addition on batter rheology and quality of deep-fat fried chicken nuggets. *Journal of Food Engineering*, 71(1), 127–132.

Du, C.-J., and Sun, D.-W. (2004). Recent developments in the applications of image processing techniques for food quality evaluation. *Trends in Food Science and Technology*, 15, 230–249.

Du, C. and Sun, D. (2006). Automatic measurement of pores and porosity in pork ham and their correlations with processing time, water content and texture. *Meat Science*, 72(2), 294–302.

Dullien, F. A. L. (1992). *Porous Media: Fluid Transport and Pore Structure* (2nd edn.). San Diego, CA: Academic Press.

Dunford, N. (2005). Not all vegetable oils are good for frying foods, http://fapc.okstate.edu/files/flash/fryingoils.pdf (accessed January 26, 2009).

El Masry, G., Wang, N., El Sayed, A., and Ngadi, M. (2007). Hyperspectral imaging for nondestructive determination of some quality attributes for strawberry. *Journal of Food Engineering*, 81(1), 98–107.

Farkas, B. E. and Singh, R. P. (1991). Physical properties of air-dried and freeze-dried chicken white meat. *Journal of Food Science*, 56(3), 611–615.

Farkas, B. E., Singh, R. P., and Rumsey, T. R. (1996). Modeling heat and mass transfer in immersion frying. I. Model development. *Journal of Food Engineering*, 29(2), 211–226.

Fernández, M. B., Tonetto, G. M., Crapiste, G. H., Ferreira, M. L., and Damiani, D. E. (2005). Hydrogenation of edible oil over Pd catalysts: A combined theoretical and experimental study. *Journal of Molecular Catalysis A: Chemical*, 237(1–2), 67–79.

Garcia, M. A., Ferrero, C., Campana, A., Bertola, N., Martino, M., and Zaritzky, N. (2004). Methylcellulose coatings applied to reduce oil uptake in fried products. *Food Science and Technology International*, 10(5), 339–346.

Giesche, H. (2006). Mercury porosimetry: A general (practical) overview. *Particle and Particle Systems Characterization*, 23(1), 9–19.

Goetz, A. F. H., Vane, G., Solomon, J. E., and Rock, B. N. (1985). Imaging spectrometry for earth remote sensing. *Science*, 228(4704), 1147–1153.

Gökmen, V., Senyuva, H. Z., Dülek, B., and Çetin, E. (2006). Computer vision based analysis of potato chips—A tool for rapid detection of acrylamide level. *Molecular Nutrition & Food Research*, 50(9), 805–810.

Gowen, A. A., O'Donnell, C. P., Cullen, P. J., Downey, G., and Frias, J. M. (2007). Hyperspectral imaging—An emerging process analytical tool for food quality and safety control. *Trends in Food Science & Technology*, 18(12), 590–598.

Hunt, R. W. G. (1991). *Measuring Color* (2nd edn.). New York: Ellis Horwood.

Huse, H. L., Mallikarjunan, P., Chinnan, M. S., Hung, Y. C., and Phillips, R. D. (1998). Edible coatings for reducing oil uptake in production of akara (deep-fat frying of cowpea paste). *Journal of Food Processing and Preservation*, 22(2), 155–165.

Huse, H. L., Hung, Y. C., and McWatters, K. H. (2006). Physical and sensory characteristics of fried cowpea (*Vigna unguiculata* L. Walp) paste formulated with soy flour and edible coatings. *Journal of Food Quality*, 29(4), 419–430.

Indira, T. N., Latha, R. B., and Prakash, M. (1999). Kinetics of deep-fat-frying of a composite product. *Journal of Food Science and Technology*, 36(4), 310–315.

Jinorose, M., Devahastin, S., Blacher, S., and Léonard, A. (2009). Application of image analysis in food drying. In C. Ratti (Ed.), *Advances in Food Dehydration* (pp. 63–95). Boca Raton, FL: CRC Press.

Karathanos, V. T., Kanellopoulos, N. K., and Belessiotis, V. G. (1996). Development of porous structure during air drying of agricultural plant products. *Journal of Food Engineering*, 29(2), 167–183.

Karmas, E. (1980). Techniques for measurement of moisture content of foods. *Food Technology*, 34(4), 52–59.

Kassama, L. S. (2003). Pore development in food during deep-fat frying, PhD thesis, Department of Bioresource Engineering, Macdonald Campus of McGill University, Montreal, Quebec, Canada.

Kassama, L. S. and Ngadi, M. O. (2003). Density, shrinkage and porosity of deep-fat fried chicken meat. Paper presented at the *American Society of Agricultural Engineers*, Las Vegas, NV.

Kassama, L. S. and Ngadi, M. O. (2004). Pore development in chicken meat during deep-fat frying. *Lebensmittel-Wissenschaft und-Technologie*, 37(8), 841–847.

Kassama, L. S. and Ngadi, M. O. (2005). Pore development and moisture transfer in chicken meat during deep-fat frying. *Drying Technology*, 23(4), 907–923.

Katz, E. E., and Labuza, T. P. (1981). Effect of water activity on the sensory crispness and mechanical deformation of snack food products. *Journal of Food Science*, 46(2), 403–409.

Kawas, M. L. and Moreira, R. G. (2001). Effect of degree of starch gelatinization on quality attributes of fried tortilla chips. *Journal of Food Science*, 66(2), 300–306.

Kazemi, S., Wang, N., Ngadi, M., and Prasher, S. O. (2005). Evaluation of frying oil quality using VIS/NIR hyperspectral analysis. *Agricultural Engineering International*, The International Commission of Agricultural Engineering (CIGR), Tsukuba, Japan, Vol. VII, Manuscript FP 05 001.

Keogh, M. K. and Auty, M. A. E. (1998). Assessment of food ingredient functionality using laser microscopy (the use of confocal laser microscopy in food ingredient evaluation), From http://www.relayresearch.ie/subapplications/teagasc/Doc (retrieved January 8, 2009).

Ketcham, R. A. and Carlson, W. D. (2001). Acquisition, optimization and interpretation of X-ray computed tomographic imagery: Applications to the geosciences. *Computers & Geosciences*, 27(4), 381–400.

Kim, M. S., Chen, Y. R., and Mehl, P. M. (2001). Hyperspectral reflectance and fluorescence imaging system for food quality and safety. *Transactions of ASAE*, 44(3), 721–729.

Kita, A., Lisinka, G., and Powolny, M. (2005). The influence of frying medium degradation on fat uptake and texture of French fries. *Journal of the Science of Food and Agriculture*, 85(7), 1113–1118.

Kroff, D. H. (1984). New rapid methods for moisture and fat analysis: A review. *Journal of Food Quality*, 6(3), 199–210.

Krokida, M. K., Oreopoulou, V., and Maroulis, Z. B. (2000). Effect of frying conditions on shrinkage and porosity of fried potatoes. *Journal of Food Engineering*, 43(3), 147–154.

Krokida, M. K., Oreopoulou, V., Maroulis, Z. B., and Marinos-Kouris, D. (2001a). Effect of pre-drying on quality of French fries. *Journal of Food Engineering*, 49(4), 347–354.

Krokida, M. K., Oreopoulou, V., Maroulis, Z. B., and Marinos-Kouris, D. (2001b). Effect of osmotic dehydration pretreatment on quality of French fries. *Journal of Food Engineering*, 49(4), 339–345.

Kumar, A. J., Singh, R. R. B., Patel, A. A., and Patil, G. R. (2006). Kinetics of colour and texture changes in Gulabjamun balls during deep-fat frying. *LWT—Food Science and Technology*, 39(7), 827–833.

Lalas, S. (2009). Quality of frying oil. In S. Sahin and S. G. Sumnu (Eds.), *Advances in Deep-Fat Frying of Foods* (pp. 57–80). Boca Raton, FL: CRC Press.

León, K., Mery, D., Pedreschi, F., and León, J. (2006). Color measurement in $L^*a^*b^*$ units from RGB digital images. *Food Research International*, 39(10), 1084–1091.

Lim, K. S. and Barigou, M. (2004). X-ray micro-computed tomography of cellular food products. *Food Research International*, 37(10), 1001–1012.

Loewe, R. (1993). Role of ingredients in batter systems. *Cereal Foods World*, 38(9), 673–677.

Mellema, M. (2003). Mechanism and reduction of fat uptake in deep-fat fried foods. *Trends in Food Science & Technology*, 14(9), 364–373.

Mendoza, F. and Aguilera, J. M. (2004). Application of image analysis for classification of ripening bananas. *Journal of Food Science*, 69(9), E471–E477.

Miri, T., Bakalis, S., Bhima, S. D., and Fryer, P. (2006). Use of x-ray micro-CT to characterize structure phenomena during frying. Paper number: 10.1051 presented at the *13th World Congress of Food Science & Technology*, Nantes, France, September 17–21, 2006.

Monteiro, S. T., Minekawa, Y., Kosugi, Y., Akazawa, T., and Oda, K. (2007). Prediction of sweetness and amino acid content in soybean crops from hyperspectral imagery. *ISPRS Journal of Photogrammetry and Remote Sensing*, 62(1), 2–12.

Moreira, R. G. and Barrufet, M. A. (1998). A new approach to describe oil absorption in fried foods: A simulation study. *Journal of Food Engineering*, 35(1), 1–22.

Moreira, R. G., Sun, X., and Chen, Y. (1997). Factors affecting oil uptake in tortilla chips in deep-fat frying. *Journal of Food Engineering*, 31(4), 485–498.

Moreira, R. G., Castell-Perrez, M. E., and Barrufet, M. A. (1999). *Deep-Fat Frying: Fundamentals and Applications*. Gaithersburg, MD: Aspen Publishers.

Ngadi, M. O., Kassama, L. S., and Raghavan, G. S. V. (2001). Porosity and pore size distribution in cooked meat patties containing soy protein. *Canadian Biosystems Engineering*, 43(3), 17–24.

Ngadi, M., Dirani, K., and Oluka, S. (2006). Mass transfer characteristics of chicken nuggets. *International Journal of Food Engineering*, 2(3), Art. 8, 1–16.

Ngadi, M. O., Li, Y., and Oluka, S. (2007). Quality changes in chicken nuggets fried in oils with different degrees of hydrogenatation. *LWT—Food Science and Technology*, 40(10), 1784–1791.

Ngadi, M. O., Wang, Y., Adedeji, A. A., and Raghavan, G. S. V. (2009). Effect of microwave pretreatment on mass transfer during deep-fat frying of chicken nugget. *LWT—Food Science and Technology*, 42(1), 438–440.

O'Brien, R. D. (1998). Fats and oils processing. In *Fats and Oils Formulating and Processing for Applications*. Lancaster, PA: Technomic Publication.

Patel, V., S., Issever A., Burghardt, A., Laib, A., Ries, M., and Majumdar, S. (2003). MicroCT evaluation of normal and osteoarthritic bone structure in human knee specimens. *Journal of Orthopaedic Research*, 21(1), 6–13.

Pedreschi, F. and Moyano, P. (2005). Oil uptake and texture development in fried potato slices. *Journal of Food Engineering*, 70(4), 557–563.

Pedreschi, F. and Zúñiga, R. N. (2009). Kinetics of quality changes during frying. In S. Sahin and S. G. Sumnu (Eds.), *Advances in Deep-Fat Frying of Foods*. Boca Raton, FL: CRC Press.

Pedreschi, F., León, J., Mery, D., and Moyano, P. (2006). Development of a computer vision system to measure the color of potato chips. *Food Research International*, 39(10), 1092–1098.

Pettinati, J. D. (1982). Introductory meat science. Current methodology and instrumentation for meat proximate analysis. Personal communication.

Piper, R. (2001). Regulation in the European union. In J. B. Rossell (Ed.), *Frying: Improving Quality*. (pp. 19–43). Cambridge, U.K.: Woodhead Publishing Ltd.

Prinyawiwatkul, W., McWatters, K. H., Beuchat, L. R., and Phillips, R. D. (1994). Physical properties of cowpea paste and akara as affected by supplementation with peanut flour. *Journal of Agricultural and Food Chemistry*, 42(8), 1750–1756.

Qiao, J., Wang, N., Ngadi, M. O., and Kazemi, S. (2007a). Predicting mechanical properties of fried chicken nuggets using image processing and neural network techniques. *Journal of Food Engineering*, 79(3), 1065–1070.

Qiao, J., Ngadi, M. O., Wang, N., Gariépy, C., and Prasher, S. O. (2007b). Pork quality and marbling level assessment using a hyperspectral imaging system. *Journal of Food Engineering*, 83(1), 10–16.

Rice, P. and Gamble, M. H. (1989). Modelling moisture loss during potato slice frying. *International Journal of Food Science and Technology*, 24(2), 183–187.

Rimac-Brnčić, S., Lelas, V., Rade, D., and Simundic, B. (2004). Decreasing of oil absorption in potato strips during deep fat frying. *Journal of Food Engineering*, 64(2), 237–241.

Roggo, Y., Edmond, A., Chalus, P., and Ulmschneider, M. (2005). Infrared hyperspectral imaging for qualitative analysis of pharmaceutical solid forms. *Analytica Chimica Acta*, 535(1–2), 79–87.

Russin, T., van de Voort, F., and Sedman, J. (2004). Rapid determination of oxidative stability of edible oils by FTIR spectroscopy using disposable IR cards. *Journal of the American Oil Chemists' Society*, 81(2), 111–116.

Saguy, I. S. and Pinthus, E. J. (1995). Oil uptake during deep-fat frying: Factors and mechanism. *Food Technology*, 49(4), 142–145,152.

Saguy, I. S., Gremaud, E., Gloria, H., and Turesky, R. J. (1997). Distribution and quantification of oil uptake in French fries utilizing a radiolabeled ^{14}C palmitic acid. *Journal of Agricultural and Food Chemistry*, 45(11), 4286–4289.

Sahin, S. and Sumnu, S. G. (2009). Preface. In S. Sahin and S. G. Sumnu (Eds.), *Advances in Deep-Fat Frying of Foods* (pp. xi). Boca Raton, FL: CRC Press.

Sanguansri, P. and Augustin, M. A. (2006). Nanoscale materials development—A food industry perspective. *Trends in Food Science & Technology*, 17(10), 547–556.

Sasov, A. and Van Dyck, D. (1998). Desktop x-ray microscopy and microtomography. *Journal of Microscopy*, 191(2), 151–158.

Sherazi, S. T. H., Kandhro, A., Mahesar, S. A., Bhanger, M. I., Talpur, M. Y., and Arain, S. (2009). Application of transmission FT-IR spectroscopy for the trans fat determination in the industrially processed edible oils. *Food Chemistry*, 114(1), 323–327.

Shyu, S.-L., Hau, L.-B., and Hwang, L. S. (2005). Effects of processing conditions on the quality of vacuum-fried carrots chips. *Journal of the Science of Food and Agriculture*, 85(11), 1903–1908.

Singh, A., Hung, Y. C., Phillips, R. D., Chinnan, M. S., and McWatters, K. H. (2004). Particle-size distribution of cowpea flours affects quality of akara (fried cowpea paste). *Journal of Food Science*, 69(7), 243–249.

Stier, R. F. (2004). Frying as a science—An introduction. *European Journal of Lipid Science and Technology*, 106(11), 715–721.

Suderman, D. R. (1983). Use of batters and breadings on food products. In D. R. Suderman and F. E. Cunningham (Eds.), *Batter and Breading Technology*. Westport, CT: AVI Publishing Co.

Sun, D. W. (2000). Inspecting pizza topping percentage and distribution by a computer vision method. *Journal of Food Engineering*, 44(4), 245–249.

Szczesniak, A. S. (1987). Correlating sensory with instrumental texture measurements—An overview of recent developments. *Journal of Texture Studies*, 18(1), 1–15.

Taiwo, K. A. and Baik, O. D. (2007). Effects of pre-treatments on the shrinkage and textural properties of fried sweet potatoes. *LWT—Food Science and Technology*, 40(4), 661–668.

Trater, A. M., Alavi, S., and Rizvi, S. S. H. (2005). Use of non-invasive x-ray microtomography for characterizing microstructure of extruded biopolymer foams. *Food Research International*, 38(6), 709–719.

Troncoso, E., Pedreschi, F., and Zúñiga, R. N. (2009). Comparative study of physical and sensory properties of pre-treated potato slices during vacuum and atmospheric frying. *LWT—Food Science and Technology*, 42(1), 187–195.

Trusell, H. J., Saber, E., and Vrhel, M. (2005). Color image processing. *IEEE Signal Processing Magazine*, 22(1), 14–22.

van Dalen, G., Blonk, H., Aalst, H., and Hendriks, C. L. L. (2003). 3-D imaging of foods using x-ray microtomography. *G. I. T. Imaging and Microscopy*, 3, 18–21.

van Dalen, G., Nootenboom, P., van Vliet, L. J., Voortman, L., and Esveld, E. (2007). 3D Imaging and analysis of porous cereal products using x-ray microtomography. *Image Analysis and Stereology*, 26, 169–177.

van de Voort, F., Ghetler, A., García-González, D., and Li, Y. (2008). Perspectives on quantitative mid-FTIR spectroscopy in relation to edible oil and lubricant analysis: Evolution and integration of analytical methodologies. *Food Analytical Methods*, 1(3), 153–163.

van Loon, W., Visser, J., Linssen, J., Somsen, D., Jan Klok, H., and Voragen, A. (2007). Effect of pre-drying and par-frying conditions on the crispness of French fries. *European Food Research and Technology*, 225(5), 929–935.

Vitrac, O., Trystram, G., and Raoult-Wack, A. (2000). Deep-fat frying of food: Heat and mass transfer, transformations and reactions inside the frying material. *European Journal of Lipid Science and Technology*, 102(8–9), 529–538.

Webb, P. A. (2001). Volume and density determinations for particle technologists: Micromeritics, http://www.micromeritics.com/Repository/Files/density_determinations.pdf (accessed August 31, 2009).

Xing, J., Saeys, W., and De Baerdemaeker, J. (2007). Combination of chemometric tools and image processing for bruise detection on apples. *Computers and Electronics in Agriculture*, 56(1), 1–13.

Yam, K. L. and Papadakis, S. E. (2004). A simple digital imaging method for measuring and analyzing color of food surfaces. *Journal of Food Engineering*, 61(1), 137–142.

Zheng, G., Chen, Y., Intes, X., Chance, B., and Glickson, J. D. (2004). Contrast-enhanced near-infrared (NIR) optical imaging for subsurface cancer detection. *Journal of Porphyrins and Phthalocyanines*, 8(9), 1106–1117.

3 Physicochemical Property Changes of Foods during Microwave-Assisted Thermal Processing

Arun Muthukumaran, Valérie Orsat, and G. S. Vijaya Raghavan

CONTENTS

3.1 INTRODUCTION

Microwaves are part of the electromagnetic spectrum in the frequency range of 300–300,000 MHz. Microwaves present many advantages compared to conventional heating methods, including rapid heating, energy conservation, and convenience (Venkatesh and Raghavan, 2004; Zhang et al., 2006). During the past few decades, a microwave oven has indeed become one of the important appliances in many homes. However, microwave cooking has not found similar popularity in industrial thermal processing applications. Apart from government regulations over the use of microwave frequencies, the major hindering factor is the poor understanding of microwaves and material interactions. Material properties, especially dielectric properties, play a major role in determining the microwave ablity of a product. Hence, understanding of the dielectric properties of food materials is the first step in developing successful microwave processing methods.

There is so far no comprehensive review on property changes of foods during microwave thermal processing. Most of the past literature either concentrates on a particular commodity or product, or a specific property change. In this chapter, an overview of physicochemical changes of selected foods during microwave thermal processing is presented. The chapter gives some important theoretical information on microwave thermal processing of foods and their property changes along with some research examples. For detailed information about a specific food product and its interaction with microwaves, the readers are suggested to use the quoted references.

3.2 DIELECTRIC PROPERTIES

Food materials are poor electric insulators and can store and dissipate energy when exposed to microwaves. Dielectric properties describe an interaction of an electromagnetic field with non- or low-conducting matter. The dielectric properties data are very limited in literature and usually available only for a few foods or food components. Wang et al. (2008), Tanaka et al. (2008), and Liao et al. (2001, 2002, 2003) have conducted experiments to get a better understanding of the dielectric properties of various food products. The dielectric properties of some food materials are shown in Table 3.1 (Regier and Schubert, 2001).

3.2.1 Dielectric Constant (ε') and Dielectric Loss Factor (ε'')

Dielectric constant (ε') and dielectric loss factor (ε'') are the two most important properties, which have a pronounced effect on the effectiveness of microwave processing of a material. The dielectric constant (ε') is a ratio of the permittivity of a substance to the permittivity of free space. The dielectric constant of a material gives the extent the material could concentrate the electric flux. It is an electrical equivalent of the relative magnetic permittivity (Regier and Schubert, 2001; Venkatesh and Raghavan, 2004).

The complex permeability (μ) and complex permittivity (ε) are two important parameters:

$$\mu = \mu' - j\mu'' \tag{3.1}$$

TABLE 3.1
Dielectric Properties of Various Materials at 2.45 GHz (Room Temperature)

Material	ε'	ε''	Remarks
Water	78.1	10.4	
Ice	3	0.003	At −2°C
1 M NaCl solution	74.8	2.14	
2 M NaCl solution	65.6	69.1	
Ethanol	7.5	7.1	
10% ethanol solution	71.5	13.8	(Weight %)
10% sucrose solution	74.5	13.1	(Weight %)
Vegetable oil	3.1	0.4	
Beef tissue	50	15	Uncooked
Cooked ham	45	25	
Mashed potatoes	65	21	
Carrot tissue	71	18	Water content 89.7%
Apple tissue	64	13	Water content 84.0%

$$\varepsilon = \varepsilon' - j\varepsilon'' \quad (3.2)$$

where
ε' is the dielectric constant
ε'' is the dielectric loss factor

Permeability and permittivity of a material can be linked by the following equation:

$$C_0\mu_0\varepsilon_0 = 1 \quad (3.3)$$

where
C_0 is the speed of light
μ_0 is the magnetic constant
ε_0 is the permittivity of vacuum

The numerical value of ε_0 is 8.854 pF m^{-1}, while the magnetic constant (μ_0) takes the value of 1.26 μH m^{-1} (Nyfors and Vainikainen, 1989; Venkatesh and Raghavan, 2004):

$$\varepsilon_r = \frac{\varepsilon_{abs}}{\varepsilon_0} \quad (3.4)$$

When the microwave energy passes through a material, part of the energy is reflected while some other part is absorbed by the material (depending on the dielectric properties of the material) and another part is transmitted. The part of the reflected energy (P_r) is a function of the dielectric constant (ε') and the angle of incidence (θ), and can be calculated using the following equation:

$$P_r(\theta) = \frac{\left(\sqrt{\varepsilon'} - 1\right)^2}{\left(\sqrt{\varepsilon'} + 1\right)^2} \tag{3.5}$$

The proportion of microwave power transmitted through the material could be easily calculated by

$$P_{trans} = (1 - P_r) \tag{3.6}$$

The quantity of energy absorbed by the material is important during microwave thermal processing. Microwave power absorbed by the material is directly correlated to the dielectric loss factor (ε'') of the material and is related to the absorbed microwave power through the following relationship:

$$P_{abs} = 5.56 \times 10^{-4} f \varepsilon'' E^2 \tag{3.7}$$

3.2.2 Penetration Depth

When microwaves pass through a lossy material, part of the energy is converted into heat and the rest of the energy decreases as the distance increases from the surface. The penetration depth can be described as the distance at which microwave power reaches 50% attenuation. Lambert's law is generally used to describe the penetration depth (d_p):

$$d_p = \frac{\lambda_0 \sqrt{\varepsilon'}}{2\pi\varepsilon''} \tag{3.8}$$

where λ_0 is the free-space microwave wavelength; at 2.45 GHz, the value for λ_0 is 12.2 cm. Most food products have dielectric loss factor of less than 25; hence, the penetration depth for most food materials is between 0.6 and 1 cm. If a food material has higher attenuation capacity, the effectiveness of the microwave processing would be limited.

The microwave power dissipated inside a material is proportional to the dielectric loss factor (ε''). The heat generated inside the material due to microwave can be expressed as

$$P_v = 2\pi f \varepsilon_0 \varepsilon'' |E|^2 \tag{3.9}$$

The amount of thermal energy generated in a dielectric material can be calculated by (Goldblith, 1966)

$$P_v = 5.56 \times 10^{-11} f \varepsilon'' E^2 \tag{3.10}$$

where

P_v is the power conversion per unit volume (W m^{-3})
f is the frequency (Hz)
ε'' is the relative dielectric loss factor
E is the electric field (V m^{-1})

3.3 MICROWAVE HEATING MECHANISM

Microwave heating could be attributed to either ionic interaction or dipolar rotation. In the case of ionic interaction, any charged particle in a microwave field experiences three orthogonal force components corresponding to the frequency of the microwaves, which results in heating. Another mechanism, which is widely used to describe microwave heating, is dipolar rotation. Water is a predominant component in food and contains dipolar molecules. When the dipolar molecules are exposed to microwaves, they would try to align with the field. This results in collision with adjacent molecules as well as friction. Collision and friction generate heat from within the food material, a mechanism that is drastically different from conventional heating. Microwaves themselves do not contain heat, and hence the physical properties of the material and its interaction with microwaves are important for heating (Buffler, 1992).

3.4 FACTORS AFFECTING DIELECTRIC PROPERTIES OF FOODS

Food is a complex mixture of different components and its dielectric properties are highly dependent on its compositions. Moisture content and salt concentration usually play a major role in determining the dielectric properties of foods. Figure 3.1 shows the relationship between the dielectric loss factor components and temperature.

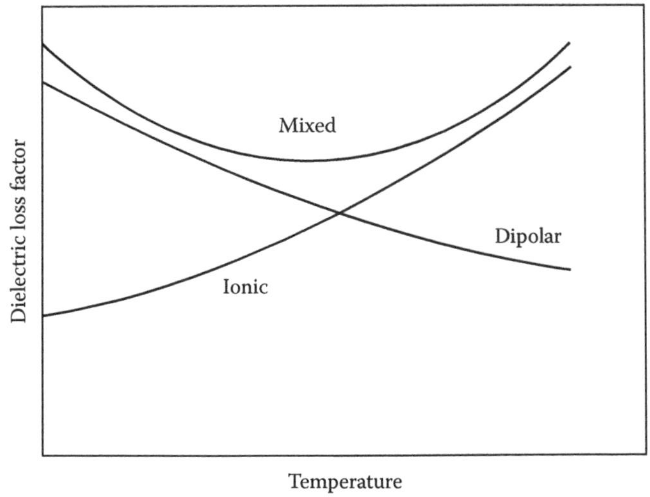

FIGURE 3.1 Temperature dependencies of the dielectric loss factor components. (From Sahin, S. and Sumnu, S.G., *Physical Properties of Foods*, Springer, New York, 2006. With permission.)

3.4.1 Effects of Temperature and Frequency

Dielectric properties of foods depend on many parameters, including microwave frequency, temperature, compositions of food, and its moisture content. Understanding the effect of each parameter and its effect on microwave processing is important in developing microwave thermal process for a particular food material. Relaxation time (τ) is an important measure, which gives the effects of temperature and frequency on dielectric properties of food. Relaxation time can be defined as the time required for a molecule to return to 1/e or 36.8% of the original condition when the static electric field is suddenly removed. Lower relaxation time is required to achieve rapid heat generation. Larger molecules have longer relaxation time in general. The relaxation time for pure water is between 0.0071 and 0.000148 ns at 20°C (Mashimo et al., 1997). Water plays a major role in determining the dielectric properties of food. Water molecules bound to the food matrix have longer relaxation time than free water molecules.

Debye developed a model to relate the relaxation time of a spherical molecule to the viscosity and temperature under Brownian movement (Von Hippel, 1954):

$$\tau = V\frac{3\nu}{kT} \tag{3.11}$$

where

τ is the relaxation time (ns)
V is the volume of the sphere (m^3)
ν is the kinematic viscosity ($m^2\ s^{-1}$)
T is the absolute temperature (K)
k is a constant

The relaxation time for nonspherical molecules could be related to the temperature and viscosity by the following equation:

$$\tau \propto \frac{\nu}{T} \tag{3.12}$$

Viscosity of all fluids decreases with the increasing temperature following Equation 3.13 (Macosko, 1994):

$$\nu = \nu_0 e^{E_a/RT} \tag{3.13}$$

where

E_a is the activation energy
R is the universal gas constant

From these equations, it is clear that an increase in temperature reduces the relaxation time. When the dielectric loss factor increases with the increasing temperature, food would experience a phenomenon known as "thermal runaway." When frozen food is thawed with higher microwave power, certain area of the food would be overheated while other areas remain much cooler. Hence, it is important to maintain low microwave power for thawing frozen food to have a uniform thawing.

3.4.2 Effect of Moisture Content

Water is a dipolar molecule; hence, it is an important factor, which determines the dielectric properties of a food material. Higher moisture content increases the dielectric constant and loss factor. The presence of free water has more significant impact on the dielectric properties than the presence of bound water. At lower moisture content, the effectiveness of microwave thermal processing decreases. Hence, it would be beneficial to use a microwave to accelerate thermal processing during an initial stage and apply different processing methods during the second or final stage of processing (Erle et al., 2000; Venkatesh and Raghavan, 2004).

3.4.3 Effect of Other Components

As mentioned earlier, presence of salt could play a major role in affecting the dielectric properties of a food material. Higher salt content reduces the availability of free water and hence reduces the dielectric constant and loss factor. It also increases the conductive loss. Sugar has a similar effect; however, sugar molecules are larger in size and hence have longer relaxation times. Hydration of sugar molecules shifts the relaxation time to lower frequencies and hence increasing the loss factor. Proteins and starch also reduce the dielectric constant and increase the loss factor (Roebuck et al., 1972).

3.5 EFFECT OF MICROWAVE THERMAL PROCESSING ON MATERIAL PROPERTIES

The following are key examples of how microwave thermal processing could affect the physical and chemical properties of a food material being processed. Each example represents major research areas in food processing, for example, fruits and vegetables, dairy products, meat products, marine products, and essential oils.

3.5.1 Fruits and Vegetables

Fruits and vegetables are highly perishable in nature. This, along with quality requirements, the need to improve energy efficiency, and lower production costs, leads to the development of newer handling and processing methods. Drying is usually a popular processing method for enhancing the shelf life of fruits and vegetables. The market for dehydrated fruits and vegetables indeed stands at around US$ 8 billion (Funebo and Ohlsson, 1998; Zhang et al., 2006). However, microwave drying alone might not be sufficient in most cases. Hybrid drying is considered as an effective method for drying of fruits and vegetables. Microwaves can be effectively combined with conventional drying methods to reduce the drying time, and improve drying efficiency and product quality. Microwave-assisted air drying, microwave-assisted drying in whole air drying process, microwave-assisted drying as final-stage air drying, microwave-assisted vacuum drying, microwave-enhanced spouted bed drying, microwave-assisted freeze drying, and microwave-assisted finish drying following osmotic dehydration are some of the newer microwave hybrid drying methods developed for fruits and vegetables. Further details on these drying methods can be found in the review work of Sunjka et al. (2004) and Zhang et al. (2006).

Information on property changes of fruits and vegetables due to microwave processing is limited in literature. Fruits and vegetables usually contain high moisture content, and hence their dielectric property changes are similar to that of pure water. Dielectric constant (ε') values of fruits and vegetables decrease as the microwave frequency increases. Ash content of fruits and vegetables is inversely proportional to ε'. Higher moisture content results in higher ε' values. The dielectric loss factor (ε'') of fruits and vegetables has negative correlation with the moisture content (Mudgett, 1995; Ryynänen, 1995; Sipahioglu and Barringer, 2003).

3.5.2 Cereal Starches

Lewandowicz et al. (1997) conducted experiments to study the effect of microwave processing on the physicochemical properties of cereal starches. They used corn, wheat, and waxy corn at an intermediate moisture content of 30%. Brabender rheological method, light microscopy, x-ray diffractometry (XRD), and differential scanning calorimetry (DSC) methods were used to test different properties of starches, both before and after microwave processing. The experimental results are shown in Figures 3.2 and 3.3. Corn and wheat had pronounced changes in their molecular structure compared to waxy corn.

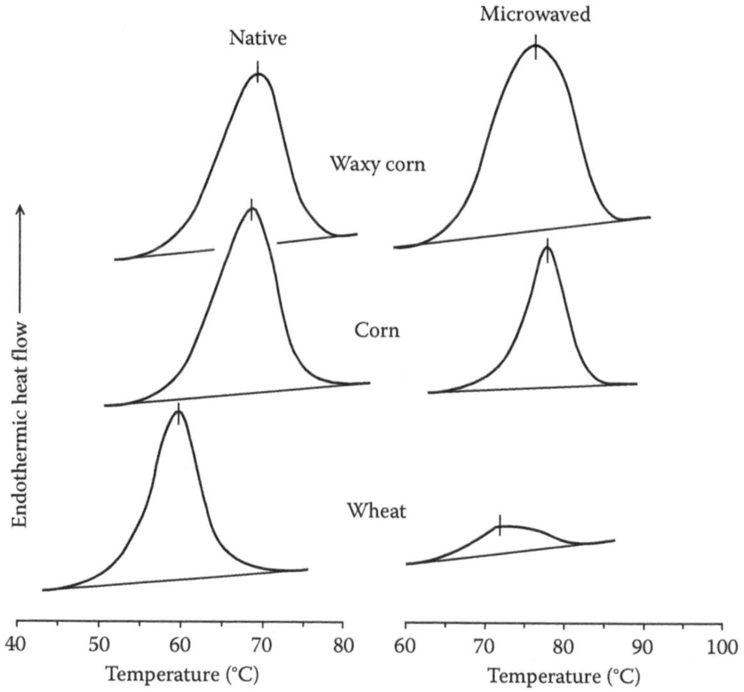

FIGURE 3.2 DSC thermograms of native and microwaved cereal starches. (From Lewandowicz, G. et al., *Carbohydr. Polym.*, 42, 193, 2000. With permission.)

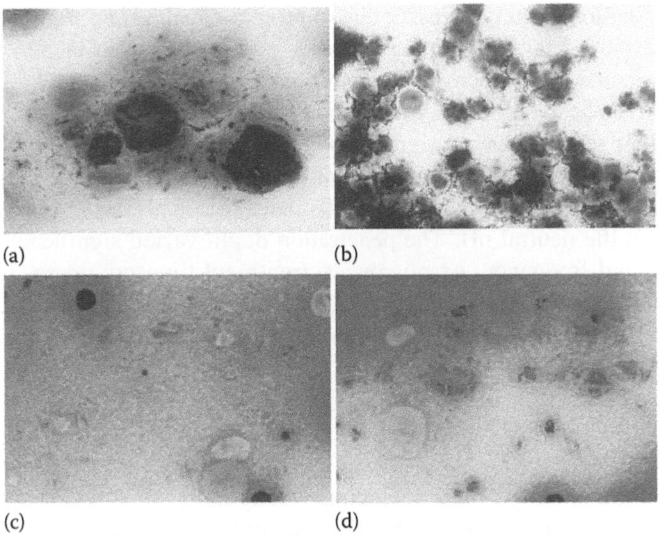

FIGURE 3.3 Light microphotographs of the native and irradiated starch samples. (a) Native corn at 95°C, (b) microwaved corn at 95°C, (c) native waxy corn at 75°C, and (d) microwaved waxy corn at 75°C. (From Lewandowicz, G. et al., *Carbohydr. Polym.*, 42, 193, 2000. With permission.)

3.5.3 Soy Protein

Soy protein is widely used as meat replacement in many new food products. Soy protein is widely available in three different formats, viz., soy concentrate, soy flour, and soy protein isolate (SPI). Ahmed et al. (2008) studied the effects of temperature and pH on the dielectric properties of SPI at different concentrations. They observed that the dielectric constant ε' decreased with an increase in temperature. The dielectric constant ε' decreased gradually as the temperature increased from 20°C to 80°C. There was a significant drop of ε' for the temperatures above 60°C. The decrease in ε' with temperature is indeed common for various foods, including proteins. Ohlsson and Bengtsson (1975), Ohlsson et al. (1974), and Wang et al. (2003) reported similar observations from their research. However, the correlation between the temperature and ε' was not linear. At temperatures above 80°C, the value of dielectric constant increased significantly. In case of proteins like SPI, the increase in dielectric constant value could be attributed to protein denaturation. Bircan and Barringer (2002) and Wang et al. (2003) reported similar increasing behavior in ε' values for meat and whey proteins, respectively.

The dielectric loss factor was found to decrease sharply at lower microwave frequencies up to 1000 MHz and then increased gradually. At lower frequencies, the change in loss factor is generally ascribed to ionic conductivity. At higher frequencies, bound water relaxation and free water relaxation are most probably responsible for dielectric loss factor changes. Dielectric constant and dielectric loss factor of SPI were affected by microwave treatment at frequencies between 200 and 2500 MHz at concentration levels of 5–10 g/100 g water (Ahmed et al., 2008). Dielectric constants generally decrease while dielectric loss factors increase with increasing temperature. At temperatures above 90°C, the loss factor decreases due to protein denaturation. Prediction models have been developed for determining dielectric parameters as a function of temperature and concentration at microwave frequencies of 915 and 2450 MHz. Decrease in both ε′ and ε′ has been reported for other foods at higher frequencies.

The pH change could have a significant effect on dielectric properties of a material. In the case of SPI, it was observed that the ε′ value increased at pH 4.5 and 10. However, there were not any significant differences in ε′ at pH 4.5 and 10 as compared to the value at the neutral pH. The penetration depth varied significantly with pH, temperature, and frequency. As microwave treatment time progresses, an increase in binding and redistribution of water around protein molecules was observed. This effect could be attributed to the transformation of macromolecules due to heating (Ahmed et al., 2008).

3.5.4 Butter

Butter is a common dairy product, which is widely used in many different cuisines. It has very short shelf life and is highly susceptible to oxidation-related changes during storage. Different processing methods have been developed to improve the storage and handling of butter. Ahmed et al. (2007) conducted experiments to study the effect of microwave frequency on the dielectric properties of butter. They also included the effects of salt and temperature on the dielectric properties in their study. This study provides a typical example for the complex interactions between different food compounds, which are often encountered during food processing.

Based on the experimental results, it was found that addition of salt changed the effect of microwave frequency on the dielectric properties of butter. For unsalted butter, the ε′ did not change with frequency, whereas for salted butter, ε′ decreased as the microwave frequency increased. A relatively constant ε′ value for the unsalted butter could be attributed to the lack of changes in the orientation of molecules in the electromagnetic field. The loss factor (ε″) of both unsalted and salted butter decreased as the frequency increased. Unsalted butter showed a significant decrease in ε″ as the frequency increased from 500 to 1500 MHz. Beyond 1500 MHz, the change in ε″ was relatively small, however. In the case of salted butter, the decrease in ε″ was gradual with the frequency increase.

Wang et al. (2005) reported that ionic conductivity played a major role in affecting the dielectric properties of butter at frequencies below 3000 MHz. Addition of salt to butter resulted in ionic conductivity changes and hence had pronounced effect on the dielectric property changes of butter under microwave frequencies. Similar results could be found in the literature (e.g., Nelson and Bartley, 2000; Everard et al., 2006).

Temperature change was found to affect the dielectric properties of both salted and unsalted butter. Increase in the processing temperature resulted in gradual decrease in ε′. Salted butter did not show a significant decrease in ε′ for processing temperatures between 30°C and 50°C. However, beyond 60°C, the ε′ value showed an upward change. The loss factor ε″ decreased for unsalted butter as the temperature increased. However, ε″ of salted butter increased as the processing temperature increased. For both salted and unsalted butter, the loss factor–frequency trend as a function of temperature followed a trend similar to that of ε′–frequency as mentioned earlier.

The lack or absence of dielectric relaxation at higher frequency range could be responsible for the decrease in dielectric constant and loss factor for unsalted butter (Nelson and Bartley, 2000). The change in ionic conduction could be attributed to the changes encountered in salted butter during microwave processing. On the other hand, longer relaxation times at lower temperatures could be responsible for the dielectric property changes in salted butter. As the temperature increased, the relaxation time and viscosity of salted butter decreased, resulting in higher ionic movements and hence the increase in dielectric parameters. Salt has a major role in affecting the loss factor, and in the case of unsalted butter, the changes could be attributed to dipole loss as the moisture content decreased during processing. Mudgett (1995) as well as Bircan and Barringer (1998) reported similar observations for salt–starch interaction on dielectric properties.

Processing temperature also plays a major role in affecting the penetration depth (d_p). For unsalted butter, d_p increased between 50°C and 70°C for the microwave frequency of 915 MHz. An increase in the temperature afterward resulted in decreased d_p. A similar trend was also observed at 2450 MHz and the variation became insignificant beyond 80°C (Wang et al., 2003; Everard et al., 2006; Ahmed et al., 2007).

3.5.5 Cheese

Cheese is another popular dairy product, which is rich in protein, calcium, and phosphorus. The dielectric properties of cheese are important in developing new microwave processing methods. Everard et al. (2006) studied the effects of different microwave frequencies (300 MHz–3 GHz) on the dielectric properties of cheese. Their results showed that the dielectric constant (ε') and the loss factor (ε'') both decreased as the processing frequency increased. For ε', the decrease with increasing frequency followed a linear path. In the case of ε'', the relationship was inversely proportional. Nelson and Bartley (2000) and Wang et al. (2003) both reported similar decrease in ε' and ε'' as the frequency increased from 27 to 1800 MHz and 27 to 915 MHz, respectively.

Temperature also has a similar effect on the dielectric properties of cheese. The effect of temperature, however, depends on the moisture content of the cheese being processed. Medium-moisture content cheese exhibits a decrease in ε' for temperatures between 5°C and 55°C. Further increase in processing temperature results in an increase in ε'. The increase in ε' value reversed again at 65°C. The decrease in ε' value between 65°C and 85°C is common for both medium- and low-moisture cheese. The dielectric property changes at higher temperatures are similar to that of soy protein. Higher processing temperatures result in protein denaturation.

The loss factor (ε'') for higher- and medium-moisture content cheese increases gradually with temperature (5°C–85°C). The trend is opposite for low-moisture cheese. The increase in ε'' for high- and medium-moisture cheese could be attributed to ionic conduction. It was reported that the effect of temperature was more pronounced at lower frequencies than at higher frequencies (above 1 GHz) (Nelson and Bartley, 2000). Models were also developed to predict the effects of moisture and salt content. These models can provide the effects of frequency, temperature, and compositions on microwave processing of cheese.

3.5.6 Edible Oils

The dielectric constant (ε') of various edible oils and fatty acids has similar frequency dependence. The changes in ε' value are minimal up to 500 kHz and then it starts decreasing. At lower frequencies, there exists an equilibrium between the oil molecules and the electromagnetic field, and hence the changes in ε' are minimal. When the frequency increases beyond 500 kHz, the ε' value gradually decreases.

The ε'' of the oils has inverted bell shape relationship with the frequency. The ε'' value decreases from 100 Hz to 13.2 kHz; after that, an increase in the frequency results in an increase in ε''. Lizhi et al. (2008) reported that the ε' values of oils were mainly affected by C18 unsaturated fatty acids. An increase in the degree of unsaturation of oils resulted in higher ε'. Hence, it could be concluded that if a fatty acid has higher number of double bonds, it would have higher ε' values. Pace et al. (1968) and Rudan-Tasič and Klofutar (1999) reported similar results in dielectric properties of oils.

3.5.7 Beef

Beef is one of the important meats in many different cuisines and is consumed all over the world. It is a rich source of selenium, zinc, iron, vitamin B, and carnitine. The dielectric constant (ε') and the loss factor (ε'') of beef both increase at higher temperatures. For example, ε' values of lean beef at 27.12 MHz were 36.0 and 68.8 at –5°C and –1°C, respectively. Temperatures in the range of –1°C and 10°C did not have a significant effect on ε' and ε'' at any microwave frequency. Higher temperature had more effect on ε'' for all meats at 27.12 MHz and ε'' increased with temperature (Ryynänen, 1995; Shukla and Anantheswaran, 2001).

Moisture content plays a major role in determining the dielectric properties of meat. Dipolar nature of water helps in microwave energy absorption. On the other hand, fat content of meat does not play a major influencing role on the microwave ablity of meat as fats have lower dielectric properties. However, higher fat content is usually associated with lower moisture content and hence indirectly affects the dielectric properties of meat. Protein has a relatively significant effect on the microwave thermal processing of meat than fats, but nevertheless has significantly lower effect than moisture content. Temperature increase significantly reduces the microwave penetration depth. For instance, d_p of lean meat was reported to decrease from 54.2 to 17.7 cm at 27.12 MHz and at –10°C and 1°C, respectively (Mudgett and Westphal, 1989; Ryynänen, 1995; Shukla and Anantheswaran, 2001; Mudgett; Farag et al., 2008). This is similar to the results obtained by Ohlsson et al. (1974) on the effect of temperature on microwave penetration depth.

3.5.8 Salmon

Wang et al. (2008) studied the effect of temperature on the dielectric properties of salmon fillets. Their results showed that between 20°C and 120°C, dielectric constant (ε') of pink salmon fillets steadily increased with temperature at low frequencies of 27 and 40 MHz. However, the ε' decreased slightly at higher frequencies up to 1800 MHz. The decrease in ε' could be attributed to higher intermolecular interaction at higher temperatures. Herve et al. (1998) and Tang et al. (2002) reported similar variations in ε'. The moisture content of salmon plays a major role in affecting the ε' and ε''. Ash content and dissolved ion content also have a significant effect on the dielectric properties and subsequent microwave thermal processing of salmon (Ohlsson et al., 1974; Nelson and Bartley, 2002; Wang et al., 2003).

3.5.9 Mirin

Mirin is a condiment with almost 40%–50% sugar and is widely used in Japanese cuisine. Dielectric loss factor of mirin is affected by both the dipolar loss component and the ionic conductivity. Ionic conductivity is lower at higher microwave frequencies. The combined effect of temperature, microwave frequency, and sugar content is complex and hard to describe. Nevertheless, the ε'' increases with frequency and temperature. The penetration depth decreases as the processing temperature increases. The effect of temperature on the d_p is significant at lower processing frequencies; at higher frequencies, temperature has only moderate effect on the d_p. Tanaka et al. (2005) reported similar results for soy sauce. It was noted by Liao et al. (2003) that this trend is distinctive for thick or complex solutions.

3.6 CONCLUDING REMARKS

Microwaves could potentially provide highly effective and versatile thermal processing methods. However, to successfully utilize the microwaves for thermal processing, it is necessary to understand the changes in food products processed with microwaves. The dielectric properties of food materials play a major role in determining the microwave ablity of the food materials. The complex nature of the food materials, combined with external factors involved in microwave thermal processing, makes it a challenge to successfully develop newer microwave processing methods. Microwaves could also be used in conjunction with other conventional thermal processing methods. More details on dielectric properties and microwave processing of the examples quoted in this chapter could be obtained from the quoted references.

ACKNOWLEDGMENTS

The authors extend their sincere gratitude for the financial support provided by the Natural Science and Engineering Research Council of Canada (NSERC) and Fonds Québécois de la recherché sur la nature et les technologies (FQRNT).

NOMENCLATURE

C_0 Speed of light
d_p Penetration depth
E Electric field, V m^{-1}

E_a Activation energy
F Frequency, Hz
J Complex number
K Constant
P_v Energy developed per unit volume, W m^{-3}
R Universal gas constant
T Absolute temperature
V Volume, m^3
V Viscosity
v_0 Initial viscosity
E Complex relative dielectric constant
ε' Dielectric constant
ε'' Dielectric loss factor
ε_0 Absolute permittivity of vacuum (8.854 pF m^{-1})
λ_0 Free space microwave wavelength
μ_0 Magnetic constant (1.26 μH m^{-1})
T Relaxation time, ns

REFERENCES

Ahmed, J., H. S. Ramaswamy, and V. G. S. Raghavan. 2007. Dielectric properties of butter in the MW frequency range as affected by salt and temperature. *Journal of Food Engineering* 82:351–358.

Ahmed, J., H. S. Ramaswamy, and G. S. V. Raghavan. 2008. Dielectric properties of soybean protein isolate dispersions as a function of concentration, temperature and pH. *LWT—Food Science and Technology* 41:71–81.

Bircan, C. and S. A. Barringer. 1998. Salt–starch interactions as evidenced by viscosity and dielectric property measurements. *Journal of Food Science* 63:983–986.

Bircan, C. and S. A. Barringer. 2002. Determination of protein denaturation of muscle foods using the dielectric properties. *Journal of Food Science* 67:202–205.

Buffler, C. R. 1992. *Microwave Cooking and Processing*. New York: Van Nostrand Reinhold.

Erle, U., M. Regier, C. Persch, and H. Schubert. 2000. Dielectric properties of emulsions and suspensions: Mixture equations and measurement comparisons. *The Journal of Microwave Power and Electromagnetic Energy* 35:185–190.

Everard, C. D., C. C. Fagan, C. P. O'Donnell, D. J. O'Callaghan, and J. G. Lyng. 2006. Dielectric properties of process cheese from 0.3 to 3 GHz. *Journal of Food Engineering* 75:415–422.

Farag, K. W., J. G. Lyng, D. J. Morgan, and D. A. Cronin. 2008. Dielectric and thermophysical properties of different beef meat blends over a temperature range of −18 to +10°C. *Meat Science* 79:740–747.

Funebo, T. and Ohlsson, T. 1998. Microwave-assisted dehydration of apple and mushroom. *Journal of Food Engineering* 39:353–367.

Goldblith, S. A. 1966. The wholesomeness of irradiated foods: Past history, present status, international aspects, and future outlook. *Food Technology* 20:93–98.

Herve, A. -G., J. Tang, L. Luedecke, and H. Feng. 1998. Dielectric properties of cottage cheese and surface treatment using microwaves. *Journal of Food Engineering* 37:389–410.

Lewandowicz, G., J. Fornal, and A. Walkowski.1997. Effect of microwave radiation on physico-chemical properties and structure of potato and tapioca starches. *Carbohydrate Polymers* 34:213–220.

Lewandowicz, G., T. Jankowski, and J. Fornal 2000. Effect of microwave radiation on physico-chemical properties and structure of cereal starches. *Carbohydrate Polymers* 42(2):193-199.

Liao, X., G. S. V. Raghavan, and V. A. Yaylayan. 2001. Dielectric properties of alcohols (C_1-C_5) at 2450 MHz and 915 MHz. *Journal of Molecular Liquids* 94:51–60.

Liao, X., G. S. V. Raghavan, and V. A. Yaylayan. 2002. Dielectric properties of aqueous solutions of α-D-glucose at 915 MHz. *Journal of Molecular Liquids* 100:199–205.

Liao, X., G. S. V. Raghavan, G. Wu, and V. A. Yaylayan. 2003. Dielectric properties of lysine aqueous solutions at 2450 MHz. *Journal of Molecular Liquids* 107:15–19.

Lizhi, H., K. Toyoda, and I. Ihara. 2008. Dielectric properties of edible oils and fatty acids as a function of frequency, temperature, moisture and composition. *Journal of Food Engineering* 88:151–158.

Macosko, C.W. 1994. *Rheology: Principles, Measurements and Applications*. Berlin, Germany: Wiley-VCH.

Mashimo, S., Kuwabara, S., Yagihara, S., and Higasi, K. 1997. Dielectric relaxation time and structure of bound water in biological materials. *Journal of Physical Chemistry* 91:6337–6338.

Mudgett, R. E. 1995. Electrical properties of foods. In M. A. Rao, and S. S. H. Rizvi (Eds.), *Engineering Properties of Foods*, pp. 389–455. New York: Marcel Dekker.

Mudgett, R. and W. Westphal. 1989. Dielectric behavior of an aqueous cation exchanger. *Journal of Microwave Power and Electromagnetic Energy* 24:33–37.

Nelson, S. O. and P. G. Bartley. 2000. Measuring frequency- and temperature-dependent dielectric properties of food materials. *Transactions of the ASAE* 43:1733–1736.

Nelson, S. O. and P. G. Bartley. 2002. Frequency and temperature dependence of the dielectric properties of food materials. *Transactions of the ASAE* 45(4):1223–1227.

Nyfors, E. and P. Vainikainen. 1989. *Industrial Microwave Sensors*. Norwood, MA: Artech House.

Ohlsson, T. and N. E. Bengtsson. 1975. Dielectric food data for microwave sterilization processing. *The Journal of Microwave Power* 10:93–108.

Ohlsson, T., N. E. Bengtsson, and P. O. Risman. 1974. The frequency and temperature dependence of dielectric food data as determined by a cavity perturbation technique. *The Journal of Microwave Power* 9:129–145.

Pace, W. E., W. B. Westphal, and S. A. Goldblith. 1968. Dielectric properties of commercial cooking oils. *Journal of Food Science* 33:30–36.

Regier, M. and Schubert, H. 2001. Microwave processing. In P. Richardson (Ed.), *Thermal Technologies in Food Processing*, pp. 178–207. Cambridge, U.K.: Woodhead Publishing Ltd.

Roebuck, B. D., S. A. Goldblith, and W. B. Westphal. 1972. Dielectric properties of carbohydrate–water mixtures at microwave frequencies. *Journal of Food Science* 37:199–204.

Rudan-Tasic, D. and C. Klofutar. 1999. Characteristics of vegetable oils of some Slovene manufacturers. *Acta Chimica Slovenica* 46:511–521.

Ryynänen, S. 1995. The electromagnetic properties of food materials: A review of the basic principles. *Journal of Food Engineering* 26:409–429.

Sahin, S. and S. G. Sumnu. 2006. *Physical Properties of Foods*. New York: Springer.

Shukla, T. P. and R. C. Anantheswaran. 2001. Ingredient interactions and product development for microwave heating. In A. K. Datta and R. C. Anantheswaran (Eds.), *Handbook of Microwave Technology for Food Applications*, pp. 355–395. Boca Raton, FL: CRC Press.

Sipahioglu, O. and S. A. Barringer. 2003. Dielectric properties of vegetables and fruits as a function of temperature, ash, and moisture content. *Journal of Food Science* 68:234–239.

Sunjka, P. S., T. J. Rennie, C. Beaudry, and G. S. V. Raghavan. 2004. Microwave-convective and microwave-vacuum drying of cranberries: A comparative study. *Drying Technology* 22:1217–1231.

Tanaka, F., K. Morita, P. Mallikarjunan, Y. C. Hung, and G. O. I. Ezeike. 2005. Analysis of dielectric properties of soy sauce. *Journal of Food Engineering* 71:92–97.

Tanaka, F., T. Uchino, D. Hamanaka, G. G. Atungulu, and Y.-C. Hung. 2008. Dielectric properties of mirin in the microwave frequency range. *Journal of Food Engineering* 89:435–440.

Tang, J., F. Hao, and M. Lau. 2002. Microwave heating in food processing. In X. H. Yang and J. Tang (Eds.), *Advances in Bioprocessing Engineering*, pp. 1–44. Singapore: World Scienctific Publishing.

Venkatesh, M. S. and G. S. V. Raghavan. 2004. An overview of microwave processing and dielectric properties of agri-food materials. *Biosystems Engineering* 88:1–18.

Von Hippel, A. 1954. *Dielectrics and Waves*. New York: Wiley.

Wang, Y., T. D. Wig, J. Tang, and L. M. Hallberg. 2003. Dielectric properties of foods relevant to RF and microwave pasteurization and sterilization. *Journal of Food Engineering* 57:257–268.

Wang, S., M. Monzon, Y. Gazit, J. Tang, E. J. Mitcham, and J. W. Armstrong. 2005. Temperature-dependent dielectric properties of selected subtropical and tropical fruits and associated insect pests. *Transactions of the ASAE* 48:1873.

Wang, Y., J. Tang, B. Rasco, F. Kong, and S. Wang. 2008. Dielectric properties of salmon fillets as a function of temperature and composition. *Journal of Food Engineering* 87:236–246.

Zhang, M., J. Tang, A. S. Mujumdar, and S. Wang. 2006. Trends in microwave-related drying of fruits and vegetables. *Trends in Food Science & Technology* 17:524–534.

4 Effect of Processing on Microbial Growth and Inactivation in Foods

Naphaporn Chiewchan

CONTENTS

4.1 INTRODUCTION

Most foods deteriorate due to action of microorganisms, which use foods as their substrates. Therefore, the knowledge of factors that favor or inhibit microbial growth is of great importance as it could be applied in food processing to ensure food safety and the prevention of food spoilage.

The contamination of microorganisms always occurs during harvesting and further processing. Food materials, plants, and animals carry their own microflora, and this can persist in processed food products. For example, although muscles of healthy animals are usually almost completely free from microorganisms, their intestines may carry a very large and diverse microflora, including human pathogens such as *Campylobacter*, *Salmonella*, and certain strains of *Escherichia coli*. These organisms may spread to other meat surfaces during postmortem processing. As a result, evisceration and dressing are regarded as key steps that need to be hygienically performed to minimize the contamination of meats and meat products. Chicken feces adhering to the outside of egg shells can be contaminated with *Salmonella*. *Vibrio parahaemolyticus* is a naturally occurring marine organism in warm coastal waters and can contaminate fishes and seafoods.

Air is also an important source of contamination. Dry air with low dust content and higher temperature has a low microbial level and vice versa. Soil contains the greatest variety of microorganisms and is an important source of heat-resistant spore-forming bacteria. Sewage can cause water contamination and transfer pathogens to shellfishes, fishes, and other seafoods. The sewage used as fertilizer in crops and plants may be contaminated with human pathogens, hence further contaminating fresh produces. Water used as an ingredient, for washing foods, cooling heated foods, and manufacturing ice may also be contaminated with foodborne pathogens, unless appropriate treatments are provided prior to processing. The selection of an appropriate location with a good water supply when establishing a plant is thus necessary.

In this chapter, a brief review of microorganisms relevant to food processing, characteristics of microbial growth, and factors affecting the microbial growth is presented. The effects of selected food processing technologies on the inactivation of microorganisms are also discussed. The reader is referred to Chapters 5 and 6 for specific discussions on the use of high-pressure processing and pulsed electric field (PEF) to treat various food products, respectively.

4.2 MICROORGANISMS RELEVANT TO FOOD PROCESSING

Raw foods are always associated with microorganisms, which may lead to spoilage or cause foodborne illness in humans. Various processing and preservation techniques have indeed been applied to eliminate, reduce, or at least control microbial growth within the safe level in order to prevent or reduce instances of foodborne illness. The microorganisms of concern in food processing and preservation are fungi and bacteria.

4.2.1 FUNGI

Microorganisms in fungi category are molds and yeasts. Molds and yeasts are common causes of food spoilage. They are relatively tolerant to low pH, low water activity, low temperature, and the presence of preservatives (Huis in't Veld, 1996). They typically spoil fruits, vegetables, and bakery products (Forsythe, 2000).

Molds are filamentous microorganisms, that is, the cells stay attached to each other to form long filaments. Many molds can form spores, which are generally different in colors, depending on the kind of mold. Species of *Aspergillus* often form black spores, while *Penicillium* is recognized by blue or green spores. Mold spores serve the purpose of reproduction, but they are not resistant to heat or chemicals.

Mold growth may cause food spoilage. The symptoms include off-flavors, discoloration, and rotting (Filtenborg et al., 1996). Important fungi that cause spoilage of fruits and vegetables include *Penicillium*, *Fusarium*, *Aspergillus*, and *Rhizopus*. For example, *Penicillium expansum*, a psychrotrophic mold, can cause blue mold rot of apple, cherry, pear, and other fruits (Marek et al., 2003; Kumar et al., 2008). *Aspergillus niger* commonly spoils garlic and onion (Filtenborg et al., 1996). Dry rot of potato is mainly caused by *Fusarium sambucinum* (Filtenborg et al., 1996). *Rhizopus stolonifer* is a serious postharvest disease of stone fruits rot (Alla et al., 2008).

Certain types of molds can produce mycotoxin. *A. flavus* and *A. parasiticus*, under favorable conditions of temperature and humidity, can produce aflatoxins. Produces that are commonly contaminated with aflatoxins are groundnut, maize, chili, spices, and cottonseed (Cotty and Jaime-Garcia, 2007; Kumar et al., 2008). Mycotoxins can also occur in milk and milk products as a result of animals consuming mycotoxin-contaminated feed (Moss, 2002; Cotty and Jaime-Garcia, 2007). *P. expansum*, causing blue mold rot, can produce toxin called patulin (Barkai-Golan and Paster, 2008; Kumar et al., 2008).

Yeasts are single-celled fungus. The reproduction of yeast is known as budding. A small bud appears on the side of a yeast cell and slowly grows until it is equal in size to the parent cell. The daughter and parent cells then separate. Yeasts can be found in plants, air, soil, and intestinal tract of animals. Yeasts are active in a very broad temperature range from 0°C to 50°C with an optimum temperature range of 20°C–30°C. Yeasts have an optimum pH of 4.0–4.5 (Garbutt, 1997). They are usually acid tolerant and are therefore associated with spoilage of acidic foods such as fruit juices and carbonated soft drinks (Martorell et al., 2007).

4.2.2 Bacteria

Bacteria are single-celled microorganisms. Some bacteria are useful. Many others, however, cause disease; for example, some species of the genus *Streptococcus* can cause sore throats and many species of *Salmonella* can cause food poisoning. When viewed under a microscope, the distinctive shapes of bacterial cells can be clearly seen.

The reproduction of bacteria is known as binary fission. The bacterial DNA molecules replicate and formation of two chromosomes occurs along with formation of cell wall. The unicellular bacterium then divides into two cells. Some bacteria, for example, *Bacillus* and *Clostridium* species, can form endospores. The endospores are not able to grow or reproduce. A spore structure enables a cell to survive environmental stresses such as cooking, freezing, drying, and high-salt and high-acid conditions. If conditions of food change, the spore then turns into a vegetative cell. The vegetative cell can grow in food and cause illness, if eaten.

Some bacteria can cause spoilage of foods. *Erwinia* and *Pseudomonas* species are the specific spoilage organisms of several ready-to-eat vegetable products (Gram et al., 2002). Pectolytic breakdown of affected tissues results in softening, liquefaction, and exudates of fresh produce (Wells and Butterfield, 1997). For example, *Pectobacterium carotovorum* (formerly *Erwinia carotovora*) can cause soft rotting of potato, tomato, pepper, lettuce, and other fresh fruits and vegetables (Wells and Butterfield, 1997; Charkowski, 2009).

Bacteria present the most important biological foodborne hazard for any food establishment; they are indeed responsible for more cases of foodborne illnesses than any other hazards. Food infection is caused by ingestion of food containing viable bacteria. This can lead to common symptoms of foodborne illnesses like nausea, vomiting, diarrhea, or fever. Sometimes, the organisms may spread to other parts of the body through the bloodstream. The bacteria causing food infection include *Salmonella* and *Listeria monocytogenes* (Sinell, 1995). Although *Salmonella* are frequently found in meat, poultry, and eggs, it can be found occasionally in fresh fruits and vegetables and their products (Lin et al., 1996; Francis et al., 1999; NACMCF, 1999). *L. monocytogenes* can be found in a wide variety of foods such as dairy products (Lundén et al., 2004), meat (including poultry), meat products (Samelis and Metaxopoulos, 1999; Berrang et al., 2000), and seafoods (Jørgensen and Huss, 1998; Farber, 2000). It is frequently reported to contaminate various fresh produces (Beuchat, 1996; Lin et al., 1996; Zagory, 1999).

Staphylococcus aureus and *Clostridium botulinum* are examples of bacteria causing food intoxication. *Staphylococcus* exists in air, dust, sewage, water, milk, or on food equipment; environmental surfaces, humans, and animals are the primary reservoirs of *Staphylococcus. S. aureus* are present in the nasal passages and throats and on the hair and skin of healthy individuals. Therefore, food handlers are usually the main source of food contamination in food poisoning outbreaks. Food poisoning arises from consuming enterotoxin produced by this organism. The staphylococcal toxin is resistant to gastrointestinal protease such as pepsin (Balaban and Rasooly, 2000). The toxin is also heat stable. The D value at 98.9°C is ≥2 h, which means that it cannot be inactivated by standard cooking temperature (Forsythe, 2000). The D values at 121°C range from 8.3 to 34 min (Bhatia and Zohoor, 2007).

Cl. botulinum is a rod-shaped and strictly anaerobic bacterium. It is of great concern for low-acid canned foods as it can produce neurotoxin. The organism is ubiquitous in nature. It can change from vegetative cell to endospore to survive under adverse conditions. The endospore possesses very high heat resistance, but the toxin is not heat stable. The D value at 121°C, to inactivate spores of *Cl. botulinum,* is 0.21 min (Adams and Moss, 2008). The toxin can be inactivated by boiling temperature. The spores of *Cl. botulinum* would not germinate and grow in food below pH 4.8 (Gavin and Weddig, 1995). Therefore, a pH of 4.6 has been selected as the dividing line between acid and low-acid foods.

4.3 MICROBIAL GROWTH CURVE

The growth pattern of microorganisms can be represented by a bell-shaped curve as shown in Figure 4.1. This curve is known as a growth curve and is most commonly plotted as the logarithm of bacterial number per unit volume of food against time.

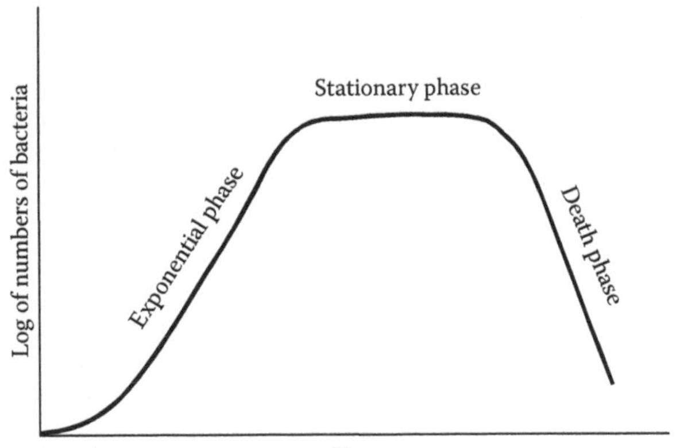

FIGURE 4.1 A typical bacterial growth curve.

The curve is ordinarily divided into different phases. Lag phase corresponds to no growth or even a decline in number. After the lag period, the growth rate continuously increases and the bacteria are in the exponential or logarithmic phase in which the multiplication rate is most rapid and can be represented by a straight line on a log-linear plot. Then, the bacteria reach the maximum cell number and the multiplication rate decreases. The number of bacteria remains constant and this phase is called the stationary phase. The death phase occurs when the cell number decreases at a faster rate than the rate new cells are formed.

4.4 FACTORS AFFECTING MICROBIAL GROWTH

Knowledge of the factors that favor or inhibit growth of microorganisms is very important in providing an understanding of the principles of food spoilage and preservation. The primary factors influencing bacterial activity are presented in Table 4.1.

TABLE 4.1
Conditions for Growth of Bacteria

Food	Foods Rich in Proteins, e.g., Meat, Poultry, Dairy Products
Acidity	Foods with pH 4.6 or higher
Temperature	5°C–60°C
Time	At least 4 h cumulative time to allow bacteria to grow to high enough numbers to cause illness
Oxygen	Aerobic, anaerobic, facultative anaerobic, microaerophilic conditions depending on type of bacteria
Moisture	Water activity (a_w) of 0.85 or higher

Source: Adapted from McSwane, D. et al., *Essentials of Food Safety and Sanitation*, Prentice Hall, Upper Saddle River, NJ, 1998.

4.4.1 Foods (Nutrients)

Microorganisms use nutrients in foods for their cellular synthesis and energy. These nutrients include carbohydrates, proteins, lipids, minerals, and vitamins. Foods rich in proteins such as milk and dairy products, meats, and egg are ideal nutrient sources of most foodborne pathogens (McSwane et al., 1998).

4.4.2 Acidity (pH)

Most microorganisms can grow over a wide range of pH. Most bacteria will not grow well at pH levels below 4.6. They can survive pH levels of 4.6–9, but optimum pH range is 6.6–7.5 (Amjadi and Hussain, 2005). However, some bacteria can grow out of this range. Yeasts and molds are generally less sensitive than bacteria and are capable of growing at a wider pH range of 2.0–9.0, with a majority favoring a pH of 4.0–6.0. *Saccharomyces cerevisiae*, for example, has a pH range of 2.35–8.6, with an optimum pH of 4.5 (Garbutt, 1997).

The internal pH of most cells is maintained near 7.0; this is the pH at which cell metabolism works best (Garbutt, 1997). When microbial cells are subject to extreme pH values, cell membranes become damaged. H^+ and OH^- ions can leak into the cells. Therefore, enzymes and nucleic acid molecules are denatured, leading to cell death. pH affects not only the rate of microbial growth, but also survival during storage, heating, drying, or other types of processing (Frazier and Westhoff, 1988).

4.4.3 Temperature

Microorganisms differ widely regarding their minimum, maximum and optimum temperature for growth. Thermophiles are bacteria that can grow at 50°C or above, while mesophiles grow best at approximately 37°C and psychrophiles can grow at 5°C or below. Examples of pathogenic bacteria belonging to the psychrotrophic group are *Aeromonas hydrophila, Cl. botulinum, L. monocytogenes,* and *Yersinia enterocolitica* (Table 4.2).

The temperature at which food is held can have a great influence on the kind, rate, and amount of microbiological changes that would take place. The growth and metabolic reactions of microorganisms depend on enzymes; the rates of enzyme reactions are directly affected by temperature. Destruction of microorganisms at high temperature is supposed to be caused by denaturation of proteins or nucleic acid and loss of cytoplasmic membrane function (Perry and Staley, 1997).

Time and temperature are the most critical factors affecting the growth of bacteria in foods. Most pathogenic bacteria can grow within a temperature rage of 5°C–60°C. This is commonly referred to as the food "temperature danger zone" (Amjadi and Hussain, 2005). Some disease-causing bacteria such as *L. monocytogenes* can grow at temperatures below 5°C, although the growth rate may be very slow.

4.4.4 Time

Bacteria can grow to dangerous levels if foods are left for a long time under suitable conditions for growth. Bacterial cells can double in number every 15–30 min

TABLE 4.2
Psychrotolerance of Some Pathogenic Bacteria

Organism	Minimum Growth Temperature
Aeromonas hydrophila	0°C–4°C
Clostridium botulinum	
Vegetative cells	3.3°C
Spore	3.3°C
Listeria monocytogenes	0°C
Yersinia enterolitica	0°C

Source: Adapted from Church, I.J. and Parsons, A.L., *Int. J. Food Sci. Technol.*, 28, 563, 1993.

under ideal conditions (McSwane et al., 1998). This implies that bacteria need about 4 h to grow to high enough number to cause foodborne illness (McSwane et al., 1998). Therefore, proper storage and handling of foods helps prevent bacterial growth.

4.4.5 Oxygen

Microorganisms vary in their requirements for oxygen and their responses to the presence of oxygen in the environment. Some microorganisms require oxygen in order to generate cellular energy in the form of adenosine triphosphate (ATP) (aerobic respiration), whereas others can generate cellular energy without oxygen (anaerobic respiration). Obligate aerobes can grow only in the presence of molecular oxygen. Facultative aerobes follow a respiratory pathway when oxygen is available, but can use alternative anaerobic energy-generating systems when oxygen is not available. Microaerophiles are organisms that prefer atmospheres with limited levels of oxygen. While they must have oxygen for respiration, they grow best at reduced levels of oxygen that range from 2% to 10%. Obligate anaerobes, on the other hand, cannot grow in the presence of oxygen and oxygen is toxic to the bacterial cells (Garbutt, 1997).

4.4.6 Moisture

All microorganisms require liquid water to enable them to grow. If there is little water present or the water present is not available to the microbes such as when water is available only in the form of ice crystal or water containing salt or sugar, the growth is slowed or even prevented. Water availability for microbial growth is often expressed in terms of the water activity (a_w):

$$a_w = \frac{p}{p_0} = \frac{1}{100}\text{ERH} \tag{4.1}$$

where

p is the partial pressure of water over the substrate or solution
p_0 is the equilibrium vapor pressure of pure water at the same temperature
ERH is the equilibrium relative humidity (Adams and Moss, 2008)

Water activity values can only fall within the range of 1 (pure water with maximum water availability) to 0 (complete absence of water). Bacteria are most sensitive to the effect of reduced water activity while yeasts and molds are more resistant. Normal yeasts grow well in the presence of glucose of approximately 20% (w/v). However, some yeasts are tolerant to low water activity caused by high concentrations of sugar (60% w/v) (Martorell et al., 2007).

4.5 EFFECT OF THERMAL PROCESSING ON MICROBIAL INACTIVATION

Thermal treatment is widely used in food processing. It is not only used to eliminate or reduce the number of microorganisms, but also for other purposes such as blanching of vegetables before freezing or drying to destroy enzymes, which also has a secondary effect of removing the majority of microbial contamination (Garbutt, 1997). The death of microorganisms is supposedly caused by protein denaturation and inactivation of enzymes required for metabolism. Damage to ribosomes has also been identified as a possible mechanism for loss of viability (Sogin and Ordal, 1967; Mackey and Derrick, 1982; Teixeira et al., 1997).

4.5.1 THERMAL RESISTANCE OF MICROORGANISMS

Vegetative cells and spores differ widely in their heat resistance. Generally, a temperature slightly above the maximum growth temperature would cause death of vegetative cells. On the other hand, bacterial spores can survive at a much higher temperature than the normal temperature range for growth of vegetative cells. Therefore, they are of primary concern during thermal processing of most foods. Moreover, even in the same population, each cell has a different heat resistance (Stephens et al., 1997).

At a constant temperature, it is assumed that a decrease in bacterial number is exponential with time; thermal process calculations are indeed based on this assumption. However, various previous works have shown that inactivation profiles are not straight lines when plotted on a semilogarithmic scale.

Typical microbial survivors found can be classified into four general types as shown in Figure 4.2 (Moats, 1971). Curve A represents a typical survivor curve with a logarithmic death rate. Curve B is a type commonly found and shows an initial lag in death rate followed by a logarithmic death rate. Curve C is similar to curve A,

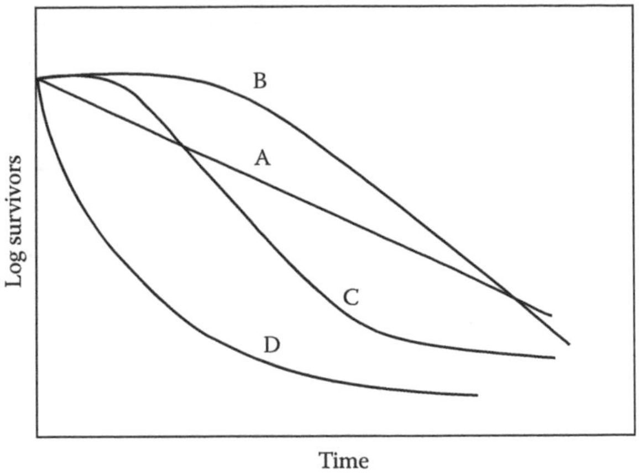

FIGURE 4.2 Some types of bacterial survival curves observed. (From Moats, W.A., *J. Bacteriol.*, 105, 165, 1971. With permission.)

but has tail, which implies a small fraction of heat-resistant cells. The shapes of curves A, B, and C are explained by assuming a uniform heat resistance of the cells (Hansen and Rieman, 1963). When curve D is found, it is assumed that cells of different heat resistance are present; the shape of the curve is determined in this case by a nonuniform distribution of heat resistance among the individual cells (Hansen and Rieman, 1963).

4.5.2 Determination of Microbial Thermal Resistance

The heat resistance of microorganisms is normally expressed in terms of thermal death time (TDT), which is the time taken at a certain temperature to kill a stated number of microorganisms (vegetative cells or spores) under specified conditions (Frazier and Westhoff, 1988). When bacteria are heated, generally, the number of surviving cells decreases exponentially with time of exposure to heat. The degree of microbial heat resistance is commonly expressed as *D*-value (decimal reduction time), which is the time required to destroy 90% of the cells or, in other words, the time required for the curve to traverse one log cycle, that is, log a/b is equal to 1 (Figure 4.3). *D*-value is frequently written with a subscript to define the temperature of the process. For example, D_{121} is the time required to kill 90% of a microbial population at 121°C. *D*-value is very useful for comparing the relative heat resistance of microorganisms and calculating thermal process time (Garbutt, 1997).

Z-value is another term used to determine the equivalent *D*-value if a product is treated at a different temperature. When the *D*-values are plotted against their temperature on a semilogarithmic scale, a straight line of decreasing slope is presented, which is called a TDT curve (Figure 4.4). *Z*-value represents degree Celsius (or Fahrenheit) change in temperature required to achieve a 10-fold change in the *D*-value (Garbutt, 1997).

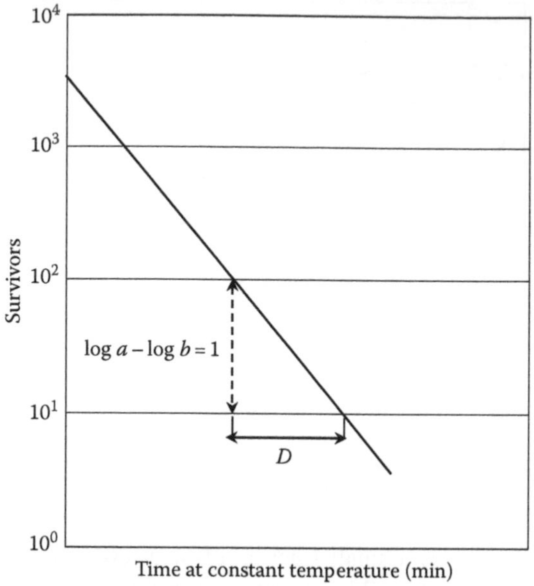

FIGURE 4.3 Logarithmic survivor curve. (Modified from Stumbo, C.R., *Thermobacteriology in Food Processing*, Academic Press, New York, 1965, 57–96.)

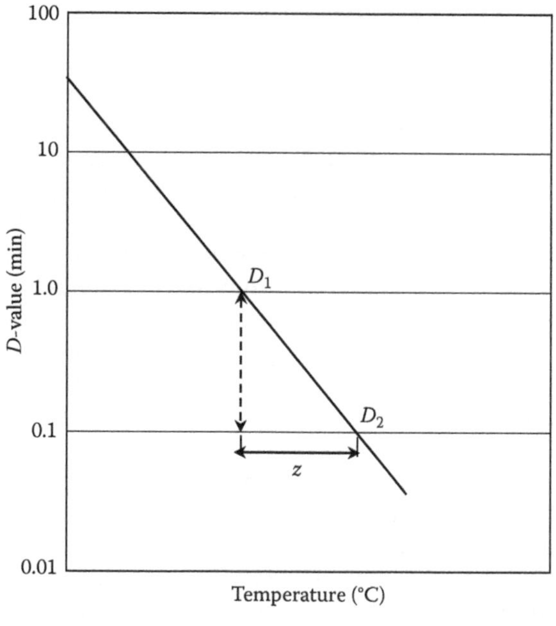

FIGURE 4.4 TDT. (Modified from Stumbo, C.R., *Thermobacteriology in Food Processing*, Academic Press, New York, 1965, 57–96.)

4.5.3 Factors Influencing Heat Resistance of Microorganisms

There are a number of factors that influence heat resistance, for example, history of microorganisms, composition of foods, pH, salt, and growth phase of microorganisms (Hansen and Riemann, 1963). It is known that cells from cultures in the logarithmic phase are less heat resistant than cells in the lag or stationary phase (Lemcke and White, 1959). Substrate or food composition has a very important effect on heat resistance. Decrease in moisture content can increase the heat resistance of microbial cells. Moist heat is much more effective in terms of microbial inactivation than dry heat since proteins, which may be destroyed during thermal processing, are more stable in a dry state (Hansen and Riemann, 1963).

Colloidal materials, especially proteins and fats, are protective against heat (Frazier and Westhoff, 1988). For example, bacteria are more heat resistant in cured ham than in phosphate buffer at the same pH (Hansen and Rieman, 1963). *L. monocytogenes* grown in butter was found to be much more heat resistant than the cells grown in half or double cream (Casadei et al., 1998). Ahmed et al. (1995) studied the heat resistance of *E. coli* O157:H7 in meats and meat products with different fat contents and found that products with higher fat levels exhibited higher *D*-values. This was believed to be because bacterial cells suspended in fat are typically more difficult to destroy than in an aqueous medium due to a reduction in water activity; increasing the fat content also altered the heat transferred into the product.

4.5.4 Effects of Canning and Pasteurization

Canning is recognized as one of the most effective methods of food preservation. The conditions used for canning must be designed to ensure that all pathogens are either inactivated or destroyed. The widely accepted minimum lethality for a thermal process applied to low-acid canned foods is that it should produce 12 decimal reductions in the number of surviving spores of *Cl. botulinum.* For example, the *D*-value, which is the time at a given temperature for the surviving population to be reduced by one log cycle, of *Cl. botulinum* spores at 121°C is 0.21 min. Therefore, the minimum process time required is $12 \times 0.21 = 2.52$ min (Adams and Moss, 2008). Canning treatments commonly double this process time, resulting in a safe but very over-processed product.

Pasteurization is a thermal process, which involves the use of treatment temperature in the range of 60°C–80°C for up to a few minutes and is usually applied to dairy products and other heat-sensitive foods such as beer and fruit juices (Perry and Staley, 1997). Pasteurization is used for two purposes, which are to eliminate specific pathogen(s) associated with a product and to eliminate a large proportion of potential spoilage organisms. As some spoilage bacteria can survive this process and can grow at ambient temperatures, refrigerated storage is often employed to improve the shelf life of pasteurized foods (Adams and Moss, 2008).

Ultra-high temperature (UHT) treatment involves thermal processing at 138°C–142°C. At this temperature, only 2–3 s, for example, is needed to achieve commercial sterility of milk. After heat treatment, the milk can then be aseptically placed in sterile containers (Garbutt, 1997). The milk can be kept unopened for

6 months without any deterioration in quality. The advantage of this thermal treatment method is that it produces very little change in flavor or loss of nutritive quality of a product (Parry and Pawsey, 1989).

4.5.5 Effect of Drying

Water availability has a great impact on the transfer of heat to microorganisms (Laroche et al., 2005). As mentioned earlier, moist heat is much more effective than dry heat in deactivating microorganisms because proteins, which may be destroyed during thermal processing, are more stable in a dry state (Hansen and Riemann, 1963). Furthermore, cells attached to a tissue are generally more heat resistant than those unattached or dispersed throughout a food or broth (Murphy et al., 2002). Microorganisms, especially bacteria, interact differently with different food matrices, so they have different degrees of sensitivity to heat. For example, Quintavalla et al. (2001) observed greater heat resistance of *Salmonella* in pork meat than in Trypticase Soy Broth. Chiewchan et al. (2007) reported that *Salmonella* Krefeld, attached on the rawhide surface with an initial a_w of 0.99 and 0.90, exhibited far higher heat resistance than in a liquid medium by 375- and 58-fold at 60°C, respectively. The results from other previous works also suggest that the heat resistance of *Salmonella* in a dry medium is much higher than in a liquid medium (Doyle and Mazzott, 2000). Therefore, more drastic heat treatment is required to eliminate bacteria (McDonough and Hargrove, 1968; Jung and Bauchat, 1999). As some microorganisms can survive in a low a_w environment, elimination of such microorganisms during drying is a critical step for the safety of a product.

4.6 EFFECTS OF LOW-TEMPERATURE AND NONTHERMAL PROCESSING ON MICROBIAL INACTIVATION

4.6.1 Chilling and Freezing

Chilling or freezing is, among many popular methods, applied to extend the shelf life of a food product. Chilled foods are foods that are stored at temperatures near, but above their freezing point, typically 0°C–5°C; the shelf life of the product would indeed depend on the storage temperature. During chilling, there may be changes in spoilage characteristics as low temperatures exert a selective effect that prevents the growth of mesophiles, leading to microflora dominated by psychrotrophs (Adams and Moss, 2008). Although mesophiles, including mesophilic pathogens, cannot grow, the ones that are injured during chilling may resume growth when the conditions become favorable. For psychrotrophs, they can grow in chilled foods, but relatively slowly, so that the onset of spoilage is delayed. Chilling may also lead to a phenomenon known as cold shock, which causes death and injury of microbial cells.

On the other hand, the temperature used for freezing is generally less than −18°C. The damage of microbial cells may be from mechanical damage to cell wall and cell membrane, which result from extra and intracellular ice crystal formation (Gieges, 1996; Archer, 2004). Bacteria in the logarithmic growth phase are more susceptible to freezing than those in the stationary phase (Geiges, 1996). However, bacterial spores are extremely resistant to the effects of freezing and repeated freeze-thawing (Georgala and Hurst, 1963; Archer, 2004), especially spores of *Bacillus* and *Clostridium* species (Geiges, 1996). The speed of freezing and thawing is another factor influencing the number of surviving microorganisms; quick freezing and thawing lead to less inactivation of microorganisms than slow freezing and thawing.

Storage temperatures above a lower limit of −10°C should not be used, as they allow the reproduction of cold-tolerant microorganisms, which can produce exoenzymes, especially proteases and lipases. The storage temperature of −18°C reliably prevents the growth of microorganisms, but permits enzymatic processes to continue (Geiges, 1996).

4.6.2 Irradiation

The electromagnetic spectrum is composed of at least six separate forms of radiation that differ in wavelength, frequency, and penetrating power. Of these forms, gamma radiation, ultraviolet (UV) radiation, and microwaves are of interest to the food industry. Ionizing radiation that is of interest for food preservation includes x-ray, beta ray, and gamma ray, which have wavelengths of 2000 Å or less. This is because these rays possess enough energy to ionize molecules in their paths and can inactivate foodborne microorganisms without increasing the temperature of the irradiated food.

The lethal effects of irradiation on microorganisms include the damage of the DNA and cytoplasmic membrane (Garbutt, 1997; Lacroix and Ouattara, 2000; Mendonca, 2002). The antimicrobial action of ionizing radiation occurs in two ways. The energy can be absorbed directly by the cell molecules, damaging cell structures, and cell biochemistry (Garbutt, 1997). The free radicals produced by the radiolysis of water in the cell as well as in the suspending menstruum also occur during irradiation. These radicals can combine with each other or oxygen molecules to give powerful oxidizing agents that can damage all cell components (Garbutt, 1997; Mendonca, 2002). The preservative effect of irradiation is a function of the absorbed radiation dose. The guidelines for dose requirements of food applications are presented in Table 4.3.

TABLE 4.3
Examples of Irradiation Dose Requirements for Food Applications

Application	Dose Requirements (kGy)
Inhibition of sprouting of potatoes and onions	0.03–0.12
Killing and sterilizing insects (disinfestation of food)	0.2–0.8
Prevention of reproduction of foodborne parasites	0.1–3.0
Decrease of after-ripening and delaying senescence of some fruits and vegetables; extension of shelf-life of food by reduction of microbial population	0.5–5.0
Elimination of viable non-sporeforming pathogenic microorganisms (other than viruses) in fresh and frozen food	1.0–7.0
Reduction or elimination of microbial population in dry food ingredients	3.0–10

Source: Adapted from Farkas, J., *Trends Food Sci. Technol.*, 17, 148, 2006.

Inactivation of foodborne microorganisms by ionizing radiation is influenced by several factors such as types of microorganisms, types of food, irradiation temperature, oxygen, and water content (Farkas, 1998). Large number of microorganisms reduce the effectiveness of a given irradiation dose. The resistance to radiation of microorganisms decreases in the following order: viruses > bacterial spores > bacterial vegetative cells > yeasts and molds (Garbutt, 1997; Mendonca, 2002). With respect to physiological state of bacteria, exponential phase cells are more sensitive to irradiation than lag phase cells or stationary phase cells (Mendonca, 2002). Low temperature increases the resistance of microorganisms to irradiation; a higher dose is required to achieve equivalent kill (Thakur and Singh, 1995). Most microorganisms exhibit increased sensitivity to irradiation in the presence of oxygen (Mendonca, 2002).

4.6.3 Hydrostatic Pressure

High hydrostatic pressure processing is nonthermal processing in which foods are subject to high hydrostatic pressure at or above 100 MPa (Devlieghere et al., 2004). Microbial growth is retarded at pressures in the range of 20–180 MPa, and these pressures can also inhibit protein synthesis (Lado and Yousef, 2002). Loss of cell viability begins at approximately 180 MPa, and the rate of inactivation increases exponentially as the pressure increases. Lethal high-pressure treatments disrupt membrane integrity and denature proteins.

Factors influencing the effect of pressure on microorganisms are the magnitude and duration of compression, stage of growth of the organisms as well as pH and temperature during compression (Hoover et al., 1989; Trujillo et al., 2002). Bacteria from the exponential phase are more pressure sensitive than cells from stationary or death phases (Trujillo et al., 2002). Microbial vegetative cells are destroyed by pressures in the range of 300–500 MPa, but spores of some bacterial species resist pressures in excess of 1000 MPa (Gould, 1996). It has been reported that it is possible to obtain milk pressurized at 400–600 MPa with a microbiological quality comparable to that of pasteurized milk (at 72°C for 15 s) (Trujillo et al., 2002).

The efficiency of hydrostatic pressure treatment may be enhanced by applying it in combination with mild temperature (30°C–50°C) and/or with bacteriocin (Trujilo et al., 2002; Devlieghere et al., 2004). Examples of commercialized products treated by high hydrostatic pressure are fruit juices, jams, sauces, and sliced cooked ham (Trujillo et al., 2002).

For more information on the use of high-pressure processing and its effect on microorganisms, the reader is referred to Chapter 5.

4.7 EFFECTS OF COMBINED PRESERVATION METHODS ON MICROBIAL INACTIVATION

Although each of the environmental factors can significantly affect the resulting microbial flora, sometimes a change in only one environmental factor is not enough to limit microbial growth. A combination of more than one factor such as pH and temperature or pH, temperature, and inhibitory substance is sometimes necessary. Moreover, foods treated with a combination of agents are often more acceptable to the consumer than those treated with a high level of a single agent (Boddy and Wimpenny, 1992).

During processing, fruits and vegetables, for example, may undergo a series of different operations, which affect the degree to which foods are preserved. Pretreatment of foods before entering the main processing steps can affect microbial population. Selection and sorting, washing with/without sanitizer(s), peeling, slicing or cutting into desired shape, as well as blanching can cause changes in the kinds and number of microorganisms associated with the food products.

Many sanitizers have been applied during washing of fresh produces, including chlorine (Beuchat and Ryu, 1997; Delaquis et al., 1999; Weissinger et al., 2000), chlorine dioxide (Du et al., 2003), ozone (Restaino et al., 1995), and organic acids (Pao and Petracek, 1997; Vijayakumar and Wolf-Hall, 2002). Use of the same sanitizer at different temperatures, concentrations, or contact time may result in different inhibitory effects (Kenny and Bauchat, 2002). For example, use of 1.0%–1.5% (v/v) acetic acid to wash cabbage slices for 5 min could reduce the *Salmonella* number by approximately 1.5 log cycles (Chiewchan and Morakotjinda, 2009). Soaking in acetic acid might have destroyed, inactivated and removed (by agitating and soaking effects) the *Salmonella* from the vegetable surface. During soaking, the acid molecules could penetrate into plant cells as well as bacterial cells. This phenomenon resulted in the accumulation of H^+, which modified the intracellular pH and disrupted the cell functions (Brul and Coote, 1999). Furthermore, acetic acid could reduce the activity of enzymes responsible for browning, that is, peroxidase and polyphenol oxidase, hence improving the appearance of the product. However, soaking vegetables in higher acetic acid concentrations and longer time may cause off-flavor and discoloration.

In recent years, there has been a trend toward reducing the levels of certain additives such as salt. High intake of salt has been linked to hypertension and stroke (He and MacGregor, 2003; Chen et al., 2009). Food additives can generally be used at lower concentrations when more than one of them is used in combination. In fermented processed meat, for example, use of added chemicals such as spices, nitrite, and salt at a suitable ratio not only gives the product a good taste and texture, but is also more effective at inhibiting microbial growth. For instance, microbial growth on cured meat could be inhibited by the use of up to 1000 ppm nitrite alone. However, the necessary concentration of nitrite fell to about 100 ppm when 5% NaCl (w/v) was also added (Boddy and Wimpenny, 1992).

Figure 4.5 illustrates the combined effect of preservation methods on the design of a food product. *Brochothrix thermosphacta* is one of several prominent food spoilage bacteria in meats and meat products; the bacteria causes flavor deterioration and surface darkening (Ayres et al., 1980). The growth zone of *B. thermosphacta* at 48 h of incubation at 25°C on the NaCl-pH gradient plate showed that the combination of more than two preservation techniques could be used to extend the shelf life of a product and could also retain the quality as well as consumer satisfaction through optimization of the combination of the applied effects (Rattanasomboon et al., 2001). The figure shows that the optimum growth occurred at neutral pH and

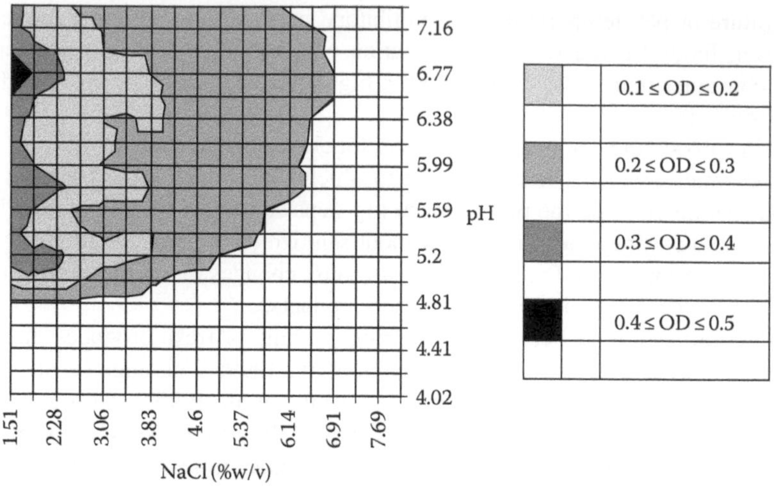

FIGURE 4.5 Contour map of the growth zone of 5 h (exponential phase) culture of *B. thermosphacta*, a meat spoilage bacterium, after (a) 0, (b) 5, (c) 10, and (d) 15 min of heat treatment. (Modified from Rattanasomboon, N. et al., *Int. J. Food Sci. Technol.*, 36, 369, 2001.)

low salt concentration (illustrations with the darkest area), and higher acidic conditions and higher salt concentrations were less favorable for bacterial growth (illustrations with the lighter area). It can be seen that growth occurred at salt concentrations up to 6.9% (w/v) and pH values as low as 4.8.

In addition to reduction of the initial contamination in foods, acid pretreatment also gives benefit during further processing. For example, the heat resistance of *Salmonella* on cabbage slices, having been pretreated with acetic acid, reduced significantly during subsequent drying (Chiewchan and Morakotjinda, 2009). By comparing to the intact cells, the bacteria might be stressed or injured during acetic acid treatment and lost the mechanism to protect the cells from subsequent thermal damage.

The reduction of oxygen in a food package is another procedure that may result in extended shelf life of foods. Modified atmosphere packaging (MAP) involves packaging of a product in an atmosphere with one-time modification of gaseous compositions to those different from the compositions of air. This modified gaseous environment can slow down respiration rate, microbial growth and reduce enzymatic degradation, with the intention of extending the shelf life. Carbon dioxide can be used in MAP as it is a noncombustible, colorless gas and leaves no toxic residues in food. In addition, CO_2 can inhibit bacterial growth by displacing the oxygen required by aerobic organisms, as well as by forming carbonic acid and thus possibly lowering the pH of the food to bacteriostatic levels.

Vacuum packaging, on the other hand, involves reduction of the amount of air from a package; the package is then hermetically sealed, so a near-perfect vacuum remains inside. The elimination of oxygen can alter spoilage organisms by preventing the growth of strict aerobes and slowing the growth of facultative anaerobes (Gould, 1996).

4.8 CONCLUDING REMARKS

Nowadays, consumer trends toward consumption of health foods and food products undergoing minimal processing are on the rise. To achieve this goal, the processes have to be well designed and consideration must be made both in terms of food quality and safety. The combined use of several preservation techniques for the production of a food product is suggested to minimize quality losses. At the same time, the processes must be designed to provide the successful control of foodborne pathogens together with the elimination of spoilage microorganisms. Knowledge of the type of microorganisms in certain foods as well as factors that favor or inhibit microbial growth can help improve the design of appropriate process operation.

REFERENCES

Adams, M.R. and Moss, M.O. 2008. *Food Microbiology*, 3rd edition. Cambridge, U.K.: The Royal Society of Chemistry.

Ahmed, N.M., Conner, D.E., and Huffman, D.L. 1995. Heat-resistance of *Escherichia coli* O157:H7 in meat and poultry as affected by product composition. *Journal of Food Science* 60: 606–610.

Alla, A., El-Sayed, M.A., Ziedan, H., and El-Mohamedy, S. 2008. Control of Rhizopus rot disease of apricot fruits (*Prunus armeniaca* L.) by some plant volatiles aldehydes. *Research Journal of Agriculture and Biological Sciences* 4: 424–433.

Amjadi, K. and Hussain, K. 2005. Integrating food hygiene into quantity food production systems. *Nutrition and Food Science* 35: 169–183.

Archer, D.L. 2004. Freezing: An underutilized food safety technology? *International Journal of Food Microbiology* 90: 127–138.

Ayres, J.C., Mundt, J.O., and Sandine, W.E. 1980. *Microbiology of Foods*. San Francisco, CA: W.H. Freeman and Company.

Balaban, N. and Rasooly, A. 2000. Staphylococcal enterotoxins. *International Journal of Food Microbiology* 61: 1–10.

Barkai-Golan, R. and Paster, N. 2008. Moldy fruits and vegetables as source of mycotoxins: Part 1. *World Mycotoxin Journal* 1: 147–159.

Berrang, M.E., Lyon, C.E., and Smith, D.P. 2000. Incidence of *Listeria monocytogenes* on pre-scald and post-chill chicken. *The Journal of Applied Poultry Research* 9: 546–550.

Beuchat, L.R. 1996. *Listeria monocytogenes*: Incidence on vegetables. *Food Control* 7: 223–228.

Beuchat, L.R. and Ryu, J.H. 1997. Produce handling and processing practices. *Emerging Infectious Disease* 3: 459–465.

Bhatia, A. and Zahoor, S. 2007. *Staphylococcus aureus* enterotoxins: A review. *Journal of Clinical and Diagnostic Research* 3: 188–197.

Boddy, L. and Wimpenny, J.W.T. 1992. Ecological concepts in food microbiology. *Journal of Applied Bacteriology Symposium Supplement* 73: 23S–38S.

Brul, S. and Coote, P. 1999. Preservative agents in foods: Mode of action and microbial resistance mechanisms. *International Journal of Food Microbiology* 50: 1–17.

Casadei, M.A., Esteves de Matos, R., Harrison, S.T., and Gaze, J.E. 1998. Heat resistance of *Listeria monocytogenes* in dairy products as affected by the growth medium. *Journal of Applied Microbiology* 84: 234–239.

Charkowski, A.O. 2009. Decaying signals: Will understanding bacterial-plant communication lead to control of soft rot? *Current Opinion in Biotechnology* 20: 178–184.

Chen, Z.Y., Peng, C., Jiao, R., Wong, Y.M., Yang, N., and Huang, Y. 2009. Antihypertensive nutraceuticals and functional foods. *Journal of Agricultural and Food Chemistry* 57: 4485–4499.

Chiewchan, N. and Morakotjinda, P. 2009. Effect of acetic acid pretreatment and hot air drying on resistance of *Salmonella* on cabbage slices. *Drying Technology* 27: 955–961.

Chiewchan, N., Pakdee, W., and Devahastin, S. 2007. Effect of water activity on thermal resistance of *Salmonella* Krefeld in liquid medium and on rawhide surface. *International Journal of Food Microbiology* 114: 43–49.

Church, I.J. and Parsons, A.L. 1993. Sous vide cook-chill technology. *International Journal of Food Science and Technology* 28: 563–574.

Cotty, P.J. and Jaime-Garcia, R. 2007. Influences of climate on aflatoxin producing fungi and aflatoxin contamination. *International Journal of Food Microbiology* 119: 109–115.

Delaquis, P.J., Stewart, S., Toivonen, P.M.A., and Moyls, A.L. 1999. Effect of warm, chlorinated water on the microbial flora of shredded iceberg lettuce. *Food Research International* 32: 7–14.

Devlieghere, F., Vermeiren, L., and Debvere, J. 2004. New preservation technologies: Possibilities and limitations. *International Dairy Journal* 14: 273–285.

Doyle, M.E. and Mazzotta, A.S. 2000. Review of studies on the thermal resistance of Salmonellae. *Journal of Food Protection* 63: 779–795.

Du, J., Han, Y., and Linton, R.H. 2003. Efficacy of chlorine dioxide gas in reducing *Escherichia coli* O157:H7 on apple surfaces. *Food Microbiology* 20: 583–591.

Farber, J.M. 2000. Present situation in Canada regarding *Listeria monocytogenes* and ready-to-eat seafood products. *International Journal of Food Microbiology* 62: 247–251.

Farkas, J. 1998. Irradiation as a method for decontamination food. *International Journal of Food Microbiology* 44: 189–204.

Farkas, J. 2006. Irradiation for better foods. *Trends in Food Science and Technology* 17: 148–152.

Filtenborg, O., Frisvad, J.C., and Thrane, U. 1996. Molds in food spoilage. *International Journal of Food Microbiology* 33: 85–102.

Forsythe, S.J. 2000. *The Microbiology of Safe Food*. Oxford, U.K.: Blackwell Science.

Francis, G.A., Thomas, C., and O'Beirne, D. 1999. The microbiological safety of minimally processed vegetables. *International Journal of Food Science and Technology* 34: 1–22.

Frazier, W.C. and Westhoff, D.C. 1988. *Food Microbiology*, 4th edition. New York: McGraw-Hill Book Company.

Garbutt, J. 1997. *Essentials of Food Microbiology*. London, U.K.: Arnold, a member of the Hodder Headline Group.

Gavin, A. and Weddig, L.M. 1995. *Canned Foods: Principles of Thermal Process Control, Acidification and Container Closure Evaluation*, 6th edition. Washington, DC: The Food Processors Institute.

Geiges, O. 1996. Microbial processes in frozen food. *Advance in Space Research* 18: 109–118.

Georgala, D.L. and Hurst, A. 1963. The survival of food poisoning bacteria in frozen foods. *Journal of Applied Bacteriology* 26: 346–358.

Gould, G.W. 1996. Methods for preservation and extension of shelf life. *International Journal of Food Microbiology* 33: 51–64.

Gram, L., Ravn, L., Rasch, M., Bruhn, J.B., Christensen, A.B., and Givskov, M. 2002. Food spoilage-interactions between food spoilage bacteria. *International Journal of Food Microbiology* 78: 79–97.

Hansen, N.H. and Riemann, H. 1963. Factors affecting the heat resistance of nonsporing organisms. *Journal of Applied Bacteriology* 26: 314–333.

He, F.J. and MacGregor, G.A. 2003. How far should salt intake be reduced? *Hypertension* 42: 1093–1099.

Hoover, D.G., Metrick, C., Papineau, A.M., Farkas, D.F., and Knorr, D. 1989. Biological effects of high hydrostatic pressure on food micro-organisms. *Food Technology* 43: 99–107.

Huis in't Veld, J.H.J. 1996. Microbial and biochemical spoilage of foods: An overview. *International Journal of Food Microbiology* 33: 1–18.

Jørgensen, L.V. and Huss, H.H. 1998. Prevalence and growth of *Listeria monocytogenes* in naturally contaminated seafood. *International Journal of Food Microbiology* 42: 127–131.

Jung, Y.S. and Bauchat, L.R. 1999. Survival of multidrug-resistant *Salmonella typhimurium* DT104 in egg powders as affected by water activity and temperature. *International Journal of Food Microbiology* 49: 1–8.

Kenny, S.J. and Beuchat, L.R. 2002. Comparison of aqueous commercial cleaners for effectiveness in removing *Escherichia coli* O157:H7 and *Salmonella* Muenchen from the surface of apples. *International Journal of Food Microbiology* 74: 47–55.

Kumar, V., Basu, M.S., and Rajendran, T.P. 2008. Mycotoxin research and mycoflora in some commercially important agricultural commodities. *Crop Protection* 27: 891–905.

Lacroix, M. and Ouattara, B. 2000. Combined industrial processes with irradiation to assure innocuity and preservation of food products—A review. *Food Research International* 33: 719–724.

Lado, B.H. and Yousef, A.E. 2002. Alternative food preservation technologies: Efficacy and mechanisms. *Microbes and Infection* 4: 433–440.

Laroche, C., Fine, F., and Gervais, P. 2005. Water activity affects heat resistance of microorganisms in food powders. *International Journal of Food Microbiology* 97: 307–315.

Lemcke, R.M. and White, H.R. 1959. The heat resistance of *Escherichia coli* cells from cultures of different ages. *Journal of Applied Bacteriology* 22: 193–201.

Lin, C.M., Fernando, S.Y., and Wei, C. 1996. Occurrence of *Listeria monocytogenes, Salmonella* spp., *Escherichia coli* and *E. coli* O157:H70. *Food Control* 7: 135–140.

Lundén, J., Tolvanen, R., and Korkeala, H. 2004. Human listeriosis outbreaks linked to dairy products in Europe. *Journal of Dairy Science* 87: E6–E11.

Mackey, B.M. and Derrick, C.M. 1982. The effect of sublethal injury by heating, freezing, drying and gamma-radiation on the duration of the lag phase of *Salmonella typhimurium. Journal of Applied Bacteriology* 53: 243–251.

Marek, R., Annamalai, T., and Venkitanarayanan, K. 2003. Detection of *Penicillium expansum* by polymerase chain reaction. *International Journal of Food Microbiology* 89: 139–144.

Martorell, P., Stratford, M., Steels, H., Fernández-Espinar, M.T., and Querol, A. 2007. Physiological characterization of spoilage strains of *Zygosaccharomyces bailii* and *Zygosaccharomyces rouxii* isolated from high sugar environments. *International Journal of Food Microbiology* 114: 234–242.

McDonough, F.E. and Hargrove, R.E. 1968. Heat resistance of *Salmonella* in dried milk. *Journal of Dairy Science* 51: 1587–1591.

McSwane, D., Rue, N., and Linton, R. 1998. *Essentials of Food Safety and Sanitation*. Upper Saddle River, NJ: Prentice Hall.

Mendonca, A.F. 2002. Inactivation by irradiation. In: Juneja, V.K. and Sofos, J.N. (eds.), *Control of Foodborne Microorganisms*, pp. 75–103. New York: Marcel Dekker, Inc.

Moats, W.A. 1971. Kinetics of thermal death of bacteria. *Journal of Bacteriology* 105: 165–171.

Moss, M.O. 2002. Mycotoxin review-1. *Aspergillus* and *Penicillium. Mycologist* 16: 116–119.

Murphy, R.Y., Duncan, L.K., Johnson, E.R., Davis, M.D., and Marcy, J.A. 2002. Thermal inactivation of *Salmonella* Senftenberg and *Listeria innocua* in beef/turkey blended patties cooked via fryer and/or air convection oven. *Journal of Food Science* 67: 1879–1885.

National Advisory Committee on Microbiological Criteria for Foods (NACMCF). 1999. Microbiological safety evaluations and recommendations on fresh produce. *Food Control* 10: 117–143.

Pao, S. and Petracek, P.D. 1997. Shelf life extension of peeled oranges by citric acid treatment. *Food Microbiology* 14: 485–491.

Perry, T.J. and Pawsey, R.K. 1989. *Principles of Microbiology for Students of Food Technology*, 2nd edn. Cheltenham, U.K.: Stanley Thornes (Publishers) Ltd.

Perry, J.J. and Staley, J.T. 1997. *Microbiology Dynamics and Diversity*. Fort Worth, TX: Saunders College Publishing.

Quintavalla, S., Larini, S., Mutti, P., and Barbuti, S. 2001. Evaluation of the thermal resistance of different *Salmonella* serotypes in pork meat containing curing additives. *International Journal of Food Microbiology* 67: 107–114.

Rattanasomboon, N., Bellara, S.R., Fryer, P.J., Thomas, C.R., and McFarlane, C.M. 2001. The gradient plate technique as a means of studying the recovery of heat injured *Brochothrix thermosphacta*. *International Journal of Food Science and Technology* 36: 369–376.

Restaino, L., Frampton, E.W., Hemphill, J.B., and Palnkar, P. 1995. Efficacy of ozonated water against various food-related micro-organisms. *Applied Environmental Microbiology* 61: 3471–3475.

Samelis, J. and Metaxopoulos, J. 1999. Incidence and principle sources of *Listeria* spp. and *Listeria monocytogenes* contamination in processed meats and a meat processing plant. *Food Microbiology* 16: 465–477.

Sinell, H.J. 1995. Control of foodborne infections and intoxication. *International Journal of Food Microbiology* 25: 209–217.

Sogin, S.J. and Ordal, Z.J. 1967. Regeneration of ribosomes and ribosomal ribonucleic acid during repair of thermal injury of *Staphylococcus aureus*. *Journal of Bacteriology* 94: 1082–1087.

Stephens, P.J., Joynson, J.A., Davies, K.W., Holbrook, R., Lappin-Scott, H.M., and Humphrey, T.J. 1997. The use of an automated growth analyser to measure recovery times of single heat-injured *Salmonella* cells. *Journal of Applied Microbiology* 83: 445–455.

Stumbo, C.R. 1965. *Thermobacteriology in Food Processing*, pp. 57–96. New York: Academic Press.

Teixeira, P., Castro, H., Mohácsi-Farkas, C., and Kirby, R. 1997. Identification of sites of injury in *Lactobacillus bulgaricus* during heat stress. *Journal of Applied Microbiology* 83: 219–226.

Thakur, B.R. and Singh, R.K. 1995. Combination processes in food irradiation. *Trends in Food Science and Technology* 6: 7–11.

Trujillo, A. J., Capellas, M., Saldo, J., Gervilla, R., and Guamis, B. 2002. Applications of high hydrostatic pressure on milk and dairy products: A review. *Innovative Food Science and Emerging Technologies* 3: 295–307.

Vijayakumar, C. and Wolf-Hall, C.E. 2002. Evaluation of household sanitizers for reducing levels of *Escherichia coli* on iceberg lettuce. *Journal of Food Protection* 65: 1646–1650.

Weissinger, W.R., Chantarapanont, W., and Beuchat, L.R. 2000. Survival and growth of *Salmonella* Baildon in shredded lettuce and diced tomatoes and effectiveness of chlorinated water as a sanitizer. *International Journal of Food Microbiology* 62: 123–131.

Wells, J.M. and Butterfield, J.E. 1997. *Salmonella* contamination associated with bacterial soft rot of fresh fruits and vegetables in the marketplace. *Plant Disease* 81: 867–872.

Zagory, D. 1999. Effects of postprocessing handling and packaging on microbial populations. *Postharvest Biology and Technology* 15: 313–321.

5 Effect of High-Pressure Food Processing on Physicochemical Changes of Foods: A Review

Navin K. Rastogi

CONTENTS

5.1 INTRODUCTION

A wide array of foods is thermally processed to inactivate microorganisms. Although these thermally processed foods are safe for consumption, this method of processing destroys the organoleptic quality of foods. On the other hand, high-pressure processing, which is an alternative technology that has recently received much attention both from academia and industry, preserves quality without compromising safety.

High-pressure processing is defined as a method of food preservation that involves subjecting food to intense pressure in the range of 300–700 MPa, with or without addition of heat, to achieve microbial inactivation while achieving consumer-desired qualities, e.g., retention of freshness and nutritive values of food products. High-pressure processing causes little changes in the "fresh" characteristics of foods; in fact, it is possible to keep many foods longer and in better condition. Small molecules such as flavor compounds, vitamins, and pigments are typically not affected by pressure. Pressure also provides a unique opportunity to create and control novel food textures in protein-based or starch-based foods. In some cases, pressure can be used to form protein gels and increase viscosity without the use of heat. Pressure-processed foods are also reported to have better texture, nutrient retention, and color compared to thermally processed foods. The technology has its roots in the material and process-engineering industry where it has been commercially used in sheet metal forming and isostatic pressing of advanced materials such as turbine components and ceramics.

Although proposed back in the 1890s, high-pressure processing was dormant technology until the late 1980s. Its use in the area of food processing began with the pioneering work of Hite (1899) for preservation of milk; its scope was then extended to the processing and preservation of fruits and vegetables (Hite et al., 1914). In the 1990s, high-pressure preserved acidified shelf-stable foods were demonstrated; however, high cost of processing equipment for most commercial applications limited their acceptability (Galazka and Ledward, 1995; Gould, 1995; Balci and Wilbey, 1999). By 2001, adequate commercial equipment was developed, so that by 2005 the process was being used for various products, ranging from juices to meats. Used since 1991 in Japan (Meidi-ya®, Meidi-Ya Co., Ltd., Tokyo, Japan) for jellies and jams, high-pressure processing met its first commercial success in the U.S. market when M/s Fresherized Foods, Texas, employed the process for guacamole dip processing. The potential of high-pressure processing is now being realized in the commercial market and in the future may lead to a wide array of appealing shelf-stable diversified food products.

Biological effects of the high-pressure inactivation of microorganisms or changing functional properties of food biopolymers are known for decades, but it is only in the last 10 years that foods preserved by high pressure have become commercial reality. Currently, there are a number of successful high-pressure-processed products on the global food market, including sliced small goods in Spain and the United States; fruit juices, jams, jellies, rice cakes, and raw squids in Japan; fruit juices in France, Italy, the United Kingdom, and the United States; salsa, guacamole meal kits, oysters in their shells, ready-to-eat meats in the United States; and apple sauce in Canada (Hugas et al., 2002). A survey of consumer opinion on high-pressure-processed foods indicated that the products were acceptable to the majority of consumers in

France, Germany, and the United Kingdom. Indeed, the French were prepared to pay a bit more for such products due to increased shelf life. However, while the health benefit of such products was the major concern for both the Germans and the British, they were reluctant to pay more (Butz et al., 2003).

High-pressure processing helps avoid undesirable alterations of foods that are caused by thermal treatment, such as vitamin loss, reduced bioavailability of essential amino acids, flavor loss, and modification of taste and color. The process thus allows processing of foods with cleaner ingredients and fewer additives. In some cases, texture has also been reported to improve as well (Balasubramaniam and Farkas, 2008). The technique is effective on most moist foods such as fruits, vegetables, sauces, and ready-to-eat meats. A high-pressure cycle generally takes no longer than several minutes, compared to traditional high-temperature processing that takes an hour or longer. In addition to its ability to destroy food-borne pathogens and spoilage microorganisms (by damaging their outer cell membrane and essential proteins in cells), high-pressure processing can also help inactivate enzymes, germinate or inactivate bacterial spores, marinate meats, shuck oysters, promote ripening of cheeses, minimize oxidative browning, etc.

A typical high-pressure processing system consists of a high-pressure vessel, a means to close the vessel off, a system for pressure generation, a system for temperature and pressure control, and a material handling system. Hydrostatic pressure is applied to a food product through a bath containing pressure-transferring medium (usually water) that surrounds the product. The hydrostatic pressure is transmitted to the product equally from all sides; this equal distribution of pressure is the reason why food is not crushed during treatment. This type of pressure also has little effect on covalent bonds and, as a result, the food being processed does not undergo significant chemical transformations.

As mentioned earlier, high-pressure processing is suitable for products with high water content and can be modified for both batch processing (for pre-packaged foods) and semi-continuous processing (for pumpable liquids). The process can also be conducted either at ambient or refrigerated temperatures, thereby eliminating thermally induced off-flavors. The technology is especially beneficial for heat-sensitive products. However, product packagings must be able to withstand a change in volume by up to 15%, followed by a return to its original size, without losing seal integrity or barrier properties.

Like any other processing methods, high-pressure processing cannot be universally applied to all types of foods. While it can be used to process both liquid and solid foods, and is especially beneficial for foods with high acid contents, the technique cannot yet be used to make shelf-stable versions of low-acid products such as vegetables, milk, or soups because of the inability of this process to destroy spores without added heat. However, it can be used to extend the refrigerated shelf life of these products and to eliminate the risk of various food-borne pathogens such as *Escherichia coli*, *Salmonella*, and *Listeria*. Another limitation is that food must contain water and not have internal air pockets. Food materials containing entrapped air such as strawberries or marshmallows would be crushed under high-pressure treatment. Dry solids, on the other hand, do not have sufficient moisture to make high-pressure processing effective for microbial destruction.

For both pasteurization and sterilization, a combined treatment of high pressure and temperature are frequently considered to be most appropriate (Farr, 1990; Patterson et al., 1995). Vegetative cells, including yeast and molds, are pressure sensitive, i.e., they can be inactivated by pressures in the range of 300–600 MPa (Knorr, 1995; Patterson et al., 1995). At high pressures, microbial death is considered to be due to permeabilization of cell membrane. For instance, it was observed that in the case of *Saccharomyces cerevisiae* at a pressure of about 400 MPa, the structure and cytoplasmic organelles were grossly deformed and large quantities of intracellular material leaked out, while at 500 MPa, the nucleus could no longer be recognized and the loss of intracellular material was almost complete (Farr, 1990). On the other hand, bacterial spores are highly pressure resistant since pressures exceeding 1200 MPa may be needed for their inactivation (Knorr, 1995). The initiation of germination or inhibition of germinated bacterial spores and inactivation of piezo-resistive microorganisms can be achieved in combination with moderate heating or other pretreatments such as ultrasound. Process temperature in the range of 90°C–121°C in conjunction with pressures of 500–800 MPa have been used to inactivate spores-forming bacteria such as *Clostridium botulinum*. Thus, sterilization of low-acid foods ($pH > 4.6$) would most probably rely on a combination of high pressure and other forms of relatively mild treatments.

In addition to food preservation, high-pressure treatment can result in food products acquiring novel structure and texture; the technique can hence be used to develop new products (Hayashi, 1990) or to increase functionality of certain ingredients. Depending on the operating parameters and the scale of operation, the cost of high-pressure treatment is typically around US$ 0.05–0.5 per liter or kilogram, the lower value being comparable to that of thermal processing (Thakur and Nelson, 1998; Balasubramaniam, 2003). Beyond the food industry, high-pressure technology could lead to processing of biological pharmaceutical products and specialized intravenous solutions or lead to development of human vaccines from pressure-inactivated viruses serving as antigens for inoculation.

The capability and limitations of high-pressure processing have been extensively reviewed (Farr, 1990; Cheftal, 1995; Knorr, 1995; Messens et al., 1997; Smelt, 1998; Thakur and Nelson, 1998; Tewari et al., 1999; Cheftel et al., 2000; Ontero and Sanz, 2000; Hugas et al., 2002; Balasubramaniam, 2003; Lakshmanan et al., 2003; Matser et al., 2004; Hogan et al., 2005; Mor-Mur and Yuste, 2005; Rastogi et al., 2007). Many of the early reviews mainly focused on the microbial efficacy or general aspects of high-pressure processing. There has been little research on the impact of high-pressure processing on the nutritional and health-promoting properties of foods to date. Most of the works also focused on juices and purees of fruits such as orange, tomato, broccoli, etc. High-pressure processing is an attractive food preservation technology and clearly offers opportunities for horticultural and food processing industries to meet the growing demand from consumers for healthier food products.

This chapter comprehensively reviews the effect of high-pressure processing on various physicochemical and microbiological changes of foods. High-pressure processing could maintain food-quality attributes such as color, flavor, and nutritional values due to its limited effect on covalent bonds. However, under pressure, chemical/biochemical reactions can also be induced and these reactions could affect quality attributes, e.g., nutritional value (Oey, 2008a,b; Sila et al., 2008).

5.2 EFFECT OF HIGH-PRESSURE PROCESSING ON QUALITY OF FRUIT AND VEGETABLE PRODUCTS

Application of high pressure to fruits and vegetables not only preserves the freshness and extend the shelf life, but also it can be used as a complimentary processing step in the chain of integrated food processing. Application of high pressure damages the plant cell wall structure, leaving cells more permeable, which leads to significant changes in tissue architecture (Farr, 1990; Dornenburg and Knorr, 1993; Rastogi et al., 1994), resulting in increased mass transfer rates during dehydration (water removal), osmotic dehydration (water removal and solute infusion), as well as solid-liquid extraction (Rastogi et al., 2007). Moreover, high pressure at ambient temperature can be used as a method of blanching, similar to hot-water or steam blanching, but without thermal degradation. Antimutagenicity of beet and tomato juices decreased by high-pressure treatment (600 MPa and 50°C or 800 MPa and 35°C), but there was no change in the antimutagenicity of carrot, leek, spinach, kohlrabi, and cauliflower juices by high-pressure processing (Butz et al., 1997). Application of high pressure during freezing results in instantaneous and homogenous formation of ice throughout the volume of a product; this eliminates internal stresses and improves the product quality. High-pressure-induced thawing reduces the loss of water-holding capacity (WHC) and improves both color and flavor of the product as well.

High-pressure-assisted thermal processing has recently emerged as a promising alternative technology for processing low-acid foods (Matser et al., 2004). The process, in general, involves simultaneous application of elevated pressure (500–700 MPa) and temperature (90°C–120°C) to a pre-heated food (Matser et al., 2004; Rajan et al., 2006a,b; Nguyen et al., 2007; Rastogi et al., 2008a). The compression heating during pressurization and rapid cooling on depressurization help reduce the severity of thermal effects encountered with conventional thermal processing (Ting et al., 2002; Rasanayagam et al., 2003; Wilson et al., 2008).

The following examples are related to the specific use of high pressure in fruits and vegetables processing as well as its effect on various organoleptic attributes such as color, flavor, texture, nutrition, and overall sensory characteristics.

5.2.1 Effect of High-Pressure Processing on Fruits and Their Products

5.2.1.1 Orange Juice

High-pressure processing (600 MPa, 20°C, 1 min), regardless of storage temperature, resulted in statistically significant reductions in aerobic bacteria as well as yeasts and molds compared with untreated Valencia and navel orange juices. High-pressure-processed Valencia orange juice stored at 4°C was microbiologically stable up to 12 weeks, while juice stored at 10°C up to 8 weeks; the yeast and mold population became detectable after 8 or 12 weeks of storage at 10°C or 4°C, respectively (Figure 5.1). Besides, 7-log reduction in *Salmonella* and partial inactivation of pectin

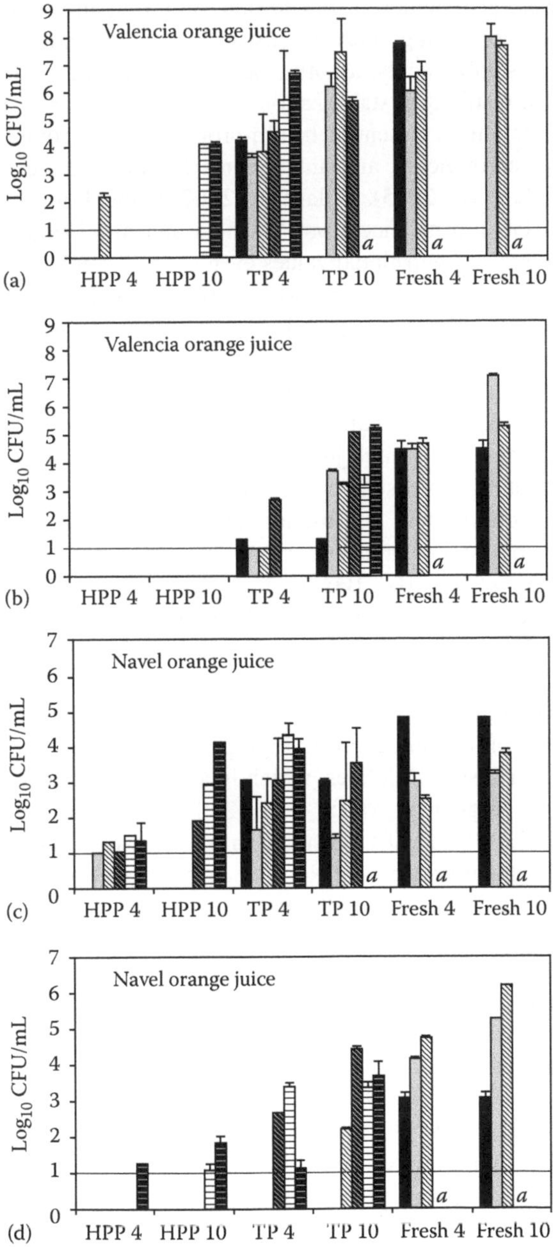

FIGURE 5.1 Mean total aerobic bacteria (a, c), yeast and mold (b, d) CFU/mL of high-pressure (600 MPa, 1 min), and thermally processed Valencia (65°C, 1 min) and navel orange juice (85°C, 25 s). HPP 4 or HPP 10 refers to high-pressure-processed juice stored at 4°C or 10°C; Thermal 4 or Thermal 10 refers to thermally processed juice stored at 4°C or 10°C; Fresh 4 or Fresh 10 refers to unprocessed juice stored at 4°C or 10°C. (■ Initial, □ week 1, ▧ week 2; ▩ week 4, ⊟ week 8, ▤ week 12). (From Bull, M.K. et al., *Innovat. Food Sci. Emerg. Technol.*, 5, 135, 2004. With permission.)

methylesterase were also achieved. In addition, high-pressure treatment resulted in no changes in color, browning index, viscosity, concentration, and titratable acidity, levels of alcohol-insoluble acids, ascorbic acid, and β-carotene (Bull et al., 2004). The treatment also resulted in stabilization of cloud (Goodner et al., 1999) as well as increased extraction of potential health-promoting compounds such as flavanones, vitamin C, carotenoids, and antioxidants in orange juice during storage at 4°C (Sanchez et al., 2003, 2005). Sellahewa (2002) pointed out that high-pressure-processed (600 MPa, 1 min) juice from Valencia and navel oranges was safe and retained its freshness and nutritional values even after 12 weeks at 4°C.

A number of researchers have reported extended shelf life of high-pressure-processed orange juice under refrigeration with increased flavor retention (Donsi et al., 1996; Parish, 1998a,b; Takahashi et al., 1998; Strolham et al., 2000; Plaza et al., 2006a). Fernandez et al. (2001a) showed that there was no significant difference in antioxidant activity, sugar, and carotene contents between high-pressure and thermally pasteurized orange juice. In contrast, Polydera et al. (2003, 2004, 2005) indicated that high-pressure-processed (500 MPa, 5 min, 35°C) orange juice had less loss of ascorbic acid and higher retention of flavor, antioxidant capacity, shelf life, sensory scores, and viscosity values as compared to conventionally pasteurized sample (80°C, 30 s). Baxter et al. (2005) showed that odor and flavor (volatile content, 20 key aroma compounds) of high-pressure-processed orange juice was acceptable to consumers even after storage for 12 weeks at lower temperatures (up to 10°C).

Butz et al. (2004) demonstrated that presence of intrinsic protective substances in freshly squeezed orange juice during high-pressure processing (600 MPa, 80°C) resulted in no major losses in folates, which is a hematopoietic vitamin of special importance during pregnancy. Ancos et al. (2002) pointed out that high-pressure processing (350 MPa, 5 min, 30°C) resulted in higher extraction of carotenoid content as compared to the untreated sample. Nienaber and Shellhammer (2001) pointed out that high-pressure processing (800 MPa, 25°C, 1 min) could be used as a method for stabilization of fresh orange juice, yielding in the lowest level of residual pectin methylesterase activity, good cloud stability, and less loss of ascorbic acid over a period of more than 2 months at 4°C or 37°C. Sampedro et al. (2008) indicated that high-pressure processing (700 MPa, 55°C, 2 min) could result in complete inactivation of pectin methylesterase.

5.2.1.2 Grapefruit and Lemon Juices

The use of a simple and effective immobilization method in combination with high pressure could be employed as a potential technique to enhance the rate of debittering of citrus juices; the bitter taste in grapefruit is due to the presence of naringin, which is a flavanone glycoside. Ferreira et al. (2008) indicated that the reduction of naringin to naringenin (a tasteless compound) using naringinase (α-rhamnopyranosidase immobilized on calcium alginate) could be increased by the application of high pressure. Under atmospheric pressure, naringin reduction was only 35% in a model system, but it was 75% under high pressure (160 MPa, 37°C, 20 min).

Donsi et al. (1998) indicated that no fungi were detected in pressure-treated lemon juices, unlike in the case of a control sample, which was spoiled by yeasts and fungi after 10 days. The study indicated that high-pressure-processed (300 MPa) lemon juice had a satisfactory shelf life without any significant change in constituents and physicochemical properties.

5.2.1.3 Pineapple

Application of high pressure resulted in an increase in the permeability of cell structure (Farr, 1990; Dornenburg and Knorr, 1993; Eshtiaghi et al., 1994; Rastogi et al., 1994), which resulted in softening of the samples. Along with the hardness, high pressure also decreased springiness and chewiness, while there was no significant effect on cohesiveness. The hardness of high-pressure treated pineapple decreased rapidly with an increase in pressure up to 300 MPa, beyond which the decrease was not significant. The compression and decompression steps during the application and release of pressure, respectively, also caused removal of a significant amount of water (Rastogi and Niranjan, 1998). Further, the reduction of hardness, which is an indication of permeabilization, resulted in increase in diffusion coefficient of water during dehydration of pineapple, which led to a reduction in drying time as inferred from the plot of moisture ratio with time (Kingsly et al., 2009a, Figure 5.2).

In the case of osmotic dehydration, high-pressure (100–800 MPa) pretreatment enhanced both water removal and solute gain due to increase in the extent of cell wall breakup with an increase in pressure. High-pressure pretreated pineapple subjected to osmotic dehydration and then air dried resulted in a product having lower solid diffusion during rehydration, and so was the release of the cellular components. The reduction in loss of soluble solids during rehydration was due to formation of a gel network between divalent ions and deesterified pectin (Eshtiaghi et al., 1994; Basak and Ramaswamy, 1998; Rastogi et al., 2000a, Figure 5.3). This may prove to be a useful technique to reduce the loss of nutrients or color from food during rehydration.

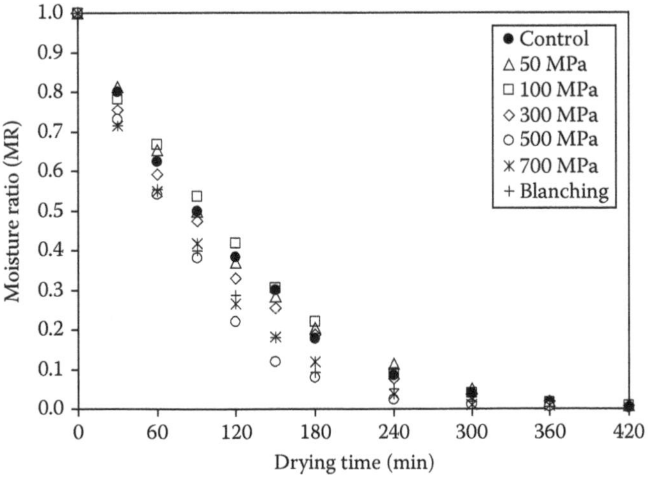

FIGURE 5.2 Variation of moisture ratio during drying of pineapple. (From Kingsly, A.R.P. et al., *J. Food Process Eng.*, 32, 369, 2009a. With permission.)

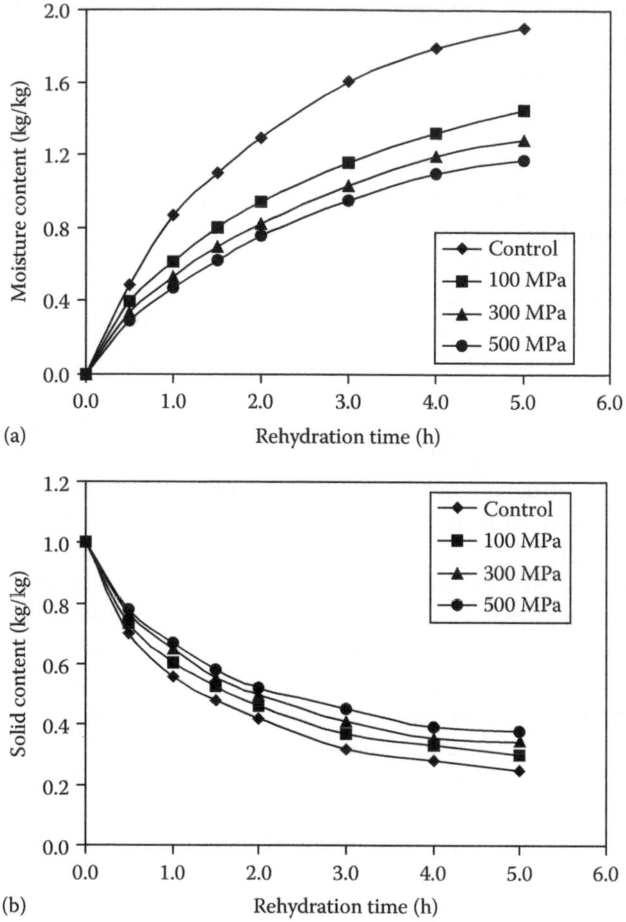

FIGURE 5.3 (a) Variation of moisture and (b) solid content with time during rehydration of high-pressure treated, partially osmotic, and air-dried pineapple sample. (From Rastogi, N.K. et al., *J. Food Sci.*, 65, 838, 2000a. With permission.)

5.2.1.4 Mango

Addition of ascorbic acid (500 ppm) and phosphoric acid (to maintain pH 3.5) prior to high-pressure processing (552 MPa, 5 min) of mango puree resulted in reduced rates of browning during storage at 3°C for 1 month. Besides, no microbial growth was observed (Guerrero-Beltran et al., 2006). Herschel–Bulkley model could well be used to describe the shear stress-shear rate data of fresh and canned high-pressure-processed mango pulp. The flow behavior index for fresh pulp decreased with pressure treatment, whereas an increasing trend was observed with the canned pulp. The consistency index of fresh pulp increased with the pressure level from 100 to 200 MPa, while a steady decrease was observed for the canned pulp (Ahmed et al., 2005).

High-pressure-processed (300 or 600 MPa for 1 min) precut mango during storage at 3°C slightly experienced reduced fresh flavor, increased off-flavor and sweetness,

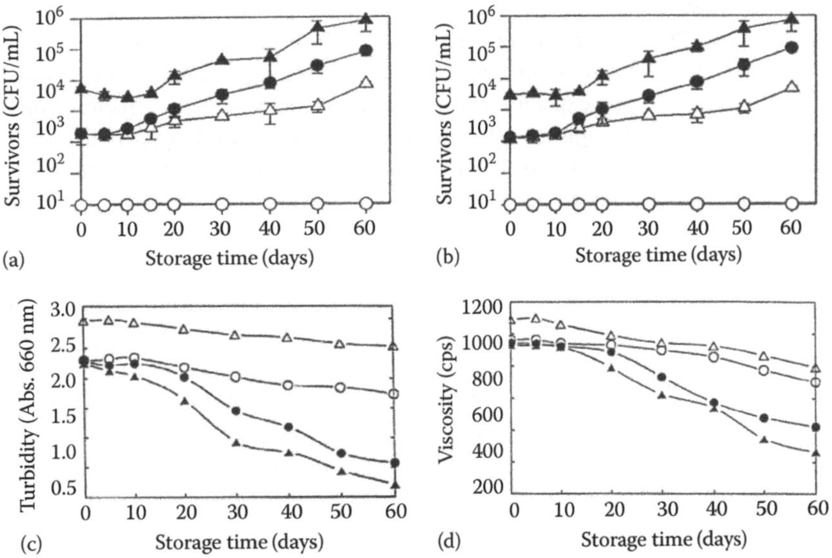

FIGURE 5.4 Effects of high-pressure treatment and thermal pasteurization on the changes in (a) total plate counts; (b) yeast and mold counts; (c) turbidity; and (d) viscosity of guava puree during storage at 4°C. (○) Pressurized at 600 MPa; (●) pressurized at 400 MPa; (△) heated at 88°C–90°C; (▲) without treatment. (From Gow, C.Y. and Hsin, T.L., *Int. J. Food Sci. Technol.*, 31, 205, 1996. With permission.)

but had improved microbial status; however, color, texture, and other sensory attributes changed very little (Boynton et al., 2002).

5.2.1.5 Guava Puree

Gow and Hsin (1996, 1999) demonstrated that high-pressure processing (600 MPa, 15 min) of guava puree effectively sterilized microbes and partially inactivated enzymes; the puree could be stored up to 40 days at 4°C without any change in color, cloudiness, ascorbic acid content, flavor distribution, and viscosity (Figure 5.4). The treatment resulted in an increase in the levels of methanol and ethanol; 2-ethylfuran was also observed during the storage period.

5.2.1.6 Peach

Kingsly et al. (2009b) indicated that pressure treatment (>300 MPa) in combination with citric acid (1–1.2%wt) was effective in inactivation of peach polyphenoloxidase, indicating that high pressure could be used as a potential alternative for conventional blanching. Sumitani et al. (1994) demonstrated that high-pressure processing (400 MPa, 20°C, 10 min) of white peach resulted in enzymatic formation of benzaldehyde, C_6 aldehydes, and alcohols by disruption of fruit tissues. The increase in benzaldehyde content in the high-pressure treated fruit during storage was caused by residual β-glucosidase activity following the high-pressure treatment.

5.2.1.7 Melon

Wolbang et al. (2008) pointed out that high-pressure processing of melon did not have any effect on the total titratable acidity and total soluble solids; however, color, ferric ion reducing capacity, and vitamin C were adversely affected. But, the β-carotene content was significantly increased. Prestamo and Arroyo (2000) indicated that high-pressure processing of melon did not cause any browning and that the texture of the fruit was acceptable.

5.2.1.8 Avocado

Lopez et al. (1998) studied the effect of high-pressure processing (345–689 MPa, 10–30 min) and initial pH values (3.9–4.3) on polyphenoloxidase activity, color, and microbial inactivation in avocado puree during storage at 5°C, 15°C, or 25°C; the results were compared with those of the untreated avocado puree. Standard plate as well as yeast and mold counts of the high-pressure treated purees were less than 10 CFU/g during 100 days of storage at 5°C, 15°C, or 25°C. Residual polyphenoloxidase activities for treatments at pH 4.1 and 689 MPa were reduced from 1078 to 266, 235, or 168 units when the process time was 10, 20, or 30 min, respectively, representing a residual activity decrease of 24.7%, 21.8%, and 15.6%, respectively (Figure 5.5a). Avocado puree with a residual polyphenoloxidase activity of less than 45% and stored at 5°C maintained an acceptable color for at least 60 days (Figure 5.5b).

Palou et al. (2000) indicated that application of high pressure (689 MPa) did not significantly affect sensory properties or color of avocado puree, i.e., guacamole. After pressure treatment, lipoxygenase and polyphenoloxidase were completely inactivated and the standard plate count was less than 10 CFU/g. High-pressure treated guacamole had an extended shelf life up to 30 days at 5°C, 15°C, or 25°C, whereas untreated samples stored at 5°C spoiled within 5 days. Storage was also associated with browning, which was attributed to a decrease in green coloration.

5.2.1.9 Apple and Cashew Apple Juices

Novotna et al. (1999) showed that sensory quality of apple juice subjected to high-pressure treatment was better than that of pasteurized juice in terms of aroma. Lavinas et al. (2008) indicated that high-pressure processing (350–400 MPa, 3 or 7 min) of cashew apple juice reduced the aerobic mesophilic bacteria count to a level below the detection limit; yeast and filamentous fungi were also inactivated. The juice could be stored for 8 weeks at 4°C without any significant change in the product quality.

5.2.1.10 Strawberry and Blackberry Purees

Lambert et al. (1999) indicated that high-pressure processing (500 or 800 MPa) of strawberry puree resulted in inactivation of enzymes responsible for degradation of food quality such as polyphenoloxidase and peroxidase. Strawberry aroma could be specifically characterized by 2 main components, viz., nerolidol and furaneol out of the 46 volatile compounds. No major changes in strawberry aroma profiles were observed up to 500 MPa, but a higher pressure (800 MPa) induced significant changes in the aroma profiles due to the synthesis of new compounds such as 3,4-dimethoxy-2-methyl-furan and γ-lactone; in addition the relative importance of the remaining

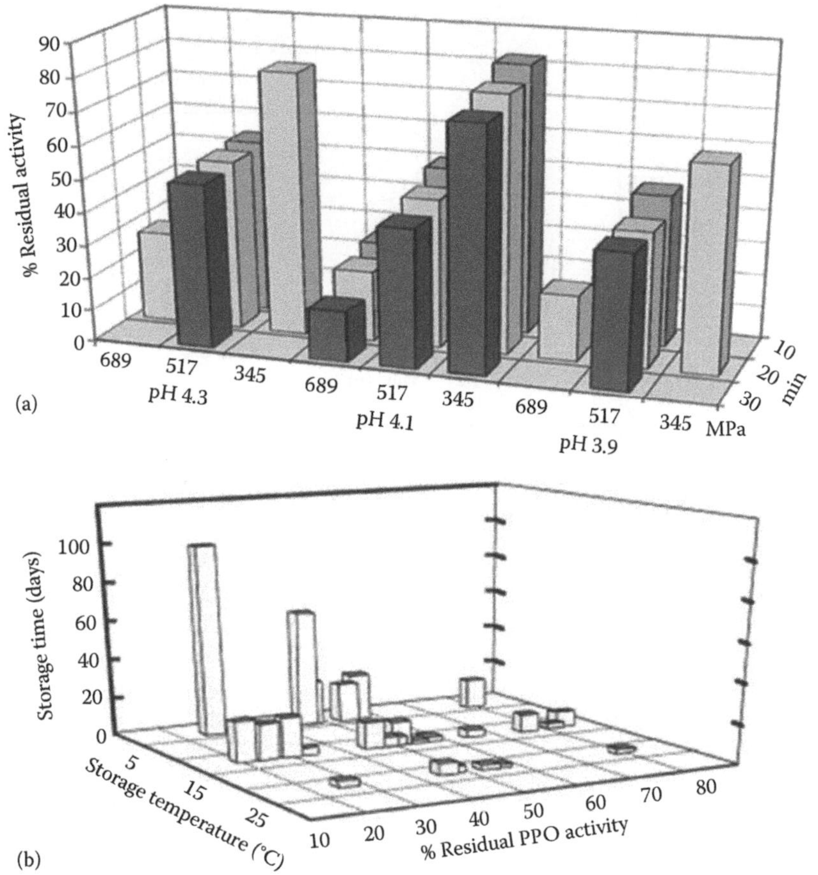

FIGURE 5.5 (a) Effect of high-pressure treatment and initial pH on residual polyphenoloxidase activity of avocado puree; (b) effect of residual polyphenoloxidase activity and storage temperature on the storage time needed to lose the green color component in avocado puree. (From Lopez, F.R. and Olano, A., *Int. Dairy J.*, 8, 623, 1998. With permission.)

compounds was also altered (Figure 5.6). Patras et al. (2009a) demonstrated that high-pressure processing could be used as an efficient method for preservation of nutritious and fresh-like strawberry and blackberry purees. High-pressure treatment (400–600 MPa, 15 min, 10°C–30°C) did not cause any significant change in the ascorbic acid and anthocyanins contents and antioxidant activities. Color changes were minor for pressurized purees, but the differences were slightly higher for thermally treated samples. Redness was well retained in high-pressure treated samples.

5.2.1.11 Grape

Rastogi et al. (1999) studied the combined effect of high pressure (100–600 MPa) with heat treatment (0°C–60°C) on inactivation of endogenous enzymes in order to develop shelf-stable red grape juice. The lowest peroxidase (55.75%) and polyphenoloxidase (41.86%) activities were found at 60°C, with pressures at 600 and 100 MPa, respectively.

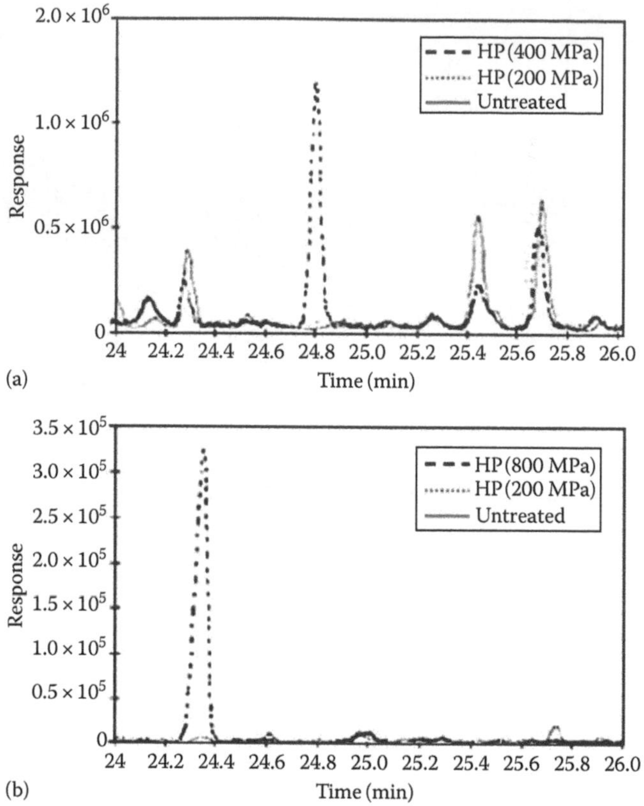

FIGURE 5.6 Volatile compounds appearing after high-pressure treatment (800 MPa, 20 min, 20°C). Peaks are characterized by retention time of (a) 12.8 min attributed to 3,4-dimethoxy 2-methyl-furan or 2-5-dimethyl 4-methoxy furan-3; (b) 24.3 min attributed to a lactone volatile compound. (From Lambert, Y. et al., *Food Chem.*, 67, 7, 1999. With permission.)

Moio et al. (1994) demonstrated that white grape must could be sterilized at 500 MPa for 3 min, with little changes in physicochemical properties. On the other hand, red grape must was not completely sterilized due to the higher stability of the natural microflora presented in it. Corrales et al. (2008a,b) pointed out that high-pressure treatment (600 MPa) resulted in increased total phenolic content, which in turn resulted in three-fold increase in antioxidant capacity along with increased extraction of acylated anthocyanins. Pressure-assisted extraction of antioxidant and anthocyanins at 600 MPa from red grape skins was maximized when the ethanol concentration and temperature of extraction were 50%, 70°C, or 100% and 50°C (Corrales et al., 2009). Addition of partially purified co-pigments (such as rosemary and thyme polyphenolic extracts) to muscadine grape juice during high-pressure processing (400 and 550 MPa, 15 min) improved color and antioxidant activity and reduced phytochemical losses during subsequent storage (Pozo et al., 2007).

5.2.1.12 Lychee

Phunchaisri and Apichartsrangkoon (2005) demonstrated that high-pressure treatment (600 MPa at 60°C for 20 min) resulted in less loss of visual quality in both fresh and syrup-processed lychee compared to thermal processing. High-pressure processing led to extensive inactivation of peroxidase and polyphenoloxidase in fresh lychee; these effects were less significant when the samples were processed in syrup.

5.2.1.13 Passion Fruit

Laboissiere et al. (2007) showed that high-pressure processing (300 MPa, 5 min, 25°C) could be used to preserve yellow passion fruit pulp, yielding a ready-to-drink juice with improved sensory quality, free from cooked and artificial flavor attributes, as compared to commercial juices. High-pressure treatment did not cause any significant modifications in compounds responsible for aroma, flavor, and consistency.

5.2.1.14 Persimmon Puree

Ancos et al. (2000) pointed out that high-pressure processing (up to 400 MPa) of persimmon puree resulted in increased amount of extractable carotenoids, which was related to an increase in vitamin A value; however, this was not correlated with an increase in antioxidant activity.

5.2.1.15 Jam

Watanabe et al. (1991) explained the method for production of high-pressure-processed jam, which involved freeze concentration to remove excess water and sterilization of the product by pressurizing. Powdered sugar, pectin, citric acid, and freeze-concentrated strawberry juice was mixed, degassed, and then pressurized (400 MPa, 5 min). The texture of jam was similar to that of conventional jam; the product had a bright red color and retained all the original flavor compounds. Gimenez et al. (2001) also pointed out that a combination of freeze concentration and high pressure yielded a strawberry jam product with better color and sensory properties compared with those of heat-treated strawberry jam. Highest stability of anthocyanins was found when the jams were stored at 4°C. A comparison of pressurized jam, traditional jam, and commercial products (Japanese jam and commercial jam) indicated that pressure-processed jam samples had better retention of anthocyanins (pelargonidin-3-rutinoside and pelargonidin-3-glucoside) concentration (Figure 5.7).

Kimura et al. (1994) showed that pressure-treated jam had better quality than heat-treated jam. The pressure-treated jam could be stored at refrigerated temperature up to 3 months. Dervisi et al. (2001) evaluated the rheological properties and color of strawberry jam produced by high pressure (400 MPa, 5 min). An increase in pectin concentration resulted in an increase in storage and loss moduli and a decrease in absorbance intensity, possibly due to degradation of anthocyanins. The optimum pectin concentration for color retention and texture quality was between 2.5% and 5%.

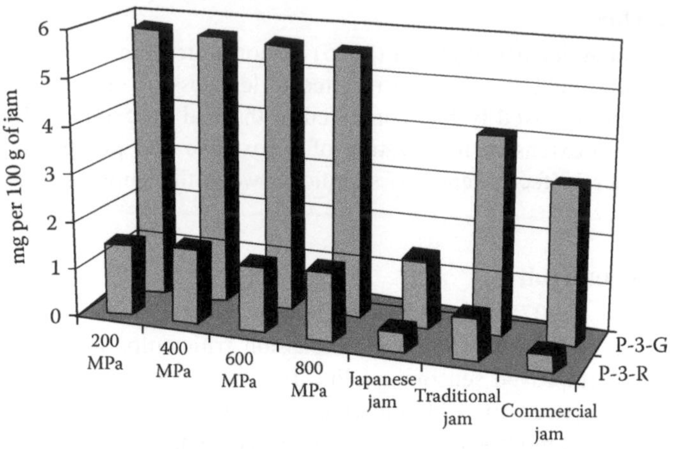

FIGURE 5.7 Pelargonidin-3-rutinoside (P-3-R) and pelargonidin-3-glucoside (P-3-G) concentrations in different strawberry jams. (From Gimenez, J. et al., *J. Sci. Food Agric.*, 81, 1228, 2001. With permission.)

5.2.2 Effect of High-Pressure Processing on Vegetables and Their Products

5.2.2.1 Potato

Application of high pressure (400 MPa, 15 min) in combination with 0.5%wt citric acid solution resulted in complete inactivation of polyphenoloxidase along with 4-log reduction in microbial count of potato samples (Eshtiaghi and Knorr, 1993).

Acceleration of mass transfer during osmotic dehydration due to application of high pressure (100–400 MPa) was demonstrated in the case of potato by Sopanangkul et al. (2002). The accelerated mass transfer was attributed to the fact that high-pressure processing opened up the tissue structure, thereby facilitating diffusion. However, pressures above 400 MPa induced starch gelatinization, which resulted in hindered diffusion. Rastogi et al. (2000b,c, 2003) demonstrated that high-pressure processing enhanced the rate of movement of the dehydration front in a potato sample subjected to subsequent osmotic dehydration. This was attributed to synergistic effect of cell permeabilization (due to high pressure) and osmotic stress (due to the concentration of the surrounding osmotic solution).

Luscher et al. (2005) indicated that pressure-shift freezing (up to 400 MPa) resulted in formation of ice of higher density than liquid water, which in turn resulted in reduced membrane damage, leading to preservation of textural properties even after freezing. Benet et al. (2006) demonstrated that optimization of high-pressure low-temperature processes (freezing and thawing) resulted in better quality and safety of whole potato. Besides, improvement in color and reduction in drip loss during thawing were indicated.

5.2.2.2 Carrot

The rapid loss of firmness of carrot due to application of high pressure was attributed to disruption of membrane, which reduced cell turgor pressure. The hardness reduced with an increase in pressure up to 300 MPa, beyond which loss of firmness was nominal (Michel and Autio, 2001; Araya et al., 2007). Kato et al. (1997) and Araya et al. (2007) showed that application of high pressure resulted in an increase in displacement with an increase in force, thereby resulting in more deformable material or rubbery-like texture, which may not sometimes be desirable. Stute et al. (1996) also illustrated that application of high pressure at ambient temperature (25°C) resulted in softening of vegetables such as carrot, potato, and green beans due to destruction of cell membrane and loss of soluble pectin along with partial liberation of cell liquor.

Upon high-pressure application, pectin methylesterase is liberated from cell wall and brought in close contact with its substrate, methylated pectin, resulting in deesterification of pectin. The reaction occurs even after release of the pressure. Deesterification of pectin led to formation of a gel network with divalent ions such as Ca^{2+} or Mg^{+}, leading to hardening of the tissue. It was also shown that high-pressure-processed vegetables no longer softened during cooking and retained textural characteristics as compared to unpressurized vegetables. Basak and Ramaswamy (1998) demonstrated that initial loss in texture was due to the instantaneous pulse action of pressure; this was followed by a more gradual change as a result of pressure-hold. The extent of the initial loss of texture was more prominent at higher pressures and partial recovery of texture was more prominent at lower pressures. The vegetables treated were firmer and brighter than raw product. The gradual change was described as texture recovery, which reached to its original value at a low pressure for long processing time (100 MPa, 30 min, 25°C).

Sila et al. (2004, 2005) indicated that high-pressure treatment in combination with $CaCl_2$ infusion improved the texture of carrot during thermal processing. Moreover, subjection of carrot to high-pressure pretreatment alone resulted in less loss of texture when these were further processed at high temperatures (100°C–125°C). Nguyen et al. (2007) demonstrated that pressure-assisted thermal process (500–700 MPa, 95°C–105°C) resulted in less quality degradation (in terms of texture, color, and carotene content) as compared to thermally processed carrots (Figure 5.8). Further, Rastogi et al. (2008a,b) suggested that the hardness during pressure-assisted thermal processing could be further improved by a combined pretreatment involving calcium infusion, heating, and pressurization. It was also demonstrated by Roeck et al. (2008, 2009) that combination of high pressure with temperature (600 MPa, 80°C) resulted in preservation of carrot texture as compared to thermally processed sample due to nonoccurrence of β-elimination reaction as well as stimulation of demethoxylation of pectin, which enhanced tissue strength by forming cross-links with calcium ions.

Fuchigami et al. (1997a,b) indicated that carrots did not freeze when subjected to high pressure (200–400 MPa) under freezing conditions (–30°C); when pressure was reduced to atmospheric pressure, quick freezing was observed. High-pressure frozen and thawed carrots had better firmness, texture, and histological structures than ordinary frozen and thawed samples.

5.2.2.3 Onion

High-pressure treatment above 100 MPa induced browning of diced onions. The rate of browning was found to increase with an increase in pressure (Butz et al., 1994). Due to the influence of pressure, onion epidermis cells and cellular components like

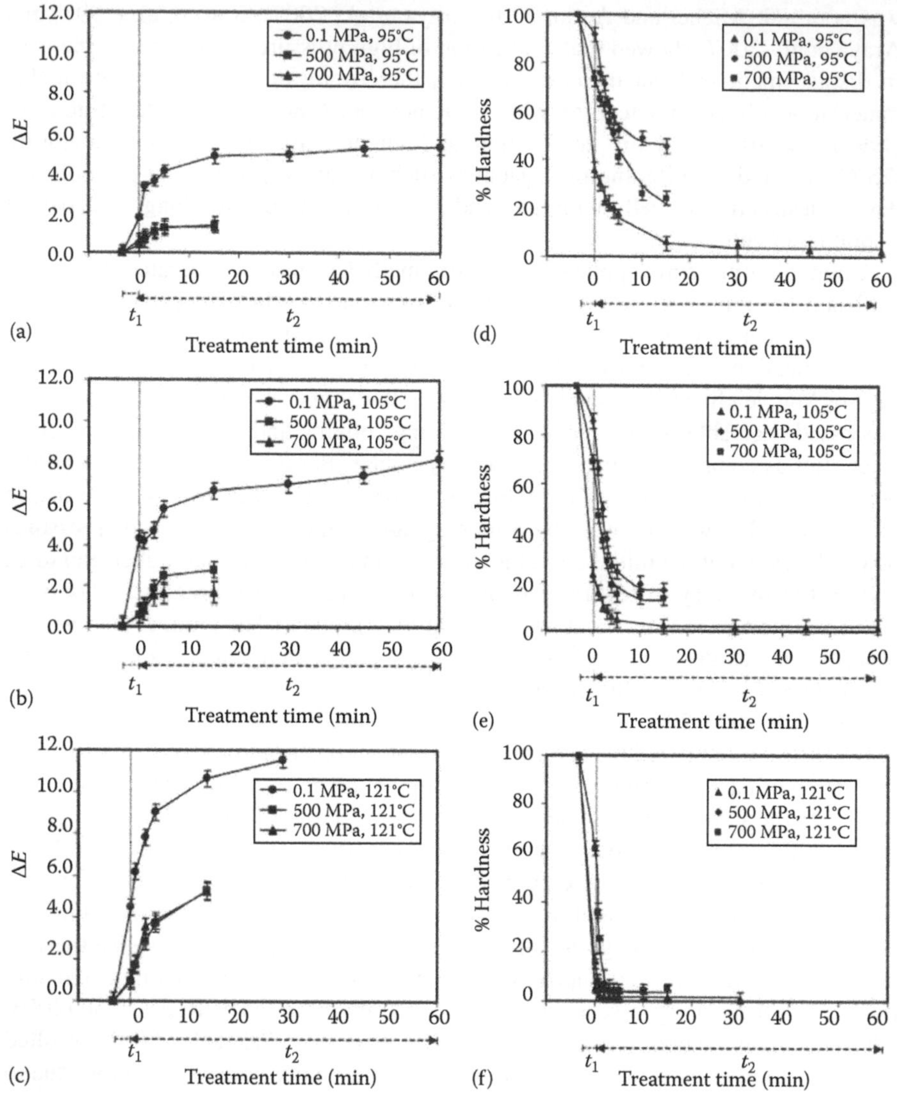

FIGURE 5.8 Changes in color (a, b, c) and texture (d, e, f) of carrot during thermal processing or pressure-assisted thermal processing at different pressures and temperatures: 95°C (a, d), 105°C (b, e), and 121°C (c, f). t_1 and t_2 are the come-up time and holding time, respectively. (From Nguyen, L.Y. et al., *J. Food Sci.*, 72, E264, 2007. With permission.)

vacuoles were affected and polyphenoloxidase was no longer separated from its phenol substrate, which was oxidized to orthoquinones and upon polymerization formed brown pigment. Roldan et al. (2009) showed that high-pressure processing (100 and 400 MPa, 5°C) of onion resulted in better extraction of flavonols (quercetin and quercetin glucosides) and increased antioxidant activity (Figure 5.9). At 400 MPa, the extraction of quercetin glucosides increased to 33%, but there was no change in the antioxidant activity.

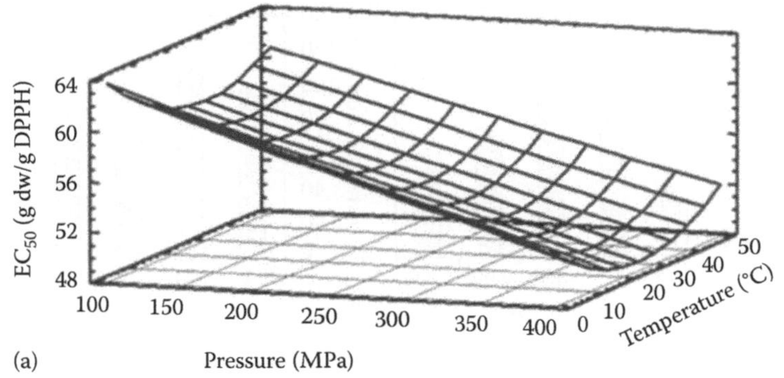

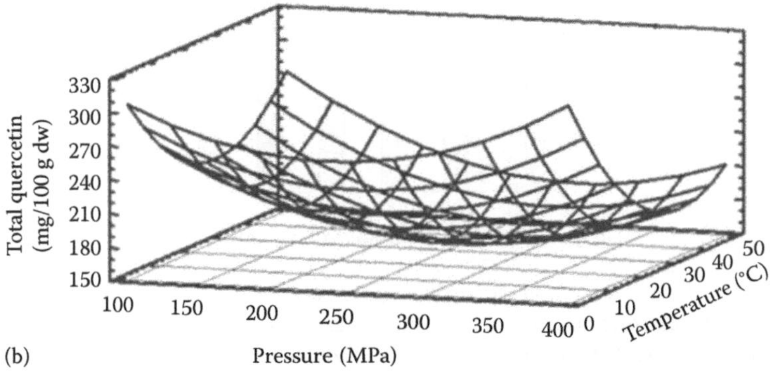

FIGURE 5.9 Response surfaces indicating the effect of high pressure and temperature on (a) onion antioxidant activity; (b) total quercetin of onion sample. (From Roldan, M.E. et al., *Lebensm. Wiss. Technol.*, 42, 835, 2009. With permission.)

5.2.2.4 Tomato

High-pressure processing of whole cherry tomato resulted in a decrease in the sample hardness up to 400 MPa. Further, an increase in pressure up to 500 or 600 MPa resulted in an increase in the hardness (as inferred from the reduction in percentage cell rupture, Figure 5.10a); however, the hardness value was less than the corresponding value for untreated tomato sample. Visual appearances of the tomato samples treated at 500 or 600 MPa were similar to the untreated samples. The increase in firmness after pressure treatment at 500 and 600 MPa was attributed to the action of pectinmethylesterase in producing low methoxypectin, which formed a gel network with divalent ions, leading to tissue hardening. The decrease in firmness below 500 MPa was due to the action of polygalacturonase, which hydrolyzed low methoxypectin to water-soluble galacturonic acid (Tangwongchai et al., 2000, Figure 5.10b and c).

Application of high pressure to tomato puree did not result in any change of water-insoluble antioxidants, lycopene, and β-carotene, but resulted in a decrease in recovery of carotenoids and an increase in water binding capacity, which was attributed to high-pressure-induced structural changes in tomato pulp tissue. However, antioxidant capacity of the water-soluble fraction increased after storage at 4°C for 21 days (Fernandez et al., 2001b). Butz et al. (2002) also pointed out that high-pressure

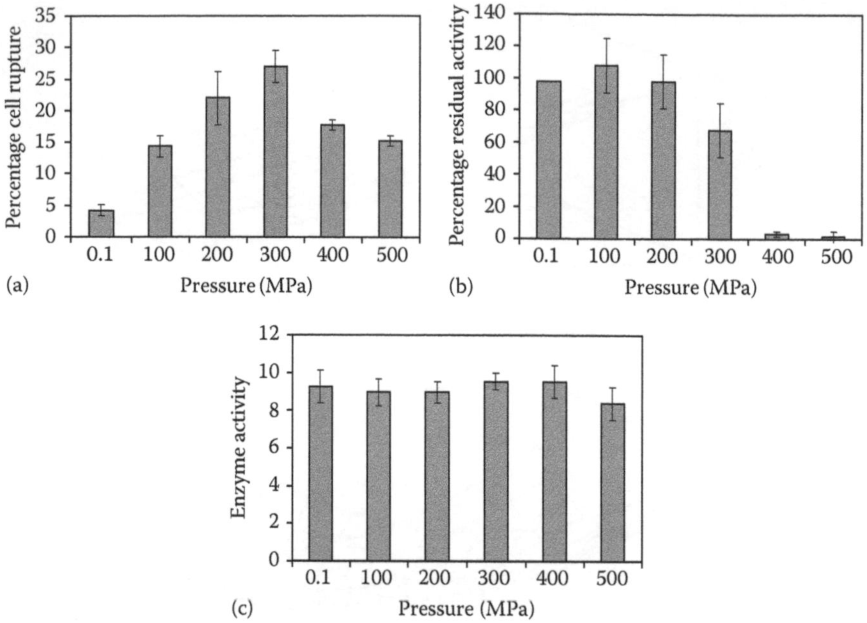

FIGURE 5.10 Effect of high-pressure treatment on (a) percentage cell rupture, the activity of crude (b) polygalacturonase and (c) pectinmethylesterase in extracts from pressure-treated whole cherry tomatoes. (From Tangwongchai, R. et al., *J. Agric. Food Chem.*, 48, 1434, 2000. With permission.)

treatment did not have a significant impact on β-carotene in tomato and antimutagenicity was also not affected. Sanchez et al. (2004) showed that high-pressure treatment increased carotenoids extractability from tomato. High-pressure treatment of the order of 200 MPa affected structure of cellular tomato matrix, such that various carotenes were released differently on the basis of their chemical features and chromoplast location. The total carotenoids content was correlated with the antioxidant activity.

Qiu et al. (2006) indicated that the highest stability of lycopene in tomato puree was obtained by high-pressure processing at 500 MPa and further storage at 4°C. Sanchez et al. (2006) verified that high-pressure-processed (400 MPa, 25°C, 15 min) tomato puree had higher redness, carotenoids, and vitamin C than low-temperature (70°C, 30 s) or high-temperature (90°C, 1 min) pasteurized samples. On the other hand, McInerney et al. (2007) indicated that the antioxidant capacity and total carotenoids content differed among vegetables, but were unaffected by high-pressure treatment.

Hsu et al. (2008) pointed out that high-pressure processing (500 MPa) of tomato juice resulted in inactivation of microorganisms and pectolytic enzymes, improvement in extractable carotenoids and lycopene contents, and retention of vitamin C compared with fresh juice. Kuo (2008) showed that high-pressure processing (200–500 MPa) of tomato retained color, extractable total carotenoids, lycopene content, and antioxidant activity. The residual activities of pectinmethylesterase and polygalacturonase were least in the lower range of pressure (200 and 400 MPa for pectinmethylesterase and polygalactronase, respectively), whereas these activities were higher at higher pressure (500 MPa).

Plaza et al. (2003) explored the possibility of using high pressure as one of the hurdles along with citric acid and sodium chloride for the manufacture of minimally processed tomato products with optimal sensory and microbiological characteristics. At 400 MPa, 4-log reduction of total microbial counts, along with a significant inactivation of food-quality-related enzymes (such as polyphenoloxidase, peroxidase, and pectin methylesterase), was obtained.

Patras et al. (2009b) demonstrated that high-pressure treatment at 600 MPa retained more than 90% of ascorbic acid in tomato purees as compared to thermal processing, but phenolic content was not affected by thermal or high-pressure treatments. Dede et al. (2007) demonstrated that high-pressure-processed (150–250 MPa) tomato juice was superior in terms of ascorbic acid retention, high antioxidant activity, pH, and minimum color loss. These quality benefits were maintained during 1 month of storage at 4°C or 25°C.

Krebbers et al. (2003) suggested that high-pressure processing (300–500 MPa, 20°C–90°C) of tomato puree might be a suitable alternative to conventional processing techniques without significant loss of color and sensory properties up to 8 weeks of storage at 4°C. Rodrigo et al. (2007) indicated that no visual color degradation of tomato puree appeared under combined thermal and high-pressure treatment (300–700 MPa, 65°C).

5.2.2.5 Green Beans

Krebbers et al. (2002) demonstrated that high-pressure processing (500 MPa, ambient temperature) had a potential to substitute conventional preservation techniques for green beans, with improved texture, nutritional quality, and appearance of the processed products. Shelf life of beans subjected to high-pressure processing could be extended for at least 1 month at 6°C or 20°C storage temperature. Leadley et al. (2008) compared high-pressure sterilization of beans with the equivalent thermal processing and concluded that the high-pressure sterilized green bean samples were darker and greener in appearance, the pressure-treated samples were twice as firm as the thermally processed samples.

5.2.2.6 Green Peas

Quaglia et al. (1996) demonstrated that high pressure could be used as an alternative to thermal blanching of green peas. High-pressure treatment (900 MPa) resulted in a higher retention of ascorbic acid (82%), in comparison with water (10%) or microwave blanching (47%). A combination of high pressure (400–900 MPa) with heat treatment (up to 60°C) did not cause any noticeable modification of pea firmness.

5.2.2.7 Mushroom

Application of high pressure (600–900 MPa) resulted in lower texture degradation of mushroom as compared to thermal blanching. However, high-pressure-induced crystallization of phospholipids in cell membrane led to permeabilization of the cell membrane. Due to increased permeability, extracellularly located polyphenoloxidase could better react with phenols and resulted in increased browning during pressure treatment. Nevertheless, product yield and color were reported to be comparable to that of thermally blanched products. The intensity of browning was less intense in the case of evacuated mushroom before pressurization (Matser et al., 2000).

5.2.2.8 Broccoli

High-pressure treatment (500 MPa, 10 min) was demonstrated to be capable of preserving nutritional substances such as sulforaphane and antimutagenic activity in apple-broccoli juices, apart from inactivating microorganisms originally present in the raw juice. Vitamin C content was not dependent of the pressure level, but it was instead dependent on the holding time. There was no difference between the sensory quality of high-pressure treated and frozen juice stored up to 70 days (Houska et al., 2006).

Butz et al. (2002) showed that high-pressure treatment did not have a significant impact on chlorophyll *a* and *b* in broccoli. On the other hand, Eylen et al. (2007, 2009) demonstrated that high concentrations of glucosinolates (GLS) present in broccoli were hydrolyzed by endogenous myrosinase and yielded isothiocyanates, which had an anticarcinogenic activity, and that at 20°C and pressures of 100–500 MPa, there was no degradation after 15 min, whereas after 35 min at 200 or 300 MPa, about 20% less GLS were present. At 40°C and pressures of 100–500 MPa, a clear effect on GLS degradation after 15 min was visible (Figure 5.11). Verlinde et al. (2008) also indicated that folylpoly-γ-glutamate present in broccoli was converted to folylmono- and folyldi-γ-glutamates due to high-pressure processing (up to 600 MPa, 25°C–45°C); this led to impaired bioavailability of dietary folate.

Pressure-shift-frozen broccoli samples were not acceptable in terms of flavor after 30 days of storage at –20°C, even though the texture remained quite firm after the treatment; however, sensory quality of samples blanched prior to high-pressure freezing was acceptable (Prestamo et al., 2004). Hence, it is necessary to blanch broccoli before it is subjected to high-pressure freezing treatment (210 MPa, –20°C). Blanched and high-pressure frozen broccoli indeed presented less cell damage, lower drip losses, and better texture than conventionally frozen samples (Fernandez et al., 2006).

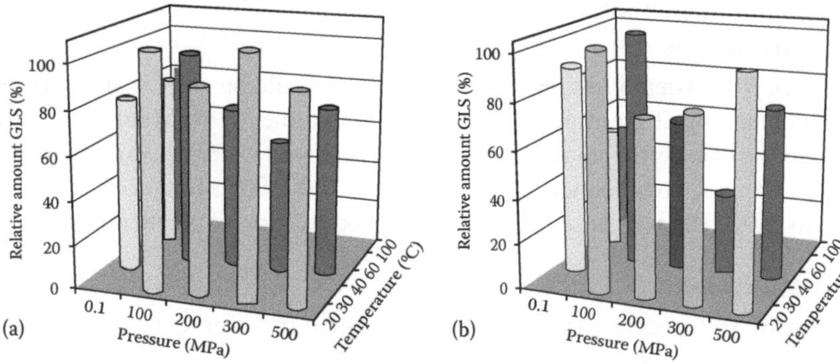

FIGURE 5.11 Relative amount of total GLS after (a) 15 min and (b) 35 min treatment at different pressures and temperatures. (From Eylen, D. et al., *Food Chem.*, 112, 646, 2009. With permission.)

5.2.2.9 Cauliflower and Spinach

A considerable loss of turgor and structural collapse was observed during high-pressure processing of cauliflower and spinach (Prestamo and Arroyo, 1998). High-pressure treatment influenced structure of spinach more extensively than cauliflower since the latter has a harder and less elastic cell structure. The soft parenchyma cell of spinach leaf was completely destroyed after high-pressure treatment (400 MPa, 30 min, 5°C).

Generally, high-pressure treatment results in membrane disruption and protein denaturation in plant tissues. High pressure also changes cell permeability and enables the movement of water from inside to outside the cells. As a result, treated vegetable tissues have a soaked or drenched appearance. However, after these changes, cauliflower maintained near-original, acceptable firmness and flavor.

5.2.2.10 White Cabbage

Wennberg and Nyman (2004) indicated that high-pressure treatment produced white cabbage with specific properties in terms of nutrition and function. The proportion of soluble fiber was reduced with an increase in pressure up to 500 MPa; however, the total fiber content remained constant. The reduction in soluble fiber content (or increase in insoluble content) due to high-pressure application was dependent on the initial levels of insoluble fiber and enzyme activity. Fuchigami et al. (1998) indicated that pressure-shift-frozen Chinese cabbage had better firmness, texture, and histological structure than the ordinary frozen samples.

5.2.2.11 Garlic

Seok and Dong (2001) showed that combined treatment, involving citric or ascorbic acid treatment (5 or 10 g/kg) and high-pressure treatment (600 MPa, 1 min) helped to retard browning of chopped garlic taken from dormant bulbs when stored under ambient conditions for 6 months. Browning of germinated bulbs was also reduced without inhibiting greening, which is a physiological disorder occurring temporarily in processed garlic as a result of external stress and is strongly affected by storage conditions.

5.2.2.12 Eggplant

Otero et al. (1998) demonstrated that high-pressure frozen eggplant samples had higher firmness and lower rupture strain and drip loss compared to those of still-air-frozen samples. The improved quality was attributed to the formation of heavy ice polymorphs due to freezing of water under high pressure (100–700 MPa), leading to volume reduction.

5.2.2.13 Pepper

Ade-Omowaye et al. (2001) demonstrated that high-pressure pretreatment (400 MPa, 10 min) resulted in an increase in drying rate of paprika during subsequent dehydration. The pretreatment could be an alternative to chemical (NaOH or HCl) pretreatments, thereby minimizing adverse consequences of chemicals.

Castro et al. (2008) suggested that high-pressure processing (100–200 MPa, 10–20 min) could be used as a pretreatment in place of blanching to produce frozen pepper with better nutritional (soluble protein and ascorbic acid) and texture (firmness) characteristics.

5.2.2.14 *Panax ginseng* and Gazpacho

Yutang et al. (2008) showed that application of combined high-pressure and microwave extraction of ginsenosides from *Panax ginseng* resulted in higher yields than other extraction methods, including soxhlet extraction, ultrasound-assisted extraction, and heat reflux extraction. The results indicated that the technique not only took a shorter time, but also allowed higher extraction yields of ginsenosides.

Plaza et al. (2006b) studied the stability of carotenoids and the antioxidant activity of Mediterranean vegetable soup (gazpacho) subjected to high-pressure treatment (up to 300 MPa, 60°C) and stored at 4°C during 40 days. The results indicated that a treatment at 150 MPa led to better preservation of both properties as compared to a treatment at 300 MPa.

5.2.2.15 Chickpea Seed, Mung Bean, and Alfalfa Seeds

Kadlec et al. (2006) pointed out that germinated chickpea seeds subjected to high-pressure treatment (500 MPa, 10 min) experienced a reduction in the total number of microorganisms without any quality and sensory changes during 21 days of storage. Doblado et al. (2007) proposed the use of high-pressure processing to produce high-quality minimally processed fresh-like cowpea sprouts. Although it was noted that the antioxidant capacity and vitamin C content, which increased during the process of germination, decreased due to high pressure (500 MPa), pressure treatment still provided a high amount of vitamin C (15–17 mg/100 g) and the antioxidant capacity (26%–59%); the values that were higher than those of raw cowpeas.

Neetoo et al. (2008) demonstrated that high-pressure processing (650 MPa, 15 min) resulted in approximately 5-log reduction in the population of *E. coli* during alfalfa seed decontamination. The germination rate of high-pressure decontaminated seeds was identical to that of untreated seeds.

Penas et al. (2008) optimized the combination of treatment time, pressure, and temperature when applied to mung bean and alfalfa seeds to reduce the native microbial load in sprouts without affecting their germination capacity. The optimal treatment pressures at 40°C were 100 and 250 MPa for alfalfa and mung bean seeds, respectively. Later, Penas et al. (2009) indicated that the counts of total aerobic mesophilic bacteria, total and fecal coliforms as well as molds and yeast were reduced with increasing pressure and concentrations of disinfectant agents (hypochlorite and carvacrol) during the production of alfalfa seed sprouts. A combination of high pressure (200 MPa) and hypochlorite (18,000 ppm) improved the microbiological quality by reducing the microbial count between 4.5- and 5-log CFU/g. Carvacrol ensured the microbial safety of sprouts, but reduced the germination percentage to unacceptable levels.

5.3 EFFECT OF HIGH-PRESSURE PROCESSING ON QUALITY OF DAIRY PRODUCTS

5.3.1 Effect on Microorganisms

Use of high pressure in dairy processing was initially explored as an alternative process to pasteurization. Hite et al. (1914) noted that high-pressure processing resulted in a significant reduction in microorganisms and combination of high pressure with temperature resulted in increased shelf life of milk. A number of researchers have since then studied inactivation of microorganisms (such as *Listeria monocytogenes*, *Staphylococcus aureus*, or *Listeria innocua*), either naturally present or introduced in milk (Styles et al., 1991; Erkman and Karatas, 1997; Gervila et al., 1997). Mussa and Ramaswamy (1997) indicated that high-pressure processing (300 MPa) resulted in 4-log reduction in microorganisms in milk; shelf life of the treated milk was 25, 18, and 12 days at 0°C, 5°C, and 10°C, respectively. High-pressure treatment (200 MPa) of pasteurized milk (63°C, 30 min) and pasteurization of high-pressure treated milk resulted in increased reduction in microbial count. Rademacher and Kessler (1997) reported increased shelf life of thermally pasteurized milk by 10 days when it was treated at 400 MPa for 15 min or at 500 MPa for 3 min. Koseki et al. (2008) indicated that a mild heat treatment (37°C, 240 min or 50°C, 10 min) inhibited the recovery of *L. monocytogenes* in high-pressure-processed milk; the product was safely stored for 70 days at 25°C (Figure 5.12).

Vachon et al. (2002) demonstrated that periodic oscillation of high pressure was very effective for the destruction of pathogens such as *L. monocytogenes*, *E. coli*, and *Salmonella enteritidis*, and that the process could be used as a promising alternative for cold pasteurization of milk. Pandey et al. (2003) studied the effect of high pressure (250–450 MPa, 0–80 min, 3°C or 21°C) on destruction kinetics of indigenous microflora and *E. coli* in raw milk. Higher pressure, longer holding time, and lower temperature resulted in greater destruction of microorganisms in raw milk. *Lactobacillus delbrueckii* subsp. *bulgaricus* showed greatest sensitivity to high pressure (480 MPa), whereas *Streptococcus thermophilus* showed greatest resistance in reconstituted skim milk. All organisms were shown to have a significant recovery in their viability after a week of refrigerated storage (Shah et al., 2008). Buzrul et al. (2008) described high-pressure (400–600 MPa, 22°C) inactivation kinetics of *E. coli* and *L. innocua* using the Weibull model. A parameter Z_p was defined as an increase in pressure resulting in 1-log reduction in microbial population.

Narisawa et al. (2008) demonstrated that protein fractions of skimmed milk provided protection against injury and inactivation of *E. coli* by high hydrostatic pressure treatment. The protective effect was found to increase with an increase in the concentration of protein in skimmed milk. Ramaswamy et al. (2009) indicated that casein and lactose present in milk were the major contributors for pressure protection of *E. coli* in milk during high-pressure treatment. Fat content in milk (0%–5%) had no significant influence on the destruction, however (Figure 5.13).

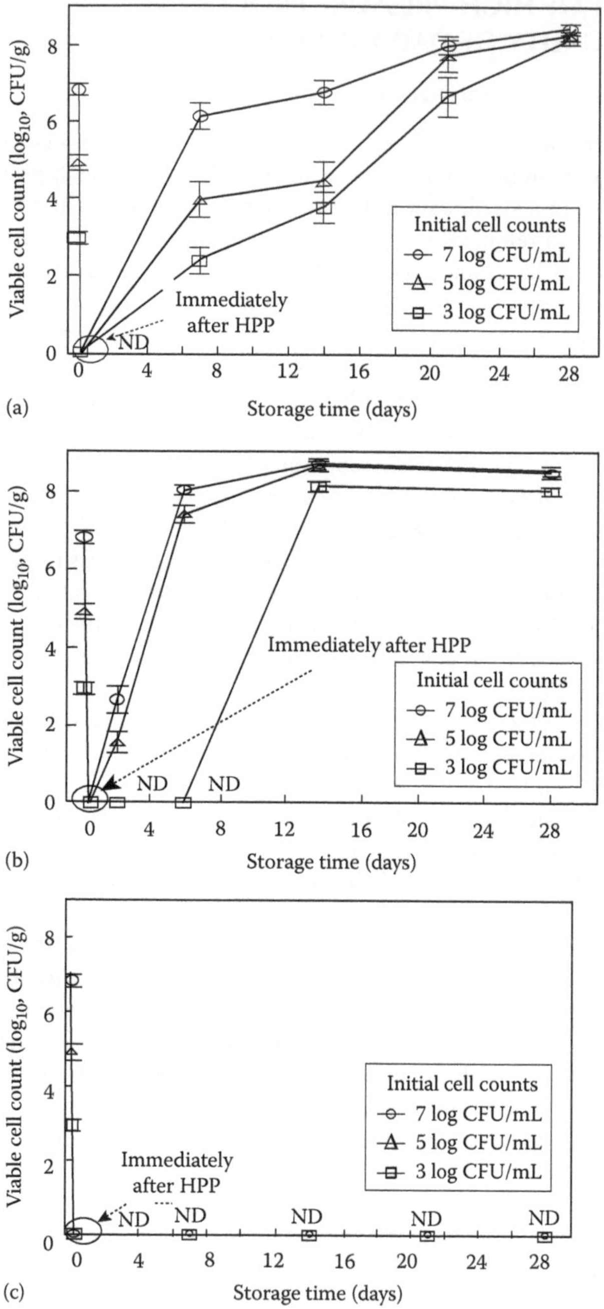

FIGURE 5.12 Changes in the number of *L. monocytogenes* cells in milk during storage at (a) 4°C; (b) 25°C; and (c) 37°C. The cells at different inoculum levels (3-, 5-, and 7-log CFU/mL) were treated at 550 MPa and 25°C for 5 min in milk. ND refers to not detected. (From Koseki, S. et al., *Food Microbiol.*, 25, 288, 2008. With permission.)

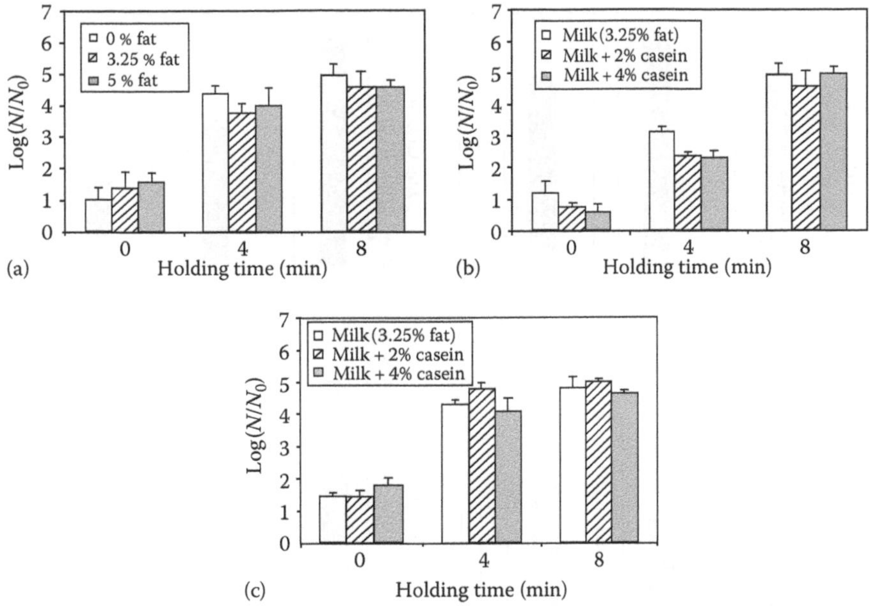

FIGURE 5.13 A comparison of logarithmic reduction of *E. coli* K12 in (a) 400 MPa treated milk with different fat contents; (b) 300 MPa treated milk (3.25% fat) with different percentages of added casein; (c) 400 MPa-treated milk (3.25% fat) with different percentages of added lactose. (From Ramaswamy, H.S. et al., *Food Bioprod. Process.*, 87, 1, 2009. With permission.)

Morgan et al. (2000) indicated that combination of high-pressure processing with bacteriocin (lacticin) resulted in a synergistic effect in controlling microbial flora of milk without significantly influencing its cheese-making properties. Masschalck et al. (2001) demonstrated synergistic effect of high pressure (155–400 MPa) and antimicrobial peptides (lactoferrin and lactoferricin at 500 mg/mL) on inactivation of bacteria; this is indeed a promising and natural method for increasing the efficiency and safety of high-pressure processing for foods. Black et al. (2005) also demonstrated that combining high-pressure processing and an antibiotic (nisin) resulted in a greater inactivation of gram-positive bacteria. The gram-negative bacteria, in this case, were found to be more sensitive to high pressure, either alone or in combination with nisin, than gram-positive bacteria. Nevertheless, this high-pressure processing–antibiotic combination allowed lower pressures and shorter processing time to be used without compromising the product safety (Figure 5.14). Black et al. (2008) later suggested that combination of high-pressure treatment and nisin might be an appealing alternative to thermal pasteurization of milk.

Sierra et al. (2000) showed that high-pressure processing (400 MPa, 30 min) was a gentler process than conventional procedures for extending shelf life of milk; no significant variations in the contents of vitamins B_1 and B_6 (pyridoxamine and pyridoxal) were observed. Viazis et al. (2008) evaluated the efficacy of high-pressure processing to inactivate pathogens in human milk without loss of any important nutritional biomolecules. Huppertz et al. (2002) reviewed the effect of high pressure on properties and contents of milk.

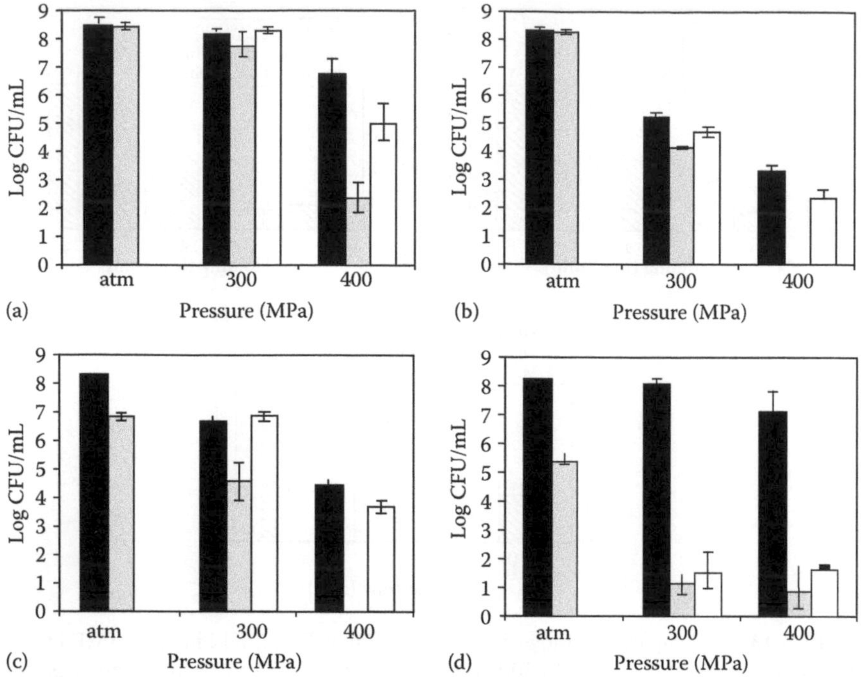

FIGURE 5.14 Effect of addition of nisin on inactivation of gram-negative bacteria (a) *E. coli* and (b) *P. fluorescens* as well as gram-positive bacteria; (c) *L. innocua*; and (d) *L. viridescens* in milk prior to (▒) or immediately after (□) HP treatment. Counts of bacteria after treatment with pressure alone (i.e., no nisin) are also shown (■). (From Black, E.P. et al., *Innovat. Food Sci. Emerg. Technol.*, 6, 286, 2005. With permission.)

5.3.2 Effect on Enzymes and Proteins

High-pressure processing up to 300 MPa was found to have little effect on β-lactoglobulin in whey, whereas further increase in pressure to above 600 MPa resulted in a decrease in the level of β-lactoglobulin due to high-pressure denaturation of protein (Brooker et al., 1998; Pandey and Ramaswamy, 1998). Anema (2008a) indicated that significant denaturation of β-lactoglobulin took place at a pressure as low as 200 MPa. At any given holding time, denaturation increased as the treatment pressure increased. Similarly, at any given pressure, denaturation increased with an increase in the holding time. At 200 MPa, and to a lesser extent at 300 MPa, the rate of denaturation of β-lactoglobulin was slower as the milk solids concentration increased. At pressures above 300 MPa, the rate of denaturation of β-lactoglobulin was similar at all milk solids concentrations. In contrast, α-lactalbumin denatured only at 400 MPa and little effect of milk solids concentration was observed (Figure 5.15).

Short time exposure to high pressure was reported to enhance activity of lipoprotein lipase and glutamyl transferase in milk. However, long exposure time (100 min) did not bring about any inactivation of lipase, while glutamyl transferase followed the first-order inactivation kinetics (Pandey and Ramaswamy, 2004). Milk enzymes

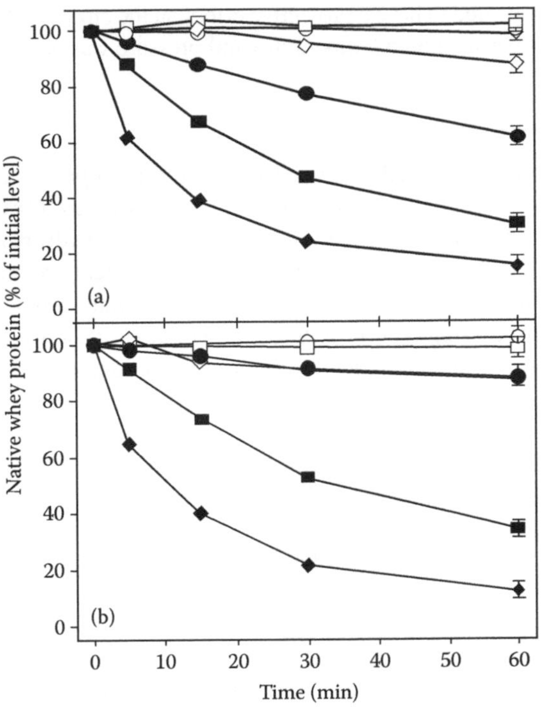

FIGURE 5.15 Denaturation of β-lactoglobulin (filled symbols) and α-lactalbumin (open symbols) in (a) 10% TS milk and (b) 20% TS skim milk following pressure treatments at 200 MPa (●, ○), 300 MPa (■, □), and 500 MPa (◆, ◇). (From Anema, S.G., *Int. Dairy J.*, 18, 228, 2008. With permission.)

were much less sensitive to pressure. Only alkaline phosphatase and proteinases were completely inactivated at 1000 MPa (Kolakowski et al., 2000).

Scollard et al. (2000) indicated that β-lactoglobulin denatured and plasmin activity decreased at pressures higher than 300 and 400 MPa, respectively. High-pressure treatment (300–500 MPa) increased proteolysis in milk. Borda et al. (2004a,b) reported that combined high-pressure and thermal inactivation of plasmin and plasminogen in milk followed the first-order kinetics. A synergistic effect of temperature and high pressure was observed in the pressure range of 300–600 MPa. However, an antagonistic effect of temperature and pressure was observed at pressures greater than 600 MPa because the enzymes were stabilized by disruption of disulfide bonds. Moatsu et al. (2008) suggested that high pressure (450 or 650 MPa) at various temperatures (20°C, 40°C, or 55°C) had a profound effect on reduction in the activity of endogenous enzymes such as plasmin, cathepsin D, and plasminogen activator, as well as denaturation of major whey proteins of ovine milk; these in turn had a significant influence on cheese yield and proteolysis during cheese ripening.

Sainz et al. (2009) examined the combined effect of pressure (300–600 MPa), temperature (40°C–60°C), and homogenization on the protease activity in milk. Inactivation of protease could extend the shelf life of milk. Protease was found to be very resistant to high pressures; pressure stability was higher in raw milk than in pasteurized milk and homogenization appeared to have a protective effect on the enzyme. A very pronounced antagonistic effect between high temperature and pressure was observed.

5.3.3 Effect on Flavor

Vazquez et al. (2006) indicated that volatile generation in pressure-treated samples (482–620 MPa, 60°C) was different from that of heat-treated (60°C) samples. Heat treatment tended to promote formation of methanethiol, hydrogen sulfide, methyl ketones, and aldehydes, whereas high-pressure treatment favored formation of hydrogen sulfide and aldehydes. Vazquez et al. (2007) later reported that pressure-assisted thermal processing of milk promoted formation of a few compounds such as 2-methylpropanal, 2,3-butanedione, and hydrogen sulfide, as well as inhibited formation of a few volatiles such as hexanal, heptanal, octanal, nonanal, and decanal. Methanethiol and methyl ketones, including 2-pentanone, 2-hexanone, 2-heptanone, 2-octanone, 2-nonanone, 2-decanone, and 2-undecanone, were the compounds that were not affected by high pressure. Due to the formation and inhibition of these compounds, the sensory acceptance of treated milk was not very high; the milk was rejected by the consumers. Flavor of the milk was termed as "cooked" milk flavor.

5.3.4 Cheese Making

High pressure induces disruption of casein micelles and denaturation of whey proteins, increases pH of milk, reduces rennet coagulation time, and increases cheese yield, thereby possessing potential application in cheese making (O'Reilly et al., 2001). San-Martin et al. (2006) discussed the application of high pressure in cheese making with reference to its effect on milk components such as casein micelles, whey proteins, milk fat globules, as well as its impact on milk color, microbial inactivation, cheese ripening, and brining.

Huppertz and Smiddy (2008) demonstrated that application of high pressure (250–300 MPa) led to an initial rapid micellar disruption; this was followed by a partial reversal of the high-pressure-induced reassociation of micellar fragments. Partial internal cross-linking of casein micelles by transglutaminase prior to pressure treatment considerably slowed down both the disruption and reassociation processes.

Degree of cross-linking and treatment at a pressure in the region 200–400 MPa can strongly influence rheological properties of the milk, which may have interesting implications for products made therefrom.

5.3.4.1 Microbial Destruction

Capellas et al. (1996) reported that samples of goat milk cheese inoculated with 10^8 CFU/g and subjected to high-pressure processing (400–500 MPa) exhibited no surviving *E. coli* even after 15, 30, or 60 days of storage at 2°C–4°C. O'Reilly et al. (2000) indeed indicated high sensitivity of *E. coli* in cheddar cheese at pressures above 200 MPa, possibly due to acid injury during cheese fermentation. Carminati et al. (2004) also showed that high-pressure treatment (400–700 MPa) was effective in reducing *L. monocytogenes* in gorgonzola cheese rinds without significantly changing its sensory properties. Evrendilek et al. (2008) indicated that in the case of

Turkish white cheese, the maximum reduction in *L. monocytogenes* counts of about 4.9-log CFU/g was achieved at 600 MPa. High-pressure processing resulted in total reduction in molds, yeasts, and *Enterobacteriaceae* counts for cheese samples produced from raw and pasteurized milk. This suggested that high-pressure processing could be effectively used to reduce the microbial load in Turkish white cheese.

Linton et al. (2008) demonstrated the potential of high pressure (500 MPa, 10 min) for treatment of raw milk to be used in the manufacture of soft cheeses. High-pressure treatment significantly reduced the level of *L. monocytogenes* in raw milk and so allowed production of safer nonthermally processed camembert-type soft cheese.

5.3.4.2 Rennet Coagulation Time

High-pressure treatment of milk indirectly affects its coagulation process and cheese-making properties through a number of effects on milk proteins, including reduction in the size of casein micelles, probably followed by interaction with micellar κ-casein. Pressures lower than 150 MPa did not have any influence on the rennet coagulation time, but at higher pressures the time was reported to decrease (Derobry et al., 1994). Many researchers have also reported that high-pressure treatment in the range of 200–400 MPa decreased the rennet coagulation time as compared to the case of untreated milk (Johnston et al., 1998; Lopez and Olano, 1998; Needs et al., 2000). A decrease in the rennet coagulation time in the case of milk was due to high-pressure-induced association of whey proteins with casein micelles (Kolakowski et al., 2000; Zobrist et al., 2005). Further increase in pressure to the range 500–600 MPa resulted, however, in an increase in the rennet coagulation time (Needs et al., 2000). Trujillo et al. (1999a) showed that the rennet coagulation time of pressure-treated (500 MPa, 15 min, 20°C) milk was higher than that of pasteurized milk (72°C, 15 s).

5.3.4.3 Rate of Curd Formation and Curd Firming

High-pressure processing accelerated the rate of curd formation and curd firming of rennet milk (Ohmiya et al., 1987). The rate of curd firming increased at pressures up to 200 MPa, beyond which it decreased (Lopez et al., 1996); the rate of curd formation was maximal at 200 MPa, but slightly decreased at 400–600 MPa (Needs et al. 2000). These investigators also observed higher curd firmness for milk treated with high pressure as compared to untreated milk. Pandey et al. (2000) indicated that a decrease in the pressure level, temperature, and holding time resulted in a decrease in WHC and an increase in gel strength of rennet curd. The change in the rate of curd formation resulted in alteration of the gel structure, which in turn affected quality and ripening characteristics of cheese.

5.3.4.4 Cheese Yield

High-pressure treatment of milk increased yield of cheese due to increased WHC as well as to denaturation of whey proteins and their association with casein. Drake et al. (1997) attributed the increased yield and higher moisture content of cheese made from high-pressure treated milk to the fact that casein molecules and fat globules might not aggregate closely and hence allow moisture to be trapped or held in cheese. Pandey and Ramaswamy (1998) reported that reduced hardness of Cheddar cheese made from high-pressure treated milk might be due to association of whey protein with casein in the pressurized milk. Arias et al. (2000) and Huppertz et al. (2004, 2005) also reported an increase in the cheese yield due to higher-pressure treatment. The yield of cheese from high-pressure treated and subsequently heated milk was greater than that from unheated and unpressurized milk. Molina et al. (2000) reported increased yield of pressurization of pasteurized milk due to improvement in the coagulation properties of proteins.

Anema (2008b) showed that application of combined process involving heat and high pressure to skim milk resulted in higher level of whey protein denaturation than that of heat or pressure treatment alone. High-pressure treatment alone decreased the casein micelle size, whereas the change in casein micelle size was not prominent for thermal or combined treatment.

5.3.4.5 Ripening of Cheese

High-pressure-induced biochemical aspects such as glycolysis, lipolysis, and proteolysis during ripening of cheese have been studied by a number of researchers. Yokohama et al. (1992) showed that high-pressure treatment resulted in considerable reduction in the ripening time due to proteolysis of milk protein, which resulted in an increase in free amino acids content and improvement in taste. The increased free amino acids content, from 21.3 to 26.5 mg/g, in cheese treated at 50 MPa for 3 days was comparable to 6-month old control cheese.

Trujillo et al. (1999b) indicated that cheese made from high-pressure treated (500 MPa) goat milk had higher pH and salt content, matured more quickly and developed stronger flavors compared to untreated milk. The presence of small peptides and free amino acids indicated a higher extent of proteolysis in cheese made from high-pressure treated milk. Buffa et al. (2001) showed that cheese prepared from raw and high-pressure treated goat milk was firmer, less fracturable and less cohesive than that prepared from pasteurized milk (72°C, 15 s). Pressure treatment resulted in more elastic, regular, and compact protein matrix with smaller and uniform fat globules, resembling the structure of cheese prepared from raw milk. Besides, the level of free amino acids increased (Saldo et al., 2002). Sheehan et al. (2005) demonstrated that high pressure (400 MPa, 5 min) did not markedly affect composition, cooking properties, pH, proteolysis, or rheological properties of reduced fat mozzarella cheese. However, WHC was markedly greater in high-pressure-processed cheese than in control cheese up to the sixth day of storage.

Daryaei et al. (2006, 2008) indicated that high-pressure processing (300–600 MPa, 5 min) is an effective process to control residual viable count of *Lactococcus* in fresh lactic curd cheese incubated aerobically or anaerobically. Spoilage yeasts and titratable acidity could also be controlled. The product shelf life could be extended up to 6–8 weeks without any adverse effects (Figure 5.16). Arques et al. (2007) indicated that the odor of La Serena cheese made from raw Merino ewe milk after second day of high-pressure treatment (300 or 400 MPa, 10 min) was scarcely affected, but aroma quality and intensity scores were lowered in comparison with those of control when stored at 8°C. However, after 50 days volatile compound profiles or the sensory characteristics were the same as those of control. Juan et al. (2007) demonstrated that high-pressure processing (300–500 MPa) altered the volatile profiles of cheeses

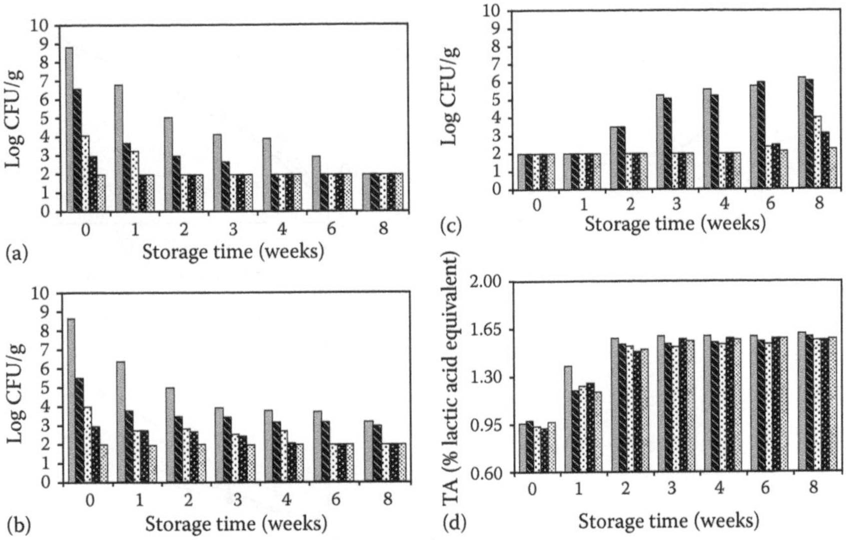

FIGURE 5.16 Residual viable count of *Lactococcus* (a) incubated aerobically; (b) incubated anaerobically; (c) spoilage yeasts; and (d) titratable acidity in lactic curd cheese. Untreated (▥) or pressure-treated at 200 (▧), 300 (▤), 400 (▦), and 600 MPa (▨) during storage for 8 weeks at 4°C. (From Daryaei, H. et al., *Innovat. Food Sci. Emerg. Technol.*, 9, 201, 2008. With permission.)

during ripening. Cheeses treated at 300 MPa after 1 day of manufacturing were characterized by higher levels of free amino acids, ethanol, ethyl esters and branched-chain aldehydes, whereas cheeses treated at 500 MPa after 1 day of manufacturing had lower microbial populations, showed the highest abundance of 2,3-butanedione, pyruvaldehyde and methyl ketones as well as the lowest abundance of alcohols.

O'Reilly et al. (2001) pointed out that high-pressure treatment accelerated degradation of α_{S1}-casein and accumulation of α_{S1}-1-casein, which enhanced the rate of ripening of commercial cheddar cheese. Later, Juan et al. (2008) pointed out that application of high pressure (300 MPa, 10 min) on the day of manufacture resulted in a decrease in α_{S1}- and β-casein and an increase in water-soluble nitrogen and free amino acids during storage up to 90 days (Figure 5.17), which resulted in a decrease in scores for taste, odor, and aroma quality compared to control; the product became softer, more elastic and less crumbly. On the other hand, pressure-treated cheese after 15 days of ripening was similar to control cheese, with more homogeneous protein network and less crumbly texture. Pressure-treated cheese had the highest percentage of short chain fatty acids and were preferred by the sensory panel.

5.3.5 Yoghurt

Jankowska et al. (2005) demonstrated that yoghurt processed at 550 MPa maintained its beneficial sensory characteristics longer than the untreated yoghurt during storage for 4 weeks at refrigerated (4°C) or room (20°C) temperature. The pressure treatment prevented post acidification of the product. The number of bacteria in the

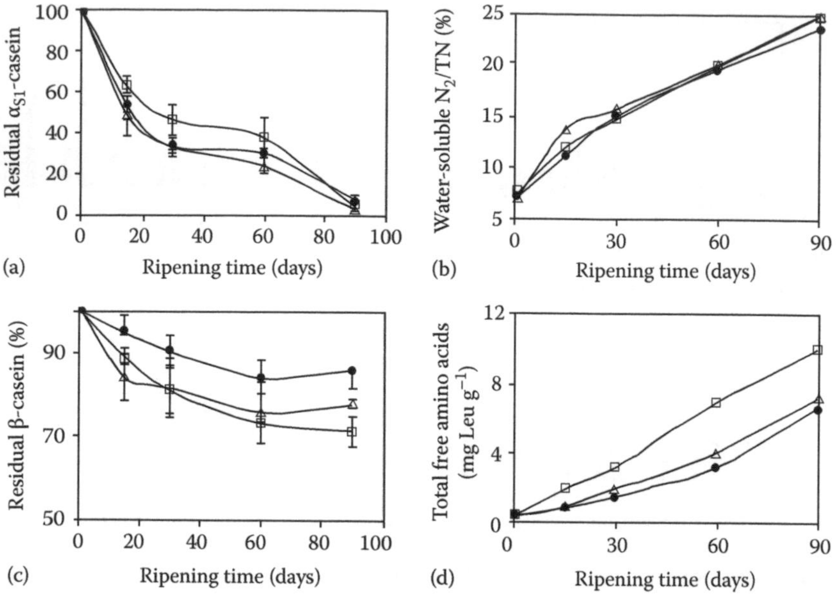

FIGURE 5.17 Residual levels of (a) α_{S1}-casein and (b) β-casein, expressed as a percentage of the amount in the corresponding 1-day-old cheese; (c) levels of water-soluble nitrogen (WSN); and (d) total free amino acids (TFAA) in control (●); 3P1 (□) and 3P15 (△) ewe milk cheese throughout ripening. (From Juan, B. et al., *Int. Dairy J.*, 18, 129, 2008. With permission.)

pressurized yoghurt stored at 4°C was maintained at less than the therapeutic minimum level of 10^6 CFU/mL. Addition of fruit preparation beneficially affected the consistency of the pressure-treated yoghurt. Walker et al. (2006) also demonstrated that high-pressure treatment (550 MPa at 4°C for 10 min) could be used to produce shelf-stable fruit yoghurt. No microbial spoilage took place in pressure-processed yoghurt even after 60 days of storage at 4.4°C or 25°C; however, the count of lactic acid bacteria decreased to <10 CFU/mL. High-pressure-processed yoghurt was thicker and smoother than untreated yoghurt and became darker in color during storage at 25°C.

5.3.6 Whey Protein

High-pressure treatment can be adopted to produce hypoallergenic hydrolysates of whey protein concentrate (WPC); the process involves combined high-pressure treatment and hydrolysis with proteinases. Nakamura et al. (1993) showed that application of high pressure (200–600 MPa) prior to enzymatic hydrolysis of WPC with proteinase resulted in a decrease in β-lactoglobulin; α-lactalbumin did not change, however. On the other hand, when thermal treatment was used in place of high-pressure treatment the amounts of both proteins decreased. Alvarez et al. (2007) pointed out that the decrease in β-lactoglobulin was attributed to the exposure of side chains of buried amino acids to solvent.

Penas et al. (2006) illustrated that high pressure (100–300 MPa) in combination with selected food-grade proteinases could be used as a treatment to remove antigenicity of whey protein hydrolysates, enabling their use as ingredients of hypoallergenic infant formulae. Chicon et al. (2009) demonstrated that pepsin and chymotrypsin under high pressure (400 MPa) produced hydrolysates in which α-lactalbumin and β-lactoglobulin were totally proteolyzed, giving rise to large and hydrophobic peptides. Such hydrolysates showed reduced antigenicity and human IgE-binding properties. The hydrolysates obtained were of improved heat stability and had superior emulsion activity to those of untreated whey proteins.

Lim et al. (2008a,b) demonstrated that application of high pressure (300 MPa, 15 min) could enhance the foaming properties of WPC, which was added to low-fat ice cream to improve its body and texture. Due to the impact of high pressure on the functional properties of whey proteins, the ice-cream mix containing the pressure-treated whey protein exhibited increased overrun and foam stability and hardness than ice cream produced with untreated whey protein.

5.4 EFFECT OF HIGH-PRESSURE PROCESSING ON QUALITY OF ANIMAL PRODUCTS

Application of high pressure induces inactivation of pathogenic and spoilage microorganisms, changes muscle enzymes and meat proteolysis, affects myofibrillar proteins, modifies muscle structure and meat texture, gelatinizes and restructures mined meat, changes myoglobin and meat color and influences lipid oxidation in meat. In terms of microbial inactivation, the extent of inactivation depends on several parameters such as type of microorganism, pressure level, process temperature and time as well as pH and composition of food or dispersion medium (Rastogi et al., 2007). The effect of high pressure on various animal products is reviewed in the following sections.

5.4.1 Beef

5.4.1.1 Inactivation of Microorganisms and Enzymes

Carlez et al. (1993, 1994) indicated that microorganisms, viz., *Pseudomonas fluorescens*, *Citrobacter freundii*, or *L. innocua* in minced meat were completely inactivated at pressures higher than 200, 280, or 400 MPa, respectively; the total flora was noted to be reduced by 3–5 log cycles. Processing of fresh minced meat at 200–300 MPa was moderately effective and delayed microbial growth by 2–6 days upon subsequent storage at 3°C. Carlez et al. (1995) later showed that high-pressure processing (200–300 MPa) of meat resulted in a whitening effect due to globin denaturation and/or heme displacement/release; at higher pressure (350 MPa) oxidation of ferrous myoglobin to ferric myoglobin took place and color of the intermediate zone of sample turned gray-brown, while cube centre and meat surface became pink (as reflected by a^* values, see Figure 5.18).

Jun et al. (1999) showed that high-pressure processing (100 MPa, 10–15 min) of beef slices within 24 h after slicing resulted in reduction in shear strength, loss of

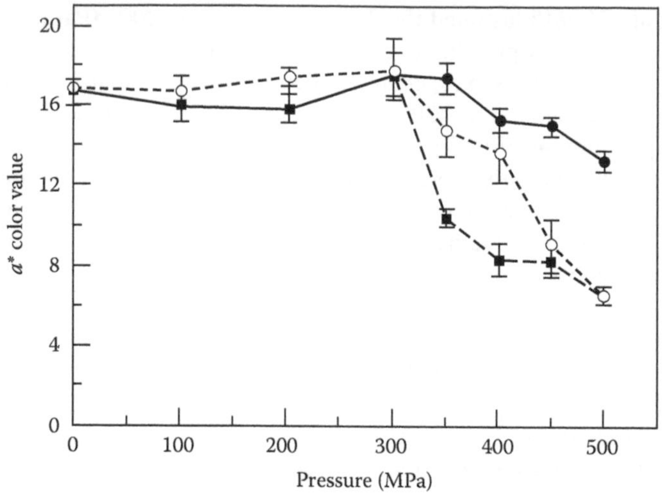

FIGURE 5.18 Changes in *a** color values at the surface (-○-), at an intermediate position (-■-), or at the centre (-●-) of meat cubes after pressurization at 10°C for 10 min in air. Results are from at least six measurements on the same *semimembranosus* meat sample. (From Carlez, A. et al., *Lebensm. Wiss. Technol.*, 28, 528, 1995. With permission.)

pink color, reduced loss of exudates, and higher score for cooked meat. However, after 7 days of storage, shear strength of the pressure-treated slices was noted to be higher than that of the control sample. Morales et al. (2008) demonstrated that four cycles of short-duration (400 MPa, 1 min) or single cycle of long-duration (400 MPa, 20 min) high-pressure treatment resulted in comparable reduction (4.38 log cycles) in microbial population in ground beef. Kramer shear force and energy were generally higher in pressurized sample than in control, which indicated high-pressure-induced hardening of the sample.

Hassan et al. (2002) compared the inactivation efficiency of gamma-irradiation (1.5–2.0 kGy) and high-pressure treatment (200–300 MPa). Both treatments resulted in 2-log decrease in the total viable cell count of bacteria and complete elimination of coliform bacteria. High-pressure treatment, however, resulted in enhancement of WHC and strong discoloration due to denaturation of muscle pigments as well as extensive denaturation of myofibrillar proteins. On the other hand, the radiation treatment caused minimal changes in appearance and texture as well as a decrease in the WHC.

Zhu et al. (2008) studied the destruction kinetics of *Clostridium sporogenes* spores in ground beef under high pressure at elevated temperatures (700–900 MPa, 80°C–100°C) and showed that bacterial spores could be destroyed within a shorter time or lower temperature than in the case of conventional thermal processing; this information can be used for development of a high-pressure-based sterilization process. Fernandez et al. (2007) indicated that high-pressure, low-temperature treatment (air-blast freezing at –30°C and 650 MPa, 10 min) significantly increased expressible moisture and did not cause any change in shear force; the color of the sample after thawing was also close to that of fresh samples. The treatment was very effective in reducing aerobic total and lactic acid bacteria counts by 2 and 2.4 log cycles, respectively. Addition of salt reduced high-pressure-induced water loss. The beef sample stored at −18°C for 45 days recovered its original color after thawing similarly to immediately-treated samples; microbial counts also remained below detection limits.

Hayman et al. (2004) demonstrated that high-pressure processing (600 MPa, 20°C, 3 min) could extend the refrigerated shelf life of ready-to-eat meats and reduce *L. monocytogenes* by more than 4-log CFU/g. Besides, counts of aerobic and anaerobic mesophiles, lactic acid bacteria, *Listeria* spp., *Staphylococci*, *Brochothrix thermosphacta*, coliforms, and fungi were also undetectable when stored at 4°C for 98 days. Moreover, consumer acceptability and sensory quality of the product was found to be very high.

5.4.1.2 Effect on Meat Hardness

Two enzymatic systems, i.e., calpain and cathepsin influence tenderization of meat. The activity of calpain reduces under high pressure. The activity of μ-calpain is also reduced during ageing (Ouali, 1990). Qin et al. (2001) showed that high-pressure treatment (100–300 MPa, 10 min) of beef resulted in a decrease in the total calpain activity; however, the acid phosphatase and alkaline phosphatase activities were not significantly reduced. Homma et al. (1995) found that the total activity of calpain in pressurized muscle increased due to a reduction in the level of calpastatin because of its pressure sensitivity; this in turn resulted in meat tenderization.

Han (2006) indicated that high-pressure treatment (200–600 MPa, 5 min, 5°C–7°C) improved the digestibility of beef extracts, which was coupled with the decrease in allergenicity of major beef antigens. It is noted that bovine serum albumin (BSA) and bovine γ-globulin (BGG) play important roles in the allergenicity of beef. Later, Han et al. (2006) confirmed that application of high pressure (600 MPa) did not affect the digestibility of BSA, whereas the digestibility of BGG was altered; this significantly led to decreased allergenicity.

Kennick et al. (1980) demonstrated that high-pressure treatment (103 MPa, 30°C–35°C) of ovine and bovine muscles resulted in increased firmness and contraction; however, after cooking meat became tenderer with higher moisture content as compared to the control sample. Suzuki et al. (1992) indicated that high-pressure processing resulted in meat tenderization without heating. High pressures up to 300 MPa caused increased myofibril fragmentation and marked modification in its ultrastructure. Suzuki et al. (1993) later reported that pressure-induced tenderization of bovine liver cells was due to improvement in actomyosin toughness.

Bai et al. (2004) showed high-pressure treatment (700 MPa, 20 min) influenced sensory, microscopic, and textural properties of cattle and mutton skeletal muscles. Color of the samples faded and extension decreased. A marked variation could be seen in the microscopic structure of myofibrils. Furthermore, shrinkage of sarcomere and reduction in shear force was observed. Han and Ledward (2004) found that, in the case of beef muscle, hardness increased with increasing pressure (200–800 MPa) at a constant temperature (up to 40°C), whereas the hardness decreased significantly with the application of pressure (200 MPa) at higher temperatures (60°C or 70°C). Accelerated proteolysis might be the major contributing factor to the loss in hardness of the beef sample.

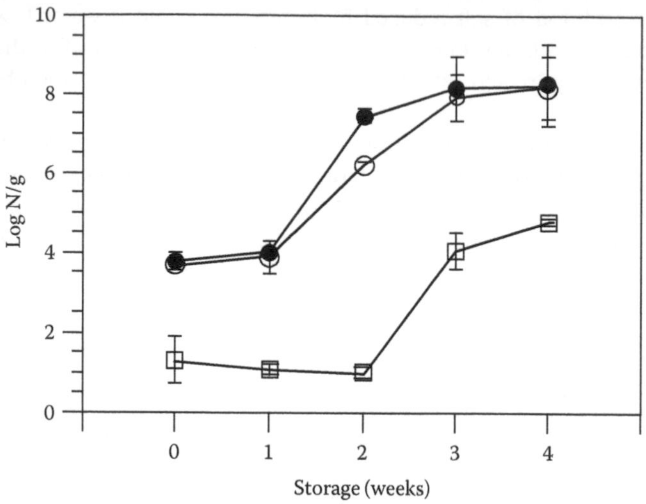

FIGURE 5.19 Total flora evolution of control and pressurized *biceps femoris* beef samples during storage at 4°C. Control sample (●); sample pressurized at 130 MPa, 260 s (○); and 520 MPa, 260 s (□). (From Jung, S. et al., *Lebensm. Wiss. Technol.*, 36, 625, 2003. With permission.)

Jung et al. (2000) pointed out that application of high pressure (520 MPa) to beef neither improved beef tenderness nor reduced the ageing period, although there was a significant increase in the activity of lysosomal enzymes such as cathepsin D and acid phosphatase. Rather, pressurization led to increased toughness due to modification in myofibrillar components, reduction in sarcomere length, and increased cooking losses. Later, Jung et al. (2003), showed that high-pressure treatment for a short time (520 MPa, 260 s) reduced total microflora and delayed microbial growth by 2 weeks (Figure 5.19); this enabled longer maturation and improved meat tenderness. Discoloration of meat occurred at pressure levels higher than 325 MPa, whereas moderate pressures (130 MPa) improved meat color by increasing its redness, which was maintained for the first 3 days of storage at 4°C.

5.4.1.3 Beef Products

Carballo et al. (1996) indicated that high-pressure treatment of finely comminuted bovine meat resulted in formation of gels with smooth cohesive texture and high water retention. Ayo et al. (2005) observed that textural properties of meat batters with walnuts were not affected by high-pressure processing. However, hardness, cohesiveness, springiness, and chewiness of the cooked products were reduced by the addition of walnut.

Canedo et al. (2009a) demonstrated that high-pressure treatment (400 MPa, 10 min, 12°C) resulted in significant changes in the levels of some volatile compounds due to microbial activity. Alcohol and aldehyde contents decreased, while other compounds such as 2,3-butanedione and 2-butanone were more abundant in high-pressure-processed meats. A significant migration of branched-chain alkanes and benzene compounds from plastic packaging material was also observed.

5.4.2 Pork

5.4.2.1 Inactivation of Microorganisms and Enzymes

Shigehisa et al. (1991) demonstrated that high-pressure treatment (400 MPa, 25°C, 10 min) of pork homogenate (pH 6.7) reduced the populations of *E. coli*, *Campylobacter jejuni*, *Pseudomonas aeruginosa*, *Salmonella* Typhimurium, *Yersinia enterocolitica*, *S. cerevisiae*, and *Candida utilis* by at least 6 log cycles when these microorganisms were inoculated at a level of 10^6–10^7 CFU/g. Noeckler et al. (2001) showed that *Trichinella spiralis* in a pork sample was completely inactivated using high-pressure processing (100–150 MPa, 5°C). Tanzi et al. (2004) demonstrated that high-pressure processing (600 MPa, 3–9 min) is a useful technique for control of *L. monocytogenes* in sliced Parma ham. The treated sample had less red color and intense salty taste, however. Garriga et al. (2004) demonstrated that, in the case of ham, high-pressure processing prevented the growth of *Enterobacteriaceae*, yeasts, and lactic acid bacteria, resulting in increased shelf life. In addition, food safety risks associated with *L. monocytogenes* and *Salmonella* were also reduced. Koseki et al. (2007) observed that high-pressure treatment (550 MPa, 10 min) of sliced cooked ham inoculated with *L. monocytogenes* reduced the microbial count below detectable limits. The bacterial count gradually increased during storage and exceeded the initial inoculum level at the end of a 70-day period.

Carpi et al. (1999) investigated the effect of high-pressure treatment (400–600 MPa, 5 min) on shelf life of sliced vacuum-packaged raw ham stored at 4°C. The control sample was spoiled by lactic acid bacteria after 15 days, while the pressure-treated samples at 400, 500, and 600 MPa were spoiled by lactic acid bacteria after 40, 60, and 74 days, respectively. Shear resistance of pressure-treated and vacuum-packaged raw ham was also lower than that of the control sample. The sensory attributes of the pressure-processed (600 MPa, 5 min) product was unchanged even after 30 days of storage at 3°C or 9°C. Ko et al. (2006) indicated that high pressure helped improve functional properties (e.g., WHC) of pork with increasing freezing rate, and that pressure-assisted freezing had more potential benefits than pressure-shift freezing over a mild pressure range. Serra et al. (2007) demonstrated that high-pressure pasteurization (400 and 600 MPa) of frozen ham resulted in lower visual color intensity as compared to control, whereas no significant effect on the flavor characteristics of the final product was observed. Higher pressures (600 MPa) led to products with significantly lower crumbliness and higher fibrousness scores than the control as well as the sample processed at 400 MPa; however, the overall sensory quality was not affected. Park et al. (2006) indicated that high-pressure thawing (50–200 MPa, 15°C) led to a significant increase in pH as well as WHC and a decrease in thawing loss than the control sample. High-pressure thawing treatment at <100 MPa improved the quality of frozen pork as the process resulted in lower cooking loss as well as no change in color and hardness.

In the case of pork, various vitamins such as riboflavin, thiamin, and thiamin monophosphate were proved to be sufficiently stable upon high-pressure and high-temperature processing (600 MPa, 25°C–100°C). Under similar conditions, vitamins decay in model solutions was up to 30 times faster, especially in the case of thiamin monophosphate (Butz et al., 2007).

Slongo et al. (2009) indicated that application of high pressure (200–400 MPa, 5–15 min) reduced growth of lactic acid bacteria, thus extended product shelf life of and better preservation of natural taste, texture, color, and vitamin content in vacuum-packaged sliced ham. Shelf life of ham subjected to 400 MPa for 15 min was extended from 19 to 85 days.

Hong et al. (2005) showed that an increase in both pressure and processing time resulted in an increase in L^* (lightness) and a^* (redness) values and a decrease in the WHC of pork; however, shear force and cooking losses were relatively unaffected. High-pressure processing (<200 MPa) of pork for 1 h was shown to maintain the quality of cooked pork.

5.4.2.2 Combined Effect of Antimicrobials and High-Pressure Processing

Jofre et al. (2008a,b) showed that antimicrobial packaging, high-pressure treatment, and refrigerated storage could be a very effective combination to obtain value-added, ready-to-eat products from ham with a safe long-term storage of up to 3 months at 6°C. A combination of high-pressure processing, nisin, and refrigeration at 6°C decreased the level of *S. aureus* by 2.4-log CFU/g after 3 months of storage.

Marcos et al. (2008a) showed that combination of high-pressure processing (400 MPa for 10 min) and natural antimicrobials (lactate diacetate) along with low-storage temperature (1°C) could reduce the level of *L. monocytogenes* during storage by 2.7-log CFU/g. Combination of high-pressure processing with another antimicrobial agent, viz., enterocins indeed reduced the population of *L. monocytogenes* to the final count of 4 MPN (most probably number)/g after 3 months of storage. Marcos et al. (2008b) indicated a possibility of combining high-pressure processing with active packaging (antimicrobial alginate films containing enterocins) to control *L. monocytogenes* growth during storage and cold chain break.

5.4.2.3 Pork Products

Zhu et al. (2004a,b) indicated that high-pressure shift freezing of pork muscle resulted in small and regular ice crystals. Near the surface, there were many fine and regular intracellular ice crystals with well-preserved muscle tissue. From midway to the center, the ice crystals were larger in size and located extracellularly. Additionally, changes in color, reduction in drip loss during thawing, considerable denaturation of myofibrillar proteins, and reduction in muscle toughness were observed as a result of high-pressure shift freezing.

Campus et al. (2008) showed that high-pressure treatments at pressures above 300 MPa resulted in an increase in lightness and a decrease in redness of dry-cured pork loins. High pressure led to a reduction in the activity of aminopeptidase as well as dipeptidyl peptidase. No significant increase in free amino acid content was noted due to the reduction in aminopeptidase activity. Oxidative stability of the pressurized dry-cured loins was also not affected. High-pressure treatment led to a reduction of several flavor compounds derived from Maillard reactions; these compounds were regenerated during vacuum storage, however.

Fulladosa et al. (2009) evaluated the effect of potassium-lactate and high-pressure processing (600 MPa) on physicochemical, instrumental color, and texture, as well as sensory characteristics of transglutaminase restructured dry-cured hams. The addition of potassium lactate did not have a negative effect on the color, flavor, or texture of restructured dry-cured hams. On the other hand, high-pressure treatment affected significantly the flavor (increasing saltiness, umami, and sweetness) and sensory texture attributes (increasing muscle binding, hardness, gumminess and fibrousness, and decreasing adhesiveness and pastiness) as well as the slice appearance (increasing brightness and iridescence and decreasing color homogeneity).

Canedo et al. (2009b) pointed out that high-pressure treatment (400 MPa, 10 min, 12°C) had a slight effect on the volatile fraction of Spanish dry-cured Serrano ham. Most compounds affected by pressurization such as alkanes (C_9–C_{12}), 2-undecene, 2-nonanone, 1-octen-3-one, 1-heptanol, 2-hexanol, 2-heptanol, ethyl pentanoate, benzaldehyde, and styrene presumably originated from the metabolism of molds. A significant effect of pressurization on migration of compounds from the plastic packaging material was also noted.

Hong et al. (2008a) evaluated combined effect of NaCl, glucono-δ-lactone (GDL), and κ-carrageenan on the binding properties of restructured pork under hydrostatic pressure. An increase in the GDL level led to a significant decrease in pH and WHC and increase in binding strength. A reduction in the NaCl level during meat restructuring could be compensated by the increase in the GDL concentration. At low GDL concentration, it was possible to achieve palatable binding properties during meat restructuring, provided that κ-carrageenan was added. Hong et al. (2008b) indicated that high pressure (>200 MPa) in combination with carrageenan (>1.5%) was beneficial in cold set binding of restructured pork as the combination resulted in increased breaking force and tensile strength. An increase in the pressure level and κ-carrageenan concentration improved the WHC and cooking losses of the restructured pork.

5.4.3 Fish

5.4.3.1 Inactivation of Microorganisms and Enzymes

Carpi et al. (1995) demonstrated that high-pressure treatment (700 MPa) could extend the shelf life of salmon spread to 60–180 days at low temperature (3°C–8°C) without significant chemical, microbiological, or sensory changes. On the other hand, Lakshmanan and Dalgaard (2004) showed that high-pressure processing up to 250 MPa could not inactivate *L. monocytogenes* in smoked salmon, but significant lag phases of 17 and 10 days were observed at approximately 5°C and 10°C, respectively. The treatment had a marked effect on both color and texture of the product. Later, Lakshmanan et al. (2005a) indicated that the activities of calpain, cathepsin B-like, and cathepsin B + L-like enzymes decreased on application of pressure. At 300 MPa, calpain was almost completely inactivated, but general protease activity was not affected by high pressure.

Wen and Kuo (2001) indicated that application of high pressure (200 MPa, 12 h) resulted in a decrease in the total plate count of Tilpia fillets from 4.7- to 2.0-log CFU/g; the freshness index of the product was also higher than the unpressurized sample. Ramaswamy et al. (2008) indicated that in the case of mackerel fish slurry, the *D*-values of *E. coli* were higher than those of *L. monocytogenes* at pressure levels equal to or greater than 350 MPa, while a reverse trend was observed at lower pressures. Corresponding Z_p values indicated that destruction rate of *L. monocytogenes* (Z_p = 103 MPa) was more sensitive to changes in pressure than *E. coli* (Z_p=185 MPa). A 10 *D* treatment followed by refrigerated storage (4°C–12°C) prevented recovery/growth of *E. coli*. Ritz et al. (2008) showed that the most effective conditions for high-pressure inactivation of *L. monocytogenes* were 200 MPa, 18°C, and pH 4.5. High-pressure treatment resulted in acceptable modifications of the physical properties of the processed fish, viz., increased *L**-value and toughness.

Cheret et al. (2005) indicated that high pressure improved safety and textural quality of fresh fish fillets. High-pressure treated sea bass exhibited slight color change, viz., an increase in lightness and a slight change of hue, which might be imperceptible in cooked fish; a decrease in exudation and WHC during storage was also noted (Figure 5.20). High-pressure treatment (>300 MPa) led to increased fish hardness after storage than in the case of the control sample. Cheret et al. (2006) later observed that calpain activity of sea bass decreased after high-pressure treatment (100 or 300 MPa for 5 min) and its evolution during storage depended on the level of pressurization. Its inhibitor, calpastatin, was not affected by high pressure, but its activity decreased during storage. Initial activity of cathepsin was increased by high-pressure treatment. Partial degradation of myofibrillar proteins occurred after treatment at 300 MPa, but these changes during storage were not revealed by SDS-PAGE profiles.

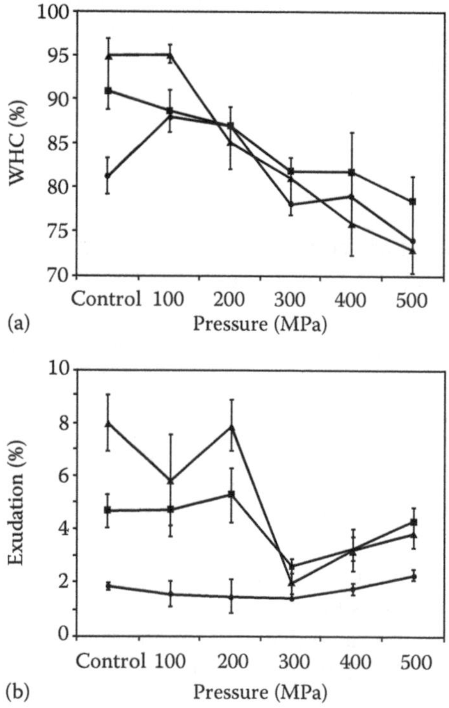

FIGURE 5.20 (a) Effect of WHC and (b) evolution of sea bass fillet exudation treated by high pressure (5 min) after storage at 4°C during 0 (◆), 7 (■), and 14 (▲) days. (From Cheret, R. et al., *J. Food Sci.*, 70, E477, 2005. With permission.)

5.4.3.2 Effect on Textural Properties

Angsupanich and Ledward (1998) showed that oxidative stability of lipids in Atlantic cod (*Gadus morhua*) muscle decreased after treatment at a pressure higher than 400 MPa due to the release of metal ions from complexes. Of the major muscle proteins, myosin denatured at 100–200 MPa, whereas actin and most sarcoplasmic proteins denatured at 300 MPa. High-pressure treated fish was harder, chewier, and gummier than both the raw and cooked products.

Caballero et al. (2005) pointed out that addition of chitosan to pressurized cod sausages (350 MPa at 7°C for 15 min) in dry form did not yield any significant differences in hardness, cohesiveness, or adhesiveness of the samples. However, addition of chitosan led to a noticeable increase in the sample elasticity (Figure 5.21).

Uresti et al. (2004) demonstrated that raw restructured fish products from arrowtooth flounder (*Atheresthes stomias*), which is an underutilized fish species abundant in Alaska, with good mechanical and functional properties could be obtained via the use of high-pressure processing at 600 MPa. Higher hardness and gel strength were obtained due to high-pressure-induced myosin aggregation (Figure 5.22). Later, Uresti et al. (2005) indicated that high-pressure treatment could also improve the mechanical properties of heat-induced gels; heat-induced gels were softer, more fragile, and had lower values of springiness, cohesiveness, and chewiness. Uresti et al. (2006) also showed that pressurization improved the mechanical properties of gels made from paste treated with transglutaminase and set at 25°C.

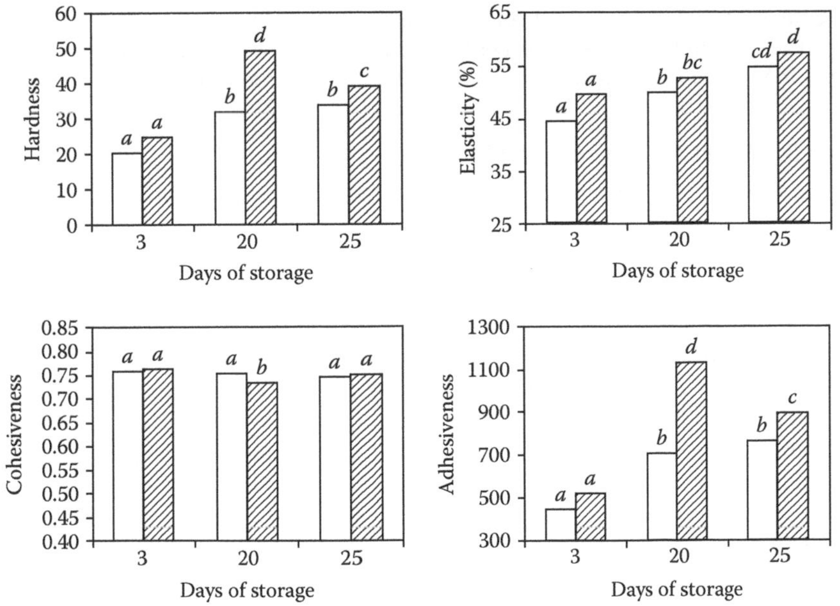

FIGURE 5.21 Hardness, cohesiveness, adhesiveness, and elasticity, as determined by texture profile analysis, of high-pressure sausages (350 MPa, 15 min, 7°C) stored at chilled temperatures (2 ± 1°C). (□) Control without chitosan; (▨) with 1.5% chitosan powder. Different letters (*a*, *b*, *c*) indicate significant differences ($p < 0.05$). (From Caballero, M.E.L. et al., *J. Food Sci.*, 10, M166, 2005. With permission.)

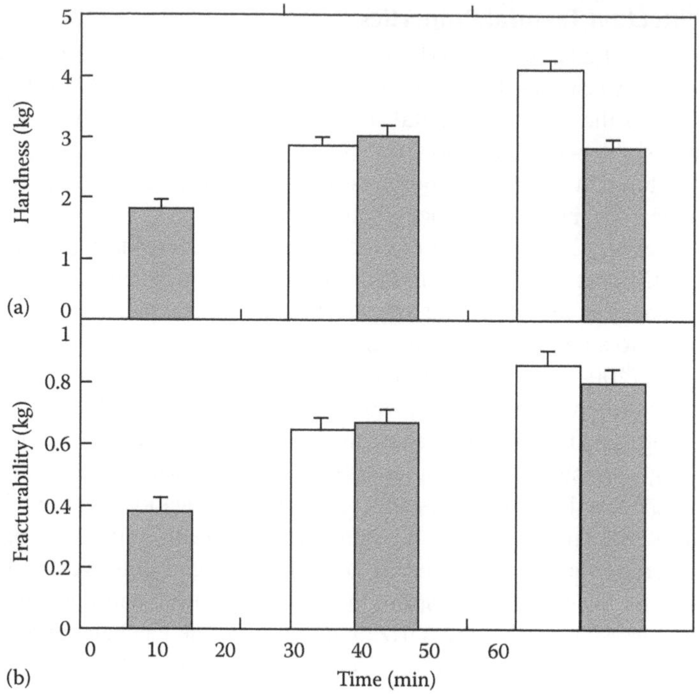

FIGURE 5.22 (a) Hardness and (b) fracturability of cooked (90°C for 15 min) pressure-treated gels. Control was cooked non-pressured fish paste (-□- 400 MPa, -■- 600 MPa). (From Uresti, R.M. et al., *J. Sci. Food Agric.*, 84, 1741, 2004. With permission.)

5.4.3.3 Effect on Freezing and Thawing

Chevalier et al. (1999) reported that high-pressure thawing of blue whiting was quicker and resulted in lower drip volume in comparison to conventional thawing. Chevalier et al. (2001), based on the analysis of physicochemical properties of fillets (such as color, lipid oxidation, and protein stability), later concluded that 140 MPa was the optimum pressure level for pressure-shift freezing of turbot fillets.

Rouille et al. (2002) showed that high-pressure thawed dogfish and scallops had better microbial quality, lower thawing time, and lower drip volume than immersion thawed products. Schubring et al. (2003) noted that high-pressure thawing (200 MPa) resulted in reduced drip loss, improved microbial status, and improved textural parameters for redfish, haddock, and whiting. However, the water-binding capacity (inferred from expressible moisture content) was higher after conventional thawing (Figure 5.23). Differential scanning calorimetry (DSC) revealed greater protein denaturation following high-pressure-assisted thawing than conventional thawing, which might be partly responsible for the product quality. Zhu et al. (2004c) showed that high-pressure treatment (150 and 200 MPa) caused marked color changes in Atlantic salmon. The texture was markedly modified at 200 MPa. Varying the freezing rate prior to thawing had no discernable effect on color and texture of the samples; however, different freezing rates resulted in different drip losses of the samples after thawing.

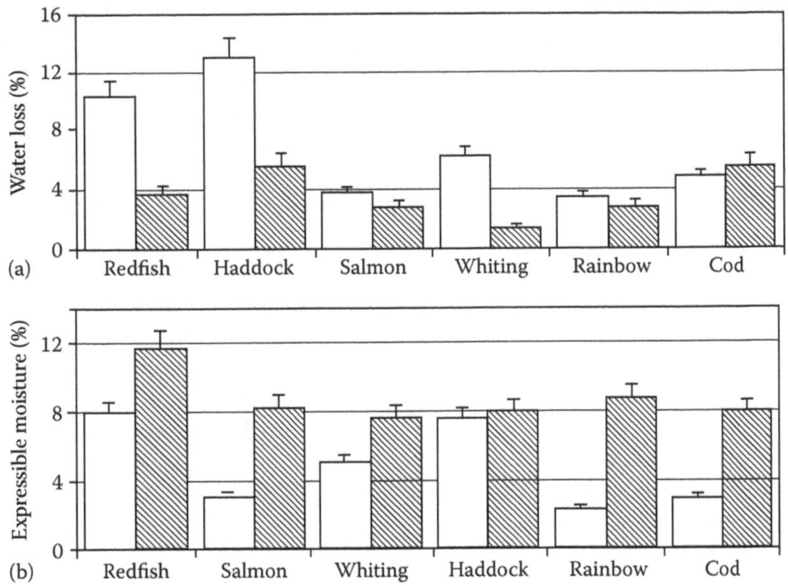

FIGURE 5.23 (a) Water loss of raw fish fillet as a function of thawing treatment; (b) thaw drip during thawing. HP and AP refer to thawing at 200 MPa and at atmospheric pressure, respectively. (From Schubring, R. et al., *Innovat. Food Sci. Emerg. Technol.*, 4, 257, 2003. With permission.)

Picart et al. (2005) demonstrated that frozen salmon mince (−28°C for 24 h) followed by pressurization and fast pressure release resulted in 2.5-log reduction for *L. innocua*. When the frozen sample was thawed using pressure-assisted thawing (207 MPa, 10°C, 23 min), a reduction of 1.2 log cycles was obtained. Alizadeh et al. (2007) demonstrated that pressure-shift freezing reduced thawing drip loss compared to air-blast freezing. The presence of small ice crystals in the pressure-shift frozen sample was probably the major reason leading to reduced drip volume. The freezing process had more influence on the quality parameters than the thawing process.

5.4.3.4 Fish Products

Lakshmanan et al. (2005b) indicated that high-pressure treatment (100–300 MPa, 20°C–30°C, 30 min) affected many quality parameters of cold-smoked salmon. However, processing for 20 min at pressures up to 200 MPa produced a product with acceptable sensory properties. Sequeira et al. (2006) demonstrated that high-pressure treatment (100–200 MPa, 4°C, 15 and 20 min) of vacuum-packed raw carp fillets (*Cyprinus carpio*) resulted in an increase in thiobarbituric acid reactive substances and free fatty acids. The color values (L^*, a^*, and b^*) of the carp fish fillets also increased with pressure and pressurization time.

Gomez et al. (2007a) indicated that the best sensory scores were achieved for cold-smoked dolphin fish when it was pressurized at 300 MPa. Lipid oxidation was prevented by phenolic compounds, which were incorporated in the process of smoking. The sample pressurized at 300 MPa was quite stable, although sensory attributes declined over storage and fell off sharply after 65 days. High pressure did not extend the product shelf life, but it was able to diminish bacterial counts during early storage period. Gomez et al. (2009) later demonstrated that application of high pressure (200–300 MPa, 7°C) is an adequate tool for obtaining high-quality desalted "bacalao" carpaccio from fish. Pressurization resulted in better microbiological quality, thereby increasing the shelf life as well as acquisition of new acceptable sensory attributes by the consumers.

Ramirez-Saurez and Morrissey (2006) indicated that high-pressure treatment increased pH and helped protect minced albacore muscle from lipid oxidation. The treatment also helped maintain low microorganism levels, change color of the muscle, and induce formation of high-molecular weight polypeptides, most likely through disulfide bonding, thus promoting texture improvement of the samples. The treatment improved shelf life of minced albacore muscle for more than 22 days at 4°C and 93 days at −20°C.

High-pressure treatment of vacuum-packed prawn extended its shelf life, although it did affect muscle color slightly, giving it a whiter appearance. The viable shelf life of 1 week for air-stored samples was extended to 21, 28, and 35 days for vacuum-packaged samples, samples treated at 200 and 400 MPa, respectively (Lopez et al., 2000).

Yagiz et al. (2007) indicated that high-pressure treatment (300 MPa) effectively reduced initial microbial population in rainbow trout and mahi-mahi up to 6- and 4-log reduction, respectively. Redness of rainbow trout was lower as compared to mahi-mahi. Optimum high pressure for influencing lipid oxidation, microbial load, and color changes were found to be 300 MPa for rainbow trout and 450 MPa for mahi-mahi.

Gomez et al. (2007b) indicated that combination of high-pressure processing and edible films yielded the best results in terms of preventing oxidation and inhibiting microbial growth in cold-smoked sardine, thereby increasing its shelf life. Coating the muscle with films enriched with oregano or rosemary extracts increased phenol content and antioxidant power when used in combination with high-pressure processing. High-pressure-induced gels (200–420 MPa) of blue whiting were found to have lower adhesiveness, higher WHC, and less yellowness than heat-induced gels (Perez and Montero, 2000).

5.4.4 Chicken, Duck, and Turkey

High-pressure treatment (400–900 MPa, 10 min) in combination with low storage temperature (4°C) was shown to be a potential technology for extension of refrigerated storage life (3–4 days for control) of fresh raw minced chicken in sealed poly-film pouches. The expected shelf life of chicken in sealed poly-film pouches processed at 408, 616, and 888 MPa was reported to be 27, 70, and more than 98 days, respectively (O'Brien and Marshall, 1996). Jimenez et al. (1998) showed that application of high pressure (200 and 400 MPa, 30 min) resulted in increased water- and fat-binding properties of chicken and pork batters, even at low ionic strength. The samples were found to be softer, more cohesive, springy, or chewy as compared to the control samples.

Trespalacios and Pla (2007) investigated the effect of simultaneous application of transglutaminase and high pressure (700–900 MPa, 40°C, 30 min) on chicken batter with addition of egg components. The gel showed marked increase in textural parameters as compared to gel without enzyme. Microstructure of the gel suggested that a higher amount and heterogeneity of cross-link was produced when meat and egg proteins were treated in the presence of transglutaminase.

Isbarn et al. (2007) reported that highly pathogenic avian influenza A virus in cell culture medium and in chicken meat could be inactivated either by thermal treatment (63°C for 2 min) or high pressure (500 MPa, 15°C, 15 s). Thus, high-pressure treatment of poultry and its products was recommended to avoid the possible health threat by avian influenza viruses. High-pressure processing has indeed a great potential for microbial control of raw chicken meat as a "fresh" chill-stored, convenience product for wok cooking. However, while raw chicken meat is oxidative stable, high-pressure treatment at 600 MPa and above induces lipid oxidation, resulting in off-flavors during subsequent cooking. Bragagnolo et al. (2007) nevertheless indicated that addition of 0.1% dried rosemary to minced chicken thighs or breasts prior to high-pressure processing inhibited lipid oxidation during subsequent cooking and could form the basis for product development. Fernandez et al. (1998) indicated that application of high pressure (200 and 400 MPa) simultaneously with cooking at 70°C resulted in formation of less compact aggregated microstructure, which had better binding properties and hence led to less harder product compared with the controls. High-pressure treatment resulted in a decrease in color parameters.

Orlien et al. (2000) suggested that 500 MPa is a critical pressure for treatment of chicken breast muscle. Up to 500 MPa, no rancidity during chilled storage was observed and the product was similar to the untreated one. Pressure treatments at 600 and 700 MPa resulted in less oxidation, but at 800 MPa lipid oxidation enhanced to the same extent as the level induced by thermal treatment. Increased lipid oxidation was probably related to membrane damage. Wiggers et al. (2004) also demonstrated that high-pressure treatment at 400 or 600 MPa led to a substantial increase in secondary lipid oxidation products in cooked breast chicken when compared to the 200 MPa treatment and the control sample. Storage period also had a considerable influence on the formation of secondary lipid oxidation products, especially in the presence of O_2 in the packs stored for 8 days. Hexanal, octanal, and nonanal were identified as products of lipid oxidation.

Ma et al. (2007) indicated that high-pressure treatment (600 and 800 MPa, 50°C–70°C) led to increased rates of lipid oxidation in chicken muscle. Addition of 1% of EDTA (ethylenediaminetetraacetic acid) disodium salt inhibited the increased rates of lipid oxidation due to release of transition metal ions from insoluble complexes that catalyzed lipid oxidation in pressure-treated muscle foods. Mariutti et al. (2008) later demonstrated that addition of sage protected minced chicken breast processed with high hydrostatic pressure (up to 800 MPa, 10 min) against lipid oxidation during subsequent chilled storage for 2 weeks. On the other hand, addition of garlic showed pro-oxidative effect at a pressure of around 300 MPa. This effect was partly counteracted by simultaneous addition of sage (Figure 5.24). Bragagnolo et al. (2006) showed that high-pressure treatment of chicken breast and thigh resulted in formation of free radicals, the extent of which increased with an increase in pressure.

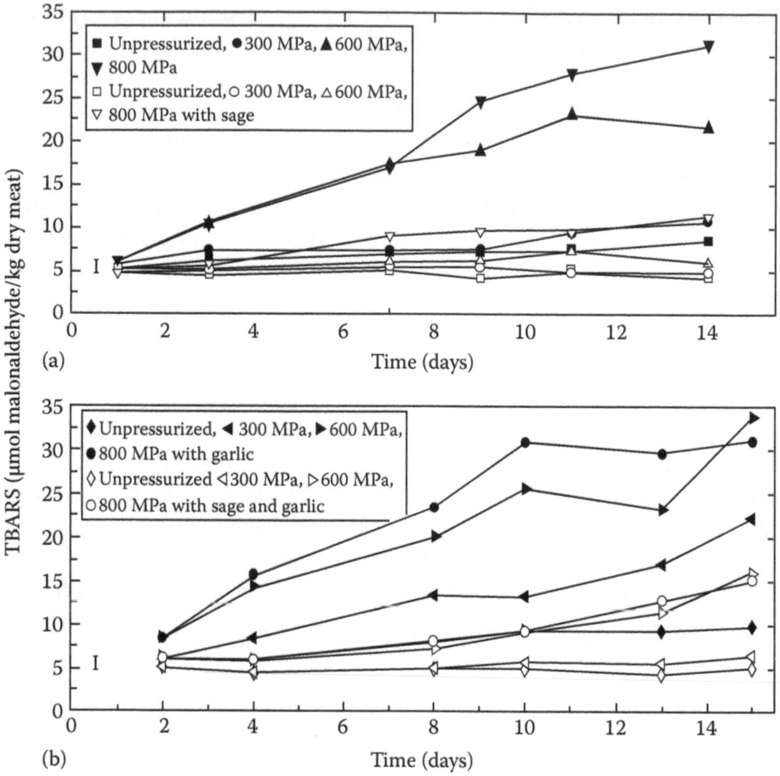

FIGURE 5.24 Formation of secondary lipid oxidation products (TBARS) in non-pressurized or pressurized minced chicken meat (a) with and without sage addition; (b) with sage and/or garlic addition during subsequent chill storage in the dark. (From Mariutti, L.R.B. et al., *Eur. Food Res. Technol.*, 227, 337, 2008. With permission.)

El Moueffak et al. (1996) demonstrated that high-pressure treatment (400 MPa, 50°C, 10 or 30 min) of *foie gras* (fatty goose or duck liver) led to a reduction in the microbial load as efficiently as the usual thermal pasteurization process; pressurization, however, maintained the unique texture as well as flavor and a high product yield. Moreover, there was no melting or separation of lipids as a result of pressure processing, in contrast to the 15% lipid loss due to thermal pasteurization. Later, El Moueffak et al. (2001) indicated that in the case of *foie gras* combined use of high pressure (350 and 550 MPa) and temperature (55°C and 65°C) could be employed to give a product of similar microbiological quality to that obtained by pasteurization. The treatment at 55°C, 550 MPa for up to 20 min resulted in 7-log reduction of total aerobic mesophilic flora and *Enterococcus faecalis*. Cruz et al. (2003) also demonstrated that total aerobic mesophilic counts decreased and no coliform bacteria were detected in fatty duck liver after high-pressure processing. High-pressure treatment (550 MPa, 55°C, 20 min) in combination with two low-oxygen-permeability films (ethylene and vinyl alcohol copolymer) resulted in a product with prolonged storage life of 90 days at 4°C, along with a significant reduction in fat loss.

Yuste et al. (2000) showed that high-pressure processing is a suitable alternative to thermal treatment for inactivating *S. enteritidis* and mesophiles in poultry sausage manufacture. Yuste et al. (2001, 2002) later indicated that the highest reduction in mesophiles (5.3-log CFU/g) and psychrotrophs (more than 7.5-log CFU/g) counts were observed in mechanically recovered poultry meat after exposure to a pressure of 450 MPa in presence of 200 ppm nisin.

Villacis et al. (2008) demonstrated that high-pressure processing could be a useful technique for salting turkey breast meat. During high-pressure come-up time, water and sodium chloride diffusion into the sample was found to be maximal at 150 MPa (Figure 5.25). Further infusion of sodium chloride into the sample was observed during holding of the sample under high pressure. High-pressure treatment resulted in minimum hardness, gumminess, chewiness, swelling of myofibrils, disappearance of the M-line, reduced difference in the density of the A-band and I-band, and breaking of segments of the Z-line.

5.4.5 Sausages

High-pressure-processed sausages were less firm, more cohesive, had lower weight loss, and higher preference scores than heat-treated samples; negligible effect of pressurization on color attributes was observed (Mor-Mur and Yuste, 2003). Hajos et al. (2004) suggested that high-pressure treatment (600 MPa, 10°C–40°C, 20 min) decreased the redness, and increased the lightness and hardness of sausage batter. Counts of *L. monocytogenes* were also reduced by 4–5 log cycles. Urea-soluble fraction of proteins of the sausage batter displayed an altered immunoreactive profile following the pressure treatment. Conformational changes in sausage batter proteins might be induced in conjunction with a certain level of epitope structure alteration, which indicated that treatment of sausage batter at 600 MPa modified the IgE immunoreactivity of proteins.

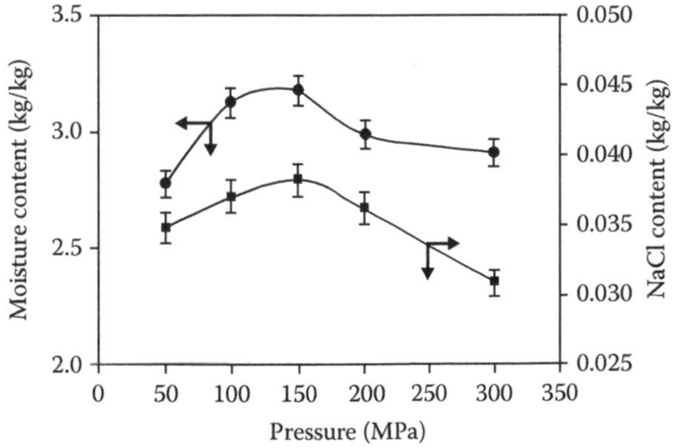

FIGURE 5.25 Effect of pressure come-up time on moisture and salt contents during salting of turkey breast at different pressures. The initial moisture and NaCl contents of the meat at ambient pressure were 2.58 kg water/kg initial dry solids and 0.038 kg NaCl/kg initial dry solids, respectively. (From Villacis, M.F. et al., *Lebensm. Wiss. Technol.*, 41, 836, 2008. With permission.)

Fermented sausage technology involves a sequence of hurdles that appear along the ripening process. A wide variety of fermented sausages are manufactured worldwide based on the concept of reduction of pH, water activity, addition of bacteriocins, or high hydrostatic pressure. Jofre et al. (2009b) demonstrated that addition of enterocins A and B to raw sausages and pressurizing them (400 MPa) at the end of ripening produced an immediate reduction in the counts of *Salmonella* but not of *L. monocytogenes* and *S. aureus*. During storage at room temperature or at 7°C, however, the counts of *Salmonella* and *L. monocytogenes* progressively decreased to less than 1 CFU/g. Ripening process, addition of the enterocins, or pressurization could also control the level of *S. aureus*. Marcos et al. (2005) showed that high-pressure treatment (300 MPa, 10 min, 17°C) could be employed as an additional hurdle to the ripening process of low-acid fermented sausages, viz., Fuet and Chorizo to yield a greater decrease in *Salmonella* population (3 MPN/g). Lower values of *L. monocytogenes* counts were noted in untreated than in pressurized sausages due to delay in pH drop caused by high-pressure inactivation of endogenous lactic acid bacteria, which may not be desirable. Discoloration of sausages was observed after high-pressure treatment, which coincided with an increase in L^* values (Figure 5.26).

Muench et al. (2005) studied the effect of high-pressure processing (400 or 800 MPa) on formation of cholesterol oxidation products in cooked sausages (cured and non-cured Bruehwurst). Application of high-pressure processing did not

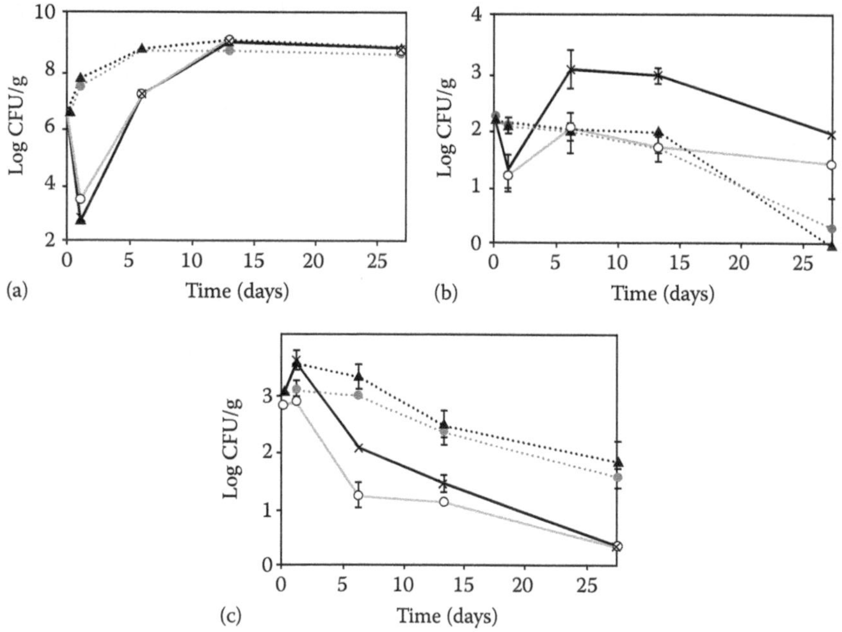

FIGURE 5.26 (a) Lactic acid bacteria; (b) *L. monocytogenes*; and (c) *Salmonella* counts of NT and high-pressure-processed sausages during ripening. (From Marcos, B. et al., *J. Food Sci.*, 70, M339, 2005. With permission.)

significantly increase cholesterol oxidation product concentration immediately after treatment or storage for up to 3 weeks at 2°C.

Dederer and Mueller (2008) demonstrated that raw dry sausages subjected to high pressure (600 MPa, 10 min, 20°C), up to 28 days after ripening, exhibited a decrease in the total count by a factor of 10^3. Lactic acid bacteria were also inactivated by high-pressure processing. High-pressure treatment resulted in increased firmness, paler color of the sausages, increased fat oxidation, and reduced fat hydrolysis. Overall sensory quality of high-pressure-processed sausages was equivalent to that of control samples. However, Mueller and Dederer (2008) later pointed out that simultaneous high-pressure processing and thermal treatment to canned Bruehwurst sausages could help inactivate almost all spores, but resulted in canned products with poor sensory quality.

Supavititpatana and Apichartsrangkoon (2007) indicated that in the case of ostrich meat sausages (yor), the amount of released and expressible water significantly decreased with increasing pressure (300–700 MPa), temperature (40°C and 60°C), and holding time (40 and 60 min). Gel strength and equilibrium stress increased with increasing severity of the treatment. DSC thermograms indicated that a pressure of 700 MPa led to formation of gel networks involving completely denatured protein with an ability to retain water. Chattong and Apichartsrangkoon (2009) later indicated that high-pressure processing (200–600 MPa) led to changes in viscoelastic properties of ostrich-meat yor (Thai sausage), which resulted in structural changes due to hydrophobic interactions and disulfide bonding. Most of the sensory attributes of pressure-treated products received higher scores than conventionally steamed products.

Colmenero et al. (1997) demonstrated that pressurization (300 MPa) of high-fat (247 g/kg) and low-fat (90 g/kg) sausages resulted in a significant decrease in microorganism activity, emulsion stability, and an increase in lightness. In the case of high-fat sausages, high pressure led to a reduction in Kramer shear force and Kramer energy. Pressurization resulted in microstructural changes, which involved formation of filamentous globular structures that differed according to the fat content.

5.4.6 Oysters

High-pressure treatment is being increasingly employed for commercial processing of oysters due to its ability to shuck oysters while keeping fresh-like characteristics of the products. Cruz et al. (2004, 2007) pointed out that high-pressure treatment (100–800 MPa, 10 min, 20°C) of oysters resulted in inactivation of pathogens, increase in moisture content and pH, as well as decrease in protein and ash contents. Oyster muscles get detached from the shells, resulting in shucking. The recovered tissue had good shape and was more voluminous and juicy than that of untreated oysters. The ΔE values, an indicator of the total color difference, indicated that there were significant ($p < 0.05$) differences in color between the samples treated by heat and pressure (Figure 5.27). One significant advantage of high-pressure treatment over thermal treatment was that the former process opened the oyster and separated muscle of the oyster from the shell. Cruz et al. (2008a) later demonstrated that high-pressure treatment (260–800 MPa) significantly changed the microflora of oysters

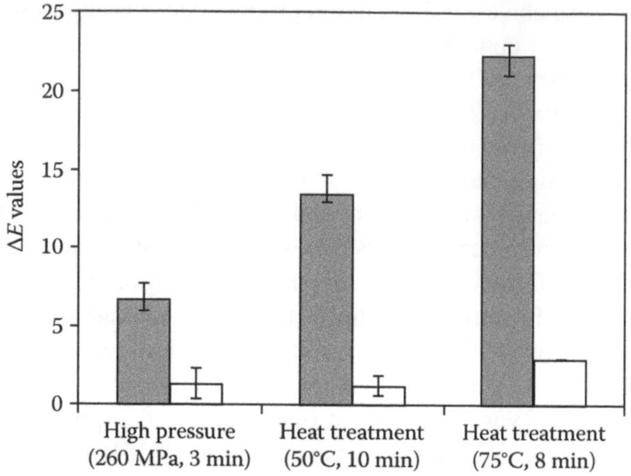

FIGURE 5.27 Effect of high-pressure or heat treatment on total color difference (ΔE) of oyster adductor muscle (dorsal side, ■) and oyster tissue (ventral side, □). (From Cruz, M.R. et al., *Innovat. Food Sci. Emerg. Technol.*, 8, 30, 2007. With permission.)

and apparently has good potential for inactivation of *Vibrio* spp., when used in combination with adequate chilled storage (14 days at 2°C). Cruz et al. (2008b) noted, however, that high-pressure treatment (260–600 MPa), apart from being able to inactivate microorganisms and delay microbial growth under chilled storage, affected various quality attributes of oysters. High pressure increased lipid oxidation; the rate of oxidation was dependent on the pressure applied. During the storage of sample at 2°C, the color of the samples turned yellow and their hardness increased. Nevertheless, Cruz et al. (2008c) indicated that high-pressure treatment did not result in significant changes of fatty acid profiles compared with those of untreated oysters. However, the level of volatile components in the headspace of fresh oysters was changed. High-pressure treated samples had higher concentrations of dimethyl sulfide, 1-penten-3-one, phenol, and 1,2,4-trimethylbenzene and lower concentrations of 1-penten-3-ol, 2,3-pentanedione, (E,E,Z)-1,3,5-octatriene, and 1,3-octadiene than the untreated samples.

Smiddy et al. (2005) confirmed higher baro-resistance of bacteria in oysters than in buffer, which indicated that studies of high-pressure-induced bacterial inactivation in buffer systems may not predict inactivation of microorganisms in foods. Calci et al. (2005) indicated that 6-log reduction of Hepatitis A virus could be achieved via application of high pressure (350–400 MPa). Li et al. (2009) demonstrated that high-pressure treatment (400 MPa, 5 min, 0°C) of oysters contaminated with murine norovirus-1 could reduce the levels of contaminant to undetectable levels.

Kural et al. (2008) pointed out that high-pressure treatment (350 MPa, 2 min, up to 35°C or 300 MPa, 2 min, 40°C) resulted in 5-log reduction of *Vibrio parahaemolyticus* in live oysters. Kural and Chen (2008) indicated that high-pressure treatment (≥250 MPa, −2°C or 1°C, ≤4 min) resulted in 5-log reduction of *Vibrio vulnificus*, which is frequently associated with oysters. Since oysters are typically consumed raw on a half shell, they can pose a threat to public health due to ingestion of this pathogenic marine microorganism. Fletcher et al. (2008) indicated that shucking of

New Zealand greenshell mussels (*Perna canaliculus*) by high-pressure processing (400 MPa) has potential benefits in product quality, increased yield, and inactivation of *L. monocytogenes*.

5.5 CONCLUDING REMARKS

Destruction of microorganisms, inactivation of enzymes at low or moderate temperatures without changing organoleptic and nutritional properties, as well as recent commercial success stories show that high-pressure technology has a great potential and can be used for development of diversified value-added food products with novel physicochemical and sensory properties. A number of high-pressure-processed products (such as fruits juices, guacamole, meat products, oysters, and dairy products) available in international markets confirm steep growth of the technology, which presents exciting opportunities for the food industry. Although high-pressure processing may not replace traditional processing methods in the near future, it may find its niche applications. Adoption of this technology would benefit consumers in terms of increased food safety, shelf life, quality, as well as availability of value-added nutritious food products at reasonable cost.

It seems clear that despite various obstacles, the benefits of high-pressure processing would eventually make it a worthwhile investment. A commercial-scale, high-pressure vessel costs between US$500,000 to US$2.5 million, depending upon equipment capacity and extent of automation. As a new processing technology with a limited market, pressure-processed products may cost 3–10 cents per pound more than thermally processed products. With two 215 L high-pressure processing units operating under typical food processing conditions, a throughput of approximately 20 million pounds per year is achievable. Factory production rates beyond 40 million pounds per year are now in operation. High capital expenditure may be offset by lower operating costs and other benefits with respect to product originality. As demand for high-pressure processing equipment grows, capital cost will continue to decrease.

Scientists are globally working to achieve breakthrough in food technology that would make great tasting, safe, and sterile foods a reality without chemical preservatives or deterioration of color and flavor due to extensive thermal processing. Food companies along with scientists are also working with regulatory authorities such as the U.S. FDA to ensure that these new products comply with their food safety requirements. Additional work also needs to be done to create guidelines for optimal processing of a variety of different products, so that companies can skip the in-house experimentation process once the machine is installed. The major challenge in the production of shelf-stable, sterilized foods is the inactivation of bacterial spores. Pressure-assisted thermal sterilization combines high pressure and thermal treatment with the instantaneous effects of adiabatic heating and cooling to give short processing time and spore inactivation.

ACKNOWLEDGMENTS

The author is grateful to Dr. V. Prakash, Director, Central Food Technological Research Institute, Mysore, India, for constant encouragement. Thanks are also due to Dr. K.S.M.S. Raghavarao, Head, Food Engineering Department, for support.

REFERENCES

Ade-Omowaye, B.I.O., Rastogi, N.K., Angersbach, A., and Knorr, D. 2001. Effect of high pressure or high electrical field pulse pretreatment on dehydration characteristics of paprika. *Innovative Food Science and Emerging Technologies* 2: 1–7.

Ahmed, J., Ramaswamy, H.S., and Hiremath, N. 2005. The effect of high pressure treatment on rheological characteristics and color of mango pulp. *International Journal of Food Science and Technology* 40: 885–895.

Alizadeh, E., Chapleau, N., Lamballerie, M. de., and LeBail, A. 2007. Effects of freezing and thawing processes on the quality of Atlantic salmon (*Salmo salar*) fillets. *Journal of Food Science* 72: E279–E284.

Alvarez, P.A., Ramaswamy, H.S., and Ismail, A.A. 2007. Effect of high-pressure treatment on the electrospray ionization mass spectrometry (ESI-MS) profiles of whey proteins. *International Dairy Journal* 17: 881–888.

Ancos, B. de., Gonzalez, E., and Cano, M.P. 2000. Effect of high-pressure treatment on the carotenoid composition and the radical scavenging activity of persimmon fruit purees. *Journal of Agricultural and Food Chemistry* 48: 3542–3548.

Ancos, B., Sgroppo, S., Plaza, L., and Cano, M.P. 2002. Possible nutritional and health-related value promotion in orange juice preserved by high-pressure treatment. *Journal of the Science of Food and Agriculture* 82: 790–796.

Anema, S.G. 2008a. Effect of milk solids concentration on whey protein denaturation, particle size changes and solubilization of casein in high-pressure-treated skim milk. *International Dairy Journal* 18: 228–235.

Anema, S.G. 2008b. Heat and/or high pressure treatment of skim milk: Changes to the casein micelle size, whey proteins and the acid gelation properties of the milk. *International Journal of Dairy Technology* 61: 245–252.

Angsupanich, K. and Ledward, D.A. 1998. High pressure treatment effects on cod (*Gadus morhua*) muscle. *Food Chemistry* 63: 39–50.

Araya, X.I.T., Hendrickx, M., Verlinden, B.E., Buggenhout, S.V., Smale, N.J., Stewart, C., and Mawson, A.J. 2007. Understanding texture changes of high pressure processed fresh carrots: A microstructural and biochemical approach. *Journal of Food Engineering* 80: 873–884.

Arias, M., Lopez, F.R., and Olano, A. 2000. Influence of pH on the effects of high pressure on milk proteins. *Milchwissenschaft* 55: 191–194.

Arques, J.L., Garde, S., Fernandez-Garcia, E., Gaya, P., and Nunez, M. 2007. Volatile compounds, odor, and aroma of La Serena cheese high pressure treated at two different stages of ripening. *Journal of Dairy Science* 90: 3627–3639.

Ayo, J., Carballo, J., Solas, M.T., and Jiménez, C.F. 2005. High pressure processing of meat batters with added walnuts. *International Journal of Food Science and Technology* 40: 47–54.

Bai, Y., Zhao, D.D., and Yang, G. 2004. Changes of microscopic structure and shear force value of bovine and mutton skeletal muscle under hydrostatic high-pressure treatment. *Food Science in China* 25: 27–31.

Balasubramaniam, V.M. 2003. High pressure food preservation. In D. Heldman (Ed.), *Encyclopedia of Agriculture, Food and Biological Engineering*, pp. 490–496. New York: Marcel Dekker, Inc.

Balasubramaniam, V.M. and Farkas, D. 2008. High pressure food processing. *Food Science and Technology International* 14: 413–418.

Balci, A.T. and Wilbey, R.A. 1999. High pressure processing of milk—The first 100 years in the development of new technology. *International Journal of Dairy Technology* 52: 149–155.

Basak, S. and Ramaswamy, H.S. 1998. Effect of high pressure processing on the texture of selected fruits and vegetables. *Journal of Texture Studies* 29: 587–601.

Baxter, I.A., Easton, K., Schneebeli, K., and Whitfield, F.B. 2005. High pressure processing of Australian navel orange juices: Sensory analysis and volatile flavor profiling. *Innovative Food Science and Emerging Technologies* 6: 372–387.

Benet, U.G., Chapleau, N., Lille, M., Bail, A., Autio, K., and Knorr, D. 2006. Quality related aspects of high pressure low temperature processed whole potatoes. *Innovative Food Science and Emerging Technologies* 7: 32–39.

Black, E.P., Kelly, A.L., and Fitzgerald, G.F. 2005. The combined effect of high pressure and nisin on inactivation of microorganisms in milk. *Innovative Food Science and Emerging Technologies*. 6: 286–292.

Black, E.P., Linton, M., McCall, R.D., Curran, W., Fitzgerald, G.F., Kelly, A.L., and Patterson, M.F. 2008. The combined effects of high pressure and nisin on germination and inactivation of Bacillus spores in milk. *Journal of Applied Microbiology* 105: 78–87.

Borda, D., Indrawati, Smout, C., Van Loey, A.M., and Hendrickx, M. 2004a. High-pressure thermal inactivation kinetics of a plasmin system. *Journal of Dairy Science* 87: 2351–2358.

Borda, D., Van Loey, A.M., Smout, C., and Hendrickx, M. 2004b. Mathematical models for combined high pressure and thermal plasmin inactivation kinetics in two model systems. *Journal of Dairy Science* 87: 4042–4049.

Boynton, B.B., Sims, C.A., Sargent, S., Balaban, M.O., and Marshall, M.R. 2002. Quality and stability of precut mangos and carambolas subjected to high-pressure processing. *Journal of Food Science* 67: 409–415.

Bragagnolo, N., Danielsen, B., and Skibsted, L.H. 2006. Combined effect of salt addition and high-pressure processing on formation of free radicals in chicken thigh and breast muscle. *European Food Research and Technology* 223: 669–673.

Bragagnolo, N., Danielsen, B., and Skibsted, L.H. 2007. Rosemary as antioxidant in pressure processed chicken during subsequent cooking as evaluated by electron spin resonance spectroscopy. *Innovative Food Science and Emerging Technologies* 8: 24–29.

Brooker, B., Ferragut, V., Gill, A., and Needs, E. 1998. Properties of rennet gel formed from high-pressure treated milk. In K. Autio (Ed.), *Proceedings of the VTT symposium, Fresh Novel Foods by High Pressure*, Helsinki, Finland, September 20–21, 1998, pp. 55–61.

Buffa, M.N., Trujillo, A.J., Pavia, M., and Guamis, B. 2001. Changes in textural, microstructural, and color characteristics during ripening of cheeses made from raw, pasteurized or high pressure treated goats' milk. *International Dairy Journal* 11: 927–934.

Bull, M.K., Zerdin, K., Howe, E., Goicoeche, G., Paramanandhan, P., Stockman, R., Sellahewa, J., Szabo, E.A., Johnson, R.L., and Stewart, C.M. 2004. The effect of high pressure processing on the microbial, physical and chemical properties of Valencia and navel orange juice. *Innovative Food Science and Emerging Technologies* 5: 135–149.

Butz, P., Koller, W.D., Tauscher, B., and Wolf, S. 1994. Ultra high pressure processing of onions: Chemical and sensory changes. *Lebensmittel-Wissenschaft und-Technologie* 27: 463–467.

Butz, P., Edenharder, R., Fister, H., and Tauscher, B. 1997. The influence of high pressure processing on antimutagenic activities of fruit and vegetable juices. *Food Research International* 30: 287–291.

Butz, P., Edenharder, R., Fernandez-Garcia, A., Fister, H., Merkel, C., and Tauscher, B. 2002. Changes in functional properties of vegetables induced by high pressure treatment. *Food Research International* 35: 295–300.

Butz, P., Needs, E.C., Baron, A., Bayer, O., Geisel, B., Gupta, B., Oltersdorf, U., and Tauscher, B. 2003. Consumer attitudes to high pressure food processing. *Food, Agriculture and Environment* 1: 30–34.

Butz, P., Serfert, Y., Fernandez-Garcia, A., Dieterich, S., Lindauer, R., Bognar, A., and Tauscher, B. 2004. Influence of high-pressure treatment at 25°C and 80°C on folates in orange juice and model media. *Journal of Food Science* 69: SNQ117–SNQ121.

Butz, P., Bognar, A., Dieterich, S., and Tauscher, B. 2007. Effect of high-pressure processing at elevated temperatures on thiamin and riboflavin in pork and model systems. *Journal of Agricultural and Food Chemistry* 55: 1289–1294.

Buzrul, S., Alpas, H., Largeteau, A., and Demazeau, G. 2008. Modeling high pressure inactivation of *Escherichia coli* and *Listeria innocua* in whole milk. *European Food Research and Technology* 227: 443–448.

Caballero, M.E.L., Guillen, M.C.G., Mateos, M.P., and Montero, P. 2005. A functional chitosan-enriched fish sausage treated by high pressure. *Journal of Food Science* 10: M166–M171.

Calci, K.R., Meade, G.K., Tezloff, R.C., and Kingsley, D.H. 2005. High-pressure inactivation of hepatitis A virus within oysters. *Applied and Environmental Microbiology* 71: 339–343.

Campus, M., Flores, M., Martinez, A., and Toldra, F. 2008. Effect of high pressure treatment on color, microbial and chemical characteristics of dry cured loin. *Meat Science* 80: 1174–1181.

Canedo, A.R., Fernandez-Garcia, E., and Nunez, M. 2009a. Volatile compounds in fresh meats subjected to high pressure processing: Effect of the packaging material. *Meat Science* 81: 321–328.

Canedo, A.R., Fernandez-Garcia, E., and Nunez, M. 2009b. Volatile compounds in dry-cured Serrano ham subjected to high pressure processing. Effect of the packaging material. *Meat Science* 82: 162–169.

Capellas, M., Mor-Mur, M., Sendra, E., Pla, R., and Guamis, B. 1996. Population of aerobic mesophiles and inoculated *E. coli* during storage of fresh goat's milk cheese treated with high pressure. *Journal of Food Protection* 59: 82–87.

Carballo, J., Fernandez, P., and Colmenero, F.J. 1996. Texture of uncooked and cooked low and high fat meat batters as affected by high hydrostatic pressure. *Journal of Agricultural and Food Chemistry* 44: 1624–1625.

Carlez, A., Rosec, J.P., Richard, N., and Cheftel, J.C. 1993. High pressure inactivation of *Citrobacter freundii*, *Pseudomonas fluorescens* and *Listeria innocua* in inoculated minced beef muscle. *Lebensmittel-Wissenschaft und-Technologie* 26: 357–363.

Carlez, A., Rosec, J.P., Richard, N., and Cheftel, J.C. 1994. Bacterial growth during chilled storage of pressure treated minced meat. *Lebensmittel-Wissenschaft und-Technologie* 27: 48–54.

Carlez, A., Veciana-Nogues, T., and Cheftel, J.C. 1995. Changes in color and myoglobin of minced beef meat due to high pressure processing. *Lebensmittel-Wissenschaft und-Technologie* 28: 528–538.

Carminati, D., Gatti, M., Bonvini, B., Neviani, E., and Mucchetti, G. 2004. High-pressure processing of Gorgonzola cheese: Influence on *Listeria monocytogenes* inactivation and on sensory characteristics. *Journal of Food Protection* 67: 1671–1675.

Carpi, G., Gola, S., Maggi, A., Rovere, P., and Buzzoni, M. 1995. Microbial and chemical shelf life of high pressure treated salmon cream at refrigeration temperatures. *Industria Conserve* 70: 386–397.

Carpi, G., Squarcina, N., Gola, S., Rovere, P., Pedrielli, R., and Bergamaschi, M. 1999. Application of high pressure treatment to extend the refrigerated shelf life of sliced cooked ham. *Industria Conserve* 74: 327–339.

Castro, S.M., Saraiva, J.A., Lopes Da Silva, J.A. et al. 2008. Effect of thermal blanching and of high pressure treatments on sweet green and red bell pepper fruits (*Capsicum annuum* L.). *Food Chemistry* 107: 1436–1449.

Chattong, U. and Apichartsrangkoon, A. 2009. Dynamic viscoelastic characterization of ostrich-meat yor (Thai sausage) following pressure, temperature and holding time regimes. *Meat Science* 81: 426–432.

Cheftel, J.C. 1995. Review: High-pressure, microbial inactivation and food preservation. *International Journal of Food Science & Technology* 1: 75–90.

Cheftel, J.C., Levy, J., and Dumay, E. 2000. Pressure-assisted' freezing and thawing: Principles and potential applications. *Food Reviews International* 16: 453–483.

Cheret, R., Chapleau, N., Delbarre-Ladrat, C., Verrez-Bagnis, V., and Lamballerie, M. de. 2005. Effects of high pressure on texture and microstructure of sea bass (*Dicentrarchus labrax* L.) fillets. *Journal of Food Science*70: E477–E483.

Cheret, R., Hernandez-Andres, A., Delbarre-Ladrat, C., Lamballerie, M. de., and Verrez-Bagnis, V. 2006. Proteins and proteolytic activity changes during refrigerated storage in sea bass (*Dicentrarchus labrax* L.) muscle after high-pressure treatment. *European Food Research and Technology* 222: 527–535.

Chevalier, D., Bail, A.L., Chourot, J.M., and Chantreau, P. 1999. High pressure thawing of fish (whiting): Influence of the process parameters on drip losses. *Lebensmittel-Wissenschaft und-Technologie* 32: 25–31.

Chevalier, D., Le Bail, A., and Ghoul, M. 2001. Evaluation of the ice ratio formed during quasi-adiabatic pressure shift freezing. *High Pressure Research* 21: 227–235.

Chicon, R., Belloque, J., Alonso, E., and Fandino, R.L. 2009. Antibody binding and functional properties of whey protein hydrolysates obtained under high pressure. *Food Hydrocolloids* 23: 593–599.

Colmenero, J.F., Carballo, J., Fernandez, P., Barreto, G., and Solas, M.T. 1997. High-pressure-induced changes in the characteristics of low-fat and high-fat sausages. *Journal of the Science of Food and Agriculture* 75: 61–66.

Corrales, M., Toepfl, S., Butz, P., Knorr, D., and Tauscher, B. 2008a. Extraction of anthocyanins from grape by-products assisted by ultrasonics, high hydrostatic pressure or pulsed electric fields: A comparison. *Innovative Food Science and Emerging Technologies* 9: 85–91.

Corrales, M., Butz, P., and Tauscher, B. 2008b. Anthocyanin condensation reactions under high hydrostatic pressure. *Food Chemistry* 110: 627–635.

Corrales, M., Fernandez, G.A., Butz, P., and Tauscher, B. 2009. Extraction of anthocyanins from grape skins assisted by high hydrostatic pressure. *Journal of Food Engineering* 90: 415–421.

Cruz, C., El Moueffak, A., Antoine, M., Montury, M., Demazeau, G., Largeteau, A., Roy, B., and Zuber, F. 2003. Preservation of fatty duck liver by high pressure treatment. *International Journal of Food Science and Technology* 38: 267–272.

Cruz, M.R., Smiddy, M., Hill, C., Kerry, J.P., and Kelly, A.L. 2004. Effects of high pressure treatment on physicochemical characteristics of fresh oysters (*Crassostrea gigas*). *Innovative Food Science and Emerging Technologies* 5: 161–169.

Cruz, M.R., Kelly, A.L., and Kerry, J.P. 2007. Effects of high-pressure and heat treatments on physical and biochemical characteristics of oysters (*Crassostrea gigas*). *Innovative Food Science and Emerging Technologies* 8: 30–38.

Cruz, M.R., Kelly, A.L., and Kerry, J.P. 2008a. Effects of high pressure treatment on the microflora of oysters (*Crassostrea gigas*) during chilled storage. *Innovative Food Science and Emerging Technologies* 9: 441–447.

Cruz, M.R., Kerry, J.P., and Kelly, A.L. 2008b. Changes in the microbiological and physicochemical quality of high pressure treated oysters (*Crassostrea gigas*) during chilled storage. *Food Control* 19: 1139–1147.

Cruz, M.R., Kerry, J.P., and Kelly, A.L. 2008c. Fatty acids, volatile compounds and color changes in high-pressure-treated oysters (*Crassostrea gigas*). *Innovative Food Science and Emerging Technologies* 9: 54–61.

Daryaei, H., Coventry, M.J., Versteeg, C., and Sherkat, F. 2006. Effects of high pressure treatment on shelf life and quality of fresh lactic curd cheese. *Australian Journal of Dairy Technology* 61: 186–188.

Daryaei, H., Coventry, M.J., Versteeg, C., and Sherkat, F. 2008. Effect of high pressure treatment on starter bacteria and spoilage yeasts in fresh lactic curd cheese of bovine milk. *Innovative Food Science and Emerging Technologies* 9: 201–205.

Dede, S., Alpas, H., and Bayindirli, A. 2007. High hydrostatic pressure treatment and storage of carrot and tomato juices: Antioxidant activity and microbial safety. *Journal of the Science of Food and Agriculture* 87: 773–782.

Dederer, I. and Mueller, W.D. 2008. High pressure induced changes in the dry sausage during ripening and storage. *Fleischwirtschaft* 88: 92–96.

Derobry, B.S., Richard, F., and Hardy, J. 1994. Study of acid and rennet coagulation of high pressurized milk. *Journal of Dairy Science* 77: 3267–3274.

Dervisi, P., Lamb, J., and Zabetakis, I. 2001. High pressure processing in jam manufacture: Effects on textural and color properties. *Food Chemistry* 73: 85–91.

Doblado, R., Frias, J., and Vidal, V.C. 2007. Changes in vitamin C content and antioxidant capacity of raw and germinated cowpea (*Vigna sinensis* var. carilla) seeds induced by high pressure treatment. *Food Chemistry* 101: 918–923.

Donsi, G., Ferrari, G., and Matteo, M.D. 1996. High pressure stabilization of orange juice: Evaluation of the effects of process conditions. *Italian Food and Beverage Technology* 8: 10–14.

Donsi, G., Ferrari, G., Matteo, M. di., and Bruno, M.C. 1998. High pressure stabilization of lemon juice. *Italian Food and Beverage Technology* 14: 14–16.

Dornenburg, H. and Knorr, D. 1993. Cellular permeabilization of cultured plant tissue by high electric field pulse and ultra high pressure for recovery of secondary metabolites. *Food Biotechnology* 7: 35–48.

Drake, M.A., Harrison, S.L., Asplund, M., Barbosa-Canovas, G., and Swanson, B.G. 1997. High pressure treatment of milk and effects on microbiological and sensory quality of cheddar cheese. *Journal of Food Science* 62: 843–845.

El Moueffak, A., Cruz, C., Antoine, M., Montury, M., Demazeau, G., Largeteau, A., Roy, B., and Zuber, F. 1996. High pressure and pasteurization effect on duck foie gras. *International Journal of Food Science and Technology* 30: 737–743.

El Moueffak, A., Cruz, C., and Antoine, M. 2001. Stabilization of duck fatty liver by high pressure treatment. Inactivation of *Enterococcus faecalis*. *Sciences des Aliments* 21: 71–76.

Erkman, O. and Karatas, S. 1997. Effect of high hydrostatic pressure on *Staphylococcus aureus* in milk. *Journal of Food Engineering* 33: 257–262.

Eshtiaghi, M.N. and Knorr, D. 1993. Potato cube response to water blanching and high hydrostatic pressure. *Journal of Food Science* 58: 1371–1374.

Eshtiaghi, M.N., Stute, R., and Knorr, D. 1994. High pressure and freezing pretreatment effects on drying, rehydration, texture and color of green beans, carrots and potatoes. *Journal of Food Science* 59: 1168–1170.

Evrendilek, G.A., Koca, N., Harper, J.W., and Balasubramanian, V.M. 2008. High-pressure processing of Turkish white cheese for microbial inactivation. *Journal of Food Protection* 71: 102–108.

Eylen, D., Oey, I., Hendrickx, M., and Loey, A. 2007. Kinetics of the stability of broccoli (*Brassica oleracea* cv. Italica) myrosinase and isothiocyanates in broccoli juice during pressure/temperature treatments. *Journal of Agricultural and Food Chemistry* 55: 2163–2170.

Eylen, D., Bellostas, N., Strobel, B.W., Oey, I., Hendrickx, M., Loey, A., Sørensen, H., and Sørensen, J.C. 2009. Influence of pressure/temperature treatments on glucosinolate conversion in broccoli (*Brassica oleraceae* L. cv Italica) heads. *Food Chemistry* 112: 646–653.

Farr, D. 1990. High pressure technology in food industry. *Trends in Food Science and Technology* 1: 14–16.

Fernandez, P., Cofrades, S., Solas, M.T., Carballo, J., and Colmenero, F.J. 1998. High pressure cooking of chicken meat batters with starch, egg white, and iota carrageenan. *Journal of Food Science* 63: 267–271.

Fernandez, G.A., Butz, P., Bognar, A., and Tauscher, B. 2001a. Antioxidative capacity, nutrient content and sensory quality of orange juice and an orange-lemon-carrot juice product after high pressure treatment and storage in different packaging. *European Food Research and Technology* 213: 290–296.

Fernandez, G.A., Butz, P., and Tauscher, B. 2001b. Effects of high-pressure processing on carotenoid extractability, antioxidant activity, glucose diffusion, and water binding of tomato puree (*Lycopersicon esculentum* Mill.). *Journal of Food Science* 66: 1033–1038.

Fernandez, P.P., Prestamo, G., Otero, L., and Sanz, P.D. 2006. Assessment of cell damage in high pressure shift frozen broccoli: Comparison with market samples. *Zeitschrift Für Lebensmittel-Untersuchung und - Forschung A* 224: 101–107.

Fernandez, P.P., Sanz, P.D., Molina-Garcia, A.D., Otero, L., Guignon, B., and Vaudagna, S.R. 2007. Conventional freezing plus high pressure-low temperature treatment: Physical properties, microbial quality and storage stability of beef meat. *Meat Science* 77: 616–625.

Ferreira, L., Afonso, C., Vila, R.H., Alfaia, A., and Ribeiro, M.H.L. 2008. Evaluation of the effect of high pressure on naringin hydrolysis in grapefruit juice with naringinase immobilized in calcium alginate beads. *Food Technology and Biotechnology* 46: 146–150.

Fletcher, G.C., Youssef, J.F., and Gupta, S. 2008. Research issues in inactivation of *Listeria monocytogenes* associated with New Zealand Greenshell mussel meat (*Perna canaliculus*) using high pressure processing. *Journal of Aquatic Food Product Technology* 17: 173–194.

Fuchigami, M., Miyazaki, K., Kato, N., and Teramoto, A. 1997a. Histological changes in high pressure frozen carrots. *Journal of Food Science* 62: 809–812.

Fuchigami, M., Kato, N., and Teramoto, A. 1997b. High pressure freezing effects on textural quality of carrots. *Journal of Food Science* 62: 804–808.

Fuchigami, M., Kato, N., and Teramoto, A. 1998. High pressure freezing effects on textural quality of Chinese cabbage. *Journal of Food Science* 63: 122–125.

Fulladosa, E., Serra, X.G., and Arnau, J.P. 2009. Effects of potassium lactate and high pressure on transglutaminase restructured dry-cured hams with reduced salt content. *Meat Science* 82: 213–218.

Galazka, V.B. and Ledward, D.A. 1995. Development in high pressure processing. *Food Technol. Int. Eur.* 12: 123–125.

Garriga, M., Grebol, N., Aymerich, M.T., Monfort, J.M., and Hugas, M. 2004. Microbial inactivation after high-pressure processing at 600 MPa in commercial meat products over its shelf life. *Innovative Food Science and Emerging Technologies* 5: 451–457.

Gervila, R., Capellas, M., Ferragur, V., and Guamis, B. 1997. Effect of high hydrostatic pressure on *Listeria innocua* 910 CECT inoculated into ewe's milk. *Journal of Food Protection* 60: 33–37.

Gimenez, J., Kajda, P., Margomenou, L., Piggott, J.R., and Zabetakis, I. 2001. A study on the color and sensory attributes of high hydrostatic pressure jams as compared with traditional jams. *Journal of the Science of Food and Agriculture* 81: 1228–1234.

Gomez, E.J., Gomez-Guillen, M.C., and Montero, P. 2007a. High pressure effects on the quality and preservation of cold-smoked dolphinfish (*Coryphaena hippurus*) fillets. *Food Chemistry* 102: 1250–1259.

Gomez, E.J., Montero, P., Gimenez, B., and Gomez-Guillen, M.C. 2007b. Effect of functional edible films and high pressure processing on microbial and oxidative spoilage in cold-smoked sardine (*Sardina pilchardus*). *Food Chemistry* 105: 511–520.

Gomez, E.J., Lopez-Caballero, M.E., Gomez-Guillen, M.C., Lopez de Lacey, A., and Montero, P. 2009. High pressure technology as a tool to obtain high quality carpaccio and carpaccio-like products from fish. *Innovative Food Science and Emerging Technologies* 10: 148–154.

Goodner, J.K., Braddock, R.J., Parish, M.E., and Sims, C.A. 1999. Cloud stabilization of orange juice by high pressure processing. *Journal of Food Science* 64: 699–700.

Gould, G.W. 1995. The microbe as a high pressure target. In D.A. Ledward, D.E. Johnston, R.G. Earnshaw, and A.P.M. Hasting (Eds.), *High Pressure Processing of Foods*, pp. 27–35. Nottingham, U.K.: Nottingham University Press.

Gow, C.Y. and Hsin, T.L. 1996. Comparison of high pressure treatment and thermal pasteurization effects on the quality and shelf life of guava puree. *International Journal of Food Science and Technology* 31: 205–213.

Gow, C.Y. and Hsin, T.L. 1999. Changes in volatile flavor components of guava juice with high-pressure treatment and heat processing and during storage. *Journal of Agricultural and Food Chemistry* 47: 2082–2087.

Guerrero-Beltran, J.A., Barbosa-Canovas, G.V., Moraga-Ballesteros, G., Moraga-Ballesteros, M.J., and Swanson, B.G. 2006. Effect of pH and ascorbic acid on high hydrostatic pressure processed mango puree. *Journal of Food Processing and Preservation.* 30: 582–596.

Hajos, G., Polgar, M., and Farkas, J. 2004. High-pressure effects on IgE immunoreactivity of proteins in a sausage batter. *Innovative Food Science and Emerging Technologies* 5: 443–449.

Han, G.D. 2006. Heat and high-pressure treatments on in vitro digestibility and allergenicity of beef extract. *Food Science and Biotechnology* 15: 523–528.

Han, J.M. and Ledward, D.A. 2004. High pressure/thermal treatment effects on the texture of beef muscle. *Meat Science* 68: 347–355.

Han, G.D., Fan, J.P., and Suzuki, A. 2006. Changes of SDS-PAGE pattern and allergenicity of BSA and BGG in beef extract treated with heat and high pressure. *Journal of Korean Society of Food Science and Nutrition* 35: 594–599.

Hassan, Y., Meszaros, L., Simon, A., Tuboly, E., Mohacsi-Farkas, C., and Farkas, J. 2002. Comparative studies on gamma radiation and high pressure induced effects on minced beef. *Acta Alimentaria* 31: 253–264.

Hayashi, R. 1990. Application of high pressure to processing and preservation: Philosophy and development. In W.E.L. Spiess and H. Schubert (Eds.), *Engineering and Food*, pp. 815–826. London, U.K.: Elsevier Applied Science.

Hayman, M.M., Baxter, I., O'Riordan, P.J., and Stewart, C.M. 2004. Effects of high-pressure processing on the safety, quality, and shelf life of ready-to-eat meats. *Journal of Food Protection* 67: 1709–1718.

Hite, B.H. 1899. The effect of pressure in the preservation of milk. *West Virginia University Agricultural Experiment Station Bulletin* 58: 15–35.

Hite, B.H., Giddings, N.J., and Weakly, C.E. 1914. The effects of pressure on certain microorganisms encountered in the preservation of fruits and vegetables. *West Virginia University Agricultural Experiment Station Bulletin* 146: 1–67.

Hogan, E., Kelly, A., and Sun, D.W. 2005. High pressure processing of foods: An overview. In D.W. Sun (Ed.), *Emerging Technologies for Food Processing*, pp. 3–32, London, U.K.: Elsevier Academic Press.

Homma, N., Ikeuchi, Y., and Suzuki, A. 1995. Effects of high pressure treatment on the proteolytic enzymes in meat. *Meat Science* 38: 219–228.

Hong, G.P., Park, S.H., Kim, J.Y., Lee, S.K., and Min, S.G. 2005. Effects of time-dependent high pressure treatment on physico-chemical properties of pork. *Food Science and Biotechnology* 14: 808–812.

Hong, G.P., Ko, S.H., Choi, M.J., and Min, S.G. 2008a. Effect of glucono-δ-lactone and κ-carrageenan combined with high pressure treatment on the physico-chemical properties of restructured pork. *Meat Science* 79: 236–243.

Hong, G.P., Min, S.G., Ko, S.H., and Choi, M.J. 2008b. Effect of high pressure treatments combined with various levels of κ-carrageenan on cold set binding in restructured pork. *International Journal of Food Science and Technology* 43: 1484–1491.

Houska, M., Strohalm, J., Kocurova, K. et al. 2006. High pressure and foods-fruit/vegetable juices. *Journal of Food Engineering* 77: 386–398.

Hsu, K.C., Tan, F.J., and Chi, H.Y. 2008. Evaluation of microbial inactivation and physicochemical properties of pressurized tomato juice during refrigerated storage. *Lebensmittel-Wissenschaft und-Technologie* 41: 367–375.

Hugas, M., Garriga, M., and Monfort, J.M. 2002. New mild technologies in meat processing: High pressure as a model technology. *Meat Science* 62: 359–371.

Huppertz, T. and Smiddy, M.A. 2008. Behaviour of partially cross linked casein micelles under high pressure. *International Journal of Dairy Technology* 61: 51–55.

Huppertz, T., Kelly, A.L., and Fox, P.F. 2002. Effects of high pressure on constituents and properties of milk. *International Dairy Journal* 12: 561–572.

Huppertz, T., Fox, P.F., and Kelly, A.L. 2004. Effects of high pressure treatment on the yield of cheese curd from bovine milk. *Innovative Food Science and Emerging Technologies* 5: 1–8.

Huppertz, T., Hinz, K., Zobrist, M.R., Uniacke, T., Kelly, A.L., and Fox, P.F. 2005. Effects of high pressure treatment on the rennet coagulation and cheese-making properties of heated milk. *Innovative Food Science and Emerging Technologies* 6: 279–285.

Isbarn, S., Buckow, R., Himmelreich, A., Lehmacher, A., and Heinz, V. 2007. Inactivation of avian influenza virus by heat and high hydrostatic pressure. *Journal of Food Protection* 70: 667–673.

Jankowska, A., Reps, A., Proszek, A., and Krasowska, M. 2005. Effect of high pressure on microflora and sensory characteristics of yoghurt. *Polish Journal of Food Nutrition and Science* 14/55: 79–84.

Jimenez, C.F., Fernandez, P., Carballo, J., and Fernandez, M.F. 1998. High pressure cooked low fat pork and chicken batters as affected by salt levels and cooking temperature. *Journal of Food Science* 63: 656–659.

Jofre, A., Aymerich, T., and Garriga, M. 2008a. Assessment of the effectiveness of antimicrobial packaging combined with high pressure to control *Salmonella* sp. in cooked ham. *Food Control* 19: 634–638.

Jofre, A., Garriga, M., and Aymerich, T. 2008b. Inhibition of *Salmonella* sp., *Listeria monocytogenes* and *Staphylococcus aureus* in cooked ham by combining antimicrobials, high hydrostatic pressure and refrigeration. *Meat Science* 78: 53–59.

Jofre, A., Aymerich, T., and Garriga, M. 2009b. Improvement of the food safety of low acid fermented sausages by enterocins A and B and high pressure. *Food Control* 20: 179–184.

Johnston, D.E., Murphy, R.J., Rutherford, J.A., and McElhone, C.A. 1998. Formation and synthesis of rennet set gel prepared from high pressure treated milk. In N.S. Isaacs (Ed.), *High Pressure Food Science, Bioscience and Chemistry*, pp. 220–226. Cambridge, U.K.: Royal Society of Chemistry.

Juan, B., Barron, L.J.R., Ferragut, V., and Trujillo, A.J. 2007. Effects of high pressure treatment on volatile profile during ripening of ewe milk cheese. *Journal of Dairy Science* 90: 124–135.

Juan, B., Ferragut, V., Guamis, B., and Trujillo, A.J. 2008. The effect of high pressure treatment at 300 MPa on ripening of ewes' milk cheese. *International Dairy Journal* 18: 129–138.

Jun, Y., Nan, X.Q., and Che, R.Z. 1999. Effects of high-pressure treatment on sensory properties in beef. *Meat Research* 4: 19–21.

Jung, S., Ghoul, M., and Lamballerie-Anton, M. De. 2000. Changes in lysosomal enzyme activities and shear values of high pressure treated meat during ageing. *Meat Science* 56: 239–246.

Jung, S., Ghoul, M., and Lamballerie-Anton, M. De. 2003. Influence of high pressure on the color and microbial quality of beef meat. *Lebensmittel-Wissenschaft und-Technologie* 36: 625–631.

Kadlec, P., Dostalova, J., Houska, M., Strohalm, J., Culkova, J., Hinkova, A., and Starhova, H. 2006. High pressure treatment of germinated chickpea *Cicer arietinum* L. seeds. *Journal of Food Engineering* 77: 445–448.

Kato, N., Teramoto, A., and Fuchigami, M. 1997. Pectic substance degradation and texture of carrot as affected by pressurization. *Journal of Food Science* 62: 359–362, 398.

Kennick, W.H., Elgasim, E.A., Holmes, Z.A., and Meyer, P.F. 1980. The effect of pressurization of pre-rigor muscle on post-rigor meat characteristics. *Meat Science* 4: 33–40.

Kimura, K., Ida, M., Yosida, Y., Okhi, K., Fukumoto, T., and Sakui, N. 1994. Comparison of keeping quality between pressure processed jam and heat processed jam: Changes in flavor components, hue and nutrients during storage. *Bioscience, Biotechnology and Biochemistry* 58: 1386–1391.

Kingsly, A.R.P., Balasubramaniam, V.M., and Rastogi, N.K. 2009a. Effect of high pressure processing on texture and drying behavior of pineapple. *Journal of Food Process Engineering* 32: 369–381.

Kingsly, A.R.P., Balasubramaniam, V.M., and Rastogi, N.K. 2009b. Influence of high pressure blanching on polyphenoloxidase activity of peach fruits and its drying behavior. *International Journal of Food Properties*. 12(3): 671–680.

Knorr, D. 1995. Hydrostatic pressure treatment of food: Microbiology. In G.W. Gould (Ed.), *New Methods for Food Preservation*, pp. 159–175. London, U.K.: Blackie Academic and Professional.

Ko, S.H., Hong, G.P., Park, S.H., Choi, M.J., and Min, S.G. 2006. Studies on physical properties of pork frozen by various high-pressure freezing process. *Korean Journal of Food Science and Animal Research* 26: 464–470.

Kolakowski, P., Reps, A., and Fetlinski, A. 2000. Microbial quality and some physicochemical properties of high pressure processed cow milk. *Polish Journal of Food Nutrition and Science* 9: 19–26.

Koseki, S., Mizuno, Y., and Yamamoto, K. 2007. Predictive modeling of the recovery of *Listeria monocytogenes* on sliced cooked ham after high pressure processing. *International Journal of Food Microbiology* 119: 300–307.

Koseki, S., Mizuno, Y., and Yamamoto, K. 2008. Use of mild heat treatment following high pressure processing to prevent recovery of pressure injured *Listeria monocytogenes* in milk. *Food Microbiology* 25: 288–293.

Krebbers, B., Matser, A.M., Koets, M., and Berg, R.W. van den. 2002. Quality and storage stability of high pressure preserved green beans. *Journal of Food Engineering* 54: 27–33.

Krebbers, B., Matser, A.M., Hoogerwerf, S.W., Moezelaar, R., Tomassen, M.M.M., and Berg, R.W. van den. 2003. Combined high pressure and thermal treatments for processing of tomato puree: Evaluation of microbial inactivation and quality parameters. *Innovative Food Science and Emerging Technologies* 4: 377–385.

Kuo, C.H. 2008. Evaluation of processing qualities of tomato juice induced by thermal and pressure processing. *Lebensmittel-Wissenschaft und-Technologie* 41: 450–459.

Kural, A.G. and Chen, H. 2008. Conditions for a 5-log reduction of *Vibrio vulnificus* in oysters through high hydrostatic pressure treatment. *International Journal of Food Microbiology* 122: 180–187.

Kural, A.G., Shearer, A.E.H., Kingsley, D.H., and Chen, H. 2008. Conditions for high pressure inactivation of *Vibrio parahaemolyticus* in oysters. *International Journal of Food Microbiology* 127: 1–5.

Laboissiere, L.H.E.S., Deliza, R., Barros, A.M., Rosenthal, A., Camargo, L.M.A.Q., and Junqueira, R.G. 2007. Effects of high hydrostatic pressure (HHP) on sensory characteristics of yellow passion fruit juice. *Innovative Food Science and Emerging Technologies* 8: 469–477.

Lakshmanan, R. and Dalgaard, P. 2004. Effects of high-pressure processing on *Listeria monocytogenes*, spoilage microflora and multiple compound quality indices in chilled cold-smoked salmon. *Journal of Applied Microbiology* 96: 398–408.

Lakshmanan, R., Piggott, J.R., and Paterson, A. 2003. Potential applications of high pressure for improvement in salmon quality. *Trends in Food Science Technology* 14: 354–363.

Lakshmanan, R., Patterson, M.F., and Piggott, J.R. 2005a. Effects of high-pressure processing on proteolytic enzymes and proteins in cold-smoked salmon during refrigerated storage. *Food Chemistry* 90: 541–548.

Lakshmanan, R., Miskin, D., and Piggott, J.R. 2005b. Quality of vacuum packed cold-smoked salmon during refrigerated storage as affected by high-pressure processing. *Journal of the Science of Food and Agriculture* 85: 655–661.

Lambert, Y., Demazeau, G., Largeteau, A., and Bouvier, J.M. 1999. Changes in aromatic volatile composition of strawberry after high pressure treatment. *Food Chemistry* 67: 7–16.

Lavinas, F.C., Miguel, M.A.L., Lopes, M.L.M., and Valente, V.L. 2008. Effect of high hydrostatic pressure on cashew apple (*Anacardium occidentale* L.) juice preservation. *Journal of Food Science* 73: M273–M277.

Leadley, C., Tucker, G., and Fryer, P.A. 2008. Comparative study of high pressure sterilization and conventional thermal sterilization: Quality effects in green beans. *Innovative Food Science and Emerging Technologies* 9: 70–79.

Li, D., Tang, Q., Wang, J., Wang, Y., Zhao, Q., and Xue, C. 2009. Effects of high-pressure processing on murine norovirus-1 in oysters (*Crassostrea gigas*) in situ. *Food Control* 20: 992–996.

Lim, S.Y., Swanson, B.G., Ross, C.F., and Clark, S. 2008a. High hydrostatic pressure modification of whey protein concentrates for improved body and texture of low-fat ice cream. *Journal of Dairy Science* 91: 1308–1316.

Lim, S.Y., Swanson, B.G., and Clark, S. 2008b. High hydrostatic pressure modifications of whey protein concentrate for improved functional properties. *Journal of Dairy Science* 91: 1299–1307.

Linton, M., Mackle, A.B., Upadhyay, V.K., Kelly, A.L., and Patterson, M.F. 2008. The fate of *Listeria monocytogenes* during the manufacture of Camembert-type cheese: A comparison between raw milk and milk treated with high hydrostatic pressure. *Innovative Food Science and Emerging Technologies* 9: 423–428.

Lopez, F.R. and Olano, A. 1998. Effect of high pressure combined with moderate temperature on rennet coagulation properties of milk. *International Dairy Journal* 8: 623–627.

Lopez, F.R., Carrascosa, A.V., and Olano, A. 1996. The effect of high pressure on whey protein denaturation and cheese making properties of raw milk. *Journal of Dairy Science* 79: 929–936.

Lopez, M.A., Palou, E., Barbosa-Cáovas, G.V., Welti-Chanes, J., and Swanson, B.G. 1998. Polyphenoloxidase activity and color changes during storage of high hydrostatic pressure treated avocado puree. *Food Research International* 31: 549–556.

Lopez, M.E., Perez-Mateos, M., Borderias, J.A., and Montero, P. 2000. Extension of the shelf life of prawns (*Penaeus japonicus*) by vacuum packaging and high-pressure treatment. *Journal of Food Protection* 63: 1381–1388.

Luscher, C., Schlueter, O., and Knorr, D. 2005. High pressure-low temperature processing of foods: Impact on cell membranes, texture, color and visual appearance of potato tissue. *Innovative Food Science and Emerging Technologies* 6: 59–71.

Ma, H.J., Ledward, D.A., Zamri, A.I., Frazier, R.A., and Zhou, G.H. 2007. Effects of high pressure/thermal treatment on lipid oxidation in beef and chicken muscle. *Food Chemistry* 104: 1575–1579.

Marcos, B., Aymerich, T., and Garriga, M. 2005. Evaluation of high pressure processing as an additional hurdle to control *Listeria monocytogenes* and *Salmonella enterica* in low-acid fermented sausages. *Journal of Food Science* 70: M339–M344.

Marcos, B., Aymerich, T., Monfort, J.M., and Garriga, M. 2008a. High pressure processing and antimicrobial biodegradable packaging to control *Listeria monocytogenes* during storage of cooked ham. *Food Microbiology* 25: 177–182.

Marcos, B., Jofre, A., Aymerich, T., Monfort, J.M., and Garriga, M. 2008b.Combined effect of natural antimicrobials and high pressure processing to prevent *Listeria monocytogenes* growth after a cold chain break during storage of cooked ham. *Food Control* 19: 76–81.

Mariutti, L.R.B., Orlien, V., Bragagnolo, N., and Skibsted, L.H. 2008. Effect of sage and garlic on lipid oxidation in high-pressure processed chicken meat. *European Food Research and Technology* 227: 337–344.

Masschalck, B., Houdt, R. van., and Michiels, C.W. 2001. High pressure increases bactericidal activity and spectrum of lactoferrin, lactoferricin and nisin. *International Journal of Food Microbiology* 64: 325–332.

Matser, A.M., Knott, E.R., Teunissen, P.G.M., and Bartels, P.V. 2000. Effects of high isostatic pressure on mushrooms. *Journal of Food Engineering* 45: 11–16.

Matser, A.M., Krebbers, B., Berg, R.W., and Bartels, P.V. 2004. Advantages of high pressure sterilization on quality of food products. *Trends in Food Science and Technology* 15: 79–85.

McInerney, J.K., Seccafien, C.A., Stewart, C.M., and Bird, A.R. 2007. Effects of high pressure processing on antioxidant activity, and total carotenoid content and availability, in vegetables. *Innovative Food Science and Emerging Technologies* 8: 543–548.

Messens, W., Van Camp, J., and Huyghebaert, A. 1997. The use of high pressure to modify functionality of food proteins. *Trends in Food Science and Technology* 8: 107–112.

Michel, M. and Autio, K. 2001. Effect of high pressure on protein and polysaccharide based structure. In M.E.G. Hendrix and D. Knorr (Eds.), *Ultra High Pressure Treatment of Foods*, pp. 189–241. New York: Kluwer Academic/Plenum Publishers.

Moatsu, G., Bakopanos, C., Katharios, D., Katsaros, G., Kandarakis, I., Taoukis, P., and Politis, I. 2008. Effect of high pressure treatment at various temperatures on indigenous proteolytic enzymes and whey protein denaturation in bovine milk. *Journal of Dairy Research* 75: 262–269.

Moio, L., Masi, P., Pietra, L. la., Cacace, D., Palmieri, L., Martino, E. de., Carpi, G., and Dall'Aglio, G. 1994. Stabilization of mustes by high pressure treatment. *Industrie delle Bevande* 23: 436–441.

Molina, E., Alvarez, M.D., Ramos, M., Olano, A., and Lopez, F.R. 2000. Use of high pressure treated milk for the production of reduced fat cheese. *International Dairy Journal* 10: 467–475.

Morales, P., Calzada, J., Avila, M., and Nunez, M. 2008. Inactivation of *Escherichia coli* O157:H7 in ground beef by single-cycle and multiple-cycle high-pressure treatments. *Journal of Food Protection* 71: 811–815.

Morgan, S.M., Ross, R.P., Beresford, T., and Hill, C. 2000. Combination of high hydrostatic pressure and lacticin 3147 causes increased killing of *Staphylococcus* and *Listeria*. *Journal of Applied Microbiology* 88: 414–420.

Mor-Mur, M. and Yuste, J. 2003. High pressure processing applied to cooked sausage manufacture: Physical properties and sensory analysis. *Meat Science* 65: 1187–1191.

Mor-Mur, M. and Yuste, J. 2005. Microbiological aspects of high pressure processing. In D.W. Sun (Ed.), *Emerging Technologies for Food Processing*, pp. 47–66. London, U.K.: Elsevier Academic Press.

Mueller, W.D. and Dederer, I. 2008. Spore inactivation in cooked sausage with combined high pressure and heat treatment. *Fleischwirtschaft* 88: 99–102.

Muench, S., Dederer, I., and Mueller, W.D. 2005. Effect of high pressure treatment on the formation of cholesterol oxides in cooked sausage cold cuts. *Fleischwirtschaft* 85: 123–126.

Mussa, D.M. and Ramaswamy, H.S. 1997. Ultra high pressure pasteurization of milk: Kinetics of microbial destruction and changes in physicochemical characteristics. *Lebensmittel-Wissenschaft und-Technologie* 30: 551–557.

Nakamura, T., Sado, H., and Syukunobe, Y. 1993. Production of low antigenic whey protein hydrolysates by enzymatic hydrolysis and denaturation with high pressure. *Milchwissenschaft* 48(3): 141–145.

Narisawa, N., Furukawa, S., Kawarai, T., Ohishi, K., Kanda, S., Kimijima, K., Negishi, S., Ogihara, H., and Yamasaki, M. 2008. Effect of skimmed milk and its fractions on the inactivation of *Escherichia coli* K12 by high hydrostatic pressure treatment. *International Journal of Food Microbiology* 124: 103–107.

Needs, E.C., Stenning, R.A., Gill, A.L., Ferragut, V., and Rich, G.T. 2000. High pressure treatment of milk: Effects on casein micelle structure and on enzyme coagulation. *Journal of Dairy Research* 67: 31–42.

Neetoo, H., Ye, M., and Chen, H. 2008. Potential application of high hydrostatic pressure to eliminate *Escherichia coli* O157:H7 on alfalfa sprouted seeds. *International Journal of Food Microbiology* 128: 348–353.

Nguyen, L.T., Rastogi, N. K., and Balasubramaniam, V.M. 2007. Evaluation of instrumental quality of pressure-assisted thermally processed carrots. *Journal of Food Science* 72: E264–E270.

Nienaber, U. and Shellhammer, T.H. 2001. High pressure processing of orange juice: Combination treatments and a shelf life study. *Journal of Food Science* 66: 332–336.

Noeckler, K., Heinz, V., Lemkau, K., and Knorr, D. 2001. Inactivation of *Trichinella spiralis* by high pressure treatment. *Fleischwirtschaft* 81: 85–88.

Novotna, P., Valentova, H., Strohalm, J., Kyhos, K., Landfeld, A., and Houska, M. 1999. Sensory evaluation of high pressure treated apple juice during its storage. *Czechoslovakia Journal of Food Science* 17: 196–198.

O'Brien, J.K. and Marshall, R.T. 1996. Microbiological quality of raw ground chicken processed at high hydrostatic pressure. *Journal of Food Protection* 59: 146–150.

Oey, I., Plancken, I. van der., Loey, A. van., and Hendrickx, M. 2008a. Does high pressure processing influence nutritional aspects of plant based food systems? *Trends in Food Science and Technology* 19: 300–308.

Oey, I., Lille, M., Loey, A. van., and Hendrickx, M. 2008b. Effect of high-pressure processing on color, texture and flavor of fruit- and vegetable-based food products: A review. *Trends in Food Science and Technology* 19: 320–328.

Ohmiya, K., Fukami, K., Shimizu, S., and Gekko, K. 1987. Milk curdling by rennet under high pressure. *Journal of Food Science* 52: 84–87.

O'Reilly, C.E., O'Connor, P.M., Kelly, A.L., Beresford, T.P., and Murphy, P.M. 2000. Use of high hydrostatic pressure for inactivation of microbial contaminants in cheese. *Applied and Environmenta*l Microbiology 66: 4890–4896.

O'Reilly, C.E., Kelly, A.L., Murphy, P.M., and Beresford, T.P. 2001. High-pressure treatment: Applications in cheese manufacture and ripening. *Trends in Food Science and Technology* 12: 51–59.

Orlien, V., Hansen, E., and Skibsted, L.H. 2000. Lipid oxidation in high pressure processed chicken breast muscle during chill storage: Critical working pressure in relation to oxidation mechanism. *European Food Research and Technology* 211: 99–104.

Otero, L. and Sanz, P.D. 2000. Modelling heat transfer in high pressure food processing: A review. *Innovative Food Science & Emerging Technologies* 4: 121–134.

Otero, L., Solas, M.T., Sanz, P.D., Elvira, C. de, and Carrasco, J.A. 1998. Contrasting effects of high pressure assisted freezing and conventional air freezing on eggplant tissue microstructure. *European Food Research and Technology* 206: 338–342.

Ouali, A.J. 1990. Meat tenderization possible causes and mechanisms: A review. *Journal of Muscle Foods* 1: 129.

Palou, E., Hernandez, S.C., Lopez, M.A., Barbosa-Canovas, G.V., Swanson, B.G., and Welti-Chanes, J. 2000. High pressure-processed guacamole. *Innovative Food Science and Emerging Technologies* 1: 69–75.

Pandey, P.K. and Ramaswamy, H.S. 1998. Effect of high pressure of milk on textural properties, moisture content and yield of Cheddar cheese. In Book of abstracts of *IFT Annual Meeting*, Atlanta, GA, pp. 173–174.

Pandey, P.K. and Ramaswamy, H.S. 2004. Effect of high-pressure treatment of milk on lipase and gamma-glutamyl transferase activity. *Journal of Food Biochemistry* 28: 449–462.

Pandey, P.K., Ramaswamy, H.S., and St. Gelasis, D. 2000. Water-holding capacity and gel strength of rennet curd as affected by high-pressure treatment of milk. *Food Research International* 33: 655–663.

Pandey, P.K., Ramaswamy, H.S., and Idziak, E. 2003. High pressure destruction kinetics of indigenous microflora and *Escherichia coli* in raw milk at two temperatures. *Journal of Food Process Engineering* 26: 265–283.

Parish, M.E. 1998a. High pressure inactivation of *Saccharomyces cerevisiae*, endogenous microflora and pectinmethylesterase in orange juice. *Journal of Food Safety* 18: 57–65.

Parish, M.E. 1998b. Orange juice quality after treatment by thermal pasteurization or isostatic high pressure. *Lebensmittel-Wissenschaft und-Technologie* 31: 439–442.

Park, S.H., Ryu, H.S., Hong, G.P., and Min, S.G. 2006. Physical properties of frozen pork thawed by high pressure assisted thawing process. *Food Science and Technology International* 12: 347–352.

Patras, A., Brunton, N.P., Pieve, S.D., and Butler, F. 2009a. Impact of high pressure processing on total antioxidant activity, phenolic, ascorbic acid, anthocyanin content and color of strawberry and blackberry purees. *Innovative Food Science and Emerging Technologies* 10: 308–313.

Patras, A., Brunton, N.P., Pieve, S.D., Butler, F., and Downey, G. 2009b. Effect of thermal and high pressure processing on antioxidant activity and instrumental color of tomato and carrot purees. *Innovative Food Science and Emerging Technologies* 10: 16–22.

Patterson, M.F., Quinn, M., Simpson, R., and Gilmour, A. 1995. Effect of high pressure on vegetable pathogens. In D.A. Ledward, D.E. Johnston, R.G. Earnshaw, and A.P.M. Hasting (Eds.), *High Pressure Processing of Foods*, pp. 47–63. Nottingham, U.K.: Nottingham University Press.

Penas, E., Gomez, R., Frias, J., and Vidal V.C. 2008. Application of high-pressure treatment on alfalfa (*Medicago sativa*) and mung bean (*Vigna radiata*) seeds to enhance the microbiological safety of their sprouts. *Food Control* 19: 698–705.

Penas, E., Gomez, R., Frias, J., and Vidal, V.C. 2009. Efficacy of combinations of high pressure treatment, temperature and antimicrobial compounds to improve the microbiological quality of alfalfa seeds for sprout production. *Food Control* 20: 31–39.

Penas, E., Restani, P., Ballabio, C., Prestamo, G., Fiocchi, A., and Gomez, R. 2006. Evaluation of the residual antigenicity of dairy whey hydrolysates obtained by combination of enzymatic hydrolysis and high-pressure treatment. *Journal of Food Protection* 69: 1707–1712.

Perez, M.M. and Montero, P. 2000. Response surface methodology multivariate analysis of properties of high pressure induced fish mince gel. *European Food Research and Technology* 211: 79–85.

Phunchaisri, C. and Apichartsrangkoon, A. 2005. Effects of ultra-high pressure on biochemical and physical modification of lychee (*Litchi chinensis* Sonn.). *Food Chemistry* 93: 57–64.

Picart, L., Dumay, E., Guiraud, J.P., and Cheftel, C. 2005. Combined high pressure–sub-zero temperature processing of smoked salmon mince: Phase transition phenomena and inactivation of *Listeria innocua*. *Journal of Food Engineering* 68: 43–56.

Plaza, L., Munoz, M., Ancos, B. de., and Cano, M.P. 2003. Effect of combined treatments of high pressure, citric acid and sodium chloride on quality parameters of tomato puree. *European Food Research and Technology* 216: 514–519.

Plaza, L., Sanchez, M.C., Elez, P., Ancos, B., Martin, B.O., and Cano, M.P. 2006a. Effect of refrigerated storage on vitamin C and antioxidant activity of orange juice processed by high-pressure or pulsed electric fields with regard to low pasteurization. *European Food Research and Technology* 223: 487–493.

Plaza, L., Sanchez-Moreno, C., De Ancos, B., and Cano, M.P. 2006b. Carotenoid content and antioxidant capacity of Mediterranean vegetable soup (gazpacho) treated by high-pressure/temperature during refrigerated storage. *European Food Research and Technology* 223: 210–215.

Polydera, A.C., Stoforos, N.G., and Taoukis, P.S. 2003. Comparative shelf life study and vitamin C loss kinetics in pasteurized and high pressure processed reconstituted orange juice. *Journal of Food Engineering* 60: 21–29.

Polydera, A.C., Stoforos, N.G., and Taoukis, P.S. 2004. The effect of storage on the antioxidant activity of reconstituted orange juice which had been pasteurized by high pressure or heat. *International Journal of Food Science and Technology* 39: 783–791.

Polydera, A.C., Stoforos, N.G., and Taoukis, P.S. 2005. Effect of high hydrostatic pressure treatment on post processing antioxidant activity of fresh navel orange juice. *Food Chemistry* 91: 495–503.

Pozo, I.D., Follo, M.A., Talcott, S.T., and Brenes, C.H. 2007. Stability of copigmented anthocyanins and ascorbic acid in muscadine grape juice processed by high hydrostatic pressure. *Journal of Food Science* 72: S247–S253.

Prestamo, G. and Arroyo, G. 1998. High hydrostatic pressure effects on vegetable structure. *Journal of Food Science* 63: 878–881.

Prestamo, G. and Arroyo, G. 2000. Preparation of preserves with fruits treated by high pressure. *Alimentaria* 318: 25–30.

Prestamo, G., Palomares, L., and Sanz, P. 2004. Broccoli (*Brasica oleracea*) treated under pressure-shift freezing process. *European Food Research and Technology* 219: 598–604.

Qin, H., Nan, Q.X., and Che, R.Z. 2001. Effects of high pressure on the activity of major enzymes in beef. *Meat Research* 3: 13–16.

Qiu, W.F., Jiang, H.H., Wang, H.F., and Gao, Y.L. 2006. Effect of high hydrostatic pressure on lycopene stability. *Food Chemistry* 97: 516–523.

Quaglia, G.B., Gravina, R., Paperi, R., and Paoletti, F. 1996. Effect of high pressure-treatments on peroxidase activity ascorbic acid content and texture in green peas. *Lebensmittel-Wissenschaft und-Technologie* 29: 552–555.

Rademacher, B. and Kessler, H.G. 1997. High pressure inactivation of microorganisms and enzymes in milk and milk products. In K. Heremans (Ed.), *High Pressure Research in Biosciences and Biotechnology*, pp. 291–293. Leuven, Belgium: Leuven University Press.

Rajan, S., Ahn, J., Balasubramaniam, V.M., and Yousef, A.E. 2006a. Combined pressure-thermal inactivation kinetics of *Bacillus amyloliquefaciens* spores in egg patty mince. *Journal of Food Protection* 69: 853–860.

Rajan, S., Pandrangi, S., Balasubramaniam, V.M., and Yousef, A.E. 2006b. Inactivation of *Bacillus stearothermophilus* spores in egg patties by pressure-assisted thermal processing. *Lebensmittel-Wissenschaft und-Technologie* 39: 844–851.

Ramaswamy, H.S., Zaman, S.U., and Smith, J.P. 2008. High pressure destruction kinetics of *Escherichia coli* (O157:H7) and *Listeria monocytogenes* (Scott A) in a fish slurry. *Journal of Food Engineering* 87: 99–106.

Ramaswamy, H.S., Jin, H., and Zhu, S. 2009. Effects of fat, casein and lactose on high-pressure destruction of *Escherichia coli* K12 (ATCC-29055) in milk. *Food and Bioproducts Processing* 87: 1–6.

Ramirez-Saurez, J.C. and Morrissey, M.T. 2006. Effect of high pressure processing (HPP) on shelf-life of albacore tuna (*Thunnus alalunga*) minced muscle. *Innovative Food Science and Emerging Technologies* 7: 19–27.

Rasanayagam, V., Balasubramaniam, V.M., Ting, E., Sizer, C.E., Bush, C., and Anderson, C. 2003. Compression heating of selected fatty food materials during high-pressure processing. *Journal of Food Science* 68: 254–259.

Rastogi, N.K. and Niranjan, K. 1998. Enhanced mass transfer during osmotic dehydration of high pressure treated pineapples. *Journal of Food Science* 63: 508–511.

Rastogi, N.K., Subramanian, R., and Raghavarao, K.S.M.S. 1994. Application of high pressure technology in food processing. *Indian Food Industry* 13: 30–34.

Rastogi, N.K., Eshtiaghi, M.N., and Knorr, D. 1999. Effect of combined high pressure and heat treatment on the reduction of peroxidase and polyphenoloxidase activity in red grape. *Food Biotechnology* 13: 195–208.

Rastogi, N.K., Angersbach, A., Niranjan, K., and Knorr, D. 2000a. Rehydration kinetics of high pressure treated and osmotically dehydrated pineapple. *Journal of Food Science* 65: 838–841.

Rastogi, N.K., Angersbach, A., and Knorr, D. 2000b. Combined effect of high hydrostatic pressure pretreatment and osmotic stress on mass transfer during osmotic dehydration. *Proceedings of the 8th International Congress of Food and Engineering* (ICFE'*8*), Pubela, Mexico, April 9–13, 2000.

Rastogi, N.K., Angersbach, A., and Knorr, D. 2000c. Synergistic effect of high hydrostatic pressure pretreatment and osmotic stress on mass transfer during osmotic dehydration. *Journal of Food Engineering* 45: 25–31.

Rastogi, N.K., Angersbach, A., and Knorr, D. 2003. Combined effect of high hydrostatic pressure pretreatment and osmotic stress on mass transfer during osmotic dehydration. In J. Welti-Chanes, J.F.V. Ruis, and G.V. Barbosa-Canovas (Eds.), *Transport Phenomena in Food Processing*, pp. 109–121. Boca Raton, FL: CRC Press.

Rastogi, N.K., Raghavarao, K.S.M.S., Balasubramaniam, V.M., Niranjan, K., and Knorr, D. 2007. Opportunities and challenges in high pressure processing of foods. *Critical Reviews in Food Science and Nutrition* 47: 69–112.

Rastogi, N.K., Nguyen, L.T., and Balasubramaniam, V.M. 2008a. Effect of pretreatments on carrot texture after thermal and pressure-assisted thermal processing. *Journal of Food Engineering* 88: 541–547.

Rastogi, N.K., Nguyen, L.T., and Balasubramaniam, V.M. 2008b. Improvement in texture of pressure-assisted thermally processed carrots using response surface methodology. *Food and Bioprocess Technology* doi.org/10.1007/s11947-008-0130-6.

Ritz, M., Jugiau, F., Federighi, M., Chapleau, N., and Lamballerie, M. de. 2008. Effects of high pressure, subzero temperature, and pH on survival of *Listeria monocytogenes* in buffer and smoked salmon. *Journal of Food Protection* 71: 1612–1618.

Rodrigo, D., Loey, A. van., and Hendrickx, M. 2007. Combined thermal and high pressure color degradation of tomato puree and strawberry juice. *Journal of Food Engineering* 79: 553–560.

Roeck, A.D., Sila, D.N., Duvetter, T., Loey, A.V., and Hendrickx, M. 2008. Effect of high pressure/high temperature processing on cell wall pectic substances in relation to firmness of carrot tissue. *Food Chemistry* 107: 1225–1235.

Roeck, A.D., Duvetter, T., Fraeye, I., Plancken, I.V., Sila, D.N., Loey, A.V., and Hendrickx, M. 2009. Effect of high-pressure/high-temperature processing on chemical pectin conversions in relation to fruit and vegetable texture. *Food Chemistry* 115: 207–213.

Roldan, M.E., Sanchez, M.C., Lloria, R., Ancos, B., and Cano, M.P. 2009. Onion high-pressure processing: Flavonol content and antioxidant activity. *Lebensmittel-Wissenschaft und-Technologie* 42: 835–841.

Rouille, J., Lebail, A., Ramaswamy, H.S., and Leclerc, L. 2002. High pressure thawing of fish and shellfish. *Journal of Food Engineering* 53: 83–88.

Sainz, C.B., Younce, F.L., Rasco, B., and Clark, S. 2009. Protease stability in bovine milk under combined thermal-high hydrostatic pressure treatment. *Innovative Food Science and Emerging Technologies* 10: 314–320.

Saldo, J., McSweeney, P.L.H., Sendra, E., Kelly, A.L., and Guamis, B. 2002. Proteolysis in caprine milk cheese treated by high pressure to accelerate cheese ripening. *International Dairy Journal* 12: 35–44.

Sampedro, F., Rodrigo, D., and Hendrickx, M. 2008. Inactivation kinetics of pectin methyl esterase under combined thermal high pressure treatment in an orange juice milk beverage. *Journal of Food Engineering* 86: 133–139.

Sanchez, M.C., Plaza, L., Ancos, B., and Cano, M.P. 2003. Effect of high-pressure processing on health-promoting attributes of freshly squeezed orange juice (*Citrus sinensis* L.) during chilled storage. *European Food Research and Technology* 216: 18–22.

Sanchez, M.C., Plaza, L., Ancos, B., and Cano, M.P. 2004. Effect of combined treatments of high-pressure and natural additives on carotenoid extractability and antioxidant activity of tomato puree (*Lycopersicum esculentum* Mill.). *European Food Research and Technology* 219: 151–160.

Sanchez, M.C., Plaza, L., Elez, M.P., de Ancos, B., Martin, B.O., and Cano, M.P. 2005. Impact of high-pressure and pulsed electric field on bioactive compounds and antioxidant activity of orange juice in comparison with traditional thermal processing. *Journal of Agricultural and Food Chemistry* 53: 4403–4409.

Sanchez, M.C., Plaza, L., de Ancos, B., and Cano, M.P. 2006. Impact of high-pressure and traditional thermal processing of tomato puree on carotenoids, vitamin C and antioxidant activity. *Journal of the Science of Food and Agriculture* 86: 171–179.

San-Martin, M.F.G., Welti-Chanes, J., and Barbosa-Canovas, G.V. 2006. Cheese manufacture assisted by high pressure. *Food Reviews International* 22: 275–289.

Schubring, R., Meyer, C., Schlueter, O., Boguslawski, S., and Knorr, D. 2003. Impact of high pressure assisted thawing on the quality of fillets from various fish species. *Innovative Food Science and Emerging Technologies* 4: 257–267.

Scollard, P.G., Beresford, T.P., Needs, E.C., Murphy, P.M., and Kelly, A.L. 2000. Plasmin activity, beta lactoglobulin denaturation and proteolysis in high pressure treated milk. *International Dairy Journal* 10: 835–841.

Sellahewa, J. 2002. Shelf life extension of orange juice using high pressure processing. *Fruit Processing*12: 344–350.

Seok, I.H. and Dong, M.K. 2001. Storage quality of chopped garlic as influenced by organic acids and high-pressure treatment. *Journal of the Science of Food and Agriculture* 81: 397–403.

Sequeira, A.M., Chevalier, D., LeBail, A., Ramaswamy, H.S., and Simpson, B.K. 2006. Physicochemical changes induced in carp (*Cyprinus carpio*) fillets by high pressure processing at low temperature. *Innovative Food Science and Emerging Technologies* 7: 13–18.

Serra, X., Grebol, N., Guardia, M.D., Guerrero, L., Gou, P., Masoliver, P., Gassiot, M., Sarraga, C., Monfort, J.M., and Arnau, J. 2007. High pressure applied to frozen ham at different process stages. 2. Effect on the sensory attributes and on the color characteristics of dry-cured ham. *Meat Science* 75: 21–28.

Shah, N.P., Tsangalis, D., Donkor, O.N., and Versteeg, C. 2008. Effect of high pressure treatment on viability of *Lactobacillus delbrueckii* ssp. *bulgaricus*, *Streptococcus thermophilus* and *L. acidophilus* and the pH of fermented milk. *Milchwissenschaft* 63: 11–14.

Sheehan, J.J., Huppertz, T., Hayes, M.G., Kelly, A.L., Beresford, T.P., and Guinee, T.P. 2005. High pressure treatment of reduced-fat mozzarella cheese: Effects on functional and rheological properties. *Innovative Food Science and Emerging Technologies* 6: 73–81.

Shigehisa, T., Ohmori, T., Saito, A., Taji, S., and Hayashi, R. 1991. Effect of high hydrostatic pressure on characteristics of pork slurries and inactivation of microorganisms associated with meat and meat products. *International Journal of Food Microbiology* 12: 207–216.

Sierra, I., Vidal, V.C., and Lopez, F.R. 2000. Effect of high pressure on the vitamin B_1 and B_6 content of milk. *Milchwissenschaft* 55: 365–367.

Sila, D.N., Smout, C., Vu, T.S., and Hendrickx, M.E. 2004. Effects of high-pressure pretreatment and calcium soaking on the texture degradation kinetics of carrots during thermal processing. *Journal of Food Science* 69: E205–E211.

Sila, D.N., Smout, C., Vu, S.T., Van Loey, A.M., and Hendrickx, M. 2005. Influence of pretreatment conditions on the texture and cell wall components of carrots during thermal processing. *Journal of Food Science* 70: E85–E91.

Sila, D.N., Duvetter, T., Roeck, A. de., Verlent, I., Smout, C., Moates, G.K., Hills, B.P., Waldron, K.K., Hendrickx, M., and Loey, A. van. 2008. Texture changes of processed fruits and vegetables: Potential use of high pressure processing. *Trends in Food Science and Technology* 19: 309–319.

Slongo, A.P., Rosenthal, A., Camargo, L.M.Q., Deliza, R., Mathias, S.P., and Aragao, G.M.F. 2009. Modeling the growth of lactic acid bacteria in sliced ham processed by high hydrostatic pressure. *Lebensmittel-Wissenschaft und-Technologie* 42: 303–306.

Smelt, J.P.P.M. 1998. Recent advances in the microbiology of high pressure processing. *Trends in Food Science and Technology* 9: 152–158.

Smiddy, M., O'Gorman, L., Sleator, D.R., Kerry, P.J., Patterson, F.M., Kelly, L.A., and Hill, C. 2005. Greater high-pressure resistance of bacteria in oysters than in buffer. *Innovative Food Science and Emerging Technologies* 6: 83–90.

Sopanangkul, A., Ledward, D.A., and Niranjan, K. 2002. Mass transfer during sucrose infusion into potatoes under high pressure. *Journal of Food Science* 67: 2217–2220.

Strolham, J., Valentova, H., Houska, M., Novotna, P., Landfeld, A., Kyhos, K., and Gree, R. 2000. Changes in quality of natural orange juice pasteurized by high pressure during storage. *Czechoslovakia Journal of Food Science* 18: 187–193.

Stute, R., Eshtiagi, M., Boguslawski, S., and Knorr, D. 1996. High pressure treatment of vegetables. In P.R. Rohr and C. Trepp (Eds.), *High Pressure Chemical Engineering*, pp. 271–276. New York: Elsevier Science.

Styles, M.F., Hoover, D.G., and Farkas, D.F. 1991. Response of *Listeria monocytogenes* and *Vibrio parahaemolyticus* to high hydrostatic pressure. *Journal of Food Science* 56: 1404–1407.

Sumitani, H., Suekane, S., Nakatani, A., and Tatsuka, K. 1994. Changes in composition of volatile compounds in high pressure treated peach. *Journal of Agricultural and Food Chemistry* 42: 785–790.

Supavititpatana, T. and Apichartsrangkoon, A. 2007. Combination effects of ultra-high pressure and temperature on the physical and thermal properties of ostrich meat sausage (yor). *Meat Science* 76: 555–560.

Suzuki, A., Kim, K., Homma, N., Ikeuchi, Y., and Saito, M. 1992. Acceleration of meat conditioning by high pressure treatment. In C. Balny, R. Hayashi, K. Heremans, and P. Mason (Eds.), *High Pressure Biotechnology*, pp. 219–227. Paris, France: Coll. Inserm.

Suzuki, A., Watanabe, M., Ikeuchi, Y., Saito, M., and Takahashi, K. 1993. Effects of high pressure treatment on the ultrastructure and thermal behavior of beef intramuscular collagen. *Meat Science* 35: 17–25.

Takahashi, F., Pehrsson, P.E., Rovere, P., and Squarcina, N. 1998. High-pressure processing of fresh orange juice. *Industria Conserve* 73: 363–368.

Tangwongchai, R., Ledward, D.A., and Ames, J.A. 2000. Effect of high pressure treatment on the texture of cherry tomato. *Journal of Agricultural and Food Chemistry* 48: 1434–1441.

Tanzi, E., Saccani, G., Barbuti, S., Grisenti, M.S., Lori, D., Bolzoni, S., and Parolari, G. 2004. High pressure treatment of raw ham. Sanitation and impact on quality. *Industria Conserve* 79: 37–50.

Tewari, G., Jayas, D.S., and Holley, R.A. 1999. High pressure processing of foods: An overview. *Sciences des Aliments* 19: 619–661.

Thakur, B.R. and Nelson, P.E. 1998. High pressure processing and preservation of foods. *Food Reviews International* 14: 427–447.

Ting, E., Balasubramaniam, V.M., and Raghubeer, E. 2002. Determining thermal effects in high-pressure processing. *Food Technology* 56: 31–35.

Trespalacios, P. and Pla, R. 2007. Synergistic action of transglutaminase and high pressure on chicken meat and egg gels in absence of phosphates. *Food Chemistry* 104: 1718–1727.

Trujillo, A.J., Royo, B., Guamis, B., and Ferragut, V. 1999a. Influence of pressurization on goat milk and cheese composition and yield. *Milchwissenschaft* 54: 197–199.

Trujillo, A.J., Royo, C., Ferragut, V., and Guamis, B. 1999b. Ripening profiles of goat cheese produced from milk treated with high pressure. *Journal of Food Science* 64: 833–837.

Uresti, R.M., Velazquez, G., Ramirez, J.A., Vazquez, M., and Antonio-Torres, J. 2004. Effect of high-pressure treatments on mechanical and functional properties of restructured products from arrowtooth flounder (*Atheresthes stomias*). *Journal of the Science of Food and Agriculture* 84: 1741–1749.

Uresti, R.M., Velazquez, G., Vazquez, M., Ramirez, J.A., and Torres, J.A. 2005. Restructured products from arrowtooth flounder (*Atheresthes stomias*) using high-pressure treatments. *European Food Research and Technology* 220: 113–119.

Uresti, M.R., Velazquez, G., Vazquez, M., Ramirez, A.J., and Torres, J.A. 2006. Effects of combining microbial transglutaminase and high pressure processing treatments on the mechanical properties of heat-induced gels prepared from arrowtooth flounder (*Atheresthes stomias*). *Food Chemistry* 94: 202–209.

Vachon, J.F., Kheadr, E.E., Giasson, J., Paquin, P., and Fliss, I. 2002. Inactivation of foodborne pathogens in milk using dynamic high pressure. *Journal of Food Protection* 65: 345–352.

Vazquez, P.A.L., Torres, J.A., and Qian, M.C. 2006. Effect of high pressure moderate temperature processing on the volatile profile of milk. *Journal of Agricultural and Food Chemistry* 54: 9184–9192.

Vazquez, P.A.L., Qian, M.C., and Torres, J.A. 2007. Kinetic analysis of volatile formation in milk subjected to pressure assisted thermal treatments. *Journal of Food Science* 72: E389–E398.

Verlinde, P., Oey, I., Hendrickx, M., and Loey, A. van. 2008. High-pressure treatments induce folate polyglutamate profile changes in intact broccoli (*Brassica oleraceae* L. cv. Italica) tissue. *Food Chemistry* 111: 220–229.

Viazis, S., Farkas, B.E., and Jaykus, L.A. 2008. Inactivation of bacterial pathogens in human milk by high-pressure processing. *Journal of Food Protection* 71: 109–118.

Villacis, M.F., Rastogi, N.K., and Balasubramaniam, V.M. 2008. Effect of high pressure on moisture and NaCl diffusion into turkey breast. *Lebensmittel-Wissenschaft und-Technologie* 41: 836–844.

Walker, M.K., Farkas, D.F., Loveridge, V., and Meunier-Goddik, L. 2006. Fruit yogurt processed with high pressure. *International Journal of Food Science and Technology* 41: 464–467.

Watanabe, M., Arai, E., Kumeno, K., and Honma, K. 1991. A new method for producing a non heated jam sample: The use of freeze concentration and high pressure sterilization. *Agricultural and Biological Chemistry* 55: 2175–2176.

Wen, C.K. and Kuo, C.H. 2001. Changes in K value and microorganisms of tilapia fillet during storage at high pressure, normal temperature. *Journal of Food Protection* 64: 94–98.

Wennberg, M. and Nyman, M. 2004. On the possibility of using high pressure treatment to modify physico-chemical properties of dietary fibre in white cabbage (*Brassica oleracea* var. capitata). *Innovative Food Science and Emerging Technologies* 5: 171–177.

Wiggers, S.B., Ohlson, M.V.K., and Skibsted, L.H. 2004. Lipid oxidation in high-pressure processed chicken breast during chill storage and subsequent heat treatment: Effect of working pressure, packaging atmosphere and storage time. *European Food Research and Technology* 219: 167–170.

Wilson, D.R., Dabrowski, L., Stringer, S., Moezelaar, R., and Brocklehurst, T.F. 2008. High pressure in combination with elevated temperature as a method for the sterilization of food. *Trends in Food Science and Technology* 19: 289–299.

Wolbang, C.M., Fitos, J.L., and Treeby, M.T. 2008. The effect of high pressure processing on nutritional value and quality attributes of *Cucumis melo* L. *Innovative Food Science and Emerging Technologies* 9: 196–200.

Yagiz, Y., Kristinsson, H.G., Balaban, M.O., and Marshall, M.R. 2007. Effect of high pressure treatment on the quality of rainbow trout (*Oncorhynchus mykiss*) and mahi mahi (*Coryphaena hippurus*). *Journal of Food Science* 72: C509–C515.

Yokohama, H., Swamura, N., and Motobyashi, N. 1992. Method for accelerating cheese ripening. European Patent Application EP 0 469 857 A1, filed July 30, 1991 and issued February 5, 1992.

Yuste, J., Pla, R., Ponce, C.E., and Mor-Mur, M. 2000. High pressure processing applied to cooked sausages: Bacterial populations during chilled storage. *Journal of Food Protection* 63: 1093–1099.

Yuste, J., Pla, R., Capellas, M., Sendra, E., Beltran, E., and Mor-Mur, M. 2001. Oscillatory high pressure processing applied to mechanically recovered poultry meat for bacterial inactivation. *Journal of Food Science* 66: 482–484.

Yuste, J., Pla, R., Capellas, M., and Mor Mur, M. 2002. Application of high pressure processing and nisin to mechanically recovered poultry meat for microbial decontamination. *Food Control* 13: 451–455.

Yutang, W., Jingyan, Y., Yong, Y., Chenling, Q., Huarong, Z., Lan, D., Hanqi, Z., and Xuwen, L. 2008. Analysis of ginsenosides in *Panax ginseng* in high pressure microwave-assisted extraction. *Food Chemistry* 110: 161–167.

Zhu, S., Le Bail, A., Ramaswamy, H.S., and Chapleau, N. 2004a. Characterization of ice crystals in pork muscle formed by pressure-shift freezing as compared with classical freezing methods. *Journal of Food Science* 69: E190–E197.

Zhu, S.M., Le Bail, A., Chapleau, N., Ramaswamy, H.S., and Lamballerie, A.M. 2004b. Pressure shift freezing of pork muscle: Effect on color, drip loss, texture, and protein stability. *Biotechnology Progress* 20: 939–945.

Zhu, S., Ramaswamy, H.S., and Simpson, B.K. 2004c. Effect of high-pressure versus conventional thawing on color, drip loss and texture of Atlantic salmon frozen by different methods. *Lebensmittel-Wissenschaft und-Technologie* 37: 291–299.

Zhu, S., Naim, F., Marcotte, M., Ramaswamy, H., and Shao, Y. 2008. High pressure destruction kinetics of *Clostridium sporogenes* spores in ground beef at elevated temperatures. *International Journal of Food Microbiology* 126: 86–92.

Zobrist, M.R., Huppertz, T., Uniacke, T., Fox, P.F., and Kelly, A.L. 2005. High pressure induced changes in the rennet coagulation properties of bovine milk. *International Dairy Journal* 15: 655–662.

6 Physicochemical Property Changes and Safety Issues of Foods during Pulsed Electric Field Processing

Malek Amiali, Michael O. Ngadi, Arun Muthukumaran, and G. S. Vijaya Raghavan

CONTENTS

6.1 INTRODUCTION

Conventional thermal pasteurizations including high-temperature-short-time (HTST), ultrahigh-temperature (UHT), and low-temperature-long-time (LTLT) methods continue to be used to increase the shelf life of food and maintain the safety of many food products. Although effective for food spoilage and pathogenic bacterial inactivation, the level of temperature (>60°C) used in these technologies could adversely affect the organoleptic properties and nutritional values of several foods. Consumer demand for foods that retain the flavor, taste, and nutritional values of the fresh form has pushed researchers to develop reduced-temperature food processing techniques.

Emerging nonthermal technologies such as pulsed electric field (PEF) is a promising technique that can replace or supplement thermal pasteurization in a hurdle process. Unlike ohmic heating and other electrical processes, which involve the passage of continuous electrical current throughout a food material in order for the product to be heated evenly, PEF is not intended for heating. On the other hand, the application of short pulses of high electric field (EF) to food placed in a treatment chamber destroys unwanted microorganisms without significant heat generation. EF intensities in the 20–80 kV cm^{-1} range are used to achieve desirable pasteurization of liquid food products (FDA, 2000a,b). The inactivation of microorganisms exposed to PEF results from electromechanical destabilization or electroplasmolysis of the cell membrane (Zimmermann, 1986; McLellan et al., 1991; Castro et al., 1993) due to the large flux of current flowing through food when a high pulse of EF is generated, which may vary in frequency from 1 to 150 pulses per second (pps). One reported advantage of PEF processing is minimal loss of volatile compounds and color during the treatment and subsequent storage compared to thermal processing (Barbosa-Cánovas et al., 1999).

PEF processing has been successful in treating a variety of food products such as fruit juices, milk, and liquid egg products (Barbosa-Cánovas et al., 1999; Amiali, 2005). In addition, PEF technologies have also been shown to be applicable for drying and microstructural modification of vegetable, fish, and meat tissues; intensification of juice yield and increase of product quality in juice production; processing of vegetable raw materials; winemaking; and sugar production (Ngadi and Bazhal, 2006).

In this chapter, a review on various PEF processing technologies with an emphasis on the safety issues of the treated foods is presented. For a review on the effects of other processing technologies on microbial growth and inactivation, the reader is referred to Chapter 4.

6.2 PULSED ELECTRIC FIELD SYSTEM

Various designs of PEF systems have been reported depending on the pulse characteristics and applications, high-voltage intensities, and types of treatment chambers (static or continuous). In general, a test apparatus for the PEF process consists of a number of components including a high-voltage power source; an energy storage capacitor bank; a switch; a treatment chamber(s); voltage, current, and temperature probes; a pump to transport food through the treatment chamber(s) (in case of a continuous system); cooling devices; and a computer or control panel to control the operation (Figure 6.1).

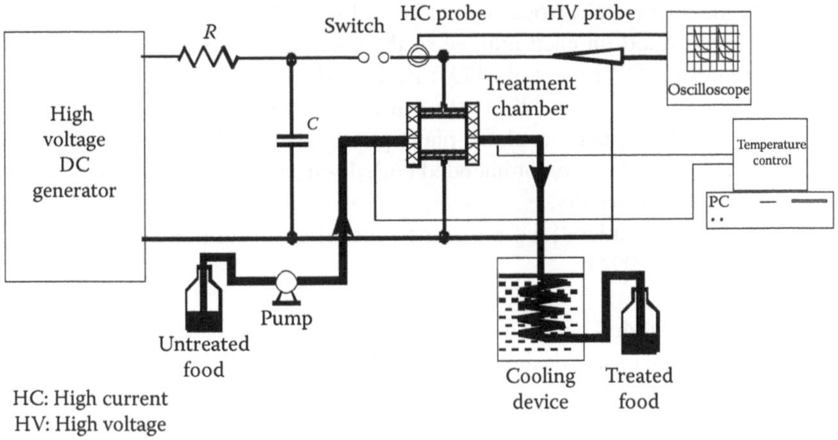

FIGURE 6.1 Major components of a pulsed electric field system. (From Amiali, M., Inactivation of *Escherichia coli* O157:H7 and *Salmonella enteritidis* in liquid egg products using pulsed electric field, PhD dissertation, McGill University, Montreal, Quebec, Canada, 2005.)

6.2.1 Test Chamber Design

A treatment chamber consists of two electrodes, held in position by insulating material that also forms an enclosure containing liquid food to be treated (Qin et al., 1996). The major function of the treatment chamber is to transfer the PEFs to the food. To obtain high field intensity, it is necessary to minimize local field enhancements since such regions increase the probability of dielectric breakdown inside the treatment chamber. An ideal chamber design ensures that all cells are exposed to the same EF intensity, number of pulses, and temperature (Qin et al., 1998). Several designs of static and continuous chambers have been suggested (Zhang et al., 1995a; Barbosa-Cánovas et al., 1999).

6.2.1.1 Static Treatment Chamber

The first systems were static and were designed for laboratory-scale experiments to treat small treatment volumes (liquids or solids). Most treatment chambers of this type have parallel electrodes of stainless steel or carbon, imbedded in different kinds of materials, with spacers made of Plexiglas, polysulfone, or nylon. The construction also requires filling and withdrawing ports and cooling jackets on the electrodes. By using different spacers, it is possible to adjust the distance between electrodes. Each treatment chamber has its own EF distribution; consequently, comparison of differing results reported in the literature is difficult.

Circular parallel electrodes have been widely used in PEF food pasteurization. This type of electrodes offers a uniform EF distribution along the gap axes and electrode surface, but creates a field enhancement problem at the edges of the electrodes. Some part of the energy used for treatment is also transformed into heat, which results in undesirable temperature increase of liquid products. These problems could be avoided by using rounded-surface electrodes in contact with the treatment region and insulate the surface electrodes by perforated polyethylene film; the EF intensity would be concentrated on each pinhole of the electrode surfaces, which raises the inactivation energy (Oshima et al., 1997; Barbosa-Cánovas et al., 1999; Sato et al., 2001; Sato, 2008). Sato et al. (2001) studied the energy efficiency of five kinds of electrodes, viz., plate–plate, insulated plate–plate, needle–plate, ring–cylinder, and coiled wire–cylinder. They recommended coiled wire system because it gave better microbial inactivation results.

The following considerations are important when designing a PEF treatment chamber (Qin et al., 1994):

1. Electrical breakdown in liquid foods is very sensitive to local enhanced EFs within a chamber. By designing a proper chamber, EF enhancement points can be located outside of the treatment region.
2. Washable and autoclavable materials must be used in constructing a treatment chamber. Polysulfone and stainless steel materials are recommended as insulation and electrode materials, respectively.
3. Filling and removal port must be easily accessible. Gas bubbles can become trigger sites for dielectric breakdown. Therefore, the filling port must facilitate air expulsion during filling.
4. When repeatedly applying high EF, the energy input must be considered. Cooling of the electrodes is required to maintain low-temperature operation, which can be done by circulating water through a jacket built onto the electrodes.

Further electrode construction recommendations were given by Bushnell et al. (1993). These include using inert materials such as platinum, gold, or metal oxides. The EF across the two electrodes must be uniform, so it is necessary to have a gap smaller than the dimensions of the electrodes. The final design of a treatment chamber should be done in conjunction with mathematical modeling of the product conductivity dependence on temperature (Qin et al., 1995b; Zhang et al., 1995a; Amiali et al., 2006a).

6.2.1.2 Continuous Treatment Chamber

Continuous treatment chambers are used for pilot or large-scale operation of liquid food pasteurization. However, the design differs according to the chamber size, capacity, and shapes of electrodes.

Coaxial and cofield are the most used PEF treatment chambers in food processing. In a coaxial treatment chamber, the EF is perpendicular to food, whereas in a cofield design, liquid food is exposed in parallel to EF strength (Min et al., 2007). The advantage of cofield treatment chamber is the ability to connect a series of chambers (up to 8 chambers) in parallel with EF intensity for a commercial-scale PEF system. A new treatment chamber design by Alkhafaji and Farid (2007) was proposed to operate with high EF intensities with limited increase in liquid temperature and fouling of electrodes. The inactivation rate of *Escherichia coli* suspended in simulated milk ultrafiltrate (SMUF) was up to 6-log reduction and treatment temperature did not exceed 38°C at a maximum EF of 49.6 kV cm^{-1}.

6.3 FACTORS AFFECTING EFFECTIVENESS OF PEF TREATMENT

6.3.1 Generation of Different Voltage Wave Shapes

PEF may be applied in the form of exponential decay, square wave, bipolar, instant charge reversal, or oscillatory pulses depending on the circuit design. In general, a direct current (DC) power supply charges a capacitor bank connected in series with a charging resistor (RC). When a trigger signal is applied, the charge stored in the capacitor flows through a food in the treatment chamber.

Oscillatory decay pulses are the least efficient in terms of microbial inactivation as they prevent a microbial cell from being continuously exposed to high-intensity EF for an extended period of time, thus preventing the cell membrane from irreversible breakdown over a large area (Jeyamkondan et al., 1999). An exponential decay voltage wave is a unidirectional voltage that rises rapidly to a maximum value and decays slowly to zero. Therefore, food is subjected to the peak voltage for a short period of time. Hence, exponential decay pulses have a long tail with a low EF, during which excess heat is generated in the food without an antimicrobial effect (Zhang et al., 1995a).

Zhang et al. (1997), while comparing square and exponential decay pulses, found that square wave pulses were most effective in extending a product shelf life. The square waveform can be obtained by using a pulse-forming network (PFN) consisting of an array of capacitors and inductors or long coaxial cable and solid-state switch devices. However, the difficulty of high-voltage square waves lies in matching the load resistance of food (R_L) to the characteristic impedance of the transmission line (Z_0). By matching impedances, a higher energy transfer to the treatment chamber can be obtained. Therefore, it is important to determine the resistance R_L in order to process food properly (Barsotti et al., 1999a).

Bipolar pulses are more lethal than any monopolar pulses (square or exponential decay) because the PEF causes movement of charged molecules in the cell membrane of microorganisms; a reversal in the orientation or polarity of the EF causes a corresponding change in the direction of charged molecules (Barbosa-Cánovas et al., 1999). The alternating changes in the movement of charged molecules in bipolar pulses cause stresses in the cell membrane and enhance cell lysis. However, the change in polarity of the pulses is obtained by alternating with the periods of relaxation, thus preventing the cells from being continuously subjected to a high EF. To overcome this problem, Ho et al. (1995) proposed instant reversal pulses where the charge was partially positive at first and partially negative immediately thereafter. The inactivation effect of the instant reversal pulses was believed to be due to a significant alternating stress on microbial cell, which caused structural fatigue. This higher killing effect of instant charge reversal pulses, compared to other pulse types, could save between 17% and 20% of the total energy and equipment costs.

Based on the reports of Ho et al. (1995) and Zhang et al. (1997) that square wave and instant reversal pulses are the most efficient pulse configurations for microbial inactivation, commercialized pulse generators made by Samtech Ltd. (Glasgow, United Kingdom), which can deliver instant reversal square waves, were tested by the Food Engineering group at McGill University (Department of Bioresource

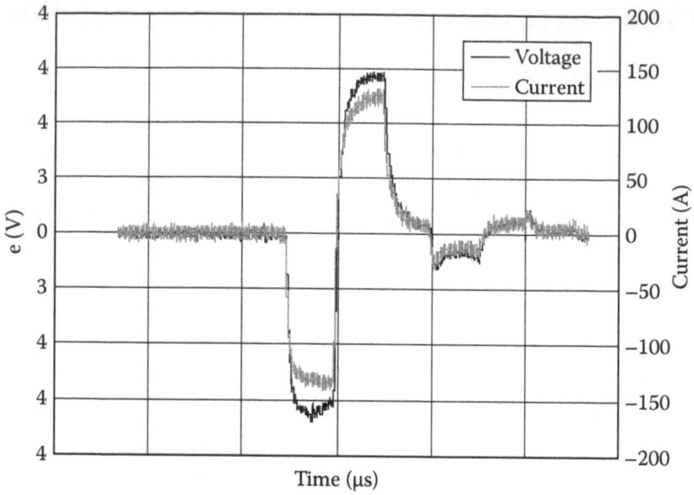

FIGURE 6.2 Instant-reversal biphase waveform used by the Food Engineering group at McGill University. (From Amiali, M., Inactivation of *Escherichia coli* 0157:H7 and *Salmonella enteritidis* in liquid egg products using pulsed electric field, PhD Dissertation, McGill University, Montreal, Quebec, Canada, 2005.)

Engineering). This kind of waveform seems to be more efficient than others in terms of liquid food pasteurization, since it combines instant reversal charge and square waveform pulses (Figure 6.2). By using this kind of waveform, Amiali et al. (2006b, 2007) reported up to 5-log reduction of *E. coli* and *Salmonella enteritidis* suspending in liquid egg product (see Figures 6.3 and 6.4).

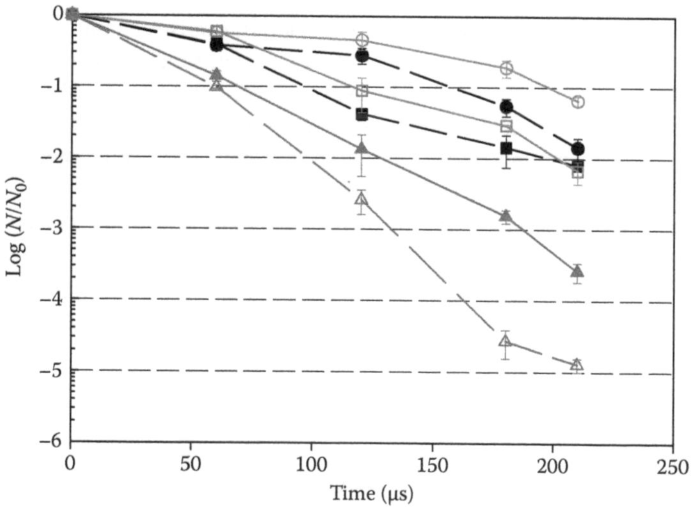

FIGURE 6.3 Survival fraction of *E. coli* O157:H7 in liquid egg yolk as a function of PEF treatment time, electric field strength, and temperature. (○) 20 kV cm^{-1} and 20°C; (●) 30 kV cm^{-1} and 20°C; (□) 20 kV cm^{-1} and 30°C; (■) 30 kV cm^{-1} and 30°C; (▲) 20 kV cm^{-1} and 40°C; (△) 30 kV cm^{-1} and 40°C. (From Amiali, M. et al., *J. Food. Eng.*, 79, 689, 2007. With permission.)

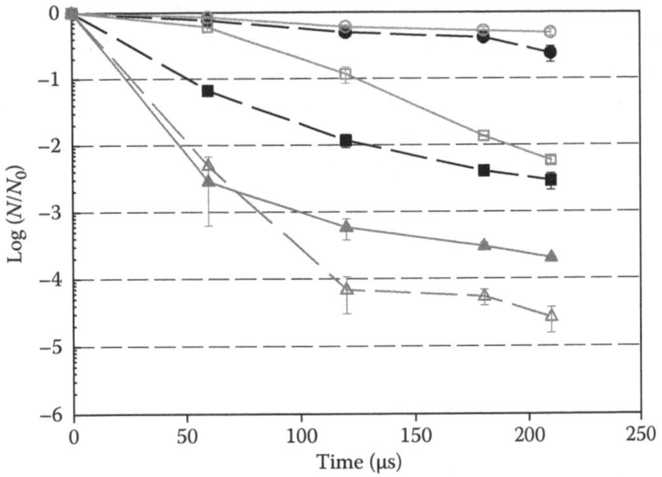

FIGURE 6.4 Survival fraction of *Salmonella enteritidis* in liquid egg yolk as a function of PEF treatment time, EF strength, and temperature. (○) 20 kV cm^{-1} and 20°C; (●) 30 kV cm^{-1} and 20°C; (□) 20 kV cm^{-1} and 30°C; (■) 30 kV cm^{-1} and 30°C; (▲) 20 kV cm^{-1} and 40°C; (△) 30 kV cm^{-1} and 40°C. (From Amiali, M. et al., *J. Food. Eng.*, 79, 689, 2007. With permission.)

6.3.2 Electric Field Strength

EF strength is defined as electric potential difference (V) between two given points in space divided by the distance (d) between them:

$$E = \frac{V}{d} \tag{6.1}$$

To achieve microbial inactivation, the applied EF needs to be greater than the critical EF for a particular microorganism (Castro et al., 1993). It is important that the EF be evenly distributed in the treatment chamber to achieve an efficient treatment. EFs of less than 4–8 kV cm^{-1} usually do not affect microbial inactivation (Peleg, 1995). An EF of 20 kV cm^{-1} or greater is usually sufficient to reduce the viability of gram-negative bacteria by up 6 log cycles (Amiali et al., 2006b, 2007; Floury et al., 2006; Alkhafaji and Farid, 2007; Ferrer et al., 2007; García et al., 2007; Mosqueda-Melgar et al., 2007; Pérez et al., 2007; Perni et al., 2007; Rojas et al., 2007; Craven et al., 2008; Shamsi et al., 2008) and gram-positive bacteria by 3–4 log cycles (Pothakamury et al., 1995; Selma et al., 2006; García et al., 2007; Mosqueda-Melgar et al., 2007; Sampedro et al., 2007). In general, the EF required to inactivate microorganisms in foods is in the range of 12–45 kV cm^{-1}. However, some studies have reported that EFs of up to 100 kV cm^{-1} could be applied to food under a continuous treatment protocol (Zhang et al., 1994a; Dunn, 1996; Smith et al., 2002; Korolczuk et al., 2006; Perni et al., 2007). The fact that microbial inactivation increases with increasing applied EF strength can be attributed to the high energy supplied to the cell suspension in a liquid food product (Pothakamury et al., 1995; Grahl and Märkl, 1996; Liu et al., 1997; Sensoy et al., 1997; Wouters et al., 1999; Amiali et al., 2004; Korolczuk et al., 2006; Alkhafaji and Farid, 2007; Charles-Rodríguez et al., 2007; Donsí et al., 2007; García et al., 2007;

Pérez et al., 2007; Rojas et al., 2007; Sampedro et al., 2007; San Martín et al., 2007; Somolinos et al., 2007; Toepfl et al., 2007; Evrendilek et al., 2008; Shamsi et al., 2008; Walkling-Ribeiro et al., 2008).

Some empirical and/or phenomenological models have been proposed to describe a relationship between EFs and microbial inactivation (Barbosa-Cánovas et al., 1999). The best known such model is that proposed by Hülsheger et al. (1981). The model gives the relationship between survival ratio ($S = N/N_0$) of microorganisms and EF:

$$\ln(S) = -b_E \cdot (E - E_c) \tag{6.2}$$

where

E_c is the critical EF obtained by extrapolating the value of E for a survival ratio of one unit

b_E is the regression coefficient (kV cm^{-1})

The value of E_c has been found to be a function of the cell size: the larger is the cell, the lower is the critical EF (Grahl and Märkl, 1996). These investigators attributed this phenomenon to the transmembrane potential experienced by the cell being proportional to its size. Hülsheger et al. (1983) found that E_c values for gram-negative bacteria were lower than those for gram-positive bacteria, which would explain the lesser resistance to PEF of gram-negative bacteria.

Survival curves of some microorganisms exposed to PEFs have a characteristic sigmoidal shape when plotted in linear coordinates. Peleg (1995) described this phenomenon by the following model:

$$S_p = \frac{100}{\left(1 + e^{(E - E_c)/a}\right)} \tag{6.3}$$

where

S_p is the percentage of surviving microorganisms

E_c is the critical EF strength when the survival level is 50% (i.e., the inflection point for S_p)

a is a parameter indicating the steepness of the survival curve around E_c

E_c and a are described by a single exponential decay model

$$E_c = E_\infty e^{(-k_1 n)} \tag{6.4}$$

$$a = a_0 e^{(-k_2 n)} \tag{6.5}$$

where

E_∞ is a constant (kV cm^{-1})

a_0 is a constant (kV cm^{-1})

k_1 and k_2 are also constants

At $E \gg E_c$, Equation 6.3 reduces to

$$S_p = \frac{100}{1 + e^{E/a}} \tag{6.6}$$

Peleg (1995) tested the model using published data (Castro et al., 1993) and found a good fit ($R^2 = 0.973$–0.999) with Equation 6.6. The experimental data of Ho et al. (1995) also supported this model although there were insufficient data to statistically quantify any parameters.

6.3.3 Treatment Time

Microbial inactivation of a liquid food increases with an increase in the number of pulses applied, up to a certain number (Sale and Hamilton, 1967; Ho et al., 1995; Zhang et al., 1995a; Marquez et al., 1997; Amiali et al., 2004, 2006a,b, 2007; Alkhafaji and Farid, 2007; Cebrián et al., 2007; Donsí et al., 2007; García et al., 2007; Mosqueda-Melgar et al., 2007; Pérez et al., 2007; Perni et al., 2007; Pizzichemi, 2007; Sampedro et al., 2007; Somolinos et al., 2007, 2008b; Evrendilek et al., 2008; Grenier et al., 2008; Walkling-Ribeiro et al., 2008). Usually, microbial inactivation is achieved during the first three or four pulses; additional pulses display a lesser lethality (Liu et al., 1997). Same observation was reported by Amiali et al. (2004, 2006b, 2007) regarding the inactivation of *E. coli* O157:H7 suspended in liquid egg products. A rapid inactivation was noticed at the beginning of the treatment (first and second batch of 15 pulses treatment), whereas a lesser lethality was observed for the rest of the treatment. The investigators attributed the phenomena to the resistance of the bacteria to the treatment. Ho et al. (1995), treating *Pseudomonas fluorescens* in various aqueous solutions using PEF (10 kV cm^{-1}, 10 × 2 μs pulses), obtained a 6-log reduction, but stated that there was no significant effect of the number of pulses on microbial inactivation. In the same study, when wider electrode gaps were used, inactivation was low, even at the highest EF intensity (20 kV cm^{-1}) and the greatest number of pulses (100). This was attributed to the nonhomogeneity of the EF in the treatment chamber.

A higher pulse frequency and shorter treatment time were reported to give a more efficient treatment of liquid foods (Barbosa-Cánovas, 1998; Alkhafaji and Farid, 2007; Charles-Rodríguez et al., 2007; Mosqueda-Melgar et al., 2007; Grenier et al., 2008). However, a very high pulse frequency tends to heat up the food product. Therefore, it is suggested to cool the electrodes of the treatment chamber(s) during the process in order to avoid any increase in the product temperature. Zhang et al. (1995b) maintained the temperature of treatment chamber electrodes within ±0.3°C by using short (2 μs) pulses (35–70 kV cm^{-1}, total of 80 pulses). Amiali et al. (2005, 2006b, 2007) kept the PEF maximum average temperature within 3°C ± 1°C by using circular parallel plate electrodes with cooling jacket system.

Marquez et al. (1997) showed that a given treatment time was more effective when the time between pulses was extended. They treated *Bacillus cereus* spores with 50 exponential decay pulses lasting 2 μs, at an EF intensity of 35 kV cm^{-1}. Reductions of 1.2 and 2.1 log cycles were obtained when 2–3 or 5–6 s were used between pulses, respectively. The results obtained were in part attributable to the pulse polarity.

Living cells, which exhibit a net charge in the presence of an EF, might migrate to the surface of electrodes. When successive unipolar pulses were applied, a shielding layer was formed. Therefore, the core became less conductive since spores had immobilized all ions presented on it. However, the ions within the cortex could penetrate the core, thus increasing the conductivity. With the pulse polarity, the ions became a shielding layer instead of penetrating the core, and therefore short electric pulses would be less effective (Marquez et al., 1997).

Different models relating the survival ratio of microorganisms to treatment time have been proposed. The first approach used to model PEF process involves the classical and widely practiced first-order kinetic model of Bigelow (Bigelow, 1921) to describe inactivation of microorganisms and enzymes. It is commonly known that the death of microorganisms is caused by inactivation of some critical enzyme systems. Since enzyme inactivation obeys first-order kinetics (van Boekel, 2002), the microbial inactivation model results in the following equation:

$$\frac{dN}{dt} = -kN \tag{6.7}$$

where

N is the microorganism population
t is the processing time at a constant rate (k) depending on its size

Integration of this expression yields

$$\ln\frac{N}{N_0} = -kt \tag{6.8}$$

where N_0 is the initial number of microorganisms. Equation 6.8 can be rearranged as follows:

$$\log\frac{N}{N_0} = \log\ S(t) = -\frac{t}{D} = S(t) = e^{-kt} \tag{6.9}$$

where

$S(t)$ is the survival fraction
t (μs^{-1}) is the treatment time
D is the decimal reduction time ($D = 2.303/k$), corresponding to the reciprocal of the first-order rate constant

The resulting semilogarithmic curve when log $S(t)$ is plotted versus time is frequently referred to as the survival curve. Martin et al. (1997) and Amiali et al. (2004, 2005, 2006b, 2007) also observed that inactivation kinetics followed first-order kinetics, i.e., a traditional exponential decay.

However, further investigations with data on microbial inactivation using PEF suggest that this model might be inadequate to describe PEF inactivation. Therefore, other models have been proposed to describe nonlinear kinetics. For example, Pruit and Kamau (1993) assumed the existence of two populations of microorganisms, which differed in their sensitivity to PEF:

$$S = S_c e^{-k_1 t} + (1 - S_c) e^{-k_2 t} \tag{6.10}$$

where
S_c is the critical survivor ratio in population 1 (PEF-sensitive)
$(1 - S_c)$ is the ratio of survivors in population 2 (PEF-resistant)
k_1 is the specific death rate of subpopulation 1 (μs^{-1})
k_2 is the specific death rate of subpopulation 2 (μs^{-1})

Amiali et al. (2004) showed that Pruit and Kamau's model adequately predicted inactivation of *E. coli* O157:H7 suspended in dialyzed liquid eggs. The inactivation kinetics followed exponential decay equation with two population sensitivity to PEF treatment. In addition, it was noted that the inactivation rate of *E. coli* O157:H7 followed exponential decay kinetic model with some tailing effect due to the resistance of survival fraction to the PEF treatment. They proposed the following model:

$$S = S_t + (1 - S_t) e^{-kt} \tag{6.11}$$

where
S_t is the tailing survival ration
$(1 - S_t)$ is the ratio of survivors in population
k is the first specific death rate (μs^{-1})

Hülsheger et al. (1981) proposed an inactivation kinetics model:

$$\ln(S) = -b_t \times \ln\left(\frac{t}{t_c}\right) \tag{6.12}$$

where
b_t is the regression coefficient
t is the treatment time (μs)
t_c is the extrapolated value of t for 100% survival

The consideration of two critical values of E_c and t_c led Hülsheger et al. (1981, 1983) to propose the following empirical equation for the calculation of the survival fraction:

$$\log(S) = \frac{(E_c - E)}{k} \times \log\left(\frac{t}{t_c}\right) \tag{6.13}$$

where
t_c is the maximum treatment time (μs) that results in an S value of 1
k is a first-order kinetics constant or microorganism constant (kV cm^{-1})

6.3.4 Treatment Temperature

The EF and treatment duration used in PEF treatment to pasteurize liquid foods significantly influence the treatment temperature; however, the heat generated has no direct effect on microbial cells. Zimmermann et al. (1974) demonstrated that temperature increase (ΔT = 3°C–5°C) induced during PEF processing had little effect on the inactivation of *E. coli*. However, since the effect was related to the temperature of the medium in which the organisms were suspended, the PEF treatment had a greater effect when processing occurred at moderate temperatures (Raso and Barbosa-Cánovas, 2003).

The temperature of the medium in which microbial (especially bacterial) cells are suspended has a significant influence in determining the membrane fluidity. The lipid bilayer of cell membrane can present different phases. At low temperatures, the well-ordered gel state occurs; fatty side chain assumes an extended conformation and nestles together with a maximum van der Waals contact. The phospholipids are assumed to be closely packed in a rigid "gel structure." However, at higher temperatures, the fatty acyl groups are less ordered and the thickness of the membrane decreases by about 15%. The phospholipids are less ordered and the membrane has a "liquid-crystalline" structure. Therefore, phase transition of the phospholipids, which occurs with shifts in temperature, can affect the physical stability of the cell membrane (Stanley, 1991; Jayaram, 2000; Russell, 2002). The phase transition temperature is about 10°C lower than the culture temperature. Since bacteria incorporate high proportions of saturated and long-chain fatty acids into phospholipids as the growth temperature increases, pulse treatments result in a greater destruction of viable cells when biological membranes are in a liquid-crystalline structure. On the other hand, below the phase transition temperature, pulse sterilization is less effective. At low temperatures, higher EF intensity is necessary to induce membrane rupture. The temperature–EF synergy can be expected to increase with increasing temperature. With increasing medium temperature, the cell membrane of bacteria becomes more fluid and its mechanical resistance decreases. Kinosita and Tsong (1977) demonstrated that the critical membrane breakdown potential decreased as the temperature of the solution increased, normally by a factor of 2 or 3 between 3°C and 25°C. Coster and Zimmermann (1975) similarly reported that the critical EF intensity for *Valonia utricularis* decreased with increasing medium temperature (decreasing from 1.0 V at 4°C to 0.64 V at 30°C).

Jayaram et al. (1992) reported that the death rate of *Lactobacillus brevis* under PEF treatment increased significantly when the temperature of the medium increased from 24°C to 60°C, even when the treatment time was short. A combination of 60°C and 25 kV cm^{-1} and a treatment time of only 12 ms resulted in 9-log reduction in *L. brevis*.

Zhang et al. (1995b) treated *E. coli* with 8 pulses of 2 μs in duration at 35 kV cm^{-1} and observed a significant increase in inactivation when the bulk temperature increased from 7°C to 20°C. At 7°C, less than 1-log reduction in viability was obtained, whereas at 20°C, more than 2-log reduction was attained. However, temperatures above 30°C (~33°C) did not result in any additional PEF inactivation.

Pothakamury et al. (1996) subjected *E. coli* to 35 and 36 kV cm^{-1} EF strengths using exponential decay and square wave pulses, the treatment temperature was varied from 3°C to 30°C. They reported greater inactivation above 20°C. For exponential decay pulses, an additional 1.5-log reduction of *E. coli* was achieved at 30°C than at 3°C. For a square wave pulse, an additional 1-log reduction was achieved at 33°C than at 7°C.

Reina et al. (1998) reported a higher inactivation rate of *L. monocytogenes* when the temperature increased from 10°C to 50°C, but not between 10°C and 25°C, 30°C, or 43°C. At 30°C and 30 kV cm^{-1} a 3.5-log reduction of *L. monocytogenes* was obtained after 600 μs of treatment, whereas at 50°C more than 4-log reduction was obtained.

Wouters et al. (1999) reported a strong synergistic effect of heat and PEF treatment. Prior to continuous treatment of *Listeria innocua* with PEF, the inlet temperature was adjusted to 20°C, 30°C, or 40°C. With an energy input of 40 J mL^{-1}, a greater inactivation rate was obtained with an inlet temperature of 40°C than at other tested temperatures (20°C and 30°C). Furthermore, less energy was needed to obtain a similar level of inactivation when using a higher medium temperature at the inlet. These investigators concluded that the inlet temperature had a substantial effect on the inactivation kinetics and on energy efficiency. Similar observations have been reported in other studies (Liu et al., 1997; Oshima et al., 1997; Sensoy et al., 1997).

Dunn and Pearlman (1987) recommended PEF treatment in combination with moderate temperatures (i.e., about 45°C or 55°C) in order to increase the antimicrobial effect of the process and to increase the product shelf life. These investigators found a combination of PEF and heat to be more efficient than conventional heat treatment alone. A higher level of inactivation was obtained using a combination of 55°C and PEF to treat milk. Dunn (1996) obtained 6-log reduction of *L. innocua* inoculated in milk after a few seconds of PEF treatment at 55°C.

Bazhal et al. (2003, 2006) showed the combined effect of heat treatment with PEF on the inactivation of *E. coli* O157:H7 in distilled water and in whole egg. The treatment conditions were 3–15 kV cm^{-1}, 300 pulses, and 0°C–80°C. They observed a synergistic effect between PEF and temperatures up to 80°C. In their second study (Bazhal et al., 2006) they investigated the inactivation of *E. coli* O157:H7 in liquid whole egg. The EF strength was varied from 9 to 15 kV cm^{-1}; the maximum treatment duration was 138 μs; and temperatures were 50°C, 55°C, or 60°C. At 60°, 2-log reduction of *E. coli* O157:H7 was obtained using thermal treatment alone, while a combination of heat and PEF resulted in 4-log reduction. These results again indicated a synergy between temperature and EF on microbial inactivation within a certain temperature range.

Amiali et al. (2005, 2006b, 2007) observed a synergistic effect of PEF and temperature when pasteurizing liquid egg products (egg white, egg yolk, and whole egg). More inactivation of *E. coli* O157:H7 and *S. enteritidis* were noticed when the EF intensity and treatment increased along with an increase in the treatment temperature. Up to 5-log reduction was reported at a field intensity of 30 kV cm^{-1}, treatment temperatures of 30°C or 40°C (depending on the egg product) and treatment time of 210 μs.

Of interest to the orange juice industry is the work of McDonald et al. (2000) who investigated the inactivation of different microorganisms with PEF treatments at 50°C and 55°C in a continuous system. At 55°C and 30 kV cm^{-1}, *E. coli*, *Leuconostoc mesenteroides*, and *L. innocua* were inactivated by 5 log cycles after 2 ms of treatment. However, only 3-log reduction was achieved for *Saccharomyces cerevisiae ascospores* after 7 pulses of 50 kV cm^{-1} at similar temperatures. In addition, Mermelstein (1998) found that after treating orange juice with a combination of PEF and mild temperature, a fresh-like quality, with no loss of organoleptic properties, was maintained for more than 8 weeks under refrigeration (4°C).

Castro et al. (1993) proposed to treat food products with a combination of PEF and heat (63°C–75°C) to extend their shelf life. However, above a certain temperature (~60°C), the texture and organoleptic qualities of certain foods were compromised. Therefore, a proper cooling device is necessary to maintain the food temperature below those generated by thermal pasteurization (Barbosa-Cánovas et al., 1999).

The temperature also has an effect on the kinetic constant and is described by using the Arrhenius-type equation (Amiali et al., 2006b, 2007).

$$k = A\exp\left(-\frac{E_a}{RT}\right) \tag{6.14}$$

where

A is an empirical constant
E_a is the activation energy
R is the gas constant
T is the treatment temperature (K)

Amiali et al. (2006b, 2007) modeled the reduction of bacterial survival fraction as a function of the treatment time at each EF treatment for inactivation of *E. coli* O157:H7 and *S. enteritidis* in liquid egg white and yolk. For liquid egg white, the determination coefficient (R^2) for the first-order kinetic model varied from 0.88 to 0.99 for *E. coli* O157:H7 and from 0.96 to 0.99 for *S. enteritidis*. The authors reported R^2 values for liquid egg yolk to vary from 0.95 to 0.99. A similar approach was also reported by Bazhal et al. (2006) for inactivation of *E. coli* O157:H7 in liquid whole egg using a combination of thermal and PEF treatment. The investigators considered the survival fraction as a function of the number of pulses (n) instead of the treatment time at each EF treatment:

$$S = S_0\exp(-k_{TE}n) \tag{6.15}$$

where

S_0 is the survival fraction after thermal treatment alone
n is the number of pulses
k_{TE} is the kinetic constant obtained using Equation 6.16,

$$k_{TE} = A\exp\left(-\frac{B}{E^2}\right) \tag{6.16}$$

where
- A and B are model constants
- E is the EF strength

The regression equations obtained from Equation 6.16 had $R^2 = 0.99$. The kinetics of *E. coli* O157:H7 inactivation estimated by the k_{TE} value depended on the treatment temperature and EF strength. Maximum k_{TE} value was achieved at 60°C and 15 kV cm^{-1}. A threshold temperature of 50°C, after which bactericidal effect of the combined treatment was significantly intensified was indicated.

6.3.5 Electrical Conductivity

The electric conductivity of a medium is defined as the ability to conduct electric current (Halden et al., 1990; Palaniappan and Sastry, 1991). It is an important variable that determines the extent of biological changes such as electropermeabilization, electrofusion, motility, and microbial inactivation induced by PEF treatment (Barbosa-Cánovas et al., 1999). The conductivity (S m^{-1}), σ, is given by

$$\sigma = \frac{d}{R \times A} = \frac{1}{\rho} \tag{6.17}$$

where
- R is the resistance of the food (Ω)
- A is the surface area (m^2)
- d is the gap between the electrodes (m)
- ρ is the resistivity (Ω-m)

In general, foods with high electrical conductivities are difficult to treat since they generate low peak EFs across a treatment chamber due to the high current that is typically generated during PEF treatment of such products (Barbosa-Cánovas et al., 1999). Therefore, it is recommended to lower the electrical conductivity to obtain greater microbial inactivation for the same applied EF. An increase in the difference between the conductivities of the medium and of the microbial cytoplasm weakens the membrane structure due to an increased flow of ionic substances across the membrane (Barbosa-Cánovas et al., 1999; Barsotti et al., 1999a). The antimicrobial effect of PEF is thus inversely proportional to the ionic strength of the suspension material (Hülsherger et al., 1981).

A product with a high resistivity (low conductivity) generally presents a higher degree of inactivation. For example, the PEF inactivation of *L. brevis* cells in a suspension of phosphate buffer solutions of different conductivities (0.17–2.23 mS cm^{-1}) was investigated by Jayaram et al. (1992). The maximum reduction in the number of viable microorganisms ($N/N_0 \approx 10^{-7}$) was obtained in the liquid possessing the lowest conductivity (0.17 mS cm^{-1}). Vega-Mercado et al. (1996) treated *E. coli* in media with KCl concentrations varying from 22.8 to 168.0 mM. The inactivation of *E. coli* treated with 0–8 pulses of 2 μs each at an EF intensity of 40 kV cm^{-1} increased with decreasing ionic strength of the KCl solutions (lowering electrical conductivity of the solution). A difference of 2.5 logs was obtained between the 168 and 22.8 mM solutions. On the other hand, for the same applied EF, Sensoy et al. (1997) reported an increase in the inactivation rate of *Salmonella dublin* with increasing conductivity of the medium. In contrast, Wouters et al. (1999), employing a field intensity of 26 kV cm^{-1}, showed that a higher electrical conductivity (7.9 mS cm^{-1} vs. 2.7 or 5.1 mS cm^{-1}) of the medium resulted in a decrease in the inactivation rate of *L. innocua*. At the outlet of the treatment chamber, for a medium temperature of 50°C, 2-log reduction was observed for the medium where $\sigma = 7.9$ mS cm^{-1}, whereas for an outlet temperature of 40°C and $\sigma = 2.7$ mS cm^{-1}, the inactivation rate was higher (4.5 log cycles). However, Ho et al. (1995) stated that a variation in σ between 0.65 mS cm^{-1} (0% NaCl w/v) and 10.2 mS cm^{-1} (0.5% NaCl w/v) had no significant effect on the microbial inhibition. Álvarez et al. (2000) found contradictory results after suspending *S. senftenberg* in citrate-phosphate buffer (pH 7.0) diluted at different concentrations to vary the electrical conductivity. They obtained lower applied EF and inactivation rate when the electrical conductivity was high. The input voltage was increased to obtain the same applied EF for different electrical conductivity levels; however, this did not influence microbial inactivation.

Some of the results presented above are contradicting to one another for the following reasons: (1) PEF equipment; (2) treatment chamber; (3) temperature of the treatment; (4) matrix; and (5) type of microorganism. Future researchers will have an opportunity to streamline these factors. Also it is important to present information about these factors while discussing conductivity and rate of inactivation of microorganisms.

Correlation of electrical conductivity and temperature for various food products are listed in Table 6.1.

Depending on the specific system and high-voltage switch used, special care may be necessary not to exceed the operating limits specified by the manufacturer. When treated food samples having too high or too low electrical conductivity with respect to the resistivity limits permitted by the electrical circuit, electrical components of the apparatus may be damaged. For example, high-intensity current

TABLE 6.1
Product Equations Correlation of Electrical Conductivity and Temperature

Product	Regression Equation	R^2
Egg white (1)	$\sigma = 0.0133T + 0.3729$	0.9983
Whole egg (II)	$\sigma = 0.0119T + 0.3482$	0.9976
Whole egg (I)	$\sigma = 0.0107T + 0.296$	0.9966
Yolk	$\sigma = 0.008T + 0.1704$	0.9978
Apple juice	$\sigma = 0.0046T + 0.1067$	0.9978
Pineapple juice	$\sigma = 0.0068T + 0.1546$	0.9978
Orange juice	$\sigma = 0.008T + 0.1912$	0.996

Source: Data from Amiali, M. et al., *International Journal of Food Properties*, 9: 533–540, 2006.

that crosses a low-resistivity food may overheat some switches unless a current limiting device is employed in the circuit. In contrast, a high-resistivity food maintains a high voltage across the switch and it may be necessary, for protection, to place a shunting resistance between the electrodes in parallel with the food sample (Barsotti et al., 1999a,b).

Several research groups have reported data base and correlations of liquid food electrical conductivity as a function of temperature that could be used to design and optimize a PEF process (Ruhlman et al., 2001; Amiali et al., 2006a).

6.4 PEF INACTIVATION OF MICROORGANISMS

Microbial challenge tests have been conducted to determine the effects of EFs on the inactivation kinetics of selected microorganisms inoculated in foods or model substrates. The tests involved applying the EF intensity and application duration/frequency, which caused inactivation of a maximum number of microorganisms without dielectric breakdown of the foods.

6.4.1 Bacteria

E. coli is the most-studied microorganism, not only in the area of PEF processing, but also among other methods of food preservation (Barbosa-Cánovas et al., 1999). Gram-negative pathogens such as *E. coli* species (Liu et al., 1997; Evrendilek et al., 1999, 2000; Iu et al., 2001; Amiali et al., 2004, 2005, 2006b, 2007; Bazhal et al., 2006; Charles-Rodríguez et al., 2007; El Zakhem et al., 2007; Ferrer et al., 2007; García et al., 2007; Mosqueda-Melgar et al., 2007; Perni et al., 2007; Grenier et al., 2008) and *Salmonella* species (Jeantet et al., 1999; Raso et al., 2000; Floury et al., 2006; Korolczuk et al., 2006; Cebrián et al., 2007; García et al., 2007; Mosqueda-Melgar et al., 2007, 2008a,b; Perni et al., 2007) have been tested and inactivated using PEF. Gram-positive microorganisms such as *Listeria* species (Cálderon-Miranda et al., 1999a,b; Wouters et al., 1999; Selma et al., 2006; García et al., 2007; Mosqueda-Melgar et al., 2007), *Bacillus* species (Heinz et al., 1999; Heinz and Knorr 2000; Pol et al., 2000), *Lactobacillus* species (Rodrigo et al., 2001; Wouters et al., 2001), *Micrococcus luteus* (Dutreux et al., 2000a; García et al., 2007; Sanpedro et al., 2007), and mixtures of bacteria have also been investigated (Dutreux et al., 2000b; McGregor et al., 2000; Russell et al., 2000; Aronsson et al., 2001; Unal et al., 2001). Acid-fast organisms, such as *Mycobacterium paratuberculosis* have also been studied (Rowan et al., 2001). Although the processing conditions differed, each study was able to confirm the effectiveness of PEF treatment.

6.4.2 Yeasts and Molds

Yeasts are one of the most important causes of food spoilage. *S. cerevisiae*, used for the leavening of bread and alcoholic fermentation, can cause spoilage in juices due to the production of alcohol and CO_2. Inactivation by PEF of several yeast and mold species (Zhang et al., 2007) including *Saccharomyces* species (Qin et al., 1994, 1995c;

Zhang et al., 1994b; Donsí et al., 2007; Somolinos et al., 2007, 2008a,b; Toepfl et al., 2007), *Candida* spp. (Hülsheger et al., 1983), *Dekkera bruxellensis* (Somolinos et al., 2007a), *Penicillium expansum* (Evrendilek et al., 2008), and *Zygosaccharomyces bailii* (Raso et al., 1998) suspended in real food systems or in food models, e.g., water, phosphate buffers, and sodium alginate solutions, has been extensively studied and reported in the literature.

6.4.3 Spores

Microbial spores are more resistant to external ambient conditions such as high temperatures and osmotic pressures, high and low pHs, and mechanical shocks than vegetative cells. Their resistance is associated to their small sizes, dehydrated cytoplasm, and mineralization. Dehydration would reduce its conductivity and make the development of a sufficiently high-voltage gradient to breach the surrounding membrane difficult (Gould, 2000). As a result of these properties, PEF treatment is of limited effectiveness in inactivation of bacterial spores. However, some studies have shown limited inactivation of spores under PEF (Marquez et al., 1997) or with PEF in conjunction with other antimicrobial treatments (Pol et al., 2001a,b).

6.4.4 Enzymes

Food spoilage can be caused by enzymes naturally present in food or by enzymes produced by certain microorganisms. According to PurePulse Technologies Inc. (San Diego, California), PEF processing did not cause any activation of endogenous alkaline phosphatase in raw milk. However, electric pulses appeared to be able to inactivate some enzymes that are detrimental to food quality and storage. The protease produced by *P. fluorescens*, which increases the risk of coagulation and enhances the bitterness of refrigerated milk, could be largely inactivated by PEF (Vega-Mercado et al., 1995). However, it should be noted that processing the same milk at 25 kV cm^{-1} and 0.6 Hz increased the sensitivity of milk proteins to this protease.

Giner-Seguí et al. (2006) studied the evolution of polygalacturonase (PG) (EC 3.2.1.15) activity in aqueous solution of commercial enzyme preparation. Up to 76.5% reduction of the PG activity could be achieved at 38 kV cm^{-1} and 1100 μs EF intensity and treatment time, respectively. However, an enhancement of PG activity at soft PEF treatment conditions (up to 110.9% at E = 15 kV cm^{-1} and 300 μs) was observed. A maximum of 80% of pectin methyl esterase activity in orange juice was inactivated at 35 kV cm^{-1} and 1500 μs EF strength and treatment time, respectively (Elez-Martinez et al., 2007).

The unfolding of lysozyme structure was reported to be induced by PEF (35 kV cm^{-1} and 300 μs); this was accompanied by the cleavage of disulfide bonds and self association aggregation when the applied PEF dosage was higher than the critical level (Zhao et al., 2007).

Table 6.2 presents a summary of treatment conditions and inactivation results obtained in several investigative studies on PEF inactivation of microorganism species.

TABLE 6.2
Inactivation Kinetics of Microorganisms by PEFs

Microorganism	Suspension Media	Process Condition	Log or % Reduction (Max)	Treatment Vessel	Source
Escherichia coli	Nutritive treatment media (NTM) pH 5, 6, and 7 a_w and 0.97, 0.985, and 1	30°C, 15–30 kV cm^{-1}, 36 mL/min, monopolar square, 4 µs duration, t = 250 µs	pH 7, 4 log cycles pH 6, 5.5 log cycles pH 5, 8 log cycles	3 × 6 cofield chamber, V = 8.5 µL, continuous	Aronsson et al. (2004)
E. coli	McIlvain buffer	<35°C, 15–28 kV cm^{-1}, square wave, 1 Hz, 2 µs	>6.0	Batch chamber, parallel plates, d = 0.25 cm, A = 2.01 cm^2	Álvarez et al. (2003)
E. coli	16 mM sodium phosphate buffer 10% ovalbumin + 20 mM sodium phosphate buffer Plate count agar Dairy cream Fish egg + 16 mM sodium phosphate buffer	25°C, ~34 kV cm^{-1}, exponential decay pulses, 10–500 pulses, 1.4–2 µs, t = 261 µs	5.0	Cylindrical chamber, D = 38 mm, parallel plates, 5 mm gap	Manas et al. (2001)
E. coli O157:H7	Whole egg	50°C–60°C, 15 kV cm^{-1}, t = 138 µs, 2 µs, 1 Hz, instant reversal charge square wave	4.0	Batch chamber, parallel plates, V = 2 mL, 1 cm gap	Bazhal et al. (2006)
E. coli O157:H7	Peptone water	0°C–80°C, 300 pulses, 3–15 kV cm^{-1}, square wave	6.8	Batch chamber, parallel plates, 1.5 mm gap, V = 2.2 mL	Bazhal et al. (2003)

(continued)

TABLE 6.2 (continued)
Inactivation Kinetics of Microorganisms by PEFs

Microorganism	Suspension Media	Process Condition	Log or % Reduction (Max)	Treatment Vessel	Source
E. coli	Orange juice	30°C, 30 kV cm^{-1}, 100 L/h, 2 ms	5.0	ColPure™ PEF system (continuous, coaxial), $d = 5$ mm	McDonald et al. (2000)
E. coli	Orange (80%) and carrot (20%) juice mixture	30°C, 25–40 kV cm^{-1}, $t = 40$–300 µs, 10^{-3} L/s, 2.5 µs wave pulse	2.6	4–6 cofield chambers, continuous	Selma et al. (2004)
E. coli	Liquid egg	<37°C, 26 kV cm^{-1}, 4 µs, 2.5 Hz, 0.5 mL/min, 100 pulses	6.0	Continuous chamber, coaxial electrodes, 11.87 mL	Martín-Belloso et al. (1997)
E. coli	Milk	17°C, 120 L/h, 41 kV cm^{-1}, 2.5 µs, 10 Hz, 63 pulses	4.0	Continuous, 28.6 mL, $d = 0.6$ cm	Dutreux et al. (2000b)
E. coli	Simulated milk ultrafiltrate (SMUF)	5–11 kV cm^{-1}, exponential pulses, 1200 IU/mL Nisin, $a_w = 0.95$	5.0	Electroporation cuvette	Terebiznik et al. (2002)
E. coli	McIlvain buffer	<35°C, 19 kV cm^{-1}, square wave, 2 µs, 2 Hz, pH 7 and 4, $t = 400$ µs	Slight sublethal injury at pH 7, 99.95% were injured at pH 4	Static parallel plate chamber, $d = 0.25$ cm, $V = 0.5$ mL	García et al. (2003)
E. coli	Orange–carrot juice	40°C, 25–40 kV cm^{-1}, $t = 40$–340 µs, square wave bipolar pulses, 2.5 µs pulse width	2.8	6 cofield treatment chamber, $D = 0.23$ cm, 0.293 cm gap	Rodrigo et al. (2003)
E. coli	SMUF	3°C–35°C, 36 kV cm^{-1}, exponential decay or square wave pulses of 2 µs width	5.0	Static chamber, parallel plate electrodes, 0.5 cm gap. $A = 27\ cm^2$, $V = 12.5\ cm^3$	Pothakamury et al. (1996)

E. coli O157:H7	Apple juice (1) Milk (2)	29°C, 24–31 kV cm^{-1}, 2.8 and 4 μs of bi- or monopolar square wave, 700 Hz, t = 141–202 μs.	2.63 (1) 1.96 (2)	6 cofield chambers, 0.137 cm gap, D = 0.29 cm, V = 0.0209 cm^3	Evrendilek and Zhang (2005)
E. coli	Apple juice	35°C–70°C, 8–40 kV cm^{-1}, 2–95 Hz, 3 kg/h, 1.5–6 μs width, exponential decay pulses, t = 6–230 μs,	>6.0	Continuous parallel treatment chamber, 2.5 mm gap, A = 2 cm^2	Heinz et al. (2003)
E. coli	SMUF	30°C, 16 kV cm^{-1}, 60 pulses, 300 μs pulse width	4.0	Electroporation cuvette, D = 0.1 cm	Pothakamury et al. (1995)
E. coli	SMUF	40°C, 36 kV cm^{-1}, exponential decay pulses, 2–3 μs pulse width	>6.0	Coaxial treatment chamber, V = 29 mL, 0.6 cm gap	Qin et al. (1998)
E. coli	16 or 40 mM sodium phosphate buffer (pH 6.6 or 7.0) Whole milk Skim milk Dairy cream	21.5°C–42°C, 17–46 kV cm^{-1}, 545 pulses, 1.1 or 100 Hz, 0.79–1.51 μs pulse width, exponential decay pulses	~6.0	Static cylinder vessel, D = 39 mm, 5 mm gap, V = 5.6 mL	Picart et al. (2002)
E. coli	Deionized water + HCl	30°C, 12.5 kV cm^{-1}, 0.3 ms pulse duration, 2–20 pulses	3.0	Electroporation cuvette, 0.2 cm gap, V = 250 μL	Geveke and Kozempel (2003)
E. coli K12	Liquid egg and milk	5–40 kV cm^{-1}, 2 μs, 60 pulses, exponential decay pulses	5.0	D = 3.7 cm	Gupta et al. (2003a)
E. coli K12	Distilled water	5°C–60°C, 16 kV cm^{-1}, 50 Hz, t = 600 μs	4.0	Static chamber, parallel plate electrodes, D = 30 mm, 10 mm cm gap	Oshima et al. (2002)
E. coli K12	Apple juice	20°C, 20 and 40 kV cm^{-1}, 0–100 pulses	5.0	Hollow container with 8.15 mm gap	Gupta et al. (2003b)
E. coli O157:H7	Gellan gums	5°C–65°C, 15–30 kV cm^{-1}, 1–20 pulses, square wave form, 4 μs pulse width	3.0	Static chamber, D = 8.4 mm, 3.4 mm gap	Ravishankar et al. (2002)

(*continued*)

TABLE 6.2 (continued)
Inactivation Kinetics of Microorganisms by PEFs

Microorganism	Suspension Media	Process Condition	Log or % Reduction (Max)	Treatment Vessel	Source
E. coli O157:H7	M9 medium	pH 3.6, 5.2, 7.0, 20–30 kV cm^{-1}, 1 mL/s, 3 µs duration, 700 Hz, bipolar square wave pulses	>6.0	6 cofield treatment chambers, $D = 0.23$ cm, 0.19 cm gap	Everendilek and Zhang (2003)
E. coli O157:H7	Peptone water (0.1% w/v)	12–20 kV cm^{-1}, 60 pulses, bipolar exponential decay pulses	2.0	—	Damar et al. (2002)
E. coli O157:H7	0.1% NaCl	35°C, 5–20 kV cm^{-1}, bipolar square, 1 mL/s, 1000 Hz, 3 µs, $t = 145.6$ µs	2.9	4 cofield treatment chambers, $V = 60$ mL, $D = 0.23$ cm, 0.292 cm gap	Unal et al. (2002)
E. coli	Distilled water (DW) 10 mM HEPES-KOH, pH 7.0 10 mM Tris–HCl, pH 7.0	35°C, 26.7 kV cm^{-1}, monopolar square pulses, 2 µs, 300 pulses, 5 Hz	Varied from 0.050% to 55% of inactivation	Electroporation cuvette of 0.2 cm gap, 2 cm^2 surface electrodes	Reyns et al. (2004)
E. coli	Bacto peptone	30°C, 30 kV cm^{-1}, 4 µs, 20 pulses, 250 Hz, 41 mL, $t = 13$ ms	5.4	Bench, continuous, 6 tubular parallel chambers, $d = 2.3$ mm	Aronsson et al. (2001)
E. coli O154:H7	0.1% NaCl	20 kV cm^{-1}, bipolar waveform, 1000 Hz, 3 µs, 1 mL/s, 12 pulses/chamber, $t = 145.6$ µs	4.5	4 chambers, continuous, $d = 0.292$ cm	Unal et al. (2001)
E. coli	Milk	17°C, 120 L/h, 41 kV cm^{-1}, 2.5 µs, 10 Hz, 63 pulses	4.0	Continuous, 28.6 mL, $d = 0.6$ cm	Dutreux et al. (2000a)
E. coli O154:H7	Glycerol	25°C, 12.5 kV cm^{-1}, 1000 Hz,	1.1–1.6	$d = 0.4$ cm	Liu et al. (1997)

E. coli O154:H7	Apple juice and apple cider	22–34 kV, t = 166 µs, 4 µs bipolar pulses, 800 Hz, 1.5 mL/s	4.5	Continuous, 6 parallel plates, d = 0.238 cm	Evrendilek et al. (2000)
E. coli O154:H7	Phosphate buffer	30°C, 5–30 kV cm^{-1}, square 4 µs, 36 mL/min, 250 Hz	5.3	Cofield, V = 8.5 µL, D = 2.0 mm, d = 2.7 mm	Aronsson et al. (2005)
E. coli O154:H7	Apple cider	42°C, 80 kV cm^{-1}, 2 µs, 30 pulses	5.35	Parallel plates, 50 mL, d = 0.3 cm	Iu et al. (2001)
E. coli O154:H7 *and E. coli* 8739	Apple juice	<35°C, 85 L/min, 18–30 kV cm^{-1}, 1000 Hz, bipolar pulses (6–3 µs), t = 86–172 µs	5.0	Continuous, 4 parallel plate chambers	Evrendilek et al. (1999)
E. coli	SMUF	49.6 kV cm^{-1}, T = 38°C, square bipolar pulse (1.7 µs), 200 Hz	6.0	4 treatment chambers, V = 0.06 cm^3	Alkhafaji and Farid (2007)
E. coli	Melon and watermelon juices	35 kV cm^{-1}, 4 µs bipolar or square, <40°C, 217 or 188 Hz, t = 1440 or 1727 µs	3.7	Continuous, 8 cofield chambers, d = 0.23 cm, 0.0415 cm^2	Mosqueda-Melgar et al. (2007)
E. coli W3110	TSA-YE	Square pulses, 2 µs, 22 kV cm^{-1}, 1 Hz, 35°C	>3.0	Static, d = 0.25 cm, parallel plate, 2.01 cm^2	Cebrian et al. (2008)
E. coli K12	Meat injection solution	12 kV cm^{-1}, exponential decay charge, T = 2°C	1.3	Cuvette d = 0.2 or 0.4 cm, V = 100 or 200 µL	Rojas et al. (2007)
E. coli	Citrate-phosphate McIlvain buffer	35°C, square wave, 2 µs pulse width, 1 Hz, 25 kV cm^{-1}, t = 400 µs	—	V = 0.5 mL, d = 0.25 cm	Garcia et al. (2007)
E. coli K12	Phosphate buffer saline (PBS)	Square wave, pulse duration = 32 ns, 100 kV cm^{-1}, 30 Hz, t = 300 s	2.0	Static cuvette, d = 1 mm, V = 100 µL	Perni et al. (2007)
Salmonella dublin	Skim milk	15–40 kV cm^{-1}, 10–50°C, t = 12–127 µs	~3.0	Cofield treatment chamber	Sensoy et al. (1997)
Salmonella enteritidis	Liquid whole egg	25 kV cm^{-1}, 1.2 mL/s, 200 Hz, 2.12 µs, t = 250 µs	4.3	4 cofield treatment chambers, D = 0.23 cm, 0.19 cm gap	Hermawan et al. (2004)

(*continued*)

TABLE 6.2 (continued)
Inactivation Kinetics of Microorganisms by PEFs

Microorganism	Suspension Media	Process Condition	Log or % Reduction (Max)	Treatment Vessel	Source
Salmonella typhimurium	Distilled water (DW) 10 mM HEPES-KOH, pH 7.0, 10 mM Tris–HCl, pH 7.0	35°C, 26.7 kV cm^{-1}, monopolar square pulses, 2 μs, 300 pulses, 5 Hz	Varied from 0.050% to 55% of inactivation	Electroporation cuvette of 0.2 cm gap, 2 cm^2 surface electrodes	Reyns et al. (2004)
S. typhimurium	Orange juice	45°C, 90 kV cm^{-1}, 50 pulses, 2 μs pulse width, nisin 100 IU/mL + lysozyme 690 IU/mL	>5.0	Batch treatment chamber, $D = 16.5$ cm, 0.3 cm gap, $V = 50$ mL	Liang et al. (2002)
S. dublin	Milk	63°C, 3.7 V/μm, 36 μs, 40 pulses	4.0	Parallel plates	Dunn and Pearlman (1987)
S. typhimurium	10 mM HEPES	15–30 kV cm^{-1}, 300 pulses of monopolar square wave, 2 μs, 1 Hz	<5.0	Electroporation cuvette, $D = 2$ mm, $A = 2\,cm^2$,	Wuytack et al. (2003)
S. enteritidis	Egg white	30, 35 kV cm^{-1}, 900 Hz, monopolar exponential decay pulses	3.5	$d = 0.2$ cm	Jeantet et al. (1999)
Salmonella senftenberg	McIlvain buffer	28 kV cm^{-1}, square wave, 15 μs, 5 Hz	~6.8	Static parallel plate chamber, $d = 0.25$ cm	Raso et al. (2000)
S. senftenberg	McIlvain buffer	Square wave pulses, 2 μs, 2 Hz, 200 pulses, 19 kV cm^{-1}	6.0	Cylinder plastic tube, 0.25 gap, $A = 2.01\,cm^2$	Álvarez et al. (2000)
S. typhimurium	Distillated water	<4°C, 10, 20 kV cm^{-1}, exponential decay pulses, 50 μs, 30 Hz	6.0	Static chamber	Russell et al. (2000)
S. typhimurium	Phosphate buffer saline (PBS)	Square wave, pulse duration = 32 ns, 100 kV cm^{-1}, 30 Hz, $t = 300$ s	1.0	Static cuvette, $d = 1$ mm, $V = 100$ μL	Perni et al. (2007)
S. enteritidis	Skim milk	47 kV cm^{-1}, 500 ns, 60 Hz, 62°C	2.3	Continuous, 5 L/h, coaxial treatment chamber, $d = 2$ mm	Floury et al. (2006)

S. senftenberg 775W	Citrate-phosphate McIlvain buffer	35°C, square wave, 2 µs pulse width, 1 Hz, 25 kV cm^{-1}, t = 400 µs	—	V = 0.5 mL, d = 0.25 cm	Garcia et al. (2007)
S. enteritidis	Melon and watermelon juices	35 kV cm^{-1}, 4 µs bipolar or square, <40°C, 217 or 188 Hz, t = 1440 or 1727 µs	3.71	Continuous, 8 cofield chambers, d = 0.23 cm, 0.0415 cm^2	Mosqueda-Melgar et al. (2007)
Listeria innocua	Milk	17°C, 120 L/h, 41 kV cm^{-1}, 2.5 µs, 10 Hz, 63 pulses	3.9	V = 28.6 mL, d = 0.6 cm	Dutreux et al. (2000b)
L. innocua	Phosphate buffer	30°C, 5–30 kV cm^{-1}, square 4 µs, 36 mL/min, 250 Hz	0.6	Cofield, V = 8.5 µL, D = 2.0 mm, d = 2.7 mm	Aronsson et al. (2005)
Bacillus cereus	5 mM potassium-HEPES buffer	30°C, 16.7 kV cm^{-1}, 50 pulses, square wave form of 2 µs duration, nisin (0.06 µg/mL)	1.8	Static chamber, V = 800 µL, D = 12.6 mm, 6 mm gap	Pol et al. (2000)
L. innocua	16 or 40 mM sodium phosphate buffer (pH 6.6 or 7.0)	21.5°C–42°C, 17–46 kV cm^{-1}, 545 pulses, 1.1 or 100 Hz, 0.79–1.51 µs pulse width, exponential decay pulses	3.0	Static cylindrical vessel, D = 39 mm, 5 mm gap, V = 5.6 mL	Picart et al. (2002)
	Whole milk, skim milk, dairy cream				
L. innocua	Bacto peptone	30°C, 30 kV cm^{-1}, 4 µs, 20 pulses, 250 Hz, 41 mL, t = 13 ms	3.0	Bench, continuous, 6 tubular parallel chambers, d = 2.3 mm	Aronsson et al. (2001)
L. innocua	Phosphate buffer	40°C, 200 L/h, 3 V/µm, 3.9 µs	6.3	ColPure™ PEF system, d = 5 mm	Wouters et al. (1999)
L. innocua	Raw skim milk	22°C–34°C, 0.5 L/min, 2 µs, 3.5 Hz, 32 pulses, 50 kV, exponential decay	2.4	Continuous, 25 mL, d = 0.6 cm	Calderon-Miranda et al. (1999a)
L. innocua	Liquid whole egg	6°C–36°C, 0.5 L/min, 2 µs, 3.5 Hz, 32 pulses, 50 kV, exponential decay	3.5	Concentric, 25 mL, d = 0.6 cm	Calderon-Miranda et al. (1999b)

(continued)

TABLE 6.2 (continued)
Inactivation Kinetics of Microorganisms by PEFs

Microorganism	Suspension Media	Process Condition	Log or % Reduction (Max)	Treatment Vessel	Source
L. innocua	Raw skim milk (0.2% milk fat)	15°C–28°C, 0.5 L/min, 100 pulses, 50 kV cm^{-1}, 2 µs, 3.5 Hz, exponential decay	2.6	Coaxial, 29 mL, d = 0.63 cm	Fernandez-Molina et al. (1999)
L. innocua	Orange juice	30°C, 30 kV cm^{-1}, 50 kV cm^{-1} (2), 100 L/h, 2 ms	5.0	ColPure™ PEF system (continuous, coaxial), d = 5 mm	McDonald et al. (2000)
L. innocua	Deionized water + HCl	30°C, 12.5 kV cm^{-1}, 0.3 ms pulse duration, 2–20 pulses	3.0	Electroporation cuvette, 0.2 cm gap, V = 250 µL	Geveke and Kozempel (2003)
Listeria monocytogenes	0.1% NaCl	20 kV cm^{-1}, bipolar waveform, 1000 Hz, 3 µs, 1 mL/s, 12 pulses/chamber, t = 145.6 µs	1.1	4 chambers, continuous, d = 0.292 cm	Unal et al. (2001)
L. monocytogenes	0.1% NaCl	35°C, 5–20 kV cm^{-1}, bipolar square, 1 mL/s, 1000 Hz, 3 µs, t = 145.6 µs	2.1	4 cofield treatment chambers, V = 60 ml, D = 0.23 cm, 0.292 cm gap,	Unal et al. (2002)
L. monocytogenes	Pasteurized whole milk (3.5% milk fat), 2% milk, 0.2% skim milk	10°C–50°C, 0.07 L/s, 30 kV cm^{-1}, 1.5 µs, 1700 Hz, bipolar pulses, t = 600 µs	3.0–4.0	Cofield flow, 20 mL	Reina et al. (1998)
L. monocytogenes	Water and skim milk	5–55 pulses, 15–30 kV cm^{-1}, 0°C–60°C	4.5	Static chamber, D = 8.4 mm, 3.4 mm gap	Fleischman et al. (2004)
L. monocytogenes	NaCl and 50% acid whey	25 kV cm^{-1}, 15°C–37°C, t = 144 µs, 1 and 2 mL/s, t = 72–144 µs, bipolar square wave pulses of 3 µs duration, 667–1000 Hz	2.5–3.7	4 cofield treatment chambers, D = 2.3 mm, 2.9 cm gap	Lado and Youcef (2003)

L. monocytogenes	Citrate-phosphate McIlvain buffer	35°C, square wave, 2 μs pulse width, 1 Hz, 25 kV cm^{-1}, $t = 400$ μs	—	$V = 0.5$ mL, $d = 0.25$ cm	Garcia et al. (2007)
L. monocytogenes	Horchata	30 kV cm^{-1}, square wave, 2.5 μs pulse width, 30°C, 1.5 10^{-3} L/s	—	Continuous, 6 cofield chambers	Selma et al. (2006)
L. monocytogenes	Distillated water	<4°C, 10, 20 kV cm^{-1}, exponential decay pulses, 50 μs, 30 Hz	4.0	Static chamber	Russell et al. (2000)
L. monocytogenes	Melon and watermelon juices	35 kV cm^{-1}, 4 μs bipolar or square, <40°C, 217 or 188 Hz, $t = 1440$ or 1727 μs	3.56	Continuous, 8 cofield chambers, $d = 0.23$ cm, 0.0415 cm^2	Mosqueda-Melgar et al. (2007)
Staphylococcus aureus	SMUF	30°C, 16 kV cm^{-1}, 60 pulses, 300 μs pulse width	4.0	Electroporation cuvette, $D = 0.1$ cm	Pothakamury et al. (1995)
S. aureus	Peptone	20°C–25°C, 30 kV cm^{-1}, 500 ns, 3000 square pulses, 5–10 Hz	3.0	Parallel plates chamber, $d = 1$ cm	McGregor et al. (2000)
S. aureus	SMUF	40°C, 36 kV cm^{-1}, exponential decay pulses, 2–3 μs pulse width	>5.0	Coaxial treatment chamber, $V = 29$ mL, 0.6 cm gap	Qin et al. (1998)
S. aureus	Skim milk	40°C, 35 kV cm^{-1}, bipolar square wave of 3.7 μs pulse duration, 250 Hz, 1 mL/s, $t = 460$ μs	3.0	6 cofield treatment chambers, $d = 2.3$ mm, 1.0 mm gap	Everendilek et al. (2004)
S. aureus	Peptone water (0.1% w/v)	12–20 kV cm^{-1}, 60 pulses, bipolar exponential decay pulses	2.0	—	Damar et al. (2002)
S. aureus	Milk	25°C, 35 kV cm^{-1}, bipolar square wave, 6 pulse width 75 Hz	6.2	$d = 0.29$ cm	Sobrino-López and Martin-Belloso (2008)
Pseudomonas aeruginosa	Peptone	20°C–25°C, 30 kV cm^{-1}, 500 ns, 3000 square pulses, 5–10 Hz	4.0	Parallel plates chamber, $d = 1$ cm	McGregor et al. (2000)
Pseudomonas fluorescens	Raw skim milk UHT skim milk	52°C–22°C, 35 kV cm^{-1}, 64 pulses, 3 μs bipolar square wave, 500 Hz, $t = 188$ μs, 1 L/min, 0.53 S/m	2.5	4 cofield treatment chambers, 0.19 cm gap, $D = 0.23$ cm	Michalac (2003)

(continued)

TABLE 6.2 (continued)
Inactivation Kinetics of Microorganisms by PEFs

Microorganism	Suspension Media	Process Condition	Log or % Reduction (Max)	Treatment Vessel	Source
Pseudomonas sp.	Milk	31 kV cm^{-1}, 55°C, 200 Hz, monopolar pulse	>5	4 cofield treatment chambers, 60 mL/min, $d = 0.2$ cm	Craven et al. (2008)
Lactobacillus plantarum	Orange (80%) and carrot (20%) juice mixture	30°C, 25–40 kV cm^{-1}, $t = 40$–300 μs, 10^{-3} L/s, 2.5 μs^2 wave pulse	1.3	4–6 cofield chambers, continuous	Selma et al. (2004)
L. plantarum	Orange–carrot juice	35.8 kV cm^{-1}, 10.6–46.3 μs, 0.5 L/min, 35.87 kV cm^{-1}	2.5	Continuous coaxial, 0.632 cm, $V = 29$ mL	Rodrigo et al. (2001)
L. plantarum	Model beer	9–18 kV cm^{-1}, 31°C, 2.2 L/h, nisin	1.5	Continuous treatment chamber of 2 mm gap	Ulmer et al. (2002)
L. plantarum	Citrate-phosphate McIlvain buffer	35°C, square wave, 2 μs pulse width, 1 Hz, 25 kV cm^{-1}, $t = 400$ μs	—	$V = 0.5$ mL, $d = 0.25$ cm	Garcia et al. (2007)
L. plantarum	Beverage formulation (orange juice, water, skimmed milk, sugar, pectin, citric acid)	35–40 kV cm^{-1}, 33 or 55°C, 6 10^{-2} L/min, bipolar square wave, 2.5 μs, $t = 180$ μs	1.5	6 cofield chambers	Sampedro et al. (2007)
Lactobacillus leichmannii	0.1% NaCl	35°C, 5–20 kV cm^{-1}, bipolar square, 1 mL/s, 1000 Hz, 3 μs, $t = 145.6$ μs	1.5	4 cofield treatment chambers, $V = 60$ mL, $D = 0.23$ cm, 0.292 cm gap	Unal et al. (2002)
Lactococcus lactis	Raw skim milk UHT skim milk	22°C–52°C, 35 kV cm^{-1}, 64 pulses, 3 μs bipolar square wave, 500 Hz, $t = 188$ μs, 1 L/min, 0.53 S/m	0.5	4 cofield treatment chambers, 0.19 cm gap, $D = 0.23$	Michalac (2003)

Leuconostoc mesenteroides	Orange juice	30°C, 30 kV cm^{-1}, 50 kV cm^{-1} (2), 100 L/h, 2 ms	5.0	ColPure™ PEF system (continuous, coaxial), $d = 5$ mm	McDonald et al. (2000)
L. mesenteroides	Bacto peptone	30°C, 30 kV cm^{-1}, 4 µs, 20 pulses, 250 Hz, 41 mL, $t = 13$ ms	3.0	Bench, continuous, 6 tubular parallel chambers, $d = 2.3$ mm	Aronsson et al. (2001)
Aerobic bacteria, molds and yeasts	Orange juice	<60°C, 35 kV cm^{-1}, 59 µs, monopolar square pulses, 600 Hz, 98 L/h, 1.4 µs	7.0	6 tubular chambers, $d = 1$ cm	Yeom et al. (2000a)
Aspergillus niger	Peptone	20°C–25°C, 30 kV cm^{-1}, 500 ns, 3000 square pulses, 5–10 Hz	2.0	Parallel plates chamber, $d = 1$ cm	McGregor et al. (2000)
Micrococcus luteus	$Na_2HPO_4 + NaH_2PO_4$	17°C, $V = 28.6$ mL, 32 kV cm^{-1}, 2 µs, 50 pulses, 0.5 L/min, $V = 1800$ mL	2.4	Concentric chamber, continuous, $d = 0.6$ cm	Dutreux et al. (2000a)
Saccharomyces cerevisiae	SMUF	40°C, 36 kV cm^{-1}, exponential decay pulses, 2–3 µs pulse width	>7	Coaxial treatment chamber, $V = 29$ mL, 0.6 cm gap	Qin et al. (1998)
S. cerevisiae	Deionized water + HCl	30°C, 12.5 kV cm^{-1}, 0.3 ms pulse duration, 2–20 pulses	3.0	Electroporation cuvette, 0.2 cm gap, $V = 250$ µL	Geveke and Kozempel (2003)
S. cerevisiae	Bacto peptone	30°C, 30 kV cm^{-1}, 4 µs, 20 pulses, 250 Hz, 41 mL, $t = 13$ ms	6.0	Bench, continuous, 6 tubular parallel chambers, $d = 2.3$ mm	Aronsson et al. (2001)
S. cerevisiae	Orange juice	30°C, 30 kV cm^{-1}, 50 kV cm^{-1} (2), 100 L/h, 2 ms	2.5	ColPure™ PEF system, $d = 5$ mm	McDonald et al. (2000)
S. cerevisiae	Apple juice	>30°C, 1.2 V μm^{-1}, 20 pulses, square wave	4.2	Bench, parallel plates	Qin et al. (1994)
S. cerevisiae	Phosphate buffer	30°C, 5–30 kV cm^{-1}, square 4 µs, 36 mL/min, 250 Hz	6.8	Cofield, $V = 8.5$ µL, $D = 2.0$ mm, $d = 2.7$ mm	Aronsson et al. (2005)

(continued)

TABLE 6.2 (continued)
Inactivation Kinetics of Microorganisms by PEFs

Microorganism	Suspension Media	Process Condition	Log or % Reduction (Max)	Treatment Vessel	Source
S. cerevisiae	Apple juice	>30°C, $V = 29$ mL, 2–10 L/min, 1 Hz, 150 exponential decay pulses	7.0	Continuous, $d = 0.6$ cm, coaxial chamber	Qin et al. (1995a)
S. cerevisiae	Peptone	20°C–25°C, 30 kV cm^{-1}, 500 ns, 3000 square pulses, 5–10 Hz	5.0–6.0	Parallel plates chamber, $d = 1$ cm	McGregor et al. (2000)
S. cerevisiae	Apple juice	20°C–30°C, 10–28 kV cm^{-1}, 8.3 pulses or 4.2–10.4 kV cm^{-1} at 20 kV cm^{-1}, bipolar square wave pulse of 2 µs duration, 84 mL/ min	4.0	6 cofield treatment chambers, 0.29 cm gap	Cserhalmi et al. (2002)
Total microorganisms	Milk	52°C, 80 kV cm^{-1}, 50 pulses, instant charge reversal pulse wave form of 2 µs, nisin (38 IU/mL), lysozyme (1638 IU/mL), pH 5 and 6.7	7.0	Circular treatment chamber, $D = 16.5$ cm, 3 mm gap, $V = 49.5$ mL	Smith et al. (2002)

6.5 APPLICATIONS OF PEF TECHNOLOGY IN FOOD PRESERVATION

The modern food industry has already been aware of the promising results of PEF treatment of apple, grape, orange juices, as well as milk and liquid egg products (Barbosa-Cánovas et al., 1999; Sampedro et al., 2006). Some of these PEF-treated foods exhibit an extension in shelf life of more than 8 weeks without refrigeration, which is one of the most attractive attributes of the technology. However, to date, PEF has been mainly applied to preserve the quality of foods such as milk, orange juice, and apple juice. Most of the results on liquid foods have shown that PEF treatment has little effect on the physicochemical, organoleptic, and nutritional properties of foods, compared to traditional thermal pasteurization and other pasteurization processes (Martín-Belloso et al., 1997; Vega-Mercado et al., 1997; Ortega-Rivas et al., 1998; Jia et al., 1999; Evrendilek et al., 2000; Yeom et al., 2000a,b; Zárate-Rodriguez et al., 2000; Cortés et al., 2006, 2008; Plaza et al., 2006; Aguilar-Rosas et al., 2007; Yongguang et al., 2007; Odriozola-Serrano et al., 2008a,b).

Hermawan et al. (2004) showed a 90% reduction in *S. enteritidis* inoculated into liquid whole egg with a circulation-mode fluid handling system using 200 × 2.12 μs pps at EF strength of 25 kV cm^{-1}. Moreover, combinations of PEF and heat treatment at 55°C showed a synergistic effect, resulting in 4.5-log reduction of *S. enteritidis*.

The PEF treatment of liquid egg products, including whole egg, egg white, and egg yolk, inoculated with *S. enteritidis* and/or *E. coli* O157:H7 was reported by Amiali et al. (2004, 2005, 2006b, 2007). Amiali et al. (2004) dialyzed liquid egg products in order to obtain high EF intensity. The maximum inactivation rates of *E. coli* O157:H7 were 1-, 2.9-, and 3.5-log reduction for egg white, egg yolk, and whole egg, respectively. The energy density required to treat egg white was 5.210 kJ L^{-1}, whereas 3.080 kJ L^{-1} was required for egg yolk and whole egg products. The results indicated that higher energy was required to process products with higher electrical conductivity (Amiali et al., 2004).

Amiali et al. (2005) inactivated *S. enteritidis* and/or *E. coli* O157:H7 in liquid whole egg using a continuous PEF system in combination with heat. The bacteria were treated at 10°C, 20°C, or 30°C using EF intensity of either 20 or 30 kV cm^{-1}. A biphasic instant reversal PEF waveform with up to 105 pulses of 2 μs in pulse width was applied. The maximum reduction of 3.9 and 3.6 log cycles were obtained for *E. coli* O157:H7 and *S. enteritidis*, respectively. Higher kinetic constant value was obtained for *S. enteritidis* (0.043 μs^{-1}), representing the more heat-PEF-sensitive bacteria compared to *E. coli* O157:H7.

Amiali et al. (2006b) reported results on a possible PEF treatment of high electrical conductivity and heat-sensitive product such as liquid egg white. The liquid egg white was inoculated with *S. enteritidis* or *E. coli* O157:H7 at a concentration of 10^8 CFU mL^{-1} and treated with up to 60 pulses (2 μs pulse width) at EF intensities of 20 and 30 kV cm^{-1}. The processing temperatures were 10°C, 20°C, and 30°C. The inactivation of 2.9 and 3.7 log cycles were obtained for *S. enteritidis* and *E. coli* O157:H7, respectively, while injured cells accounted for 0.9 and 0.5 log cycle for *S. enteritidis* and *E. coli* O157:H7, respectively. A synergy between EF intensity and processing temperature was noted. The inactivation rate constants on both selective (VRBA or SSA) and nonselective (TSA) agars for *E. coli* O157:H7 were 8.2×10^{-3} and $6.6 \times 10^{-3}\,\mu s^{-1}$, whereas the values for *S. enteritidis* were 16.2×10^{-3} and $12.6 \times 10^{-3}\,\mu s^{-1}$, respectively. This again indicated that *E. coli* O157:H7 was more resistant to heat-PEF treatment compared to *S. enteritidis*.

Synergistic effect of temperature and PEF on inactivation of *S. enteritidis* or *E. coli* O157:H7 in liquid egg yolk was investigated by Amiali et al. (2007). The flow process temperatures and EF intensities were 20°C, 30°C, and 40°C and 20 and 30 kV cm^{-1}, respectively. The Arrhenius-type equation was used to determine the changes of the kinetics constant with respect to temperature. At 30 kV cm^{-1} and 40°C, the population of both bacteria were reduced by 5 log cycles. The rate constant increased from 0.004 to 0.098 μs^{-1} for *S. enteritidis*, whereas for *E. coli* O157:H7 the constant increased from 0.009 to 0.039 μs^{-1} as the processing temperature increased from 20°C to 40°C. In this case, *S. enteritidis* was more resistant to heat-PEF inactivation than *E. coli* O157:H7 at lower processing temperatures. The data of kinetic rate constants with respect of temperature yielded Equations 6.18 and 6.19 for *E. coli* O157:H7 and *S. enteritidis*, respectively.

$$k_T = 2.46\times10^6 \cdot e^{\left[\frac{-47}{8.31\cdot10^{-3}\cdot T}\right]} \tag{6.18}$$

$$k_T = 9.24\times10^6 \cdot e^{\left[\frac{-48.3}{8.31\cdot10^{-3}\cdot T}\right]} \tag{6.19}$$

6.6 CONCLUDING REMARKS

Research conducted in the past has clearly demonstrated the effectiveness of PEF in food processing applications, especially for microbial inactivation. PEF can replace many conventional thermal processing methods to produce higher quality food products, which today consumers expect from food processors.

ACKNOWLEDGMENTS

The authors extend their sincere gratitude for the financial support provided by the Natural Science and Engineering Research Council of Canada (NSERC) and Fonds Québécois de la recherché sur la nature et les technologies (FQRNT).

REFERENCES

Aguilar-Rosas, S.F., Ballinas-Casarrubias, M.L., Nevarez-Moorillon, G.V., Martin-Belloso, O., and Ortega-Rivas, E. 2007. Thermal and pulsed electric fields pasteurization of apple juice: Effects on physicochemical properties and flavour compounds. *Journal of Food Engineering* 83: 41–46.

Alkhafaji, S.R. and Farid, M. 2007. An investigation on pulsed electric fields technology using new treatment chamber design. *Innovative Food Science and Emerging Technologies* 8: 205–212.

Alvarez, I., Raso, J., Palop, A., and Sala, F.J. 2000. Influence of different factors on the inactivation of *Salmonella senftenberg* by pulsed electric fields. *International Journal of Food Microbiology* 55: 143–146.

Alvarez, I., Virto, R., Raso, J., and Condon, S. 2003. Comparing predicting models for *Escherichia coli* inactivation by pulsed electric fields. *Innovative Food Science and Emerging Technologies* 4: 195–202.

Amiali, M. 2005. Inactivation of *Escherichia coli* O157:H7 and *Salmonella enteritidis* in liquid egg products using pulsed electric field. PhD dissertation, McGill University, Montreal, Quebec, Canada.

Amiali, M., Ngadi, M.O., Smith, J.P., and Raghavan, V.G.S. 2004. Inactivation of *Escherichia coli* O157:H7 in liquid dialyzed egg using pulsed electric fields. *Food and Bioproducts Processing* 82: 151–156.

Amiali, M., Ngadi, M.O., Smith, J.P., and Raghavan, V.G.S. 2005. Inactivation of *Escherichia coli* O157:H7 and *Salmonella enteritidis* in liquid whole egg using continuous pulsed electric field system. *International Journal of Food Engineering* 1(5): Article 8.

Amiali, M., Ngadi, M.O., Raghavan, V.G.S., and Nguyen, D.H. 2006a. Electrical conductivities of liquid egg products and fruits juices exposed to high pulsed electric fields. *International Journal of Food Properties* 9: 533–540.

Amiali, M., Ngadi, M.O., Smith, J.P., and Raghavan, V.G.S. 2006b. Inactivation of *Escherichia coli* O157:H7 and *Salmonella enteritidis* in liquid egg white using pulsed electric field. *Journal of Food Science* 71(3): M88–M94.

Amiali, M., Ngadi, M.O., Smith, J.P., and Raghavan, V.G.S. 2007. Synergistic effect of temperature and pulsed electric field on inactivation of *Escherichia coli* O157:H7 and *Salmonella enteritidis* in liquid egg yolk. *Journal of Food Engineering* 79: 689–694.

Aronsson, K., Lindgren, M., Johansson, B.R., and Rönner, U. 2001. Inactivation of microorganisms using pulsed electric fields: The influence of process parameters on *Escherichia coli, Listeria innocua, Leuconostoc mesenteroides* and *Saccharomyces cerevisiae. Innovative Food Science and Emerging Technologies* 2: 41–45.

Aronsson, K., Borch, E., Stenlöf, B., and Rönner, U. 2004. Growth of pulsed electric field exposed *Escherichia coli* in relation to inactivation and environmental factors. *International Journal of Food Microbiology* 93: 1–10.

Aronsson, K., Rönner, U., and Borch, E. 2005. Inactivation of *Escherichia coli, Listeria innocua* and *Saccharomyces cerevisiae* in relation to membrane permeabilization and subsequent leakage of intracellular compounds due to pulsed electric field processing. *International Journal of Food Microbiology* 99: 19–32.

Barbosa-Cánovas, G.V., Góngora-Nieto, M.M., Pothakamury, U.R., and Swanson, B.G. 1999. *Preservation of Foods with Pulsed Electric Fields*. San Diego, CA: Academic Press.

Barsotti, L., Merle, P., and Cheftel, J.C. 1999a. Food processing by pulsed electric fields. I. Physical aspects. *Food Review International* 15: 163–180.

Barsotti, L., Merle, P., and Cheftel, J.C. 1999b. Food processing by pulsed electric fields. II. Biological aspects. *Food Review International* 15: 181–213.

Bazhal, M.I., Ngadi, M.O., Smith, J.P., and Raghavan, V.G.S. 2003. Combined effect of pulsed electric field and temperature on *Escherichia coli* O157:H7 inactivation. *ASAE Annual International Meeting*, Las Vegas, NV, July 13–16, 2003.

Bazhal, M.I., Ngadi, M.O., Smith, J.P., and Raghavan, V.G.S. 2006. Inactivation of *Escherichia coli* in liquid whole egg using combined PEF and thermal treatments. *LWT-Food Science and Technology* 39: 419–425.

Bigelow, W.D. 1921. The logarithmic nature of thermal death time curves. *Journal of Infectious Diseases* 29: 528–536.

Bryant, G. and Wolfe, J. 1987. Electromechanical stress produced in plasma membranes of suspended cell by applied electrical fields. *Journal of Membrane Biology* 96: 129–139.

Bushnell, A.H., Dunn, J.E., and Clark, R.W. 1993. High pulsed voltage systems for extending the shelf life of pumpable food products. U.S. Patent 5235905.

Calderón-Miranda, M.L., Barbosa-Cánovas, G.V., and Swanson, B.G. 1999a. Transmission electron microscopy of *Listeria innocua* treated by pulsed electric fields and nisin skimmed milk. *International Journal of Food Microbiology* 51: 31–38.

Calderón-Miranda, M.L., Barbosa-Cánovas, G.V., and Swanson, B.G. 1999b. Inactivation of *Listeria innocua* in liquid whole egg by pulsed electric fields and nisin. *International Journal of Food Microbiology* 51: 7–17.

Castro, A., Barbosa-Cánovas, G.V., and Swanson, B.G. 1993. Microbial inactivation of foods by pulsed electric fields. *Journal of Food Preservation* 17: 47–73.

Cebrián, G., Sagarzazu, N., Pagán, R., Condón, S., and Mañas, P. 2007. Heat and pulsed electric field resistance of pigmented and non-pigmented enterotoxigenic strains of *Staphylococcus aureus* in exponential and stationary phase of growth. *International Journal of Food Microbiology* 118: 304–311.

Cebrián, G., Sagarzazu, N., Pagán, R., Condón, S., and Mañas, P. 2008. Resistance of *Escherichia coli* grown at different temperatures to various environmental stresses. *Journal of Applied Microbiology* 105: 271–278.

Charles-Rodríguez, A.V., Nevárez-Moorillón, G.V., Zhang, Q.H., and Ortega-Rivas, E. 2007. Comparison of thermal processing and pulsed electric fields treatment in pasteurization of apple juice. *Food and Bioproducts Processing* 85: 93–97.

Cortés, C., Esteve, M.J., and Frígola, A. 2008. Color of orange juice treated by high intensity pulse electric fields during refrigerated storage and comparison with pasteurized juice. *Food Control* 19: 151–158.

Cortés, C., Esteve, M.J., Rodrigo, D., Torregrosa, F., and Frígola, A. 2006. Changes of colour and carotenoids contents during high intensity pulsed electric field treatment in orange juices. *Food and Chemical Toxicology* 44: 1932–1939.

Coster, H.G.L. and Zimmerman, U. 1975. The mechanisms of electrical breakdown in the membrane of *Valonia utricularis*. *Journal of Membrane Biology* 22: 73–90.

Craven, H.M., Swiergon, P., Ng, S., Midgely, J., Versteeg, C., Coventry, M.J., and Wan, J. 2008. Evaluation of pulsed electric field and minimal heat treatments for inactivation of pseudomonads and enhancement of milk shelf-life. *Innovative Food Science and Emerging Technologies* 9: 211–216.

Cserhalmi, Zs., Vidács, I., Beczner, J., and Czukor, B. 2002. Inactivation of *Saccharomyces cerevisiae* and *Bacillus cereus* by pulsed electric fields technology. *Innovative Food Science and Emerging Technologies* 3: 41–45.

Damar, S., Bozoğhi, F., Hizal, M., and Bayindirh, A. 2002. Inactivation and injury of *Escherichia coli* O157:H7 and *Staphylococcus aureus* by pulsed electric fields. *World Journal of Microbiology and Biotechnology* 18: 1–6.

Donsì, G., Ferrari, G., and Pataro, G. 2007. Inactivation kinetics of *Saccharomyces cerevisiae* by pulsed electric fields in a batch treatment chamber: The effect of electric field unevenness and initial cell concentration. *Journal of Food Engineering* 78: 784–792.

Dunn, J. 1996. Pulsed light and pulsed electric field for foods and eggs. *Poultry Science* 75: 1133–1136.

Dunn, J.E. and Pearlman, J.S. 1987. Methods and apparatus of extending the shelf-life of fluid food products. U.S. Patent 4695472.

Dutreux, N., Notermans, S., Góngora-Nieto, M.M., Barbosa-Cánovas, G.V., and Swanson, B.G. 2000a. Effects of combined exposure of *Micrococcus luteus* to nisin and pulsed electric fields. *International Journal of Food Microbiology* 60: 147–162.

Dutreux, N., Notermans, S., Wijtzes, T., Góngora-Nieto, M.M., Barbosa-Cánovas, G.V., and Swanson, B.G. 2000b. Pulsed electric fields inactivation of attached and free-living *Listeria innocua* under several conditions. *International Journal of Food Microbiology* 54: 91–98.

El Zakhem, H., Lanoisellé, J.L., Lebovka, N.I., Nonus, M., and Vorobiev, E. 2007. Influence of temperature and surfactant on *Escherichia coli* inactivation in aqueous suspensions treated by moderate pulsed electric fields. *International Journal of Food Microbiology* 120: 259–265.

Elez-Martínez, P., Suárez-Recio, M., and Martín-Belloso, O. 2007. Modeling the reduction of pectin methyl esterase activity in orange juice by high intensity pulsed electric fields. *Journal of Food Engineering* 78: 184–193.

Evrendilek, G.A. and Zhang, Q.H. 2003. Effect of pH, temperature, and prepulsed electric field and heat inactivation of *Escherichia coli* O157:H7. *Journal of Food Protection* 66: 735–759.

Evrendilek, G.A. and Zhang, Q.H. 2005. Effects of pulse polarity and pulse delaying time on pulsed electric fields-induced pasteurization on *Escherichia coli* O157:H7. *Journal of Food Engineering* 68: 271–276.

Evrendilek, G.A., Zhang, Q.H., and Richter, E.R. 1999. Inactivation of *Escherichia coli* O157:H7 and *Escherichia coli* 8739 in apple juice by pulsed electric fields. *Journal of Food Protection* 62: 793–796.

Evrendilek, G.A., Jin, Z.T., Ruhlman, K.T., Qiu, X., Zhang, Q.H., and Richter, E.R. 2000. Microbial safety and shelf-life of apple juice and cider processed by bench and pilot scale PEF systems. *Innovative Food Science and Emerging Technologies* 1: 77–86.

Everendilek, G.A., Zhang, Q.H., and Richter, E.R. 2004. Application of pulsed electric fields to skim milk inoculated with *Staphylococcus aureus*. *Biosystems Engineering* 87: 137–144.

Evrendilek, G.A., Tok, F.M., Soylu, E.M., and Soylu, S. 2008. Inactivation of *Penicillium expansum* in sour cherry juice, peach and apricot nectars by pulsed electric fields. *Food Microbiology* 25: 662–667.

FDA. 2000a. Kinetics of microbial inactivation for alternative food processing technologies, Washington, DC.

FDA. 2000b. Pulsed electric field, Report, Washington, DC.

Fernández-Molina, J.J., Barkstrom, E., Torstensson, P., Barbosa-Cánovas, G.V., and Swanson, B.G. 1999. Shelf-life extension of raw skim milk by combining heat and pulsed electric fields. In: Barbosa-Cánovas, G.V. and Lombardo, S. (Eds.), *The 6th Conference of Food Engineering (CoFE&rlenis;99), AIChE*, Dallas, TX, pp. 349–355.

Ferrer, C., Rodrigo, D., Pina, M.C., Klein, G., Rodrigo, M., and Martínez, A. 2007. The Monte Carlo simulation is used to establish the most influential parameters on the final load of pulsed electric fields *E. coli* cells. *Food Control* 18: 934–938.

Fleischman, G.J., Ravishankar, S., and Balasubramaniam, V.M. 2004. The inactivation of *Listeria monocytogenes* by pulsed electric field (PEF) treatment in static chamber. *Food Microbiology* 21: 91–95.

Floury, J., Grosset, N., Lesne, E., and Jeantet, R. 2006. Continuous processing of skim milk by a combination of pulsed electric field and conventional heat treatments: Does a synergetic effect on microbial inactivation exist? *Lait* 86: 203–211.

Gálvez, A., López, R.L., Abriouel, H., Valdivia, E., and Omar, N.B. 2008. Application of bacteriocins in the control of foodborne pathogenic and spoilage bacteria. *Critical Reviews in Biotechnology* 28: 125–151.

García, D., Gómez, N., Condón, S., Raso, J., and Pagán, R. 2003. Pulsed electric field cause sublethal injury in *Escherichia coli*. *Letters in Applied Microbiology* 36: 140–144.

García, D., Gómez, N., Mañas, P., Raso, J., and Pagán, R. 2007. Pulsed electric fields cause bacterial envelopes permeabilization depending on the treatment intensity, the treatment medium pH and the microorganism investigated. *International Journal of Food Microbiology* 113: 219–227.

Geveke, D.J. and Kozempel, M.F. 2003. Pulsed electric field effects on bacteria and yeast cells. *Journal of Food Processing and Preservation* 27: 65–72.

Giner-Seguí, J., Bailo-Ballarín, E., Gorinstein, S., and Martín-Belloso, O. 2006. New kinetics approach to the evolution pf polygalacturonase (EC 3.2.1.15) activity in a commercial enzyme preparation under pulsed electric fields. *Journal of Food Science* 71: E262–E269.

Gould, G.W. 2000. Preservation: Past, present and future. *British Medical Bulletin* 56: 84–96.

Grahl, T. and Märkl, H. 1996. Killing of microorganisms by pulsed electric fields. *Applied Microbiology and Biotechnology* 45: 148–157.

Grenier, J.R., Jayaram, S.H., Kazerani, M., Wang, H., and Griffiths, M.W. 2008. MOSFET-based pulse power supply for bacterial transformation. *IEEE Transactions on Industry Applications* 44: 25–31.

Gupta, B.S., Masterson, F., and Magee, T.R.A. 2003a. Application of high voltage pulsed electric field in pasteurization of liquid egg and milk. *Indian Chemistry Ingredients Section A* 45: 31–34.

Gupta, B.S., Masterson, F., and Magee, T.R.A. 2003b. Inactivation of *Escherichia coli* K12 in apple juice by high voltage pulsed electric field. *European Food Research and Technology* 217: 434–437.

Halden, K., De Alwis, A.A.P., and Fryer, P.J. 1990. Change in electric conductivity of foods during ohmic heating. *International Journal of Food Science and Technology* 25: 9–25.

Heinz, V. and Knorr, D. 2000. Effect of pH, ethanol addition and high hydrostatic pressure on the inactivation of *Bacillus subtilis* by pulsed electric fields. *Innovative Food Science and Emerging Technology* 1: 151–159.

Heinz, V., Phillips, S.T., Zenker, M., and Knorr, D. 1999. Inactivation of *Bacillus subtilis* by high intensity pulsed electric fields under close to isothermal conditions. *Food Biology* 13: 155–168.

Heinz, V., Toepfl, S., and Knorr, D. 2003. Impact of temperature on lethality and energy efficiency of apple juice pasteurization by pulsed electric fields treatment. *Innovative Food Science and Emerging Technologies* 4: 167–175.

Hermawan, G.A., Evrendilek, W.R., Zhang, Q.H., and Richter, E.R. 2004. Pulsed electric field treatment of liquid whole egg inoculated with *Salmonella enteritidis*. *Journal of Food Safety* 24: 1–85.

Ho, S.Y., Mittal, G.S., Cross, J.D., and Griffiths, M.W. 1995. Inactivation of *Pseudomonas fluorescens* by high voltage electric pulses. *Journal of Food Science* 60: 1337–1340, 1343.

Hülsheger, H., Potel, J., and Niemann, E.G. 1981. Killing of bacteria with electric pulses of high field strength. *Radiation and Environmental Biophysics* 20: 53–65.

Hülsheger, H., Potel, J., and Niemann, E.G. 1983. Electric field effects on bacteria and yeast cells. *Radiation and Environmental Biophysics* 22: 149–162.

Iu, J., Mittal, G.S., and Griffiths, M.W. 2001. Reduction in levels of *Escherichia coli* O157:H7 in apple cider by pulsed electric fields. *Journal of Food Protection* 64: 964–969.

Jayaram, S.H. 2000. Sterilisation of liquid foods by pulsed electric fields. *IEEE Electrical Insulation Magazine* 16: 17–25.

Jayaram, S., Castle, G.S.P., and Margaritis, A. 1992. Kinetics of sterilization of *Lactobacillus brevis* cells by the application of high voltage pulses. *Biotechnology Bioengineering* 40: 1412–1420.

Jeantet, R., Baron, F., Nau, F., Roignant, M., and Brulé, G. 1999. High intensity pulsed electric fields applied to egg white: Effect on *Salmonella enteritidis* inactivation and protein denaturation. *Journal of Food Protection* 62: 1381–1386.

Jeantet, R., Mc Keag, J.R., Fernández, J.C., Gosset, N., Baron, F., and Korolczuk, J. 2004. Pulsed electric field continuous treatment of egg products. *Sciences des Aliments* 24: 137–158.

Jeyamkondan, S., Jayas, D.S., and Holley, R.A. 1999. Pulsed electric field processing of foods: A review. *Journal of Food Protection* 62: 1088–1096.

Jia, M., Zhang, Q.H., and Min, D.B. 1999. Pulsed electric field processing effects on flavor compounds and microorganisms of orange juice. *Food Chemistry* 65: 445–451.

Kinosita, K. Jr. and Tsong, T.Y. 1977. Hemolysis of erythrocytes by transient electric field. *Proceeding of the National Academy of Science of the USA* 74: 1923–1927.

Korolczuk, J., Rippoll Mc Keag, J., Carballeira Fernandez, J., Baron, F., Grosset, N., and Jeantet, R. 2006. Effect of pulsed electric field processing parameters on *Salmonella enteritidis* inactivation. *Journal of Food Engineering* 75: 11–20.

Lado, B.H. and Youcef, A.E. 2003. Selection and identification of a *Listeria monocytogenes* target strain for pulsed electric fields process optimization. *Applied and Environmental Microbiology* 4: 2223–2229.

Liang, Z., Mittal, G.S., and Griffiths, M.W. 2002. Inactivation of *Salmonella typhimurium* in orange juice containing antimicrobial agents by pulsed electric field. *Journal of Food Protection* 65: 1081–1087.

Liu, X., Youcef, A.E., and Chism, G.W. 1997. Inactivation of *Escherichia coli* O157:H7 by the combination of organic acids and pulsed electric field. *Journal of Food Safety* 16: 287–299.

Manãs, P., Barsotti, L., and Cheftel, J.C. 2001. Microbial inactivation by pulsed electric fields in a batch treatment chamber: Effect of some electrical parameters and food constituents. *Innovative Food Science and Emerging Technologies* 2: 239–249.

Marquez, V.O., Mital, G.S., and Griffiths, M.W. 1997. Destruction and inhibition of bacterial spores by high voltage pulsed electric fields. *Journal of Food Science* 62: 399–409.

Martín, O., Qin, B.L., Chang, F.J., Barbosa-Cánovas, G.V., and Swanson, B.G. 1997. Inactivation of *Escherichia coli* in skim milk by high intensity pulsed electric fields. *Journal of Food Process Engineering* 20: 317–336.

Martín-Belloso, O., Vega-Mercado, H., Qin, B.L., Chang, F.J., Barbosa-Cánovas, G.V., and Swanson, B.G. 1997. Inactivation of *Escherichia coli* suspended in liquid egg using pulsed electric fields. *Journal of Food Processing and Preservation* 21: 193–208.

McDonald, C.J., Lloyd, S.W., Vitale, M.A., Petersson, K., and Innings, E. 2000. Effects of pulsed electric fields on microorganisms in orange juice using electric field strengths of 30 and 50kV/cm. *Journal of Food Science* 65: 984–989.

McGregor, S.J., Farish, O., Fouracre, R., Rowan, N.J., and Anderson, J.G. 2000. Inactivation of pathogenic and spoilage microorganisms in test liquid using pulsed electric fields. *IEEE Transactions on Plasma Science* 28: 144–149.

McLellan, M.R., Kime, R.L., and Lind, K.R. 1991. Electroplasmolysis and other treatments to improve apple juice yield. *Journal of the Science of Food and Agriculture* 57: 303–306.

Mermelstein, N.H. 1998. Processing paper cover wide range topics. *Food Technology* 52: 50–53.

Michalac, S. 2003. Inactivation of selected microorganisms and properties of pulsed electric field processed milk. *Journal of Food Processing and Preservation* 27: 137–151.

Min, S., Evrendilek, G.A., and Zhang, H.Q. 2007. Pulsed electric fields: Processing system, microbial and enzyme inhibition, and shelf life extension of foods. *IEEE Transactions on Plasma Science* 35: 59–71.

Mosqueda-Melgar, J., Raybaudi-Massilia, R.M., and Martín-Belloso, O. 2007. Influence of treatment time and pulse frequency on *Salmonella enteritidis*, *Escherichia coli* and *Listeria monocytogenes* populations inoculated in melon and watermelon juices treated by pulsed electric fields. *International Journal of Food Microbiology* 117: 192–200.

Mosqueda-Melgar, J., Raybaudi-Massilia, R.M., and Martín-Belloso, O. 2008a. Combination of high-intensity pulsed electric fields with natural antimicrobials to inactivate pathogenic microorganisms and extend the shelf-life of melon and watermelon juices. *Food Microbiology* 25: 479–491.

Mosqueda-Melgar, J., Raybaudi-Massilia, R.M., and Martín-Belloso, O. 2008b. Inactivation of *Salmonella enterica* ser. Enteritidis in tomato juice by combining of high-intensity pulsed electric fields with natural antimicrobials. *Journal of Food Science* 23: M47–M53.

Ngadi, M.O. and Bazhal, M. 2006. Pulsed electric field assisted juice extraction. *Stewart Postharvest Review* 2(4): 1–8.

Odriozola-Serrano, I., Soliva-Fortuny, R., and Martín-Belloso, O. 2008a. Changes of health-related compounds throughout cold storage of tomato juice stabilized by thermal or high intensity pulsed electric field treatments. *Innovative Food Science and Emerging Technologies* 9: 272–279.

Odriozola-Serrano, I., Soliva-Fortuny, R., Gimeno-Añó, V., and Martín-Belloso, O. 2008b. Modeling changes in health-related compounds of tomato juice treated by high-intensity pulsed electric fields. *Journal of Food Engineering* 89: 210–216.

Ortega-Rivas, E., Zárate-Rodriguez, E., and Barbosa-Cánovas, G.V. 1998. Apple juice pasteurization and pulsed electric fields. *Transactions IChemE*, Part C 76: 193–198.

Oshima, T., Sato, K., Terauchi, M., and Sato, M. 1997. Physical and chemical modifications of high-voltage pulse sterilization. *Journal of Electrostatics* 42: 159–166.

Oshima, T., Okuyama, K., and Sato, M. 2002. Effect of cultural temperature on high voltage pulse sterilization of *Escherichia coli*. *Journal of Electrostatics* 55: 227–235.

Palaniappan, S. and Sastry, S.K. 1991. Electrical conductivity of selected juices: Influence of temperature, solids content, applied voltage, and particle size. *Journal of Food Process Engineering* 14: 247–260.

Peleg, M. 1995. A model of microbial survival after exposure to pulsed electric fields. *Journal of the Food science and Agriculture* 67: 93–99.

Pérez, M.C.P., Aliaga, D.R., Bernat, C.F., Enguidanos, M.R., and López, A.M. 2007. Inactivation of *Enterobacter sakazakii* by pulsed electric field in buffered peptone water and infant formula milk. *International Dairy Journal* 17: 1441–1449.

Perni, S., Chalise, P.R., Shama, G., and Kong, M.G. 2007. Bacterial cells exposed to nano-second pulsed electric fields show lethal and sublethal effects. *International Journal of Food Microbiology* 120: 311–314.

Picart, L., Dumay, E., and Cheftel, J.C. 2002. Inactivation of *Listeria innocua* in dairy fluids by pulsed electric fields: Influence of electric parameters and food composition. *Innovative Food Science and Emerging Technologies* 3: 357–369.

Pizzichemi, M. 2007. Application of pulsed electric fields to food treatment. *Nuclear Physics B (Proceedings Supplements)* 172: 314–316.

Plaza, L., Sánchez-Moreno, C., Elez-Martínez, P., de Ancos, B., Martín-Belloso, O., and Pilar Cano, M. 2006. Effect of refrigerated storage on vitamin C and antioxidant activity of orange juice processed by high-pressure or pulsed electric fields with regard to low pasteurization. *European Food Research and Technology* 223: 487–493.

Pol, I.E., Mastwijik, H.C., Bartel, P.V., and Smid, E.J. 2000. Pulsed electric field treatment enhances the bactericidal action of nisin against *Bacillus cereus*. *Applied and Environmental Microbiology* 1: 428–430.

Pol, I.E., Van Arendonk, W.G.C., Mstwijik, H.C., Krommer, J., Smid, E.J., and Moezelaar, R. 2001a. Sensitivities of germinating spores and carvacrol-adapted vegetative cells and spores of *Bacillus cereus* to nisin and pulsed-electric-field treatment. *Applied and Environmental Microbiology* 4: 1693–1699.

Pol, I.E., Mastwijik, H.C., Slump, R.A., Popa, M.E., and Smid, E.J. 2001b. Influence of food matrix on inactivation of *Bacillus cereus* by combination of nisin, pulsed electric field treatment, and carvacrol. *Journal of Food Protection* 64: 1012–1018.

Pothakamury, U.R., Monsalve-González, A., Barbosa-Cánovas, G.V., and Swanson, B.G. 1995. Inactivation of *Escherichia coli* and *Staphylococcus aureus* in model foods by pulsed electric field technology. *Food Research International* 28: 167–171.

Pothakamury, U.R., Vega, H., Zhang, Q., Barbosa-Cánovas, G.V., and Swanson, B.G. 1996. Effect of growth stage and processing temperature on the inactivation of *Escherichia coli* by pulsed electric fields. *Journal of Food Processing* 59: 1167–1171.

Pruit, K. and Kamau, D.N. 1993. Mathematical models of bacteria growth, inhibition and death under combined stress conditions. *Journal of Industrial Microbiology* 12: 221–231.

Qin, B.L., Barbosa-Cánovas, G.V., Swanson, B.G., and Pedrow, P.D. 1994. Inactivation of microorganisms by pulsed electric field of different voltage waveforms. *IEEE Transactions on Dielectrics and Electrical Insulation* 1: 1047–1057.

Qin, B., Zhang, Q., Barbosa-Cánovas, G.V., Swanson, B.G., and Pedrow, P.D. 1995a. Pulsed electric field treatment chamber design for liquid food pasteurization using a finite element method. *Transactions of the ASAE* 38: 557–565.

Qin, B.L., Chang, F.J., Barbosa-Cánovas, G.V., and Swanson, B.G. 1995b. Nonthermal inactivation of *Saccharomyces cerevisiae* in apple juice using pulsed electric fields. *LWT-Food Science and Technology* 28: 564–568.

Qin, B.-L., Pothakamury, U.R., Barbosa-Cánovas, G.V., and Swanson, B.G. 1996. Nonthermal pasteurization of liquid foods using high-intensity pulsed electric field. *Critical Reviews in Food Science & Nutrition* 36: 603–627.

Qin, B.L., Barbosa-Cánovas, G.V., Swanson, B.G., and Pedrow, P.D. 1998. Inactivating microorganisms using a pulsed electric field continuous treatment system. *IEEE Transactions on Industrial Applications* 34(1): 43–49.

Raso, J. and Barbosa-Cánovas, G.V. 2003. Nonthermal preservation of foods using combined processing techniques. *Critical Reviews in Food Science & Nutrition* 43(3): 265–285.

Raso, J., Calderón, M.L., Góngora-Nieto, M.M., Barbosa-Cánovas, G.V., and Swanson, B.G. 1998. Inactivation of *Zygosaccharomyces bailii* in fruit juices by heat, hydrostatic pressure and pulsed electric fields. *Journal of Food Science* 63: 1042–1044.

Raso, J., Alvarez, I., Condón, S., and Sala, F.J. 2000. Predicting inactivation of *Salmonella senftenberg* by pulsed electric fields. *Innovative Food Science and Emerging Technologies* 1: 21–30.

Ravishankar, S., Fleischman, G.J., and Balasubramaniam, V.M. 2002. The inactivation of *Escherichia coli* O157:H7 during pulsed electric field (PEF) treatment in static chamber. *Food Microbiology* 19: 351–361.

Reina, L.D., Jin, Z.T., Youcef, A.E., and Zhang, Q.H. 1998. Inactivation of *Listeria monocytogenes* in milk by pulsed electric fields. *Journal of Food Protection* 61: 1203–1206.

Reyns, K.M.F.A., Diels, A.M.J., and Michiels, C.W. 2004. Generation of bacterial and mutagenic components by pulsed electric treatment. *International Journal of Food Microbiology* 93: 165–173.

Rodrigo, D., Martínez, A., Harte, F., Barbosa-Cánovas, G.V., and Rodrigo, M. 2001. Study of inactivation of *Lactobacillus plantarum* in orange-carrot juice by means of pulsed electric fields: Comparison of inactivation kinetics models. *Journal of Food Protection* 64: 259–263.

Rodrigo, D., Barbosa-Cánovas, G.V., Martínez, A., and Rodrigo, M. 2003. Weibull distribution function based on empirical mathematical model for inactivation of *Escherichia coli* by pulsed electric fields. *Journal of Food Protection* 66: 1007–1012.

Rojas, M.C., Martin, S.E., Wicklund, R.A., Paulson, D.D., Desantos, F.A., and Brewer, M.S. 2007. Effect of high-intensity pulsed electric fields on survival of *Escherichia coli* K-12 suspended in meat injection solutions. *Journal of Food Safety* 27: 411–425.

Rowan, N.J., MacGregor, S.J., Anderson, J.G., Cameron, D., and Farish, O. 2001. Inactivation of *Mycobacterium paratuberculosis* by pulsed electric fields. *Applied and Environmental Microbiology* 6: 2833–2836.

Ruhlman, K.T., Jin, Z.T., and Zhang, Q.H. 2001. Physical properties of liquid foods for pulsed electric fields treatment. In: Barbosa-Cánovas, G.V. and Zhang, Q.W. (Eds.), *Pulsed Electric Field in Food Processing: Fundamental Aspect and Application*, pp. 45–56. Lancaster, PA: Technomic Publishing.

Russell, N.J. 2002. Bacterial membranes: The effect of chill storage and food processing. An overview. *International Journal of Food Microbiology* 79: 27–34.

Russell, N.J., Colley, M., Simpson, R.K., Trivett, A.J., and Evans, R.I. 2000. Mechanism of action of pulsed high electric fields (PHEF) on the membranes of food-poisoning bacteria is an 'all-or-nothing' effect. *International Journal of Food Microbiology* 55: 133–136.

Sale, A.J.H. and Hamilton, W.A. 1967. Effects of high electric fields on microorganisms. I. Killing of bacteria and yeasts. *Biochemica et Biophysica Acta* 143: 781–788.

Sampedro, F., Rodrigo, D., Martínez, A., Barbosa-Cánovas, G.V., and Rodrigo, M. 2006. Review: Application of pulsed electric fields in egg and egg derivatives. *Food Science and Technology International* 12: 397–405.

Sampedro, F., Rivas, A., Rodrigo, D., Martínez, M., and Rodrigo, M. 2007. Pulsed electric fields inactivation of *Lactobacillus plantarum* in an orange juice-milk based beverage: Effect of process parameters. *Journal of Food Engineering* 80: 931–938.

San Martín, M.F., Sepúlveda, D.R., Altunakar, B., Góngora-Nieto, M.M., Swanson, B.G., and Barbosa-Cánovas, G.V. 2007. Evaluation of selected mathematical models to predict the inactivation of *Listeria innocua* by pulsed electric fields. *LWT—Food Science and Technology* 40: 1271–1279.

Sato, M. 2008. Environmental and biotechnological applications of high-voltage pulsed discharges in water. *Plasma Sources Science and Technology* 17: 1–7.

Sato, M., Ishida, N.M., Sugiarto, A.T., Oshima, T., and Tanigushi, H. 2001. High efficiency sterilizer by high voltage pulse using concentrated field electrode system. *IEEE Transactions on Industry Applications* 37: 1646–1650.

Selma, M.V., Salmerón, M.C., Valero, M., and Fernández, P.S. 2004. Control of *Lactobacillus plantarum* and *Escherichia coli* by pulsed electric fields in MRS broth, nutrient broth and orange-carrot juice. *Food Microbiology* 21: 519–525.

Selma, M.V., Salmerón, M.C., Valero, M., and Fernández, P.S. 2006. Efficacy of pulsed electric fields for *Listeria monocytogenes* inactivation and control in horchata. *Journal of Food Safety* 26: 137–149.

Sensoy, I., Zhang, Q.H., and Sastry, S.K. 1997. Inactivation kinetics of *Salmonella dublin* by pulsed electric field. *Journal of Food Process Engineering* 20: 367–381.

Shamsi, K., Versteeg, C., Sherkat, F., and Wan, J. 2008. Alkaline phosphatase and microbial inactivation by pulsed electric field in bovine milk. *Innovative Food Science and Emerging Technologies* 9: 217–223.

Smith, K., Mittal, G.S., and Griffiths, M.W. 2002. Pasteurization of milk using pulsed electrical field and antimicrobials. *Food Microbiology and Safety* 67: 2304–2308.

Sobrino-López, A. and Martín-Belloso, O. 2008. Use of nisin and other bacteriocins for preservation of dairy products. *International Dairy Journal* 18: 329–343.

Somolinos, M., Garcia, D., Condon, S., Manas, P., and Pagan, R. 2007. Relationship between sublethal injury and inactivation of yeast cells by the combination of sorbic acid and pulsed electric fields. *Applied and Environmental Microbiology* 73: 3814–3821.

Somolinos, M., Garcia, D., Condon, S., Manas, P., and Pagan, R. 2008a. Biosynthetic requirements for the repair of sublethally injured *Saccharomyces cerevisiae* cells after pulsed electric fields. *Journal of Applied Microbiology* 105: 166–174.

Somolinos, M., Mañas, P., Condón, S., Pagán, R., and García, D. 2008b. Recovery of *Saccharomyces cerevisiae* sublethally injured cells after pulsed electric fields. *International Journal of Food Microbiology* 125: 352–353.

Somolinos, M., Garcia, D., Pagan, R., Condon, S., and Manas, P. 2008a. Relationship between sublethal injury and inactivation of yeast cells by the combination of sorbic acid and pulsed electric fields. *Applied and Environmental Microbiology* 73: 3814–3821.

Somolinos, M., Mañas, P., Condón, S., Pagán, R., and García, D. 2008b. Recovery of *Saccharomyces cerevisiae* sublethally injured cells after pulsed electric fields. *International Journal of Food Microbiology* 125: 352–356.

Stanley, D.W. 1991. Biological membrane deterioration and associated quality losses in food tissues. In: Clydesdale, F.M. (Ed.), *Critical Reviews in Food Science and Nutrition*, pp. 235–245. New York: CRC Press.

Terebiznik, M., Jagus, R., Cerrutti, P., de Huergo, M.S., and Pilosof, A.M.R. 2002. Inactivation of *Escherichia coli* by combination of nisin, pulsed electric field, and water activity reduction by sodium chloride. *Journal of Food Protection* 65: 1253–1258.

Toepfl, S., Heinz, V., and Knorr, D. 2007. High intensity pulsed electric fields applied for food preservation. *Chemical Engineering and Processing* 46: 537–546.

Ulmer, H.M., Heinz, V., Gänzie, M.G., Knorr, D., and Vogel, R.F. 2002. Effects of pulsed electric fields on inactivation and metabolic activity of *Lactobacillus plantarum* in model beer. *Journal of Applied Microbiology* 93: 326–355.

Unal, R., Kim, J.G., and Youcef, A.E. 2001. Inactivation of *Escherichia coli* O157:H7, *Listeria monocytogenes*, and *Lactobacillus leichmannii* by combinations of ozone and pulsed electric field. *Journal of Food Processing* 64: 777–782.

Unal, R., Yousef, A.E., and Dunne, P.C. 2002. Spectrophotometric assessment of bacterial cell membrane damage by pulsed electric field. *Innovative Food Science and Emerging Technologies* 3: 247–254.

van Boekel, M. 2002. On the use of the Weibull model to describe thermal inactivation of microbial vegetative cells. *International Journal of Food Microbiology* 74: 139–159.

Vega-Mercado, H., Powers, J.R., Barbosa-Cánovas, G.V., and Swanson, B.G. 1995. Plasmin inactivation with pulsed electric fields. *Journal of Food Science* 60: 1143–1146.

Vega-Mercado, H., Martín-Belloso, O., Chang, F.J., Barbosa-Cánovas, G.V., and Swanson, B.G. 1996. Inactivation of *Escherichia coli* and *Bacillus subtilis* suspended in pea soup using pulsed electric fields. *Journal of Food Processing and Preservation* 20: 501–510.

Vega-Mercado, H., Martín-Belloso, O., Qin, B.L., Chang, F.J., Góngora-Nieto, M.M., Barbosa-Cánovas, G.V., and Swanson, B.G. 1997. Non-thermal food preservation: Pulsed electric fields. *Trends in Food Science and Technology* 8: 151–157.

Walkling-Ribeiro, M., Noci, F., Cronin, D.A., Riener, J., Lyng, J.G., and Morgan, D.J. 2008. Reduction of *Staphylococcus aureus* and quality changes in apple juice processed by ultraviolet irradiation, pre-heating and pulsed electric fields. *Journal of Food Engineering* 89: 267–273.

Wouters, P.C., Bos, A.D.P., and Ueckert, J. 2001. Membrane permeabilization in relation to inactivation kinetics of *Lactobacillus* species due to pulsed electric fields. *Applied and Environmental Microbiology* 7: 3092–3101.

Wouters, P.C., Dutreux, N., Smelt, J.P.M., and Lelieveld, H.L.M. 1999. Effects of pulsed electric fields on inactivation kinetics of *Listeria innocua. Applied and Environmental Microbiology* 65: 5364–5371.

Wuytack, E.Y., Phuong, D.T., Aertsen, A., Reyns, K.M.F., Marquenie, D., De Ketelaere, B., Masschalk, B., Van Opstal, I., Diels, A.M.J., and Michiels, C.W. 2003. Comparison of sublethal injury induced *Salmonella enterica* serovar Typhimurium by heat and by different nonthermal treatments. *Journal of Food Protection* 66: 31–37.

Yeom, H.W., Streaker, C.B., Zhang, Q.H., and Min, D.B. 2000a. Effects on the quality of orange juice and comparison with heat pasteurization. *Journal of Agricultural and Food Chemistry* 48: 4597–4605.

Yeom, H.W., Streaker, C.B., Zhang, Q.H., and Min, D.B. 2000b. Effects of pulsed electric fields on the activities of microorganisms and pectin methyl esterase in orange juice. *Journal of Food Science* 65: 1359–1363.

Yin, Y., Han, Y., and Liu, J. 2007. A novel protecting method for visual green color in spinach puree treated by high intensity pulsed electric fields. *Journal of Food Engineering* 79: 1256–1260.

Yongguang, Y., Yong, H., and Jingbo, L. 2007. A novel protecting method for visual green color in spinach puree treated by high intensity pulsed electric fields. *Journal of Food Engineering* 79: 1256–1260.

Zárate-Rodríguez, E., Ortega-Rivas, E., and Barbosa-Cánovas, G.V. 2000. Quality changes in apple juice as related to nonthermal processing. *Journal of Food Quality* 23: 337–349.

Zhang, Q., Monsalve-González, A., Qin, B.L., Barbosa-Cánovas, G.V., and Swanson, B.G. 1994a. Inactivation of *Saccharomyces cerevisiae* in apple juice by square-wave and exponential-decay pulsed electric fields. *Journal of Food Process Engineering* 17: 469–478.

Zhang, Q.H., Chang, F.J., Barbosa-Cánovas, G.V., and Swanson, B.G. 1994b. Inactivation of microorganisms in a semisolid model food using high voltage pulsed electric fields. *LWT-Food Science and Technology* 27: 538–543.

Zhang, Q.H., Chang, F.J., Barbosa-Cánovas, G.V., and Swanson, B.G. 1995a. Engineering aspects of pulsed electric field pasteurization. *Journal of Food Engineering* 25: 261–291.

Zhang, Q.H., Qin, B.-L., Barbosa-Cánovas, G.V., and Swanson, B.G. 1995b. Inactivation of *Escherichia coli* for food pasteurization by high-strength pulsed electric fields. *Journal of Food Processing and Preservation* 19: 103–118.

Zhang, Q.H., Qiu, X., and Sharma, S.K. 1997. *Recent development in pulsed processing.* Washington, DC: National Food Processors Association. New Technologies Yearbook.

Zhao, W., Yang, R., Lu, R., Tang, Y., and Zhang, W. 2007. Investigation of the mechanisms of pulsed electric fields on inactivation of enzyme: Lysozyme. *Journal of Agricultural and Food Chemistry* 55: 9850–9858.

Zimmermann, U. 1986. Electrical breakdown, electropermeabilization and electrofusion. *Reviews of Physiology, Biochemistry and Pharmacology* 105: 176–256.

Zimmermann, U., Pilwat, G., and Riemann, F. 1974. Dielectric breakdown of cells membranes. *Biophysical Journal* 14: 881–899.

7 Physicochemical Changes of Foods during Freezing and Thawing

Adriana E. Delgado and Da-Wen Sun

CONTENTS

7.1 INTRODUCTION

Freezing is a well-known long-term preservation process widely used in the food industry. This is because changes in the nutritional or sensory characteristics of foods are small if appropriate freezing and storage procedures are followed. The freezing of foods normally consists of pre-freezing treatments, freezing, frozen storage, and thawing, each of which must be properly conducted to obtain optimum results (Fennema, 1977). When a product is frozen, the formed ice crystals may cause cell rupture and alterations in the transport properties of cell membranes, which have practical consequences in terms of leaching of cellular substances from tissues as well as water loss, leading normally to disappointing consequences in terms of texture (Delgado and Rubiolo, 2005). There is a general acceptance that high freezing rates retain the quality of a food product better than lower freezing rates since evidence tends to show that relatively slow freezing causes large ice crystals to form exclusively in extracellular areas, while high freezing rates produce small uniformly distributed ice crystals (Partman, 1975; Mallet, 1993). However, ultrarapid freezing may produce, in some cases, unfavorable effects as well (Mascheroni and Agnelli, 1999; Chambers et al., 2006). In fact, the control of the crystal size has been identified by the food freezing industry as critical; therefore, extensive research has been carried out to control the crystal size and to minimize the time of the water–ice transition phase.

Freezing and thawing are complex unit operations, involving heat and mass transfer and the possibility of a series of physical and chemical modifications that decrease the food quality. Consequently, when selecting a freezing or thawing method, care should be exercised to ensure that the process cost is minimized and that the process could provide a broad range of conditions to maintain the product quality. Other important aspects such as energy usage and environmental impact should also be taken into account.

There is a significant body of literature on the physical and chemical changes that the major groups of commercially frozen foods, namely, fruits, vegetables, seafoods, meats, baked goods, and prepared foods (pizzas, desserts, ice cream, complete meals, and cook-freeze dishes) undergo during freezing and thawing processes. Bald (1991), Jeremiah (1996), Kennedy (2000), Hui et al. (2004), Sun (2006), and Evans (2008), among others, have reviewed the different aspects of freezing and thawing as well as the effects of these processes on the various changes of foods. For detailed information on the effects of freezing and frozen storage on the quality changes of seafoods and red meats, the reader is referred to Chapters 9 and 10, respectively.

This chapter is broadly divided into three sections. The first section is related to the changes that the main food constituents undergo during freezing; the second section is dedicated to the factors influencing the quality of frozen foods such as freezing and thawing method and rate, storage conditions, and temperature abuse; finally, a general overview of the recent developments in freezing and thawing technology is given by focusing on their effects on the food quality.

7.2 INFLUENCE OF FREEZING ON PRINCIPAL CONSTITUENTS OF FOODS

The effect of freezing on food components is diverse and some components are affected more than others. For example, proteins can be irreversibly denatured by freezing, whereas carbohydrates are generally more stable (Lim et al., 2004). Other common chemical changes that can proceed during freezing and frozen storage are lipid oxidation, enzymatic browning, flavor deterioration, and the degradation of pigments and vitamins (Zaritzky, 2000). The main goal of the freezing process is to extend the shelf life of a raw material or product beyond that achievable at temperatures above the initial freezing point of the material. Therefore, it is important to understand the modifications that can occur during freezing in food components and that can further lead to quality degradation (James, 2006; Lim et al., 2006).

7.2.1 Water

Water is an essential constituent of most foods. It is present in a very wide range, varying, for example, from 4% in milk powder up to 95% in tomato and lettuce. Water may exist as an intracellular or extracellular component in vegetable and animal products, as a dispersing medium or solvent in a variety of products, as the dispersed phase in some emulsified products such as butter and margarine, or as a minor constituent in other foods (deMan, 1999). The conversion of water into ice during freezing has the advantage of fixing the tissue structure and separating the water fraction in the form of ice crystals in such a way that water is not available as a solvent or cannot take part in deterioration reactions. On the other hand, ice crystals formed during freezing can affect quality parameters such as color, texture, and flavor. Meanwhile, in the remaining unfrozen portion, the concentration of dissolved substances increases, while the water activity of a product decreases (Zaritzky, 2000).

For water activity up to about 0.2, most water molecules are strongly sorbed and almost immobilized and thus behave in many respects like a solid (Van den Berg, 1991). Usually, this part of water is nonfreezable and, therefore, not available for chemical reactions or as plasticizers (Okos et al., 1992). The water that does not freeze is normally considered to correspond to the monomolecular layer of adsorbed water and has been suggested to be the critical water content above which deteriorative changes may occur. The nonfreezable water content, in percentage (%) of total water for various foodstuffs, is given in Table 7.1.

TABLE 7.1
Nonfreezable Water in % of Total Water for Various Foodstuffs

Product	Total Water Content (%)	Nonfreezable Water in % of Total Water
Lean beef	74.0	12
Haddock	83.5	8
Cod	80.5	9
Whole eggs liquid	74.0	7
White bread	40.0	46
Fruit juice	88.0	3
Spinach	90.0	2

Source: Adapted from Zaritzky, N.E., Factors affecting the stability of frozen foods, in *Managing Frozen Foods*, ed. C.J. Kennedy, Woodhead Publishing Limited, Boca Raton, FL, pp. 111–135, 2000.

Water is a rather unusual substance having high boiling and freezing points, high specific heat, high latent heats of fusion and vaporization, high surface tension, high polarity, and unusual density changes (Zaritzky, 2006). The considerable difference in the densities of water and ice may result in structural damage to foods when they are frozen, being more likely in plant tissue with its rigid structure and poorly aligned cells than in muscle with its pliable consistency and the parallel arrangement of cells (Fennema, 1973). When water freezes at atmospheric pressure, it expands nearly 9%. The degree of expansion varies considerably owing to the following factors (Fellows, 2000):

- Moisture content, as higher moisture contents produce greater changes in volume.
- Cell arrangement, i.e., intercellular air spaces, which are common in plant tissue. These spaces can probably accommodate growing crystals, and thereby minimize changes in the specimen exterior dimensions (Fennema, 1973; deMan, 1999); for example, whole strawberries increase in volume by 3%, whereas coarsely ground strawberries increase by 8.2%, when both are frozen to −20°C (Leniger and Beverloo, 1975).
- Concentration of solutes, since high concentrations reduce the freezing point and do not freeze or expand at commercial freezing temperatures.
- Freezer temperature, which determines the amount of unfrozen water and hence the degree of expansion.
- Crystallized components, including ice, fats, and solutes, which contract when they are cooled; this reduces the volume of food.

Table 7.2 shows volume changes in some fruit products and water–sucrose solutions. It can be observed that highly concentrated sucrose solutions do not show expansion.

As mentioned earlier, crystallization occurs during freezing. It is well known that the crystallization of ice has two steps: the formation of nuclei and the later growth of the nuclei to a specific crystal size, with the final crystal size being a function of the rates of nucleation and crystal growth and also of the final temperature (Price, 1997; Martino et al., 1998). Nucleation is the first decisive step in crystal formation, beginning with molecules colliding with each other due to their random movement in the solution, which leads to the formation of pre-nucleating clusters. As the population of these clusters increases, the clusters begin to associate to form an embryo. Some embryos, through additional collisions, grow into nuclei (tiny crystallites of the smallest size capable of independent existence), contributing to the formation of macroscopic crystals in the process termed *crystal growth* (Banga et al., 2004).

The driving force for crystallization, the supersaturation, is composed of two zones, the metastable and unstable zones (Figure 7.1). The solid line in Figure 7.1 represents a saturation or solubility curve (S), while the dashed line corresponds to the supersolubility or supersaturation curve (SS). Below S, crystallization is impossible. Above S, in the metastable zone, the system is supersaturated and crystallization is possible with the aid of agitation or seeding. On the other hand, in the unstable region, crystallization is spontaneous and crystals appear after nucleation

TABLE 7.2
Volume Change of Fruit Products and Aqueous Sucrose Solutions during Freezing

Product	Volume Increase during Temperature Change from 21.1°C to −17.8°C (%)
Apple juice	8.3
Orange juice	8.0
Whole raspberries	4.0
Crushed raspberries	6.3
Whole strawberries	3.0
Crushed strawberries	8.2
Sucrose (%)	
0	8.6
10	8.7
20	8.2
30	6.2
40	5.1
50	3.9
60	None
70	−1.0 (decrease)

Source: Adapted from deMan, J.M., Water, in *Principles of Food Chemistry*, 3rd edn., ed. J.M. deMan, Aspen Publishers, Inc., Gaithersburg, MD, pp. 1–30, 1999.

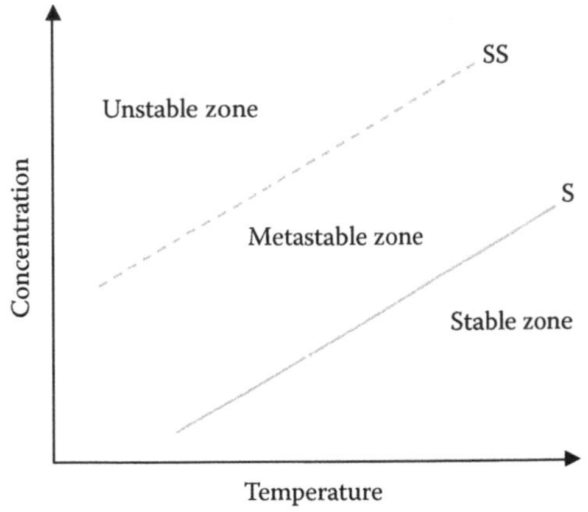

FIGURE 7.1 Schematic representation of the saturation–supersaturation curves (S = saturation or solubility; SS = supersolubility or supersaturation). (Adapted from Banga, S. et al., *Business Brief. Pharm.*, 1, 2004.)

(Banga et al., 2004; Lawler and Dimick, 2008). Nucleation is more energetically demanding than crystal growth; accelerating this stage can thus have a remarkable influence on the crystallization rate. As the boundary of supersolubility (SS) is crossed, nucleation rate increases rapidly (Banga et al., 2004).

Two distinct processes are identified in the nucleation of crystals viz. primary and secondary or contact nucleation. Primary nucleation involves the formation of a crystal in a solution containing no existing crystals (Chow et al., 2005). Nucleation, which takes place in pure systems, is referred to as homogenous or three-dimensional nucleation. If a solid interface, whether a containing wall or a pre-existent crystal, is involved, the nucleation is called heterogeneous, catalytic, or two-dimensional (Fennema, 1973; Chow et al., 2003). Of the two kinds of nucleation, the heterogeneous type is believed to predominate in foods and living matters (Fennema, 1973). Secondary nucleation involves the production of new crystals in a solution containing pre-existing crystals. It can occur either by the crystals acting as templates for new crystals nuclei to be formed or by the crystals fragmenting to produce more nucleation sites (Chow et al., 2003). Figure 7.2 shows the classification of nucleation.

New approaches are currently being analyzed to induce crystallization and control the crystal size. An example of these novel approaches is crystallization in the presence of an ultrasonic wave or sonocrystallization, which is discussed in Section 7.4.1.

There are 11 different forms of crystalline ice, among which the hexagonal form, known as ice Ih, is the normal form of ice in frozen food at atmospheric pressure and is the only ice form that is less dense than water (Préstamo et al., 2005; Zaritzky, 2006). Ice Ih expands in volume upon freezing; this increase in volume seems to be the cause of tissue damage. The volume increment in liquid-ice III, liquid-ice V, or liquid-ice VI phase changes is negative; these ice polymorphs are thus expected to cause less damage upon freezing (Sanz and Otero, 2005). Figure 7.3 shows that by manipulating ambient temperature and/or pressure, various pathways of changing the physical state of food can be followed. This will be discussed further in Section 7.4.3.

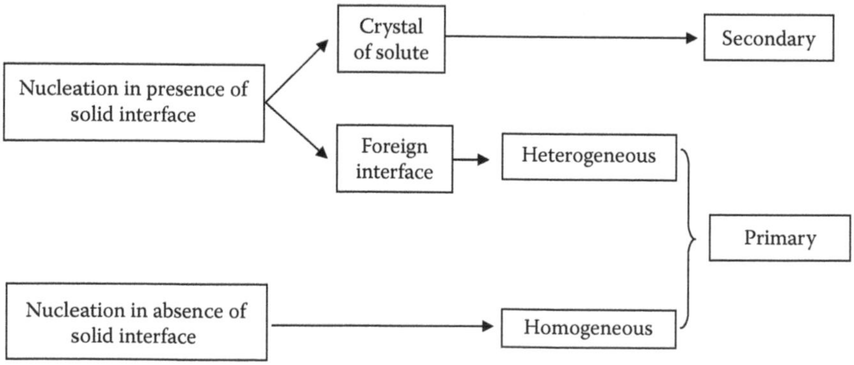

FIGURE 7.2 Classification of nucleation. (Adapted from Chow, R.K. et al., *Ultrasonics*, 41, 595, 2003. With permission.)

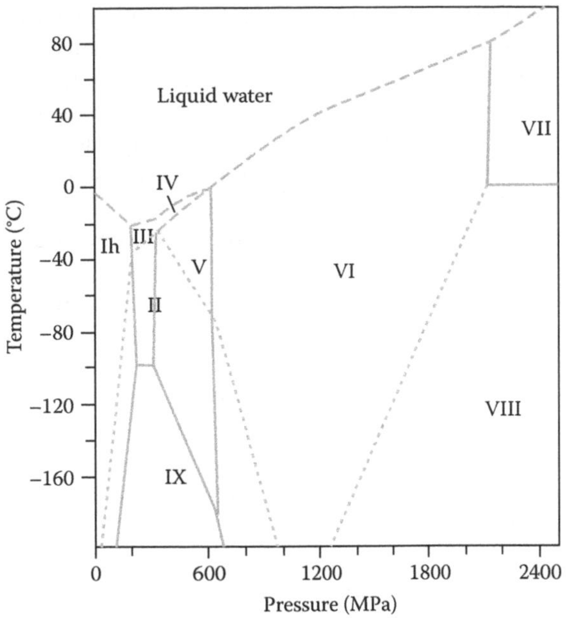

FIGURE 7.3 Phase diagram of water. (Adapted from Akyurt, M. et al., *Energy Convers. Manage.*, 43, 1773, 2002.)

7.2.2 Proteins

Proteins may undergo changes during freezing and frozen storage, primarily because of denaturation (Lim et al., 2004). *Denaturation* can be defined as a major change in the native structure that does not involve alteration of the amino acid sequences and usually involves the loss of biological activity and significant changes in some physical or functional properties such as solubility (deMan, 1999).

Oxidative processes during storage can also contribute to protein denaturation; oxidizing agents (e.g., enzymes, transition metals) can react with proteins via lipid and nonlipid radicals (Zaritzky, 2000). For example, the addition of malonaldehyde, a commonly occurring product of lipid oxidation, to trout myosin solutions during storage at −4°C was found to accelerate protein denaturation (Lim et al., 2004). Fish protein is particularly sensitive to denaturation where the protein develops cross-links between adjacent protein molecules that effectively stop the thawed fish protein to reabsorb water to recreate the pre-frozen gel structure (Lavety, 1991). This denatured protein has a much tougher and rubbery texture than the native protein. The textural changes that occur in fish proteins have been attributed to changes in the myofibrils, as has been reported for cod and Alaska Pollack muscle (Hedges and Nielsen, 2000). The rate at which fish or beef muscle is frozen also influences the degree of protein denaturation. Although rapid freezing generally results in less denaturation than slower freezing, intermediate freezing rates can be more detrimental than slow freezing, as judged by textural changes and the solubility of actomyosin. For example, cod fillets frozen at intermediate rates developed intracellular ice crystals large enough to damage the cellular membranes (Foegeding et al., 1996).

In the case of beef tissue, studies by differential scanning calorimetry (DSC) and the measurements of ATPase (adenosine triphosphatase) activity have also shown that the freezing rate is related to protein denaturation, with the denaturation effect on myofibrillar proteins being higher for slower freezing rates (Zaritzky, 2000). Myofibrillar proteins could be stabilized during freezing and frozen storage by the infusion of cryoprotectants such as sugars, sugar alcohols, or starch hydrolysis products (Foegeding et al., 1996).

Freezing and frozen storage do not significantly affect the nutritional value of meat and fish proteins. However, on thawing frozen meat and fish, substantial amounts of intra- and extracellular fluids and their associated water-soluble proteins and other nutrients may be lost (the so-called drip loss). The volume of drip loss on thawing of meat and fish is highly variable, usually of the order of 2%–10% of net weight; however, in exceptional circumstances, up to 15% of the weight of the product may be lost (Fletcher, 2002). Nevertheless, it was observed for fish that if the product is stored for an appropriate short time and at a sufficiently low temperature, the subsequently thawed fish would rehydrate with the protein returning to its original gel condition (Lavety, 1991).

The caseinate micelles of milk, which are quite stable to heat, may also be destabilized by freezing. On frozen storage of milk, the stability of caseinate progressively decreases and this may lead to complete coagulation (deMan, 1999). Enzymes have also been linked to protein denaturation, as it is known that low temperature decreases the activity of enzymes in tissue, but does not inactivate them (Lim et al., 2004; Zaritzky, 2006).

7.2.3 Lipids

Lipids in food exhibit unique physical and chemical properties. Their compositions, crystalline structure, melting properties, and ability to associate with water and other nonlipid molecules are especially important to their functional properties in many foods. During processing, storage, and handling of foods, lipids undergo complex chemical changes and react with other food constituents, producing numerous compounds both desirable and deleterious to food quality (Nawar, 1996). The process of autooxidation and the resulting deterioration in flavor of fats and fatty foods are often described by the term rancidity (deMan, 1999). In particular, the unsaturated bonds present in all fats and oils represent active centers that, among other things, may react with oxygen. This reaction leads to the formation of primary, secondary, and tertiary oxidation products that may make the fats or fat-containing foods unsuitable for consumption (deMan, 1999).

Lipids can degrade in frozen systems by means of hydrolysis and oxidation. Lipid oxidation is indeed one of the major causes of food spoilage. It is of great economic concern to the food industry because it leads to the development of various off-flavors and off-odors. In addition, oxidative reactions can decrease the nutritional quality of foods (Nawar, 1996).

Lipids in foods can be oxidized by both enzymatic and nonenzymatic mechanisms. One of the enzymes that is considered important in lipid oxidation is lipoxygenase, which has recognition for its off-flavor development in vegetables and legumes and has also been found in fish gill tissue, chicken muscle, and sardine skin (Erickson, 2002). Lipoxygenase is the main enzyme responsible for pigment bleaching and off-odors in frozen vegetables; if the enzyme is not inactivated before freezing by blanching, it can generate offensive flavors and loss of pigment color (Zaritzky, 2006).

At temperatures below −10°C, both enzymatic and nonenzymatic reactions associated with lipid oxidation are decelerated. However, in the range from 0°C to −10°C, decreased oxidative stabilities have been noted (Erickson, 2002). Unless the rate is very slow, the rate of freezing has been found to have little influence on the oxidative stability of frozen products (Tomas, 1988). Instead, storage temperatures play a dominant role in dictating the stability of food products, including muscle foods. The order of time/temperature holding treatments, on the other hand, markedly influences the development of rancidity. Lamb held at temperatures of −5°C to −10°C before storage at −35°C developed more rancidity than lamb first stored at −35°C and followed by storage at −5°C or −10°C (Hagyard et al., 1993).

The hydrolysis of lipids or lipolysis results in the release of free fatty acids. Aubourg and Gallardo (2005) studied the effect of brine freezing on quality changes of small pelagic fish species. These investigators followed the rancidity development during storage at −18°C for up to 9 months; the quality change results were compared to those observed at common freezing conditions. Fish samples treated under brine freezing conditions showed higher lipid oxidation development (as indicated by higher peroxide value and thiobarbituric acid index) and worse scores on some sensory attributes (i.e., general aspect, odor, and color) than the control fish. However, samples treated under brine freezing conditions provided lower lipid hydrolysis development (free fatty acid formation) and better scores for consistency.

Freezing can facilitate lipid oxidation, partly because the competing reactions of microbiological spoilage are avoided and partly because of the concentration effects. Thus, lipid oxidation is relatively more important in frozen muscle tissue than in fresh tissue (Foegeding et al., 1996).

Lipid degradation can be reduced in frozen foods by lowering the storage temperature, excluding oxygen (e.g., use of vacuum packaging), adding antioxidants (e.g., butylated hydroxytoluene or BHT as well as natural vitamin E), and supplementing the diet of animals with antioxidants (Lim et al., 2004). Studies involving dietary modifications have consistently shown that oxidative stability reflects unsaturation in the muscle tissue. When the level of unsaturation in the diet increases, lipid oxidation occurs to a greater extent (Erickson, 2002).

7.2.4 Vitamins

Freezing is considered as one of the best food preservation methods when judged on the basis of nutrients retention. However, it is well known that significant amounts of some vitamins can be lost from processing prior to freezing (e.g., peeling and trimming), leaching (especially during blanching), chemical degradation, and thawing (Fennema, 1977; Lim et al., 2004). The stability of vitamins in foods is generally influenced by pH and the presence of oxygen, light, metals, reducing agents, and heat (Mandigo and Osburn, 1996).

It has been reported that for some frozen foods such as strawberries, the total and biologically active ascorbic acid remain at essentially the same level for a year or longer if the foods are stored below −18°C, although vitamin C losses have also been found to occur at temperatures as low as −23°C (Ottaway, 2002). The conversion to the partially active dehydroascorbic acid and the totally inactive 2,3-diketogulonic acid increases with increasing storage temperature; complete conversion practically occurs in 8 months at −10°C and in less than 2 months at −2°C. Such findings were instrumental in establishing −18°C as the upper limit for frozen food storage and for using biologically active ascorbic acid as a general indicator of quality deterioration during frozen storage (Kramer, 1979). For peaches and boysenberries, a 10°C rise in the temperature from −18°C to −7°C caused the rate of vitamin C degradation to increase by a factor of 30–70 (Fennema, 1977).

Vitamin C and thiamine (vitamin B_1) have been studied extensively since they are water soluble, highly susceptible to chemical degradation, and present in many foods; they are also required in the diet and are sometimes deficient in the diet (Fennema, 1977). Therefore, it is generally assumed that if these vitamins are retained, all other nutrients would also be well retained.

Table 7.3 shows vitamin C losses in several important vegetables during the entire freezing process (blanching, freezing, storage, and thawing). It can be observed that average losses are about 50%, with the exception of small loss values for asparagus and large loss values for spinach. The losses are mainly due to the water solubility of vitamin C during blanching. When water or steam is used for heating, the leaching of vitamins, flavors, color, carbohydrates, and other water-soluble components occurs (Cano, 1996). The relative content of certain vitamins in peas at various stages, from production to consumption, is listed in Table 7.4. Blanched, frozen, and then cooked peas showed the greatest losses of vitamins C and B_1, mainly during blanching and cooking (Cano, 1996).

Thiamine in tissues, which is bound to proteins, is more stable to thermal destruction than free thiamine. The B vitamins are less affected by process and more by loss through drip following a freeze–thaw cycle (Kennedy, 2000).

7.2.5 Carbohydrates and Minerals

Carbohydrates occur in plant and animal tissues in many different forms and levels. In animal organisms, the main sugar is glucose and the storage carbohydrate is glycogen; in milk, the main sugar is almost exclusively the disaccharide lactose (deMan, 1999). In plant organisms, approximately 75% of the solid matter is carbohydrate. The total carbohydrate content can be as low as 2% of the fresh weight in some fruits or nuts, more than 30% in starchy vegetables, and over 60% in some pulses and cereals. In plants, the storage carbohydrate is starch, while the structural polysaccharide is cellulose (Haard and Chism, 1996).

The nutritive value of carbohydrates is not significantly affected during handling of fresh foods and the subsequent processing and distribution of frozen foods (Fennema, 1977). In general, carbohydrates are susceptible to hydrolysis during frozen storage, which can still occur at temperatures as low as −22°C (Lim et al., 2004).

TABLE 7.3
Loss of Vitamin C from Vegetables and Fruits during the Entire Freezing Process

Product	Storage Time at −18°C (Months)	Loss of Vitamin C (% Mean and Range)
Asparagus	6–12	12 (12–13)
Green beans	6–12	45 (30–68)
Lima beans	6–12	51 (39–64)
Broccoli	6–12	49 (35–68)
Cauliflower	6–12	50 (40–60)
Green peas	6–12	43 (32–67)
Spinach	6–12	65 (54–80)
Strawberries		
Seventeen varieties, sliced, sugared, in metal cans	5	17 (0–44)
Puree, 5 + 1 or 3 + 1 sugar	6	16
Whole, no syrup or sugar, in polyethylene bags	10	34
Partially sliced, 6 + 1 sugar, in polyethylene boxes	10	42
Citrus products		
Orange juice concentrate, 42 Brix	9	1
Orange juice, unconcentrated	6	32
Grapefruit juice concentrate, 42 Brix	9	5
Cantaloupes		
In syrup	5–9	9–44
In syrup with added vitamin C	5	23
Plain	9	65–85
Cherries		
Sweet, pitted, in syrup, with or without added vitamin C and citric acid	10	19 (11–28)
Peaches		
Twelve varieties, sliced, in syrup, with added vitamin C	8	23 (12–40)
Twelve varieties, sliced, in syrup, in moisture-proof containers	8	69 (38–82)
Sliced, in syrup, in glass jars	5	29

Source: Adapted from Fennema, O.R., *Food Technol.*, 12, 32–38, 1977.

Like B vitamins and proteins, carbohydrates are less affected by process and more by loss through drip following a freeze–thaw cycle (Kennedy, 2000).

Sugar hydrolysis increases the number of solutes in the food matrix, resulting in a reduction in the amount of ice in the product, which may alter certain physical properties; for example, the firmness of ice cream was found to inversely relate to the degree of hydrolysis (Lim et al., 2004).

TABLE 7.4
Relative Content of Some Vitamins in Peas at Various Steps from Production to Consumption

	Percentage Remaining of Original Amount			
Step in Process	**Vitamin C**	**Vitamin B_1**	**Vitamin B_2**	**Niacin**
Newly harvested	100	100	100	100
Blanched	67	95	81	90
Blanched, frozen	55	94	78	76
Blanched, frozen, heated	38	63	72	79

Source: Adapted from Cano, M.P., Vegetables, in *Freezing Effects on Food Quality*, ed. L.E. Jeremiah, Marcel Dekker, Inc., New York, pp. 247–298, 1996.

Blanching and freezing can cause changes in texture and the pectic composition of certain foods. Both treatments produce a gradual breakdown in the protoplasmic structure organization, with a subsequent loss of turgor pressure, release of pectic substances, and final softening effect (Préstamo et al., 1998). Préstamo et al. (1998) studied the influences of blanching and freezing on the microstructure of carrot in order to determine if the softening of carrot was related to changes in pectic substances or to firmness. These investigators observed that blanching and freezing slightly reduced the firmness compared with the freezing treatment on its own; it was clear that the most softness in the tissue was due to the freezing process. These investigators also observed that the pectin contents were higher in the frozen and frozen blanched samples than in the raw and blanched carrot samples, respectively. The formation of a gel in the blanched sample decreased pectin extraction; this gave an idea of how cells were damaged in the middle lamella, i.e., the main damage being due to freezing rather than blanching (Préstamo et al., 1998). The results of Préstamo et al. (1998) with respect to the changes of the total pectic substances during freezing are, in general, in accordance with a similar study carried out by Fuchigami et al. (1995).

Minerals present in any form (e.g., chemical compounds, molecular complexes, and free ions) can dramatically affect the color, texture, flavor, and stability of foods (Lim et al., 2004). Minerals are chemically stable under typical conditions of handling and processing, and nutrient losses are negligible, provided that losses by physical means (e.g., leaching) are avoided (Fennema, 1977). Nevertheless, no changes were observed in six mineral elements (Ca, Cu, Mg, Mn, Ni, and Zn) between fresh and frozen artichokes, green beans, and peas; boiled fresh vegetables and boiled frozen vegetables also exhibited similar mineral contents (Lim et al., 2006).

7.3 CHANGES IN FOODS DURING FREEZING AND THAWING

Freezing method and freezing rate, storage conditions and temperature abuse, as well as thawing are factors influencing the quality of frozen foods (Chambers et al., 2006). As mentioned earlier, slow freezing produces larger ice crystals, while fast freezing produces a greater number of smaller crystals. Whether large or small crystal size is preferable depends on the purpose of freezing (Nesvadba, 2008). In ice cream, the ice crystals must be as small as possible, so as to make the product as creamy and smooth as possible; however, to concentrate liquid food products, large crystals are easier to separate from the freeze concentrate; or in freeze drying, it is usually desirable to produce a small number of large crystals in order to accelerate the subsequent sublimation process (Nesvadba, 2008). Storage conditions, e.g., total storage time, the type of arrangement of product containers in a storage room, temperature fluctuations/freeze–thaw cycles, permeability, and integrity of packaging, are also critical parameters for maintaining frozen food quality (Chambers et al., 2006). Many frozen foods are thawed before consumption or further processing, and although superficially thawing is the reversal of freezing, in several ways the thawing process may be as important and difficult to control as the freezing process (Bogh-Sorensen, 2000; Nesvadba, 2008).

7.3.1 Effects of Freezing Method and Rate

Air blast, plate contact, immersion freezing, cryogenic freezing, and their combinations are the most common methods used for food freezing (Sun, 2001). Air-blast and multi-plate freezers are most widespread, while air fluidizing systems are used for individual quick freezing (IQF) of small products. The application of cryogenic IQF is, however, still very restricted because of the high price of the liquefied gases used (Fikiin, 2008). During the past few years, new methods are being analyzed in order to accelerate and/or improve the freezing process. Alongside the emerging and novel freezing techniques (e.g., hydrofluidization, immersion freezing with smart agitation modes, application of ice slurries or air impingement, air-cycle-based freezing, flash-freezing cryogenic methods, high-pressure shift freezing (HPSF), and magnetic resonance freezing), a number of promising freezing process innovations have also been launched, such as ultrasonic freezing, dehydrofreezing, and applications of antifreeze protein and ice nucleation protein (Sun and Zheng, 2006; Fikiin, 2008). Some of these new approaches are presently small scale and still unlikely to be implemented for commercial refrigeration in a short-term perspective (Fikiin, 2008).

It is widely recognized that the quality of frozen products is largely dependent on the rate of freezing (Ramaswamy and Tung, 1984). Terms such as slow, sharp, rapid, quick, and ultrarapid are frequently used; although they are sometimes adequately defined, the definitions and methodology associated with these terms are varied. For example, a freezing velocity described as slow for cellular suspensions may be defined as ultrarapid when dealing with larger specimens such as foods (Fennema, 1973).

An average freezing rate (w) has been suggested by Leniger and Beverloo (1975) to determine the nature of the freezing process. If w is greater than 5 cm/h, the freezing process is considered fast; if between 1 and 5 cm/h, it is moderately fast, and at less than 1 cm/h, freezing is slow. In most cases, the aim is to achieve an average freezing rate of at least 2 cm/h (Ramaswamy and Tung, 1984). Another typical classification of the freezing rate is to consider the process as slow freezing when the freezing rate is around 1°C–10°C/h, as commercial freezing when the rate is between

10 and 50°C/h, and as rapid freezing for freezing rates above 50°C/h (Brown, 1991). Many researchers (Mascheroni, 1977; Bevilacqua et al., 1979; Añon and Calvelo, 1980; Tomas, 1988) have considered the freezing rate for meat products in terms of a local freezing time (characteristic time), defined as the time necessary to cool the center temperature from −1.1°C (freezing point of water of the muscle tissue) to −7°C (temperature at which 80% of the tissue water is frozen).

High cooling rates coupled with low final temperatures would nearly always result in the severe cracking of products containing large percentages of water (Fennema, 1973; Pham, 2008). Water expands by about 9% in volume when turning into ice, causing considerable stresses in foods during freezing. In cryogenic freezing, this expansion is followed by a significant thermal contraction, of the order of 0.5% in linear terms or 1.5% in volumetric terms (Rabin and Steif, 1998). The rapid temperature drop down to −196°C in liquid nitrogen, for example, can induce stress as high as 1.5 MPa, whereas freezing in a medium at −40°C yields stress in the range of 1–0.5 MPa (Shi et al., 1999; Le Bail, 2004). As frozen food is brittle, these stresses may cause cracking (Pham, 2008). It was shown, for instance, that the cryomechanical freezing of mushrooms, even for short time of immersion in liquid nitrogen, always led to cracks; cryomechanical freezing is thus not recommended for mushrooms (Mascheroni and Agnelli, 1999). Systems with high void spaces show a higher probability that internal stress would dissipate, instead of accumulate, thus reducing the possibility of freeze-cracking (Zaritzky, 2000).

The combinations of freezing rate and storage conditions can contribute to the ultimate food quality. Yi and Kerr (2009) observed that dough frozen at relatively slow freezing rates and stored at higher temperatures (−10°C to −20°C) was softer, had higher specific volume, and was lighter in color, yet had greater propensity for staling. Rapidly frozen egg white was reported to have a viscosity, foam stability, and native protein content closest to those of unfrozen eggs, while lower freezing rates resulted in greater denaturation, lower viscosity, and less foam stability (Kerr, 2006).

Freezing rate can also influence the color of animal tissues in the frozen state. Rapid freezing causes tissues to become very pale. This is undesirable for red meats but desirable for poultry. This effect is apparently related to the number and size of ice crystals produced and the thickness and nature of the tissue matrix. Once thawed, the appearance of rapidly frozen and slowly frozen tissues cannot be distinguished, however (Foegeding et al., 1996). In the case of meat, there is some disagreement in the literature about which one, fast freezing or slow freezing, has more benefits although the method of freezing clearly affects the ultrastructure of the muscle (James, 2008).

The critical zone of water crystallization (from −1°C to −8°C) should be quickly passed through, so as to ensure achieving a fine ice crystal structure that prevents cellular tissues from perceptible damage (Fikiin, 2008). The location of ice crystals in tissues and cellular suspensions is a function of the nature of the cells, freezing rate, and specimen temperature (Zaritzky, 2006). It is generally accepted that crystallization, regardless of the freezing rate, is initiated in the extracellular fluid (Zaritzky, 2006). Figure 7.4 shows an example of the typical structure of a plant cell (Figure 7.4a) and consequences of extracellular ice formation (Figure 7.4b). In the typical plant cell, the intact cell membrane (PM) is pushed back against a rigid

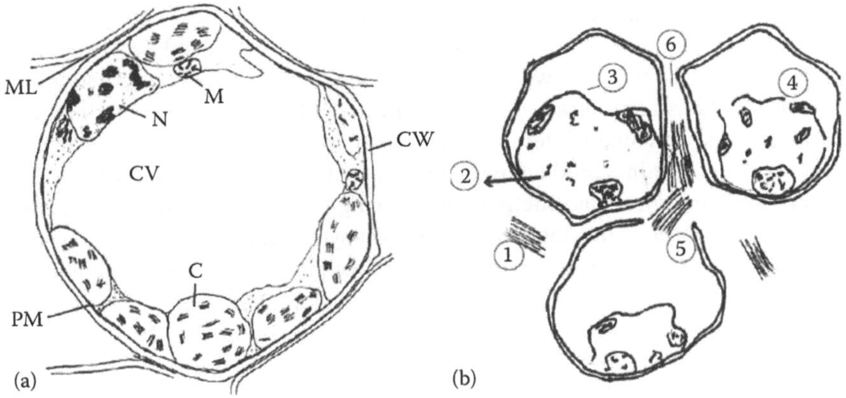

FIGURE 7.4 (a) Structure of a plant cell and (b) effects of formation of extracellular ice on the cell structure. PM, cell membrane; CW, cell wall; C, chloroplasts; M, mitochondria; N, nucleus; CV, central vacuole; ML, middle lamella. ① Movement of water, ② shrinkage of PM; ③ concentration of solutes, ④ membrane disruption, ⑤ damage of CWs ⑥ separation of cells. (From Kerr, W.L., Frozen food texture, in *Handbook of Food Science, Technology and Engineering*, vol. 2, ed. Y.H. Hui, Taylor & Francis Group, LLC, Boca Raton, FL, pp. 61-1/61-15, 2006. With permission.)

cell wall (CW). Internal organelles such as chloroplasts (C), mitochondria (M), and nucleus (N) are occluded by a large central vacuole (CV) (Kerr, 2006). It can be observed in Figure 7.4b that the formation of extracellular ice causes movement of water from cells, resulting in the shrinkage of the plasma membrane and concentration of internal solutes as well as disruption of the membrane. Ice may also penetrate CWs or separate adjacent cells at the middle lamella (Kerr, 2006). With the exception of bread dough, most food groups suffer the least changes in textural quality when frozen at a rapid rate (Kerr, 2006).

7.3.2 Effects of Frozen Storage and Temperature Abuse

Even if a food product is adequately frozen, physicochemical and biochemical changes during storage can lead to degradation of its quality, with this deterioration during cold storage being a slow, continuous, cumulative, and irreversible process (Blond and Le Meste, 2004; Zaritzky, 2008). Physical changes that can be produced during frozen storage are moisture migration and ice recrystallization (Zaritzky, 2006). Apart from these physical changes, chemical modifications including enzymatic reactions, lipid oxidation, protein denaturation, degradation of pigments and vitamins, and flavor deterioration are also produced (see Section 7.2). Therefore, investments in improved freezing technologies can be wasted during the storage and distribution chain (Oliveira et al., 1999).

7.3.2.1 Moisture Migration and Recrystallization

Major physical changes that occur during frozen storage result from water migration; moisture loss or gain from one region or food component to another region would continuously occur since there is always some difference in water vapor pressure due to a temperature gradient or a surface energy difference (Blond and Le Meste, 2004). As a result, changes in either water content or ice crystal size viz. recrystallization are produced. Moisture migration can be minimized by maintaining small temperature fluctuations and small internal temperature gradients and by the inclusion of internal barriers within a product and within a packaging (Zaritzky, 2006).

Water loss through surface dehydration during frozen storage (freezer burn) can occur if improper packaging techniques are employed. Freezer burn occurs when the equilibrium water pressure above the surface of the product is greater than that in the air, causing ice crystals to sublime. In meat, for example, the oxidation and darkening of heme pigments usually accompany freezer burn (Foegeding et al., 1996).

There are different types of recrystallization, although the most common ones in foods are the following (Fellows, 2000):

- Isomass recrystallization. This is a change in surface shape or internal structure, usually resulting in a lower surface-area-to-volume ratio. The ice crystals of irregular shape and large surface-to-volume ratio adopt a more compact configuration and tend to become more spherical and smoother over time (Zaritzky, 2006).
- Accretive recrystallization or sintering. In this case, two adjacent ice crystals join together to form a larger crystal, causing an overall reduction in the number of crystals. The proposed mechanism of crystal aggregation is surface diffusion (Zaritzky, 2008). In systems where the ice volume is important, this mechanism is dominant since the ice crystal contacts make migration easier (Blond and Le Meste, 2004).
- Migratory recrystallization or grain growth, also known as Ostwald ripening. This involves an increase in the average size and a reduction in the average number of crystals caused by the growth of larger crystals at the expense of smaller crystals.
- Additional ice formation can occur by irruptive recrystallization. Under conditions of very fast freezing, aqueous specimens solidify in a partially noncrystalline state, and not all freezable water is converted into ice. Upon warming to some critical temperature, the crystallization of ice occurs abruptly, which is referred to as irruptive recrystallization or devitrification (Zaritzky, 2008).

Among the above types of recrystallization, migratory recrystallization is the most important in most foods and is largely caused by fluctuations in the storage temperature. If there is a gradual reduction in the number of small crystals and an increase in the size of larger crystals, it would result in the loss of quality similar to that observed in slow freezing (Fellows, 2000). A common practice to control ice crystallization and ice crystal growth during freezing and storage is through the addition of stabilizers. Soukoulis et al. (2009) studied the use of dietary fibers (a group of heterogeneous substances such as celluloses, hemicelluloses, lignins, pectins, and seeweed or bacteria-derived gums) as crystallization and recrystallization phenomena controllers in frozen dairy products and found that these stabilizers offered great potential.

Unfavorable conditions can occur not only during frozen storage, but also in commercial display cases or during the distribution of frozen foods. In fact, conditions qualifying as temperature abuse of frozen items may also occur in the consumer's home freezer or during transport by the consumer from the retail store to the consumer's home (Chambers et al., 2006). Products should therefore be tested under these conditions to assess the loss of sensory quality and shelf life that each episode of temperature fluctuation may produce.

7.3.2.2 Storage Conditions and Stability of Frozen Foods

For many decades, the concept of water activity (a_w) has been sufficient to describe the stability of food products. However, as a_w was proved to be inadequate in some cases, the concept of glass transition emerged as a parameter for quantifying water mobility and food stability, which became popular in the late 1980s (Oliveira et al., 1999). Glass transition is a transformation in amorphous materials from a solid glassy state to a supercooled viscous liquid or rubbery state. Glass transition occurs over a range of temperature, although it is often referred to a single temperature value, known as the glass transition temperature T_g, which is defined as the onset or midpoint temperature of the glass transition temperature range. For practical purposes, the freezing process is considered complete when the temperature at the center reaches −18°C and most of the water has been converted into ice. Thus, a food system can be considered as being composed of a crystalline phase of pure water and an amorphous domain, which contains solutes and residual water (Blond and Le Meste, 2004). As temperature decreases further, there would be a point where the unfrozen solution is so cryo-concentrated that its freezing temperature equals its T_g; the whole unfrozen matrix would then undergo glass transition. This point corresponds to the maximum concentration for ice formation and is designated as T_g', i.e., glass transition temperature of the maximum cryo-concentrated solution (Oliveira et al., 1999). T_g' is often regarded as a useful indicator of the temperature below which food would be well protected from deteriorative reactions that are diffusion-limited (Slade and Levine, 1991; Goff, 1994). Unfortunately, most food products have such a large amount of water that their T_g' values are below normal frozen storage temperatures. This implies that most frozen foods are stored with a more or less significant amount of unfrozen cryo-concentrated aqueous solution, and there is a fair degree of molecular mobility (Oliveira et al., 1999).

Glass transition temperatures are most often determined by DSC and by thermomechanical analysis (TMA) (Goff, 1994). Table 7.5 shows the T_g' of some food products. T_g' is strongly affected by the composition of the matrix; T_g' also depends on the procedure used to determine its value. In the case of ice creams, for example, T_g' depends on the recipe, particularly on the level of sugars. As Brake and Fennema (1999) pointed out, standardized procedures for using DSC to measure T_g' have not been developed, and different procedures in terms of the cooling rate, warming rate, holding time and temperature, annealing conditions (or lack thereof), and sample size have been reported. These different procedures can affect the degree to which the maximum freeze-concentration is attained and thus the determination of T_g'; furthermore, complex products such as plant or animal tissues have been known to exhibit more than one T_g' (Brake and Fennema, 1999). The knowledge of the

TABLE 7.5
Glass Transition (T_g') Values of Different Foods

Food	(T_g')(°C)	Reference
Fruits and fruit products		
Apple	−41 to −42	Fennema (1996)
	−37	Guegov (1981)
	−71	Sa et al. (1999)
Banana	−35	Fennema (1996)
Peach	−36	Fennema (1996)
Strawberry	−33 to −41	Fennema (1996)
Tomato	−41	Fennema (1996)
	−21	Guegov (1981)
Blueberry (flesh)	−41	Fennema (1996)
White grape juice	−42	Fennema (1996)
Pineapple juice	−37	Fennema (1996)
Strawberry juice	−45	Torreggiani et al. (1999)
Lemon juice	−38	Maltini (1974)
Orange juice	−34	Moreira and Simatos (1977)
Raspberry juice	−31 to −35	Moreira and Simatos (1977)
Vegetables, fresh or frozen		
Sweet corn (fresh)	−15	Fennema (1996)
Potato (fresh)	−12	Fennema (1996)
Pea (frozen)	−25	Fennema (1996)
Broccoli, head (frozen)	−12	Fennema (1996)
Spinach (frozen)	−17	Fennema (1996)
Carrot	−32	Guegov (1981)
Egg		
Egg white	−38	Simatos et al. (1975)
Egg yolk	−32	Simatos et al. (1975)
Fish		
Cod (whole muscle)	−11.7 ± 0.6	Brake and Fennema (1999)
Mackerel (whole muscle)	−13.3 ± 0.5	Brake and Fennema (1999)
Tuna	−70	Inoue and Ishikawa (1997)
Beef muscle	−12.0 ± 0.3	Brake and Fennema (1999)
Desserts		
Commercial ice creams	−31 to −33	Fennema (1996)
	−27.5 to −40	Levine and Slade (1990)
	−34	Blond (1996)
Cheese		
Cheddar	−24	Fennema (1996)
Provolone	−13	Fennema (1996)
Cream cheese	−33	Fennema (1996)

glassy-state condition during commercial storage is important since very low storage temperatures may result in higher costs.

7.3.3 Effect of Thawing

Thawing is considered as the reverse process of freezing, except for the different thermal properties of water and ice. Water has a higher heat capacity and lower thermal conductivity than ice; hence, for an identical driving force, the thawing time would be longer than the freezing time. Similar to the freezing process, the thawing time is a parameter of interest since it influences the quality of a food product.

Figure 7.5 shows a typical thawing curve where three parts in which the thawing process can be divided are depicted viz. tempering (AB), thawing (BC), and heating of food above its thawing plateau (above C). After the initial rapid rise in temperature (AB), there is a long period where the temperature of the food is close to the melting temperature of ice (BC). During this period, any cellular damage caused by slow freezing or recrystallization would result in a release of cell constituents to form drip losses (Fellows, 2000). This causes a loss of water-soluble nutrients. For example, beef can lose 12% thiamine, 10% riboflavin, 14% niacin, 32% pyridoxine, and 8% folic acid; or 30% of vitamin C can be lost in fruits (Fellows, 2000).

Most of the textural changes due to freezing and frozen storage would only become apparent after thawing. These textural changes are more noticeable in fruits and vegetables that have high water content and especially those that are eaten raw such as tomato, celery, and lettuce, which turn into mush and liquid after thawing. This explains why celery, lettuce, or tomato is not usually frozen, or why frozen fruits are best served before they have completely thawed (James, 2006). Microstructural changes in strawberry after freezing and thawing processes were studied by Delgado and Rubiolo (2005) using scanning electron microscopy (SEM). The influences of different thawing rates, as obtained by varying the temperature and air velocity, were

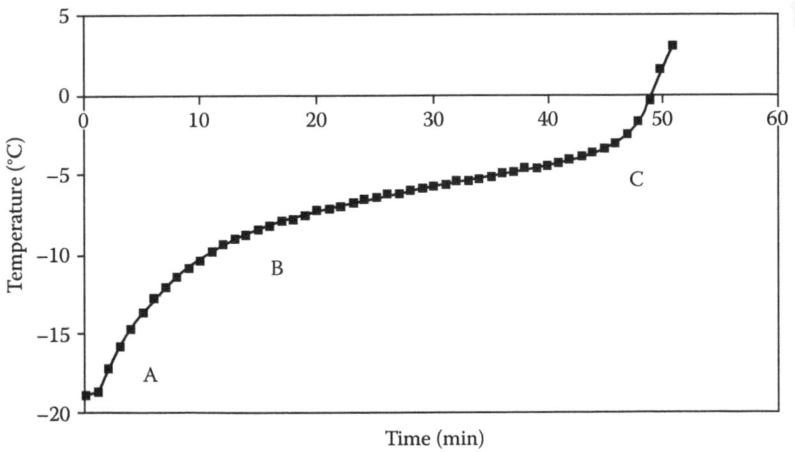

FIGURE 7.5 Characteristic temperature–time thawing curve. (Adapted from Delgado, A.E. and Sun, D.-W., *J. Food Eng.*, 57, 383, 2003.)

analyzed on samples frozen under the same freezing conditions. These investigators observed that lower air velocities caused greater tissue damage; the higher the ambient temperature, the higher the deterioration.

The method of thawing adopted in practice is dependent on the further use of a product. Among the many possible variants, three typical situations can be distinguished (Daudin, 1992):

- Thawing/cooking. When a product is thawed for immediate consumption, as is the case at home or in a public restaurant, the product is cooked directly from the frozen state. This practice combines speed with bacteriological safety.
- Partial thawing. This entails the preparation of individual frozen portions (e.g., steaks and fish fillets) from frozen half-products. In this case, the temperature is raised to about −5°C, with 60%–70% of the water still being in the form of ice, but the consistency of the product is sufficiently pliable for the operations of filleting and shaping to be performed. The product is then refrozen.
- Thawing-transformation. A typical example is the boning of carcasses followed by cutting into joints for sale in the "fresh" state, which necessitates complete defrosting for these operations to be performed without difficulty. The same applies to joints of meat, which make up finished products such as cooked ham defrosted before curing.

In the case of meats in particular, failure to thaw frozen meat adequately is considered a contributory factor in sporadic outbreaks. During the freezing operation, surface temperatures are reduced rapidly and bacterial multiplication is severely limited. During the thawing operation, these same surface areas are the first to rise in temperature and bacterial multiplication can recommence (James and Bailey, 1984). On large objects subjected to long uncontrolled thawing cycles, surface spoilage can occur before the center regions have fully thawed; for these and other reasons, the thawing process is not a simple operation and in several ways it may be as important as the freezing process (James and Bailey, 1984). The loss of juice from foods due to dripping and evaporation as well as microbial growth on the surface due to the use of high temperatures and long exposure time are listed as primary deteriorative effects that should be minimized during thawing (Mannapperuma and Singh, 1988).

Thawing methods can be divided into two general groups: surface heating methods and electrical (internal heating) methods (e.g., microwaves and radio waves) (Bogh-Sorensen, 2000). The U.S. Food and Drug Administration (2005) (FDA) Model Food Code recommends several thawing methods for raw meat products: thawing under refrigeration (≤5°C), thawing under cold running water (≤21°C), and thawing as part of a cooking process, e.g., in the case of microwave thawing (Shrestha et al., 2009). Each of these thawing methods presents some disadvantages to the foodservice operator; e.g., thawing under refrigeration or running water can be time-consuming. Shrestha et al. (2009) proposed that submersion in hot water could be used to rapidly thaw chicken breast portions, while retaining quality and safety. The hot-water thawing temperature was chosen as 60°C, i.e., an approximate temperature setting for foodservice hot holding equipment. This temperature is also not expected to cause the localized or surface overheating of the meat as it could occur during microwave thawing.

In the food industry, air thawing is a common method, but for fish in particular, thawing in circulating water is normally used (Bogh-Sorensen, 2000). In the case of air thawing of unwrapped products, it is also important to control the relative humidity of air in order to minimize weight loss. When food is thawed by microwave or dielectric heaters, the main considerations are to avoid overheating, to minimize thawing time, and to avoid the excessive dehydration of the food (Fellows, 2000).

In selecting a thawing system for industrial use, a balance must be made between the thawing time, appearance and bacteriological condition of a product, processing problems such as effluent disposal, as well as capital and operating costs of the respective systems (James, 2006). Of these factors, thawing time is the principal criterion that governs the selection of the system, while appearance, bacteriological condition, and weight loss are important if the material is to be sold in the thawed condition, but are less so if it is for further processing (James, 2006).

7.4 EFFECTS ON FOOD QUALITY OF RECENT DEVELOPMENTS IN FREEZING AND THAWING TECHNOLOGY

Changing lifestyles and work patterns as well as the raising levels of disposable income have resulted in more nutritional and health-conscious consumers, which demand products of higher quality. Currently, chemical and physical aids to freezing and thawing processes are researched and developed with the consideration of energy saving and/or quality improvement (Li and Sun, 2002b). As a result, numerous innovations have taken place and significant improvements in product quality have been achieved (Sun and Zheng, 2006).

7.4.1 Ultrasound-Assisted Freezing

The use of ultrasound has attracted considerable interest in food science and technology due to its promising effects in food processing and preservation (Knorr et al., 2004). The sound ranges employed can be broadly divided into high-frequency, low-energy, diagnostic ultrasound in the MHz range and low-frequency, high-energy, power ultrasound in the kHz range (Mason and Chemat, 2003). Power ultrasound, in particular, has proved to be extremely useful in crystallization processes since it is possible to modify both the nucleation and crystal growth stages of solidification (Acton and Morris, 1992; Mason, 1998). Compared to other nucleation methods (e.g., use of chemicals, amino acids, ice nucleation bacteria, and seed crystals), power ultrasound offers several advantages: the initial nucleation temperature of the liquid can be adjusted, the technique is chemically noninvasive and does not require direct contact with the product to be frozen, and the use of power ultrasound does not present legislative difficulties (Acton and Morris, 1992). The controlled crystallization of sugar solutions, hardening of fats, and manufacture of chocolate and margarine are examples of food processes that can be improved by the application of power ultrasound (Leadley and Williams, 2006). Although ultrasound has been successfully used for many years, the use of power ultrasound to directly improve processes and products has not yet been sufficiently exploited in food manufacturing (Zheng and Sun, 2005, 2006; Delgado et al., 2009).

As mentioned earlier, the advantages of food preservation by freezing (e.g., volatile retention, minimum change of organoleptic properties, and absence of microbial growth) are to a certain degree counterbalanced by the risk of damage caused by the formation of ice within tissues. This is because ice formation can mechanically affect cell membranes and distort the tissue structure (Martino et al., 1998; Sanz et al., 1999). However, as previously pointed out, minimizing the time to cross the zone of maximum ice crystal formation or "thermal arrest time" could help better maintain the quality of a food product. Under the influence of ultrasound, food materials can be frozen with a much shorter thermal arrest time (Mason et al., 1996; Leadley and Williams, 2006). Research related to the use of ultrasound to assist immersion freezing showed that the freezing rate, as represented by the characteristic time, could be significantly improved from 8% up to 20% as compared to that of conventional immersion freezing. This makes the ultrasound-assisted immersion chilling and freezing (ICF) process very attractive from the cost point of view (Li and Sun, 2002a; Sun and Li, 2003; Delgado et al., 2009).

Low-frequency power ultrasound acts in a number of ways during crystallization: (1) it initiates seeding because the cavitation bubbles tend to act like crystal nuclei and so enhances the rate and uniformity of seeding, and (2) ultrasound can break up any large crystalline agglomerates (Mason et al., 1996). It is also possible to "tailor" a crystal size distribution between the extreme cases of a short burst of ultrasound to nucleate at lower levels of supersaturation and allow growth to large crystals and the production of small crystals via continuous sonication throughout the duration of the process, which can facilitate prolific nucleation at higher levels of supersaturation at the expense of some crystal growth (Ruecroft et al., 2005). A pulsed or an intermittent application of ultrasound can give intermediate effects. In any event, optimum conditions need to be determined by experimental investigation (Ruecroft et al., 2005; Luque de Castro and Priego-Capote, 2007). While high-power ultrasonic systems become more and more standardized, the way the energy is applied to the medium is unique for every application; therefore, the potential to obtain patent protection is relatively large (Patist and Bates, 2008).

Sun and Li (2003), via the use of cryo-SEM, showed that the microstructure of potato tissue exhibited a better cellular structure under specific conditions of ultrasound application (ultrasonic power of 15.85 W and a total acoustic treatment time of 2 min). In an experimental study carried out by Acton and Morris (1992), the acoustic effect on crystal size distribution was analyzed during the freezing of a sucrose solution (Figure 7.6). These investigators observed that in the ultrasonically treated sample, there were crystals of smaller mean diameter, together with a reduction in the percentage of crystals with a diameter greater than 50 μm. In the control sample, 55% of the water existed in crystals of 50 μm or larger, while this fraction was only 33% in the ultrasonically treated sample (Acton and Morris, 1992).

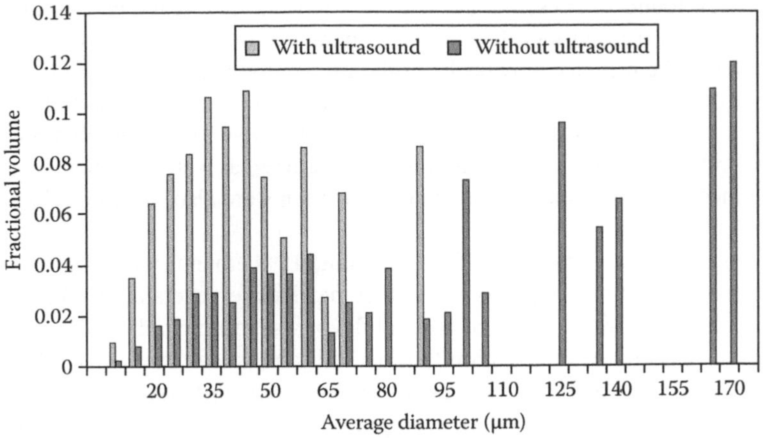

FIGURE 7.6 Acoustic effect on crystal size distribution in frozen sucrose solution. (Adapted from Acton, E. and Morris, G.J., Method and apparatus for the control of solidification in liquids. WO/1992/020420, U.S. Patent Application (Granted: 12/03/1997, Withdrawn: 19/10/1999, 1992.)

7.4.2 Dehydrofreezing

As it is generally accepted, the cause of undesirable physicochemical modifications during freezing is the crystallization of water and sometimes solutes. The use of prefreezing treatments can help reduce (or even, in some cases, avoid) these changes, either by inactivating deterioration reactions directly or by reducing the water content in the material (Torreggiani et al., 2000). Currently, there is a renewed interest in implementing partial dehydration prior to freezing; the reason for this is the versatility of the technique, which makes it possible to reduce the water content, improve quality, and/or develop new products (Torreggiani et al., 2000).

Dehydrofreezing has been successfully applied to fruits and vegetables. The advantages over conventional freezing include energy savings, since the water load to the freezer is reduced; better quality and stability (color, flavor); as well as better thawing behavior (lower drip loss) (Kennedy, 2003).

Partial dehydration can be achieved either by partial air drying, osmotic dehydration, or ICF in the solutions of high sugar or salt concentration. To avoid browning during air drying, blanching or other treatments such as dipping in antioxidant solutions (e.g., ascorbic or citric acid) can be used (Giangiacomo et al., 1994; Kennedy, 2003). ICF is quite similar to osmotic dehydration in that both processes involve cross transfer of water and solute, but differ in the operating temperature range; ICF is carried out at low temperatures, while for osmotic dehydration, the operating temperatures range from 30°C to 80°C. After dehydration, freezing would then follow.

As mentioned earlier, one of the partial dehydration methods is ICF. Binary brine solutions are generally used, most often sodium chloride (eutectic ≈ −20°C) or calcium chloride (eutectic ≈ −50°C), although ternary solutions (e.g., water + NaCl + ethanol, or sugars) have also been studied (Lucas and Raoult-Wack, 1996). The ICF process was found to preserve the texture of fruits and vegetables tissues more successfully and caused less dehydration during the freezing process (Torreggiani et al., 2000). The primary drawbacks of the immersion applications are the adverse effect on the product flavor, which is caused by residual freezant due to the uncontrolled solute uptake by the refrigerated products. Operational problems with the immersion liquids (high viscosity at low temperatures) as well as difficulty to maintain the medium at a defined constant concentration were also reported (Elias, 1978; Fikiin, 2003).

A formulation composing of 15% table salt, 15% ethanol, and 70% water was used to freeze carrot dices, peas, and other particulate vegetable products; it was observed that the adverse effect on product flavor was minimized. At the same time, those characteristics of the solution that make it a practical freezing medium were also maintained (Elias, 1978). For carrots in particular, a pretreatment (e.g., dipping the product in a 5% sucrose solution) might be necessary to reduce the salt flavor. Strawberries can be pretreated with sugars or syrups before freezing because these additives not only reduce the amount of water to freeze, but also sweeten the product, contribute to retaining volatile compounds, and decrease browning by acting as a barrier to oxygen. Garrote and Bertone (1989) found that strawberry halves osmotically treated with solutions of glycerol, glucose, and sucrose of varying concentrations sustained a significantly smaller exudate loss ($p < 0.05$) than untreated fresh strawberry halves (Li and Sun, 2002b). It is emphasized that the ICF process would produce higher quality results, especially with small pieces of food (Lucas and Raoult-Wack, 1998). The possible employment of pumpable ice slurries (known under different trade names such as Flo-Ice, Binary Ice, Slurry-Ice, Liquid Ice, Pumpable Ice, or Fluid Ice) as refrigerating media is another important reason for the renewed interest in immersion freezing techniques (Fikiin, 2003).

The so-called hydrofluidization method (HFM) for the fast freezing of foods has been suggested and patented to bring together the advantages and to overcome the drawbacks of both air fluidization and immersion food freezing techniques (Fikiin, 2003). The HFM uses a circulating system that pumps the refrigerating liquid upward, through orifices or nozzles, in a refrigerating vessel, thereby creating agitating jets. These form a fluidized bed of highly turbulent liquid and moving products; extremely high surface heat transfer coefficients can thus be obtained (Fikiin et al., 2003). The HFM technology presents a series of advantages over the conventional IQF modes, which can be summarized as follows (Fikiin et al., 2003; Fikiin, 2008):

- High frozen food quality due to high freezing rate: fine-grain crystal structure; sharp reduction of the surface mass transfer; easy incorporation of antioxidants, flavors, aromas, and micronutrients; and improved nutritional value and sensory properties.
- High energy and economic efficiency due to reduced investment, maintenance and power costs, high throughput, and cost efficiency.
- Environmental friendliness due to the use of environmentally friendly secondary coolants (syrup-type solutions or ice slurries) and primary refrigerant in a small isolated system. The use of product-friendly refrigerant

media makes it possible to exploit the osmotic phenomena that occur at the product–media interface; therefore, products with added value can be formulated. As dehydrofrozen foods would require rehydration prior to consumption, establishing correlation between processing conditions and rehydration characteristics is also important for maintaining good sensory characteristics and texture (Sun and Zheng, 2006).

7.4.3 High-Pressure Freezing and Thawing

In order to reduce the chemical and mechanical damages to food systems during freezing, new technologies have been developed to freeze foods under high pressure. Under high pressure, there is a nonfrozen liquid state of water below 0°C and the freezing point can be reduced to –21°C at 210 MPa; the opposite effect is observed above this pressure level, as shown in the phase diagram of water (Figure 7.3). This phenomenon allows carrying out different high-pressure processes of freezing and thawing of foods (Le Bail et al., 2002). Increasing freezing/thawing rates, ice polymorphs of higher density, and smaller and uniform ice crystal formation are among the benefits of this technique. Furthermore, high-pressure treatments from 100 MPa prevent microbial growth, although protein modifications may take place from 100 to 200 MPa (Le Bail et al., 2002).

High-pressure freezing processes such as pressure-assisted freezing (PAF) and HPSF are novel technologies that have been recently studied. In the former process, phase transitions occur under constant pressure, higher than atmospheric pressure, and as the latent heat of crystallization is reduced when pressure increases, reduction in phase transition time can be achieved. On the other hand, releasing the pressure once the temperature of food reduces to the modified freezing point results in a high supercooling effect, leading to an enhanced rate of ice nucleation. This forms a basis for HPSF, with its main advantage of the instantaneous and homogeneous initial formation of ice throughout the whole volume of a product (Norton et al., 2009). Therefore, HPSF can be especially useful to freeze foods with large dimensions where the effects of freeze cracking caused by thermal gradients can become pronounced (Martino et al., 1998).

The applications of HPSF are still under development, but the amount of data available is increasing (Norton et al., 2009). Most of the studies carried out to date focus on the advantageous effects of HPSF on the texture and structure of various products. Koch et al. (1996), for example, observed that the HPSF of potato cubes resulted in less damage to the cell structure, less drip loss on thawing, and less enzymatic browning than conventional freezing. Préstamo et al. (2005) studied the effects of the different conditions of HPSF on protein content as well as peroxidase and polyphenoloxidase activities in potato samples. The results showed that none of the conditions analyzed was enough to inactive the enzymes, with peroxidase being more resistant than polyphenoloxidase; the protein content decreases by about 50% after the high-pressure treatment (210 MPa and –20°C). Fuchigami et al. (1997) compared the texture of raw or thermally blanched carrots subjected to high-pressure freezing and concluded that high-pressure freezing at 200, 340, and 400 MPa was effective in improving both the texture and histological structure of the frozen carrots. In the work of Otero et al. (1998), who compared the processes of conventional air freezing and HPSF, it was found that high-pressure shift frozen samples had the appearance of fresh samples, and no differences between the center and surface cell structures were observed, indicating that uniform ice nucleation had been achieved (Norton et al., 2009).

HPSF has also been applied in the processing of fruits, pork, lobster, and tofu. Martino et al. (1998) observed that a uniform and instantaneous ice nucleation within the whole volume of a pork sample was achieved only by HPSF when compared to classical freezing methods. Otero et al. (2000) confirmed the beneficial effects of HPSF on whole peach and mango as compared to air-blast frozen samples.

Pressure-assisted thawing (PAT) also offers many advantages in comparison to atmospheric thawing, with the most important benefit of reduced thawing time (Le Bail et al., 2002). Limitations on the application of PAT are mainly high cost, pressure-induced protein denaturation, and meat discoloration (Li and Sun, 2002b). However, reduced drip loss and enhanced microbial reduction appear as motivation criteria for valuable foods; these advantages should justify the higher cost when compared to conventional thawing at atmospheric pressure (Le Bail et al., 2002).

An interdisciplinary approach toward providing a scientific and technical basis of high-pressure low-temperature (HPLT) processes was that of the European Commission funded project "SAFE ICE: Low-Temperature Pressure Processing of Foods: Safety and Quality Aspects, Process Parameters and Consumer Acceptance" (Urrutia et al., 2007). The project covered many angles of the subject such as evaluation of thermo-physical properties under pressure, modeling freezing and thawing at high pressure, effect of high pressure on activity of food spoilage enzymes, consumers' perception, impact of the technology on quality-related parameters (e.g., texture, color, and microstructure), as well as the development of high-pressure measuring techniques. The results showed that high-pressure freezing and thawing offer unique opportunities for product development and the improvement of food quality. However, the applications of the process in industrial-scale facilities as well as studies on further products of interest (besides potato, salmon, pork, starch gels, dough, strawberry, carrot, and broccoli) along with the economic study of the applicability of such process are still challenging (Urrutia et al., 2007).

Additional information on high-pressure assisted freezing and thawing can be found in Chapter 5.

7.4.4 Ohmic, Acoustic, and Microwave Thawing

The process of thawing bulky frozen foodstuffs is intrinsically slow; this can have expensive consequences in industrial food processing and can be inconvenient in large and small catering and even at home (Miles et al., 1999). Besides time inefficiency, microbial contamination, inhomogeneous temperature distribution due to low conductivity, and prohibitive cost involved are problems that need to be solved (Miao et al., 2007). In order to address these concerns, new technologies have been proposed and analyzed.

Ohmic heating, also known as Joule heating, electric resistance heating, direct electric resistance heating, electro heating, or electro conductive heating, is a thermal treatment process in which an electric current passes at an appropriate voltage through conducting food with high electrical resistance; heat is internally generated within the food, thus increasing the temperature of the food material (Li and Sun, 2002b). Ohmic heating rate is principally dependent on the electric conductivity of food material as well as the frequency of the electric current (Halden et al., 1990; Imai et al., 1995). Provided that the electric conductivities of all components in food are similar, ohmic technology has the advantages of high heating rate, homogeneous temperature distribution, and energy economization (Miao et al., 2007).

The most suitable freezing rate and thawing method for raw and blanched carrots were investigated by Fuchigami et al. (1994). The results showed that frozen carrot disks were defrosted comparatively fast even at −3°C by electrostatic thawing; drip, cell damage, and softening were also prevented by ohmic heating (Fuchigami et al., 1994). The use of an alternating electric field might be more beneficial than an electrostatic field, since the latter might cause electrolysis and thus require expensive electrodes (Li and Sun, 2002b).

Miao et al. (2003) developed an ohmic thawing system in which frozen food material was electrically connected to two electrodes by an electrode solution (e.g., salt solution with a concentration in the range of 0.1%–8.0%); an alternating current was then passed for thawing. These investigators used the system in the thawing of minced-meat, Japanese white radish, and recently in the thawing of frozen saline surimi cubes. The results of their work showed that temperature distribution and thawing time are related to the applied voltage and the concentration of electrode solution. In the thawing of frozen surimi, for example, and for a voltage of 20 V and frequency of 60 Hz, a homogeneous temperature distribution could be achieved when the concentration of electrode solution was below 4%. Miao et al. (2007) showed that ohmic thawing could be well applied with frozen surimi since the process gave higher thawing rates and resulted in stronger gels when compared to conventional thawing.

Ultrasound has also been used to assist the thawing process as there is reliable evidence on the ability of power ultrasound to improve heat and mass transport operations. In terms of heat transfer, the potential of ultrasonic energy has been demonstrated during sterilization and other thermal treatments where the fluid-to-particle heat transfer coefficient is significantly enhanced. In the case of thawing, ultrasound could supply all the energy required to thaw meat samples (Sastry et al., 1989). Miles et al. (1999) indeed used high-intensity ultrasound to supply all the energy required for thawing frozen meats (beef and pork) and fish (cod) samples. These investigators clearly distinguished their approach from previous attempts by other investigators, where ultrasound was used merely to assist thawing of materials immersed in warm water. Frequencies and intensities around 500 kHz and 0.5 W/cm^2 were found to offer effective thawing for beef, pork, and cod samples while minimizing surface heating.

Since effective acoustic or ultrasound thawing depends on the way the energy is applied to the medium, the selected frequency and acoustic intensity should be appropriate in order to avoid excessive heating near the surface. Some rapid thawing techniques can cause excessive heating at the product surface, leading to the loss of product quality. However, surface overheating might be overcome by combining acoustic and water bath thawing.

It has been proposed that the absorption of ultrasound energy depends on the thermo-elastic relaxation of ice crystals in food and is affected by ice crystal orientation and size, impurities present in the ice crystals, as well as temperature (Kissam et al., 1981). Miles and Shore (1978) and Shore et al. (1986) found that ultrasound was more attenuated in frozen meats than in unfrozen tissues and that the attenuation increased markedly with temperature, reaching the maximum near the initial freezing points of the foods before decreasing rapidly at higher temperatures, thus making ultrasound particularly suitable for the controlled thawing of foods (Miles et al., 1999; Torley and Bhandari, 2007).

Microwaves can easily penetrate a whole frozen product, thus effectively reaching the inner regions within a short time. For example, microwave tempering can be performed in 5–10 min for a large amount of frozen products (20–40 kg) (Ahmed and Ramaswamy, 2007). Maximum homogeneity is achieved with temperatures slightly above zero; after that the inhomogeneity rises again. Therefore, it is advantageous to reduce the thawing process to plain tempering, i.e., to stop the heating at the temperatures around −5 to −2°C (Fu, 2006).

Microwave thawing requires shorter thawing time and smaller space for processing and could reduce drip loss, microbial contamination, and chemical deterioration (Li and Sun, 2002b). However, there are still severe restrictions on the rate at which thawing can be accomplished. A main difficulty is the formation of wide temperature gradients (runaway heating) within a product. It is not uncommon for a frozen food heated in a microwave oven to boil around the edges, while the center remains frozen (Yam and Lai, 2004). The preferential absorption of microwaves by liquid water over ice is a major cause for runaway heating (Fu, 2006). Thus, if, for any reason, the temperature of a product is nonuniform, warm regions become centers for more power dissipation and the nonuniformity is amplified. When this happens, runaway heating may result in regions of cooked product embedded within substantially frozen product, little warmer than it was initially (Miles et al., 1999).

An irregular shape of a food can also cause nonuniform heating; the thin parts tend to overcook, while the thick parts tend to undercook. Another cause of nonuniform heating is that different foods have different dielectric and thermal properties. For example, when microwave heating a frozen meal consisting of meat and vegetable, the vegetable often becomes overheated and dried out before the meat reaches the serving temperature (Yam and Lai, 2004). The temperature uniformity during microwave thawing can be improved when appropriate sample thickness, microwave power level, frequency, and/or surface cooling are applied (Fu, 2006).

7.5 CONCLUDING REMARKS

Freezing and thawing are complex processes involving physical and chemical changes that might greatly affect the food quality. Satisfying consumers' demands for high-quality products has over recent years led to the development of new approaches for improving the freezing/thawing processes with consideration also to energy use and environmental impact besides quality enhancement. Food products processed by novel methods constitute an emerging niche market, which creates opportunities and challenges to the food industry. Due to the large variety of foods and formulations and since the determination of the optimum process parameters is highly product dependent, more evidence of the benefits as well as the development of processing equipment is still needed for the commercialization of these novel processes.

REFERENCES

Acton, E. and Morris, G.J. 1992. Method and apparatus for the control of solidification in liquids. WO/1992/020420, U.S. Patent Application (Granted: 12/03/1997, Withdrawn: 19/10/1999).

Ahmed, J. and Ramaswamy, H.S. 2007. Microwave pasteurization and sterilization of foods. In *Handbook of Food Preservation*, 2nd edn., ed. M.S. Rahman, pp. 691–711. Boca Raton, FL: CRC Press.

Akyurt, M., Zaki, G., and Habeebullah, B. 2002. Freezing phenomena in ice-water systems. *Energy Conversion and Management* 43: 1773–1789.

Añon, M.C. and Calvelo, A. 1980. Freezing rate effects on the drip losses of frozen beef. *Meat Science* 4: 1–14.

Aubourg, S.P. and Gallardo, J.M. 2005. Effect of brine freezing on the rancidity development during the frozen storage of small pelagic fish species. *European Food Research and Technology* 220: 107–112.

Bald, W.B. 1991. *Food Freezing: Today and Tomorrow*. Berlin, Germany: Springer-Verlag.

Banga, S., Chawla, G., and Bansal, A.K. 2004. New trends in the crystallization of active pharmaceutical ingredients. *Business Briefing: Pharmagenerics* 1–5.

Bevilacqua, A., Zaritzky, N.E., and Calvelo, A. 1979. Histological measurements of ice in frozen beef. *Journal of Food Technology* 14: 237–251.

Blond, G. 1996. Bases theoriques de la structure des glaces: Influence du procede de fabrication et de la formulation. *Colloque Alliance 7—Cedus "La Texture des Produits Sucres"*, pp. 59–68. Paris: Cedus.

Blond, G. and Le Meste, M. 2004. Principles of frozen storage. In *Handbook of Frozen Foods*, eds. Y.H. Hui, P. Cornillon, I. Guerrero Legaretta, M.H. Lim, K.D. Murrell, and W.-K. Nip, pp. 25–53. New York: Marcel Dekker, Inc.

Bogh-Sorensen, L. 2000. Maintaining safety in the cold chain. In *Managing Frozen Foods*, ed. C.J. Kennedy, pp. 5–26. Boca Raton, FL: Woodhead Publishing Limited.

Brake, N.C. and Fennema, O.R. 1999. Glass transition values of muscle tissue. *Journal of Food Science* 64(1): 10–15.

Brown, M.H. 1991. Microbiological aspects of frozen foods. In *Food Freezing: Today and Tomorrow*, ed. W.B. Bald, pp. 15–25. Berlin, Germany: Springer-Verlag.

Cano, M.P. 1996. Vegetables. In *Freezing Effects on Food Quality*, ed. L.E. Jeremiah, pp. 247–298. New York: Marcel Dekker, Inc.

Chambers, E. IV, McGraw, S., and Smiley, K. 2006. Sensory analysis of frozen foods. In *Handbook of Frozen Food Processing and Packaging*, ed. D.-W. Sun, pp. 561–576. Boca Raton, FL: CRC Press.

Chow, R.K., Blindt, R., Chivers, R., and Povey, M. 2003. The sonocrystallisation of ice in sucrose solutions: Primary and secondary nucleation. *Ultrasonics* 41(8): 595–604.

Chow, R.K., Blindt, R., Chivers, R., and Povey, M. 2005. A study on the primary and secondary nucleation of ice by power ultrasound. *Ultrasonics* 43: 227–230.

Daudin, J.D. 1992. Freezing. In *Technology of Meat and Meat Products*, ed. J.P. Girard, pp. 5–31. Chichester, U.K.: Ellis Horwood Limited.

deMan, J.M. 1999. Water. In *Principles of Food Chemistry*, 3rd edn., ed. J.M. deMan, pp. 1–30. Gaithersburg, MD: Aspen Publishers, Inc.

Delgado, A.E. and Rubiolo, A.C. 2005. Microstructural changes in strawberry after freezing and thawing processes. *LWT—Food Science and Technology* 38(2): 135–142.

Delgado, A.E. and Sun, D.-W. 2003. One-dimensional finite difference modelling of heat and mass transfer during thawing of cooked cured meat. *Journal of Food Engineering* 57(4): 383–389.

Delgado, A.E., Zheng, L., and Sun, D.-W. 2009. Influence of ultrasound on freezing rate of immersion-frozen apples. *Food and Bioprocess Technology* 2(3): 263–270.

Elias, S. 1978. Direct immersion freezing...(cold) wave of the future? *Food Engineering International* 12: 44–45.

Erickson, M.C. 2002. Lipid oxidation of muscle foods. In *Food Lipids—Chemistry, Nutrition and Biotechnology*, eds. C.C. Akoh and D.B. Min, pp. 383–429. New York: Marcel Dekker, Inc.

Evans, J.A. 2008. *Frozen Food Science and Technology*. Oxford, U.K.: Blackwell Publishing Ltd.

Fellows, P. 2000. Freezing. In *Food Processing Technology, Principles and Practice*, 2nd edn., ed. P. Fellows, pp. 418–440. Cambridge, U.K.: Woodhead Publishing Limited.

Fennema, O. R. 1973. Nature of the freezing process. In *Low-Temperature Preservation of Foods and Living Matter*, eds. O.R. Fennema, W.D. Powrie, and E.H. Marth, pp. 150–239. New York: Marcel Dekker, Inc.

Fennema, O.R. 1977. Loss of vitamins in fresh and frozen foods. *Food Technology* 12: 32–38.

Fennema, O.R. 1996. Water and ice. In *Food Chemistry*, 3rd edn., ed. O.R. Fennema, pp. 17–94. New York: Marcel Dekker, Inc.

Fikiin, K. 2003. Novelties of food freezing research in Europe and beyond. Flair-Flow 4 Synthesis Report. Paris, France: Institut National de la Recherche Agronomique.

Fikiin, K. 2008. Emerging and novel freezing processes. In *Frozen Food Science and Technology*, ed. J.A. Evans, pp. 101–123. Oxford, U.K.: Blackwell Publishing Ltd.

Fikiin, K., Tsvetkov, O., Laptev, Y., Fikiin, A., and Kolodyaznaya, V. 2003. Thermophysical and engineering issues of the immersion freezing of fruits in ice slurries based on sugar-ethanol solution. *Ecolibrium* 2(7): 10–15.

Fletcher, J. M. 2002. Freezing. In *Nutrition Handbook for Food Processors*, eds. C.J.K. Henry and C. Chapman, pp. 331–341. Cambridge, U.K.: Woodhead Publishing Ltd.

Foegeding, E.A., Lanier, T.C., and Hultin, H.O. 1996. Characteristics of edible muscle tissues. In *Food Chemistry*, 3rd edn., ed. O.R. Fennema, pp. 879–942. New York: Marcel Dekker, Inc.

Fu, Y.-C. 2006. Microwave heating in food processing. In *Handbook of Food Science, Technology and Engineering*, Vol. 2, ed. Y.H. Hui, pp. 125-1/125-15. Boca Raton, FL: Taylor & Francis Group, LLC.

Fuchigami, M., Hyakuumoto, N., Miyazaki, K., Nomura, T., and Sasaki, J. 1994. Texture and histological structure of carrots frozen at a programmed rate and thawed in an electrostatic field. *Journal of Food Science* 59(6): 1162–1167.

Fuchigami, M., Hyakumoto, N., and Miyazaki, K. 1995. Programmed freezing effects on texture, pectic composition and electron microscopic structure of carrots. *Journal of Food Science* 60(1): 137–141.

Fuchigami, M., Kato, N., and Teramoto, A. 1997. High-pressure-freezing effects on textural quality of carrots. *Journal of Food Science* 62(4): 804–808.

Garrote, R. L. and Bertone, R. A. 1989. Osmotic concentration at low temperature of frozen strawberry halves. Effect of glycerol, glucose, and sucrose solution on exudate loss during thawing. *Food Science and Technology* 22: 264–267.

Giangiacomo, R., Torreggiani, D., Erba, M.L., and Messina, G. 1994. Use of osmodehydrofrozen fruit cubes in yoghurt. *Italian Journal of Food Science* 3: 345–350.

Goff, H.D. 1994. Measuring and interpreting the glass transition in frozen foods and model systems. *Food Research International* 27: 187–189.

Guegov, Y. 1981. Phase transitions of water in some products of plant origin at low and superlow temperatures. In *Advances in Food Research*, ed. C.O. Chichester, pp. 297–360. New York: Academic Press.

Haard, N.F. and Chism, G.W. 1996. Characteristics of edible plant tissues. In *Food Chemistry*, 3rd edn., ed. O.R. Fennema, pp. 943–1012. New York: Marcel Dekker, Inc.

Hagyard, J., Keiller, A.H., Cummings, T.L., and Chrystall, B.B. 1993. Frozen storage conditions and rancid flavour development in lamb. *Meat Science* 35: 305–312.

Halden, K., De Alwis, A.A.P., and Fryer, P. J. 1990. Changes in the electrical conductivity of foods during ohmic thawing. *International Journal of Food Science and Technology* 25: 9–25.

Hedges, N. and Nielsen, J. 2000. The selection and pre-treatment of fish. In *Managing Frozen Foods*, ed. C. Kennedy, pp. 95–110. Boca Raton, FL: Woodhead Publishing Limited.

Hui, Y.H., Cornillon, P., Guerrero Legaretta, I., Lim, M.H., Murrell, K.D., and Nip, W.-K. 2004. *Handbook of Frozen Foods*. New York: Marcel Dekker, Inc.

Imai, T., Uemura, K., Ishida, N., Yoshizaki, S., and Noguchi, A. 1995. Ohmic heating of Japanese white radish *Rhaphanus sativus* L. *International Journal of Food Science and Technology* 30: 9–25.

Inoue, C. and Ishikawa, M. 1997. Glass transition of tuna flesh at low temperature and effects of salt and moisture. *Journal of Food Science* 62: 496–499.

James, S. 2006. Principles of food refrigeration and freezing. In *Handbook of Food Science, Technology and Engineering*, Vol. 2, ed. Y.H. Hui, pp. 112-1/112-13. Boca Raton, FL: Taylor & Francis Group, LLC.

James, S. 2008. Freezing of meat. In *Frozen Food Science and Technology*, ed. J.A. Evans, pp. 124–150. Oxford, U.K.: Blackwell Publishing Ltd.

James, S.J. and Bailey, C. 1984. The theory and practice of food thawing. In *Thermal Processing and Quality of Foods*, ed. P. Zeuthen, pp. 566–578. London, U.K.: Elsevier.

Jeremiah, L.E. 1996. *Freezing Effects on Food Quality*. New York: Marcel Dekker, Inc.

Kennedy, C.J. 2000. Freezing processed foods. In *Managing Frozen Foods*, ed. C.J. Kennedy, pp. 137–158. Boca Raton, FL: Woodhead Publishing Limited.

Kennedy, C. 2003. Developments in freezing. In *Food Preservation Techniques*, eds. P. Zeuthen and L. Bøgh-Sørensen, pp. 228–240. Cambridge, U.K.: CRC Press.

Kerr, W.L. 2006. Frozen food texture. In *Handbook of Food Science, Technology and Engineering*, Vol. 2, ed. Y.H. Hui, pp. 61-1/61-15. Boca Raton, FL: Taylor & Francis Group, LLC.

Kissam, A.D., Nelson, R.W., Ngao, J., and Hunter, P. 1981. Water-thawing of fish using low frequency acoustics. *Journal of Food Science* 47: 71–75.

Knorr, D., Zenker, M., Heinz, V., and Lee, D.-U. 2004. Applications and potential of ultrasonics in food processing. *Trends in Food Science and Technology* 15(5): 261–266.

Koch, H., Seyderhelm, I., Wille, P., Kalichevsky, M.T., and Knorr, D. 1996. Pressure-shift freezing and its influence on texture, color, microstructure and rehydration behaviour of potato cubes. *Nahrung-Food* 40(3): 125–131.

Kramer, A. 1979. Effects of freezing and frozen storage on nutrient retention of fruits and vegetables. *Food Technology* 2: 58–65.

Lavety, J. 1991. Physico-chemical problems associated with fish freezing. In *Food Freezing: Today and Tomorrow*, ed. W.B. Bald, pp. 123–131. Berlin, Germany: Springer-Verlag.

Lawler, P.J. and Dimick, P.S. 2008. Crystallization and polymorphism of fats. In *Food Lipids: Chemistry, Nutrition, and Biotechnology*, 3rd edn., eds. C.C. Akoh and D.B. Min, pp. 245–263. New York: CRC Press.

Leadley, C.E. and Williams, A. 2006. Pulsed electric field processing, power ultrasound and other emerging technologies. In *Food Processing Handbook*, ed. J.G. Brennan, pp. 201–236. Weinheim, Germany: Wiley-VCH Verlag GmbH & Co. kGaA.

Le Bail, A. 2004. Freezing processes: Physical aspects. In *Handbook of Frozen Foods*, eds. Y.H. Hui, P. Cornillon, I. Guerrero Legaretta, M.H. Lim, K.D. Murrell, and W.-K. Nip, pp. 1–11. New York: Marcel Dekker, Inc.

Le Bail, A., Chevalier, D., Mussa, D.M., and Ghoul, M. 2002. High pressure freezing and thawing of foods: A review. *International Journal of Refrigeration* 25: 504–513.

Leniger, H. A. and Beverloo, W. A. 1975. *Food Process Engineering*. Dordrecht, the Netherlands: D. Reidel Publishing Co.

Levine, H. and Slade, L. 1990. Cryostabilization technology: Thermoanalytical evaluation of food ingredients and systems. In *Thermal Analysis of Foods*, eds. V.R. Harwalkar and C.Y. Ma, pp. 221–305. London, U.K.: Elsevier Applied Science.

Li, B. and Sun, D.-W. 2002a. Effect of power ultrasound on freezing rate during immersion freezing of potatoes. *Journal of Food Engineering* 55(3): 277–282.

Li, B. and Sun, D.-W. 2002b. Novel methods for rapid freezing and thawing of foods—A review. *Journal of Food Engineering* 54: 175–182.

Lim, M.H., McFetridge, J.E., and Liesebach, J. 2004. Frozen foods components and chemical reactions. In *Handbook of Frozen Foods*, eds. Y.H. Hui, P. Cornillon, I. Guerrero Legaretta, M.H. Lim, K.D. Murrell, and W.-K. Nip, pp. 67–81. New York: Marcel Dekker, Inc.

Lim, M.H., McFetridge, J.E., and Liesebach, J. 2006. Frozen food: Components and chemical reactions. In *Handbook of Food Science, Technology and Engineering*, Vol. 3, ed. Y.H. Hui, pp. 114-1/114-10. Boca Raton, FL: Taylor & Francis Group, LLC.

Lucas, T. and Raoult-Wack, A.-L. 1996. Immersion chilling and freezing: Phase change and mass transfer in model food. *Journal of Food Science* 61: 127–132.

Lucas, T. and Raoult-Wack, A.-L. 1998. Immersion chilling and freezing in aqueous refrigerating media: Review and future trends. *International Journal of Refrigeration* 21(6): 419–429.

Luque de Castro, M.D. and Priego-Capote, F. 2007. Ultrasound-assisted crystallization (sonocrystallization). *Ultrasonics Sonochemistry* 14(6): 717–724.

Mallet, C.P. 1993. *Frozen Food Technology*. London, U.K.: Chapman & Hall.

Maltini, E. 1974. Thermophysical properties of frozen lemon juice related to freeze-drying problems. Current studies on the *Thermophysical Properties of Foodstuffs, IIR-IIFC1-2*. Bressanone, Italy, pp. 201–207.

Mandigo, R.W. and Osburn, W.N. 1996. Cured and processed meats. In *Freezing Effects on Food Quality*, ed. L.E. Jeremiah, pp. 135–182. New York: Marcel Dekker, Inc.

Mannapperuma, J.D. and Singh, R.P. 1988. Prediction of freezing and thawing times of foods using a numerical method based on enthalpy formulation. *Journal of Food Science* 53(2): 626–630.

Martino, M.N., Otero, L., Sanz, P.D., and Zaritzky, N.E. 1998. Size and location of ice crystals in pork frozen by high-pressure-assisted freezing as compared to classical methods. *Meat Science* 50(3): 303–313.

Mascheroni, R.H. 1977. Transferencia de calor con simultáneo cambio de fase en tejidos cárneos. PhD dissertation, National University of La Plata, Argentina.

Mascheroni, R.H. and Agnelli, M.E. 1999. Cryomechanical freezing of foodstuffs: Quality improvements. *Proceedings of 20th International Congress of Refrigeration*, IIR/IIF, Sydney, Australia, p. 380.

Mason, T.J. 1998. Power ultrasound in food processing—The way forward. In *Ultrasound in Food Processing*, eds. J.W. Povey and T.J. Mason, pp. 105–126. London, U.K.: Blackie Academic & Professional.

Mason, T.J. and Chemat, F. 2003. Ultrasound as a preservation technology. In *Food Preservation Techniques*, eds. P. Zeuthen and L. Bøgh-Sørensen, pp. 303–337. Cambridge, U.K.: CRC Press.

Mason, T.J., Paniwnyk, L., and Lorimer, J.P. 1996. The uses of ultrasound in food technology. *Ultrasonics Sonochemistry* 3: S253–S260.

Miao, Y., Kawai, M., Horibe, K., and Noguchi, A. 2003. Ohmic thawing of frozen food. *Proceedings of the 4th Annual Meeting of Japan Society for Food Engineering*, Vol. 30. Otsu City, Japan (in Japanese).

Miao, Y., Chen, J.Y., and Noguchi, A. 2007. Studies on the ohmic thawing of frozen surimi. *Food Science and Technology Research* 13(4): 296–300.

Miles, C.A. and Shore, D. 1978. Changes in the attenuation of ultrasound during freezing. *Proceedings of the 24th European Meeting of Meat Research Workers*, pp. D.4:3–D.4:6. Germany, Kulmbach.

Miles, C.A., Morley, M.J., and Rendell, M. 1999. High power ultrasonic thawing of frozen foods. *Journal of Food Engineering* 39(2): 151–159.

Moreira, T. and Simatos, D. 1977. Quelques données sur les relations entre l'aptitude à la lyophilisation des jus de fruits et leur composition chimique. Freezing, Frozen Storage and Freeze-Drying. IIR-IIF C1-2, 487–493.

Nawar, W.W. 1996. Lipids. In *Food Chemistry*, 3rd edn., ed. O.R. Fennema, pp. 225–320. New York: Marcel Dekker, Inc.

Nesvadba, P. 2008. Thermal properties and ice crystal development in frozen foods. In *Frozen Food Science and Technology*, ed. J.A. Evans, pp. 51–80. Oxford, U.K.: Blackwell Publishing Ltd.

Norton, T., Delgado, A., Hogan, E., Grace, P., and Sun, D.-W. 2009. Simulation of high pressure freezing processes by enthalpy method. *Journal of Food Engineering* 91: 261–268.

Okos, M.R., Narsimhan, G., Singh, R.K., and Weitnauer, A.C. 1992. Food dehydration. In *Handbook of Food Engineering*, eds. R. Heldman and D.B. Lund, pp. 437–562. New York: Marcel Dekker.

Oliveira, J.C., Pereira, P.M., Frias, J.M., Cruz, I.B., and MacInnes, W.M. 1999. Application of the concepts of biomaterials science to the quality optimization of frozen foods. In *Processing Foods—Quality Optimization and Process Assessment*, eds. F.A.R. Oliveira and J.C. Oliveira, pp. 107–130. Boca Raton, FL: CRC Press LLC.

Otero, L., Solas, M.T., and Sanz, P.D. 1998. Contrasting effects of high-pressure-assisted freezing and conventional air-freezing on eggplant tissue microstructure. *Zeitschrift für Lebensmittel Untersuchung und Forschung* A206: 338–342.

Otero, L., Martino, M., Zaritzky, N., Solas, M., and Sanz, P.D. 2000. Preservation of microstructure in peach and mango during high-pressure-shift freezing. *Journal of Food Science* 65(3): 466–470.

Ottaway, P.B. 2002. The stability of vitamins during food processing. In *Nutrition Handbook for Food Processors*, eds. C.J.K. Henry and C. Chapman, pp. 247–264. Cambridge, U.K.: Woodhead Publishing Ltd.

Partman, W. 1975. The effects of freezing and thawing on food quality. In *Water Relations of Foods*, ed. R.B. Duckworth, pp. 505–537. London, U.K.: Academic Press Inc.

Patist, A. and Bates, D. 2008. Ultrasonics innovations in the food industry: From the laboratory to commercial production. *Innovative Food Science and Emerging Technologies* 9: 147–154.

Pham, Q.T. 2008. Modelling of freezing process. In *Frozen Food Science and Technology*, ed. J.A. Evans, pp. 1–25. Oxford, U.K.: Blackwell Publishing Ltd.

Préstamo, G., Fuster, C., and Risueño, M.C. 1998. Effect of blanching and freezing on the structure of carrots cells and their implications for food processing. *Journal of the Science of Food and Agriculture* 77: 223–229.

Préstamo, G., Palomares, L., and Sanz, P. 2005. Frozen foods treated by pressure shift freezing: Proteins and enzymes. *Journal of Food Science* 70(1): S22–S27.

Price, C.J. 1997. Take some solid steps to improve crystallization. *Chemical Engineering Progress*, September, 93(9): 34–43.

Rabin, Y. and Steif, P.S. 1998. Thermal stresses in a freezing sphere and its application in cryobiology. *Transactions of the ASME* 65: 328–333.

Ramaswamy, H.S. and Tung, M.A. 1984. A review on predicting freezing times of foods. *Journal of Food Process Engineering* 7: 169–203.

Ruecroft, G., Hipkiss, D., Ly, T., Maxted, N., and Cains, P.W. 2005. Sonocrystallization: The use of ultrasound for improved industrial crystallization. *Organic Process Research & Development* 9(6): 923–932.

Sa, M.M., Figueiredo, A.M., and Sereno, A.M. 1999. Glass transition and state diagrams for fresh and processed apple. *Thermochimica Acta* 329: 31–38.

Sanz, P.D. and Otero, L. 2005. High-pressure freezing. In *Emerging Technologies for Food Processing*, ed. D.-W. Sun, pp. 627–652. London, U.K.: Elsevier Academic Press.

Sanz, P.D., de Elvira, C., Martino, M., Zaritzky, N., Otero, L., and Carrasco, J.A. 1999. Freezing rate simulation as an aid to reducing crystallization damage in foods. *Meat Science* 52(3): 275–278.

Sastry, S.K., Shen, G.Q., and Blaisdell, J.L. 1989. Effect of ultrasonic vibration on fluid-to-particle convective heat transfer coefficients. *Journal of Food Science* 54(1): 229–230.

Shi, X., Datta, A.K., and Mukherjee, S. 1999. Thermal fracture in a biomaterial during rapid freezing. *Journal of Thermal Stresses* 22: 275–292.

Shore, D., Woods, M.O., and Miles, C.A. 1986. Attenuation of ultrasound in post rigor bovine skeletal muscle. *Ultrasonics* 24: 81–87.

Shrestha, S., Schaffner, D., and Nummer, B.A. 2009. Sensory quality and food safety of boneless chicken breast portions thawed rapidly by submersion in hot water. *Food Control* 20: 706–708.

Simatos, D., Faure, M., Bonjour, E., and Couach, M. 1975. Differential thermal analysis and differential scanning calorimetry in the study of water in foods. In *Water Relations of Foods*, ed. R.B. Duckworth, pp. 193–209. New York: Academic Press.

Slade, L. and Levine, H. 1991. Beyond water activity: Recent advances on an alternative approach to the assessment of food quality and safety. *Critical Review in Food Science and Nutrition* 30(2–3): 115–360.

Soukoulis, C., Lebesi, D., and Tzia, C. 2009. Enrichment of ice cream with dietary fibre: Effects on rheological properties, ice crystallisation and glass transition phenomena. *Food Chemistry* 115: 665–671.

Sun, D.-W. 2001. *Advances in Food Refrigeration*. Surrey, U.K.: Leatherhead Publishing, LFRA Ltd.

Sun, D.-W. 2006. *Handbook of Frozen Food Processing and Packaging*. Boca Raton, FL: CRC Press.

Sun, D.-W. and Li, B. 2003. Microstructural change of potato tissues frozen by ultrasound-assisted immersion freezing. *Journal of Food Engineering* 57(14): 337–345.

Sun, D.-W. and Zheng, L. 2006. Innovations in freezing process. In *Handbook of Frozen Food Processing and Packaging*, ed. D.-W. Sun, pp. 175–195. Boca Raton, FL: CRC Press.

Tomas, M.C. 1988. Oxidación lipídica en carnes refrigeradas y congeladas. PhD dissertation, Nacional University of La Plata, Buenos Aires, Argentina.

Torley, P.J. and Bhandari, B.R. 2007. Ultrasound in food processing and preservation. In *Handbook of Food Preservation*, 2nd edn., ed. M.S. Rahman, pp. 713–740. Boca Raton, FL: CRC Press.

Torreggiani, D., Forni, E., Guercilena, I. et al. 1999. Modification of glass transition temperature through carbohydrates additions: Effect upon color and anthocyanin pigment stability in frozen strawberry juices. *Food Research International* 32: 441–446.

Torreggiani, D., Lucas, T., and Raoult-Wack, A.L. 2000. The pre-treatment of fruits and vegetables. In *Managing Frozen Foods*, ed. C. Kennedy, pp. 54–80. Boca Raton, FL: Woodhead Publishing Limited.

United States Food and Drug Administration 2005. Food code 3.501.13. http://www.fda.gov/Food/FoodSafety/RetailFoodProtection/Foodcode/FoodCode2005/default.htm (accessed August 3, 2009).

Urrutia, G., Arabas, J., Autio, K. et al. 2007. SAFE ICE: Low-temperature pressure processing of foods: Safety and quality aspects, process parameters and consumer acceptance. *Journal of Food Engineering* 83: 293–315.

Van den Berg, C. 1991. Food-water relations: Progress and integration, comments and thoughts. In *Water Relationships in Foods—Advances in the 1980s and Trends for the 1990s*, eds. H. Levine and L. Slade, pp. 21–28. New York: Plenum Press.

Yam, K.L. and Lai, C.C. 2004. Microwable frozen food or meals. In *Handbook of Frozen Foods*, eds. Y.H. Hui, P. Cornillon, I. Guerrero Legaretta, M.H. Lim, K.D. Murrell, and W.-K. Nip, pp. 581–593. New York: Marcel Dekker, Inc.

Yi, J. and Kerr, W.L. 2009. Combined effects of dough freezing and storage conditions on bread quality factors. *Journal of Food Engineering* 93: 495–501.

Zaritzky, N.E. 2000. Factors affecting the stability of frozen foods. In *Managing Frozen Foods*, ed. C.J. Kennedy, pp. 111–135. Boca Raton, FL: Woodhead Publishing Limited.

Zaritzky, N. 2006. Physical-chemical principles in freezing. In *Handbook of Frozen Food Processing and Packaging*, ed. D.-W. Sun, pp. 3–31. Boca Raton, FL: Taylor & Francis Group, LLC.

Zaritzky, N. 2008. Frozen storage. In *Frozen Food Science and Technology*, ed. J.A. Evans, pp. 224–247. Oxford, U.K.: Blackwell Publishing Ltd.

Zheng, L. and Sun, D.-W. 2005. Ultrasonic assistance of food freezing. In *Emerging Technologies for Food Processing*, ed. D.-W. Sun, pp. 603–626. San Diego, CA: Elsevier Academic Press.

Zheng, L. and Sun, D.-W. 2006. Innovative applications of power ultrasound during food freezing process—A review. *Trends in Food Science and Technology* 17(1): 16–23.

Part II

Physicochemical Changes of Foods: A Focus on Products

8 Structural Changes of Multiphase Food Systems Generated by Proteins

Parichat Hongsprabhas

CONTENTS

8.1 INTRODUCTION

From a structural standpoint, foods are composed of macro- and microstructural elements (e.g., gas bubbles, crystals, oil droplets, and hydrated macromolecules) arranged in three dimensions, such as those found in baked products, butter, fluid milk, puddings, and sausages. The creation of multiphase food structures has become considerably significant for the food industry. However, such a task requires understanding the roles played by various physicochemical properties of food ingredients, at both the molecular level (structure at nano length scale) and the colloidal level (micro length scale), in the formation of a multiphase structure. This understanding would allow food processors to generate texture; and stabilize and control releases of tastes, flavors, nutrients, and bioactive compounds as well as postprandial metabolic responses of the created multiphase structure.

This chapter reviews the influences of food proteins in structure creation at the molecular, microscopic, and macroscopic levels. This information is important in view of the fact that proteins exhibit multiple effects on various important properties of foods, including surface properties (solubility, wettability, dispersibility, emulsifying, emulsion stabilizing, foaming, and foam stabilizing) and hydrodynamic properties (viscosifying, thickening, gelling, water-holding, and texturizing) (Damodaran, 1996). These multifunctionalities of proteins allow food processors to engineer the structure of foods with different characteristics by controlling the assembling process of the structural elements in each product.

In addition to size, shape, distribution, and volume fraction of structural elements generated by proteins, physical and chemical interactions among proteins and other food ingredients also play significant roles in determining the sensory attributes such as texture, flavor, and taste; stability; satiety; and bioactivity of foods (Aguilera and Stanley, 1999; Parada and Aguilera, 2007; Lundin et al., 2008; McClements et al., 2008). Apart from their roles in structure creation, proteins also play an important role in providing satiety through amino acids released during stomach digestion and lipolysis (Lundin et al., 2008; McClements et al., 2008; Na Nakornpanom and Hongsprabhas, 2008). Understanding the physicochemical basis of structure generated by proteins is thus crucial in the fabrication of foods for specific functions. The insights in structural changes of food proteins during processing and storage in colloidal food systems are reviewed in this chapter.

8.2 MULTIPHASE FOOD STRUCTURE

8.2.1 Creating Phase Separation—Mixing of Biopolymers

Normally, mixing of polymer A and polymer B could result in either synergistic interactions or phase separation, depending on the thermodynamic compatibility of the two components as well as the governing mechanisms of mixing (Aguilera, 1992; Matveev et al., 2000; Tanabe et al., 2002; Cheng et al., 2003).

Ledward (1993) summarized that the behavior of mixed macromolecules that do not chemically interact is controlled by the enthalpy of segment–segment interactions. Therefore, mixing of macromolecules could either yield a single phase containing different macromolecules in which the two polymers mutually exclude one another, or a single phase containing a soluble complex, or a liquid two-phase system in which the two macromolecules are primarily in separated phases, or an insoluble associative complex dispersed in a soluble serum phase that is depleted in both macromolecules. Such physical interactions depend on the chemical potentials, which determine whether the interactions between are thermodynamically favorable or not.

If interactions between macromolecules are not favorable, the mutual exclusion of each macromolecule from the domain of the others might take place. Examples are those found during mixing macromolecules of a similarly charged group (e.g., gelatin–whey proteins and agar–gelatin), or when mixing a charged macromolecule with a polar macromolecule (e.g., gelatin–maltodextrin mixture). Such thermodynamic phase separation is called segregative phase separation. On the other hand, if interactions between macromolecules are attractive or thermodynamically favorable (e.g., mixing of oppositely charged macromolecules), a soluble and/or insoluble coacervate complex could be formed. Mixing of negatively charged gelatin with positively charged gum Arabic or mixing of positively charged proteins and negatively charged polysaccharides are thermodynamically favorable, for example. Phase separation caused by this latter kind of interaction could be called associative phase separation or coacervation (Ledward, 1993).

A phase diagram for macromolecular mixing (polymer A and polymer B) is illustrated in Figure 8.1. The solid curve is the binodal or cloud point curve. Compositions of macromolecular mixtures lying under the cloud point curve would coexist as a single-phase solution. On the other hand, when the concentration region is above this curve, macromolecules form a two-phase system or water-in-water dispersion (w/w), since the system spontaneously separates (Antonov et al., 1977; Tolstoguzov, 1991, 2003). A w/w dispersion is composed of droplets of an immiscible polymer dispersed in a continuous phase of another polymer. Schorsch et al. (1999) illustrated phase separation of locust bean gum (LBG) suspension mixed with skim milk (SM) suspension at different volume ratios (Figure 8.2). At a higher LBG ratio, the gum

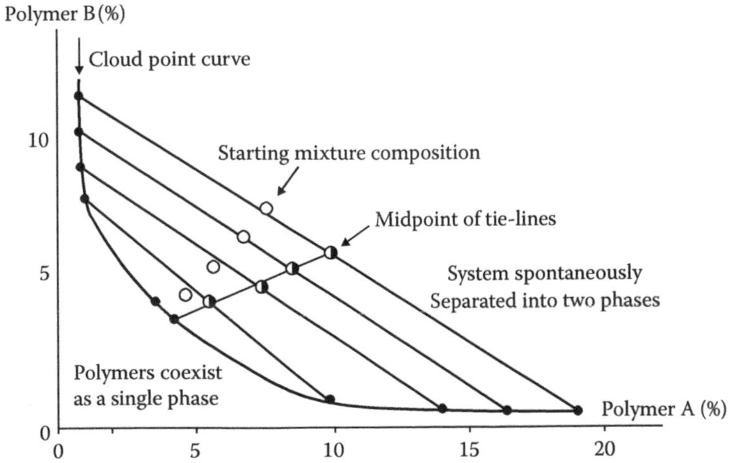

FIGURE 8.1 Phase diagram of mixed macromolecules (polymer A and polymer B). (Modified from Antonov, Y.A. et al., *Colloid Polym. Sci.*, 255, 937, 1977. With permission.)

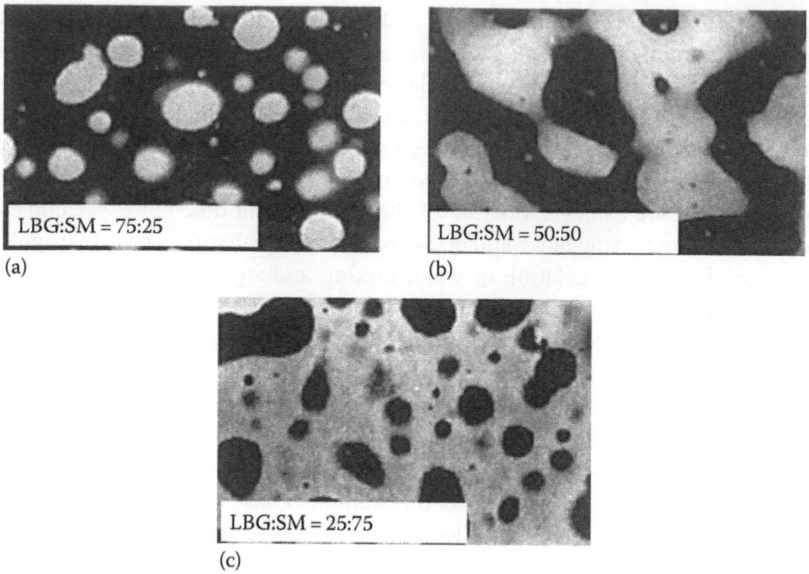

FIGURE 8.2 Mixing of LBG and SM. Proteins fluoresced in light color. (From Schorsch, C. et al., *Food Hydrocolloids*, 13, 89, 1999. With permission.)

existed as a continuous dark area. Increasing the SM ratio could generate a bicontinuous phase of LBG and SM; the inversion of a continuous phase from LBG to SM occurred at an LBG to SM ratio of 25:75.

8.2.2 Factors Affecting Phase Separation

An interplay between food formulation, process, and storage conditions results in a nonequilibrium nature of processed foods. The behavior of a food system indeed depends on (a) the volume fraction of the assembled structural elements, (b) types and strength of the interactions, and (c) the length scale or distance to which the elements are structured (Mezzenga, 2007). It is inevitable to consider these regimes simultaneously in order to understand the multiphase structure of foods.

8.2.2.1 Volume Fraction of Each Phase

In a dilute protein solution, the nano length scale or the molecular structure of protein molecules determines the thermodynamic equilibrium between protein–protein and protein–water interactions. The consequent surface and hydrodynamic properties of proteins are resulted from the proportion of hydrophobic, hydrophilic, and charged amino acid residues. For example, caseins could adopt a random coil structure due to their flexible structure as a result of phosphorylated serine residues; caseins indeed lack the ordered structures of α-helix, β-sheet, and β-turn found in globular proteins. This gives rise to better multifunctionality of caseins over globular proteins.

On the other hand, globular food proteins, such as whey proteins, egg albumen (EA) proteins, and soy proteins, possess an ordered structure, which develops from the folding of polypeptides to bury the hydrophobic residues inside and to expose the polar and charged residues to the exterior; this is to keep the net free energy to the minimum (Damodaran, 1996). Nevertheless, globular proteins may have a hydrophobic patch on the surface, together with the hydrophobic residues. These hydrophobic patches could influence the solubility characteristics and the surface properties of the globular proteins in a dilute solution. Therefore, globular proteins need to be unfolded prior to being adsorbed at an interface to have better emulsifying capability than the native proteins.

Proteins adsorbed at an oil–water interface may stabilize the oil droplets by the Derjaguin, Landau, Verwey, and Overbeek (DLVO) interactions and/or the steric stabilization mechanism. The proteins may possess or be capable of adopting extended structures, which protrude into a solution for a considerable distance from the interface. This extended hydrated layer may form the basis for steric stabilization of the emulsion. Interactions between the adsorbed protein layers can involve a reduction in configurational entropy as molecular chains overlap (Darling and Birkett, 1987). In addition, hydration of adsorbed hydrophilic components can lead to an enthalpic repulsion when two particles are in close proximity. This tends to force the oil droplets apart (Darling and Birkett, 1987).

At a higher volume fraction of protein, a multiphase structure may be formed as a viscous suspension, gel, glass, or crystalline (Mezzenga, 2007). The formation of different microstructures, such as liquid crystal, solid foam, high internal phase emulsion (HIPE), multiple emulsion, and glassy matrix of polymer may bring protein molecules in close proximity. Interactions among food components can subsequently occur irreversibly and generate new phases. The micro-length-scale (100 nm–100 μm) structural element generated by protein exists in many forms and positions, e.g., microparticles, particulate network, oil–water interface, and air–water interface. This concentrate regime is usually achieved during normal food processing, such as concentration and dehydration.

The protein molecules within close proximity could further react and form a new structure with different characteristics. For instance, Na Nakornpanom et al. (2009) found that the addition of calcium (Ca) lactate at an acidic pH of 3.0 to heated mixed proteins of soy protein isolate (SPI) and sodium caseinate (SCN) (80°C, 30 min) increased the aggregation of the heated mixed proteins (Figure 8.3). The Ca lactate-mediated aggregation was higher at higher ratios of SCN. This is probably due to the sensitivity of phosphorylated caseins to ionic calcium. Raising the Ca lactate concentration increased the aggregation of heated mixed proteins; the aggregation reached a plateau of around 25 mM of added Ca lactate. It is likely that a sulfhydryl (SH)–disulfide (SS) interchange was also involved in such an aggregation of soy and phosphorylated caseins. Lowering the pH of protein suspension and the addition of Ca lactate led to the lower charge dispersion of phosphorylated caseins, allowing the polypeptides to come close enough for Ca-crosslinking and subsequent SH–SS interchange. In the presence of the sulfhydryl-blocking agent *N*-ethylmaleimide (NEM), the aggregation of heated mixed proteins was quite low despite the increase in Ca lactate concentration. It was apparent that the metastable behavior of colloidal proteins attained in the concentrated regime could further alter their surface and interfacial characteristics.

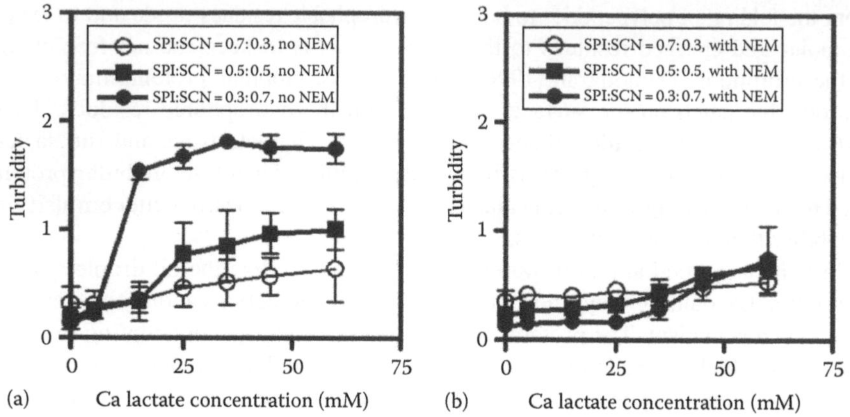

FIGURE 8.3 Effect of Ca lactate concentration on aggregation of heated SPI and SCN at a pH of 3.0 in (a) absence or (b) presence of NEM. (Modified from Na Nakornpanom, N. et al., *Kasetsart J. (Nat. Sci.)*, 43, 780–790, 2009. With permission.)

8.2.2.2 Phase Transition Temperature

Generally, starch gelation involves phase transition or gelatinization from a state of starch granules with a semicrystalline structure to a disordered hydrated state at an elevated temperature; this is followed by granule swelling and leaching of linear-chain amylose, leading to the formation of a three-dimensional network outside the granules. As a result, a starch gel is composed of granular starch gel particles embedded in a continuous amylose matrix. However, when starch and protein are heated together, a mixed structure could be formed.

Heating of mixed cassava starch (CS) and whey protein isolate (WPI) suspension at a pH of 5.75, which is close to the isoelectric pH of whey proteins, to 85°C was found to generate a composite gel when the starch fraction was below 0.7 (Aguilera and Rojas, 1996). At low starch fractions, i.e., 0.2–0.3, the elastic modulus and storage modulus of the mixed starch–protein gel were higher than those of pure whey protein gel (Figure 8.4). Thermal transition and microstructure of the mixed suspension indicated that gelatinization of starch fraction and denaturation of protein fraction was independent of each other. At starch fractions of higher than 0.7, the gel could not be formed due to excessive gelatinization of CS and its long paste characteristics. When the starch fraction increased, a continuous protein network phase filled with gelatinized starch inclusions was reported to invert to a continuous gelatinized starch matrix embedded with particulate protein inclusions. Consequently, the mixed starch–protein gel had weaker mechanical properties than whey protein gels due to the inversion of the continuous phase as shown in Figure 8.4 (Aguilera, 2000).

The inversion of the continuous protein phase to the starch phase could nevertheless be avoided. This could be done by using a salt-induced gelation method of globular proteins described by Barbut and Foegeding (1993) and Doi (1993) to induce the formation of a protein matrix prior to starch gelatinization. The reheating step of such a composite structure could limit starch swelling and leaching of amylose

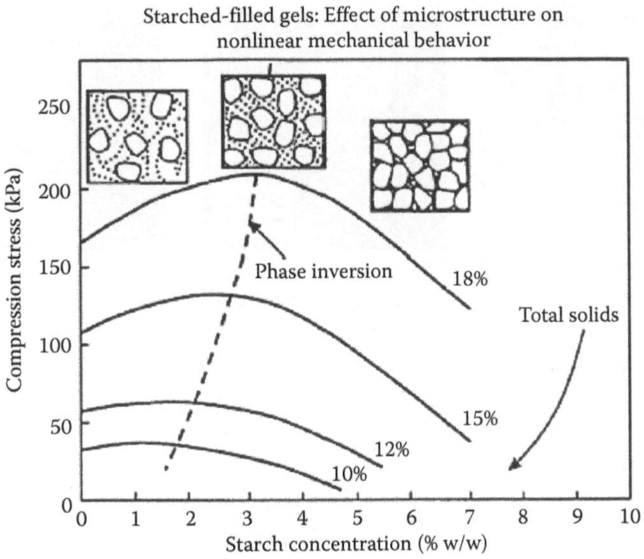

FIGURE 8.4 Effect of starch concentration on compressive stress of cassava-WPI mixed gel at different total solid concentrations. (From Aguilera, J.M., *Food Technol.*, 54, 56, 2000. With permission.)

(Hongsprabhas and Dit-udom-po, 2006). The use of two-stage gelation of the protein matrix made it possible to avoid the inversion of the continuous protein phase into the gelatinized starch phase when the starch-to-protein ratio was 2:1 (Hongsprabhas and Dit-udom-po, 2006). Such a starch-to-protein ratio led to a close-packed composite structure where the swollen starch granules are packed densely within the continuous protein matrix. It was also reported that EA-flour composite gels containing rice flour (RF) were more rigid than the ones containing CS.

Apart from the fraction of starch filler, structural reinforcement of a protein–starch composite also relies on the interactions between swollen starch granules and proteins at an interphase. Confocal laser scanning microscopy (CLSM) showed that swollen mungbean starch (MB) granules were bound by the protein network of waxy rice flour (WF) after thermal treatment (Figure 8.5). The resistance to shear during pasting of mixed WF and MB suspension, compared to that of waxy flour or MB alone, was likely due to the binding of swollen MB granules via the interactions between the WF protein network and the protein at the envelope of the swollen MB granules (Israkarn and Hongsprabhas, 2008). The induction of protein interactions at the interphase between the protein and starch phases in w/w dispersion could play an important role in preventing phase inversion at higher starch fractions in protein–starch mixes after the continuous protein phase was formed, regardless of the volume fraction of starch.

8.2.2.3 Physicochemical Properties of Macromolecules

Interactions between protein–protein, protein–small molecular weight (MW) emulsifier, and protein–polysaccharide at an interface or in each phase may be

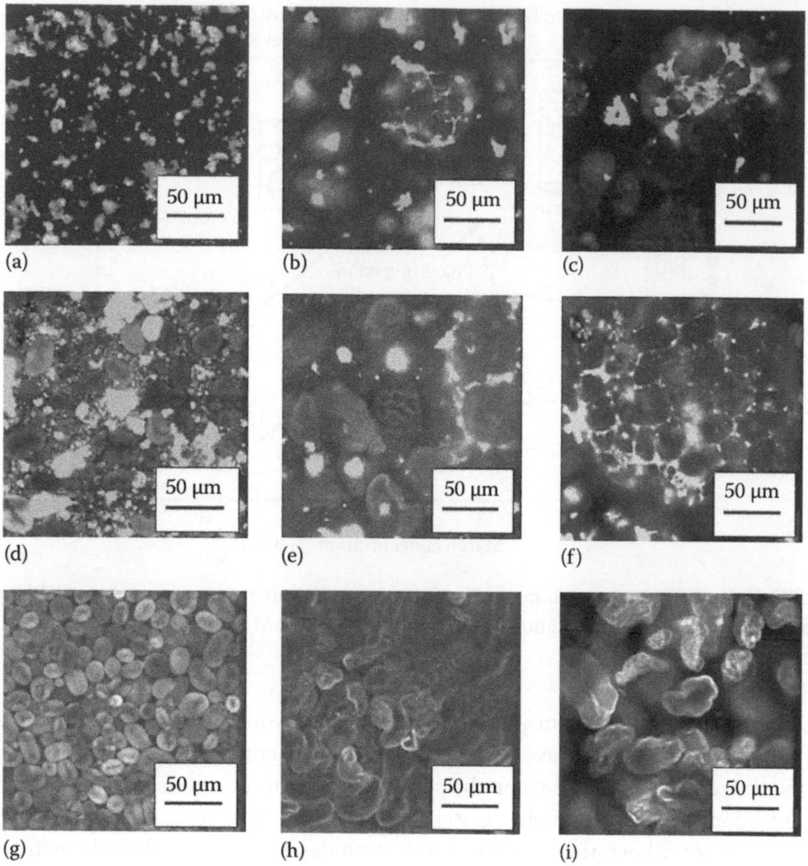

FIGURE 8.5 CLSM of WF and MB. (a) Unheated WF; (b) heated WF; (c) retrograded WF; (d) unheated WF:MB (0.5:0.5); (e) heated WF:MB (0.5:0.5); (f) retrograded WF:MB (0.5:0.5); (g) unheated MB; (h) heated MB; and (i) retrograded MB. Starch/flour suspensions (0.67% w/v) were heated at 80°C for 30 min and retrograded at 4°C for 48 h under quiescent condition. Proteins are fluoresced in light color by rhodamine B. (Modified from Israkarn, K. and Hongsprabhas, P., *Kasetsart J. (Nat. Sci)*, 42, 376, 2008. With permission.)

enhanced or delayed during processing by manipulating the molecular cooperation and competition through alterations of such parameters as temperature, pH, and ionic strength.

A concentrate regime is usually attained in foods and should always be under close consideration. The daily reference values (DRV) for a person with normal metabolism suggest a caloric intake of carbohydrate-to-lipid-to-protein of 45–65:20–35:10–35 per day. It is apparent that the physicochemical characteristics of normal foods are laid under the concentrate regime rather than the dilute regime. Therefore, the length scale of microstructural elements as well as the strength and types of interactions involved in the protein fraction are of important consideration when designing and controlling the multiphase structure of foods, as is discussed below.

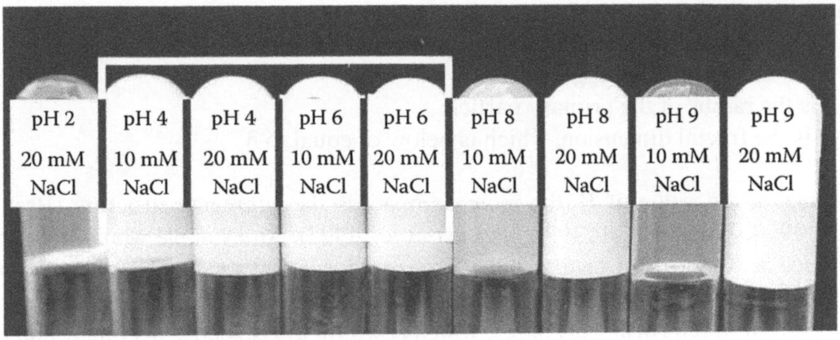

FIGURE 8.6 Appearance of heat-set WPI gels formed at different pH and NaCl concentrations. Heat-set WPI gel prepared within the isoelectric pH range (pH 4–6) had low gel strength and poor water-holding capacity.

8.3 PROTEINS AS MICROSTRUCTURAL ELEMENTS IN FOODS

8.3.1 Particulate Protein Gel

Gels can be defined as soft, deformable elastic solids consisting of a connected network of either small particles or large molecules (Dickinson, 1990). The network structure can be divided into two major types: the first is built of particles, e.g., protein aggregates, and the second is built by flexible macromolecules, e.g., gelatin and polysaccharides (Dickinson, 1990; van Vliet et al., 1991). A particle or particulate protein gel can either be translucent or opaque, depending on the size and arrangement of the particles forming its structure. Examples of particle gels include EA, dairy, and soy protein gels, which are commonly formed during heating.

With the exception of gelatin, which forms macromolecular gel, food proteins can form gels by either random or linear aggregation, depending on such factors as pH and salt concentration (Tombs, 1974; Doi, 1993). The random process results in the formation of an opaque gel, while linear aggregation results in a more transparent gel. On either the acidic or alkali side of the isoelectric pH, high electrostatic repulsion would slow down the aggregation rate, thus inducing linear aggregation. In contrast, at a pH close to the isoelectric pH, random aggregation occurs; as shown in Figure 8.6, the gels had different clarity. Overall, network formation can be controlled by factors such as protein concentration, salt type and concentration as well as heating conditions.

The arrangement of protein in a three-dimensional network can be characterized quantitatively using fractal geometry approach. A fractal object possesses a characteristic dimension, which is smaller than the dimensionality of the space within which it is embedded. Thus, a fractal object has the property of being less dense with increasing size. An aggregate of fractal geometry can be described by the following equation (Walstra et al., 1991):

$$N_p = \left(\frac{R_a}{a}\right)^D \tag{8.1}$$

where

- N_p is the number of particles in a growing aggregate
- R_a is the radius of an aggregate
- a is the radius of the primary particle
- D is the fractal dimension, which is below or equal to 3

The higher the value of D, the more compact is the aggregate structure (Bremer et al., 1990; Walstra et al., 1991; Vreeker et al., 1992). An example would be a fine-stranded heat-induced β-lactoglobulin (β-lg) gel formed at a pH of 7.5 in the absence of salt, which has a D value of approximately 2.91. On the other hand, a random aggregate would form at a pH of 5.3, which is within the isolectric pH range, and has a D value of approximately 2.46 (Stading et al., 1993).

Not only the size of an aggregate, but the pore size, strand arrangement, type of particulate network (i.e., random aggregation or linear aggregation process), and the relative strength of the links between aggregates to those within the aggregates also have an impact on the rheological and functional properties of protein gels (Hongsprabhas et al., 1999; Marangoni et al., 2000). The strength of the inter-aggregate, in comparison with the strength within the aggregate, is dependent on the nature of the aggregation process. In a diffusion-limited aggregation (DLA), aggregation is fast and limited by the diffusional motion of the aggregating particles, while in a reaction-limited aggregation (RLA), the aggregation process is slow due to the presence of a repulsive energy barrier between approaching particles. The fractal dimension of aggregates originated from DLA is around 1.7–1.8, whereas RLA aggregates have a fractal dimension in the range of 2.0–2.1 (Meakin, 1988).

The dependence of the elastic modulus on protein concentration has been used to establish the framework of fractal geometry. Bremer et al. (1990) indicated that a cluster of protein molecules would possess a fractal nature if the power-law dependence exists between the amount of flocs in the cluster and the radius of the cluster. In addition, the magnitude of this power, signified as the fractal dimension, D, would be below 3. The elastic constant of protein aggregates could be described as a function of the aggregate volume fraction:

$$\text{Elastic constant } (K) \sim \phi^{\mu} \tag{8.2}$$

where

- ϕ is the aggregate volume fraction
- μ is a constant, which can be graphically determined as the slope of a log ϕ versus log K plot

The models proposed by Bremer et al. (1990) and Shih et al. (1990) were summarized by Vreeker et al. (1992). Briefly, in the Shih et al. (1990) model, the mechanical properties of the network are determined by the elastic constants of individual aggregates and their backbone as well as the strong-linked inter-aggregate and weak-linked inter-aggregate. In this model, μ is related to the fractal dimension by either of the following equations:

$$\text{Weak-linked regime; } \mu = \frac{d-2}{d-D} \tag{8.3}$$

$$\text{Strong-linked regime; } \mu = \frac{d+x}{d-D} \tag{8.4}$$

where

- d is the Euclidian dimension (i.e., 3, for a close-packed space-filling aggregate, which is non-fractal)
- x is the backbone dimension of the aggregate, which could increase from 1 to 1.3 as the concentration is lowered

Food formulation and process parameters can be used to change the nature of protein aggregates and their network arrangement. The resulting three-dimensional network could thus exhibit different mechanical or textural properties that define their end use. The relative strength of the chemical interactions within the aggregates, in comparison to those between the aggregates, the network density as well as the overall characters of the microstructural elements within the three-dimensional network could be altered by using different protein types and concentration, salt types and concentration, and thermal or pressure treatments.

Micrographs shown in Figure 8.7 illustrate that whey protein aggregates could have different particle sizes and particulate networks possessing different porosities when prepared at different protein and salt concentrations. An increase in the protein concentration led to the formation of a translucent gel with small-sized aggregates, compared to the opaque gel with large-sized aggregates at constant fractal dimension formed at lower concentration. However, an increase in the $CaCl_2$ concentration led to a more opaque gel due to an increase in the diameter of the aggregates (Hongsprabhas and Barbut, 1997; Hongsprabhas et al., 1999; Marangoni et al., 2000). The fractal dimension of such a $CaCl_2$-induced cold-set WPI gel, as determined by Marangoni et al. (2000) (Table 8.1), indicated that the fast aggregation rate of heat-denatured whey proteins was due to the Ca^{2+}-crosslinking effect within the aggregate (weak-linked regime).

The fractal characteristics of protein aggregates making up the network determine the elastic properties caused by enthalpic changes (i.e., deformation of a gel structure involving breakage of chemical linkages). Thus, an understanding of the nature of protein aggregation or mechanisms involved in the three-dimensional network formation may help food scientists in designing the microscopic structure to provide a desirable texture and mouthfeel, or to control the releases of tastes or flavors, which are the characteristics of food proteins at the macroscopic length scale (Bremer et al., 1990; Dickinson, 1990; Walstra et al., 1991; Vreeker et al., 1992; Hongsprabhas et al., 1999; Marangoni et al., 2000; Renkema and van Vliet, 2004; Maltais et al., 2008).

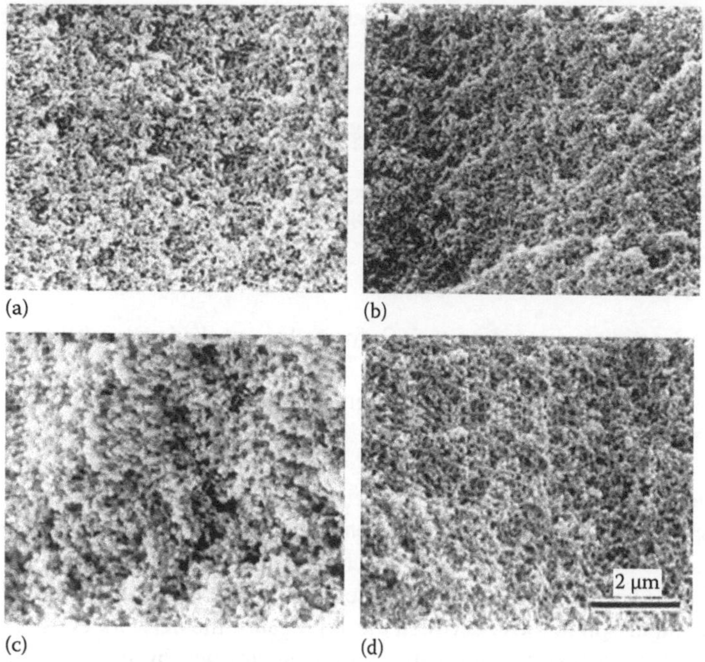

FIGURE 8.7 Scanning electron micrographs of $CaCl_2$-induced cold-set WPI gels. (a) WPI 10% protein, 10 mM $CaCl_2$, (b) WPI 10% protein, 120 mM $CaCl_2$, and (c) WPI 6% protein, 10 mM $CaCl_2$, and (d) WPI 6% protein, 120 mM $CaCl_2$. (Modified from Hongsprabhas, P. et al., *LWT-Food Sci. Technol.*, 32, 196, 1999.)

TABLE 8.1
Fractal Dimension (*D*) of $CaCl_2$-Induced Cold-Set Whey Protein Isolate Gel

	Fractal Dimension (*D*) Determined by	
[$CaCl_2$] mM	Rheology	Microscopy
10	2.54	2.25
30	2.57	2.45
120	2.62	2.61

Source: Marangoni, A.G. et al., *Food Hydrocolloids*, 14, 61, 2000. With permission.

8.3.2 Composite Structure

A composite structure is defined as a matrix, constituting of at least two phases with different properties. This kind of structure creates a novel property, which is different from that of the original. A continuous phase, often appearing in a large quantity, is called a matrix, whereas the other is referred to as a filler. The volume fraction and rheological characteristics of the filler, as well as the interactions at the filler and matrix interphase, can strengthen or weaken the overall mechanical property of the matrix (Aguilera, 1992). Examples of composite food structures involving proteins are described below.

8.3.2.1 Protein-Stabilized Emulsion

In any food emulsion, there are many different molecular species capable of lowering the interfacial tension between the aqueous phase and the oil phase. Of the types of surfactants used in food emulsions, it is proteins that play an important role as emulsifiers and long-term stabilizers. The driving force for protein adsorption at an interface involves the removal of nonpolar amino acid residues from the H-bonding environment of the bulk aqueous phase and the simultaneous displacement of vicinal water molecules from the hydrophobic environment of the oil–water interface (Dickinson, 1992). Following adsorption, protein forms a macromolecular barrier, which protects oil droplets against flocculation and coalescence. A favorable feature of a protein emulsifier is the ease of unfolding at an interface. The extent of unfolding depends on the surface area availability, the amount of time spent in contact with the interface, and the macromolecular structure prior to adsorption (Dickinson, 1992). The crucial aspect of emulsion formation and stability is the attainment of the balance between molecular cooperation and competition among emulsifiers at the oil–water interface during and after emulsification.

The functional requirements of practical food emulsions are not complete stability, but rather controlled instability. Destabilizing reactions of food emulsions involve creaming, flocculation, and coalescence. An emulsion would cream or sediment if the dispersed phase is sufficiently different in density from the continuous phase. Creaming can be reduced by increasing the viscosity of the aqueous phase or be enhanced by increasing the particle size of oil droplets or lowering the density of the oil phase.

Apart from the protein concentration at an interface, the structure of protein at the interface also plays an important role in stabilizing an emulsion (Friberg et al., 1990). Proteins used in the oil-in-water (o/w) emulsions include milk proteins (e.g., casein micelles, caseinates, and whey proteins), egg proteins (EA and egg yolk granules), animal proteins (e.g., plasma and gelatin), and plant proteins (e.g., soy proteins). From the molecular flexibility point of view, casein could produce a more stable emulsion; this is followed by whey proteins and soy proteins (Britten and Giroux, 1991a); among caseins, β-casein adsorbs at the highest extent (Dickinson et al., 1988).

A segment perpendicular to a surface results in an effective steric barrier, while the number of contact points with the interface influences the strength of adsorption. For example, flexible caseins have numerous proline residues, so they have little ordered secondary structure and no intramolecular crosslink. As a result, caseins are able to adopt a number of different conformational states when being adsorbed at the oil–water interface. They are usually adsorbed at the interface in such a way that considerable portions of their structures protruding into the aqueous phase are available (Dickinson, 1992). On the other hand, serum milk proteins, such as β-lg and α-lactalbumin (α-la), bind relatively close to the interface and do not protrude into a solution to a marked extent (Fang and Dalgleish, 1993). The thickness of an adsorbed layer increases at higher surface concentration of these proteins, suggesting multilayer formation of globular whey proteins at the interface.

The ability to adopt a number of conformations of caseins when they are adsorbed provides a possibility of monolayer formation of caseins at an interface. β-Casein was suggested to adsorb at the interface by the train/loop/tail model, with more hydrophobic train segments coming in direct contact with the surface and more hydrophilic loop and tail regions extending into the bulk aqueous phase (Dalgleish and Leaver, 1991; Dickinson, 1994). Approximately 70% of the adsorbed β-casein monolayer was found to closely associate with the hydrocarbon oil–water interface in the form of trains and small loops. In contrast, an adsorbed layer of globular proteins was reported to be thinner, but denser than that of the flexible β-casein. A monolayer of adsorbed globular proteins was a rather dense two-dimensional assembly of interacting deformable particles. The close-packed globular protein layer was consistent with slow and limited unfolding of the molecules after adsorption. The strong molecular force (attractive and repulsive) of globular proteins led to severely restricted molecular mobility at the interface (Dalgleish and Leaver, 1991).

The dissociation of a quaternary structure or denaturation of proteins is required prior to emulsification. Therefore, casein micelles are adsorbed at an interface in a semi-intact form (Oortwijn et al., 1977). The thermal denaturation of globular proteins prior to emulsification was reported to improve the emulsifying properties. The high level of the thermally denatured whey protein fraction in mixed proteins (of denatured and undenatured proteins) increased the emulsion viscosity and coalescence stability compared with the low-level denatured fraction (Britten et al., 1994).

It is apparent that the molecular structure of proteins plays an important role in determining the structure of proteins at an interface as well as the subsequent surface characteristics of oil droplets dispersed in an aqueous phase. Physicochemical changes of proteins at the oil–water interface are discussed in detail in Section 8.4.

8.3.2.2 Protein–Starch Composite Structure

A composite structure of protein–starch naturally occurring in grain endosperms is commonly found in cooked or steamed grains. Cooking or steaming cereals and legumes results in simultaneous changes of protein and starch phases where both protein denaturation and starch gelatinization occur at the same time.

Proteins, particularly granule-bound starch synthesis enzymes entrapped in dried starch granules, are also involved in the determination of the starch-pasting characteristics, even when they are present at less than 0.2% in dried starch. Figure 8.8 illustrates a separated protein phase in an envelope of swollen CS after heating (shown in light color by rhodamine B, which bound specifically to the amine group of proteins) (Hongsprabhas et al., 2007). In heated CS granules (80°C for 30 min), the protein fraction progressed to the circumference of the granules due to the thermodynamic incompatibility with the hydrated starch fraction. This protein fraction could help maintain the integrity of the granule envelope after heat treatment. The alterations of these granule-bound proteins could change the mechanical properties

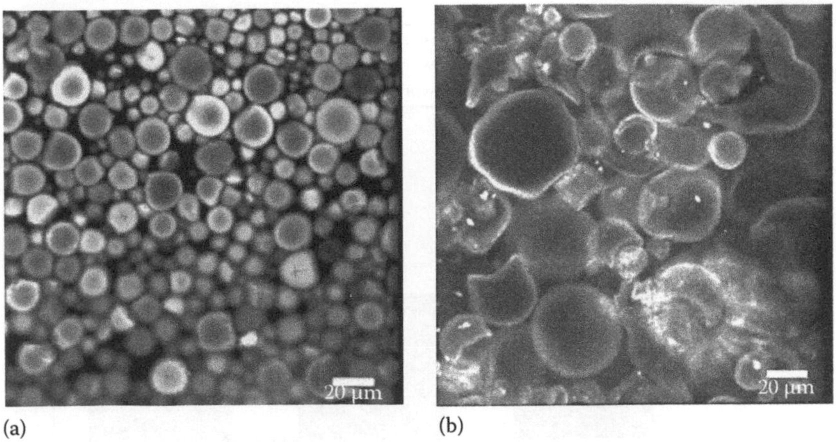

FIGURE 8.8 CLSM of unheated and heated MB and CS. Starch suspensions (0.67% w/v in distilled water) were heated at 80°C for 30 min under quiescent condition. Proteins were stained with rhodamine B, which appeared in light color. (a) Unheated CS and (b) heated CS. Bars = 20 µm. (Modified from Hongsprabhas, P. et al., *Carbohydr. Polym.*, 67, 614, 2007. With permission.)

and the glass transition temperature of thermoplastic starch, the raw material used in the fabrication of extruded starch-based bioplastic films, even if it is present in minute amounts (Hongsprabhas and Israkarn, 2008).

Starches and proteins are main ingredients in many food formulations. The mechanism of starch reinforcement in a protein–starch composite matrix relies upon the fractions of both macromolecules as well as the water availability, gelation temperature of a continuous matrix, volume fraction of the separated phase, or filler and mechanical strength of the fillers in relation to the continuous matrix, as previously described. Generally, the reinforcement could be obtained by increasing the volume fraction of the filler (Richardson et al., 1981; Ross-Murphy and Todd, 1983).

An ability to form a composite structure between heated globular proteins and starch granules may expand the application of the separated phase under some conditions. The presence of CS in an EA composite film, for example, was noted to provide a means to regulate the thermo-mechanical properties of the composite film containing sunflower oil at both low and high relative humidity (RH) (Wongsasulak et al., 2006). Such a composite structure can be used as a controlled release film or coating material for food ingredients. CS was used in this case as the reinforcing filler and regulator of the thermo-mechanical characteristics and effective diffusivity (D_{eff}) of oil-soluble core materials at different RH values (Wongsasulak et al., 2007).

The release of a carotenoid compound, which is oil soluble, from composite films with a continuous protein phase and CS as the filler to an external oil phase was found to be mainly driven by Fickian diffusion (Wongsasulak et al., 2007). The D_{eff} values of the carotenoid compound through the EA–CS composite film were between 1.38×10^{-12} and 5.46×10^{-13} m^2/s (Figure 8.9). The EA network having high strength and elasticity (formed at a pH of 11 with 150 mM NaCl) resulted in composite films that could delay the release of the core material. On the other hand, reducing

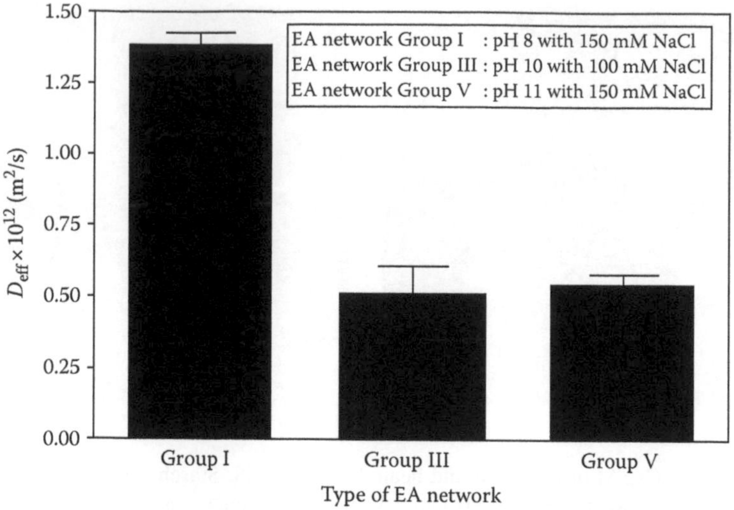

FIGURE 8.9 Effect of EA network type on D_{eff} of oil-soluble core material (carotenoid compound diffused from the composite film into the bulk oil phase). (From Wongsasulak, S. et al., *Food Res. Int.*, 40, 249, 2007. With permission.)

the strength and elasticity of the EA protein matrix resulted in a faster release of the core material from the EA–CS composite structure. However, increasing the RH during aging of the composite films reduced the D_{eff} ($p \leq 0.05$), as shown in Figure 8.10. The film formed at a pH of 11 with 150 mM NaCl and aged at 100% RH showed no release of the core material after 2.5 h.

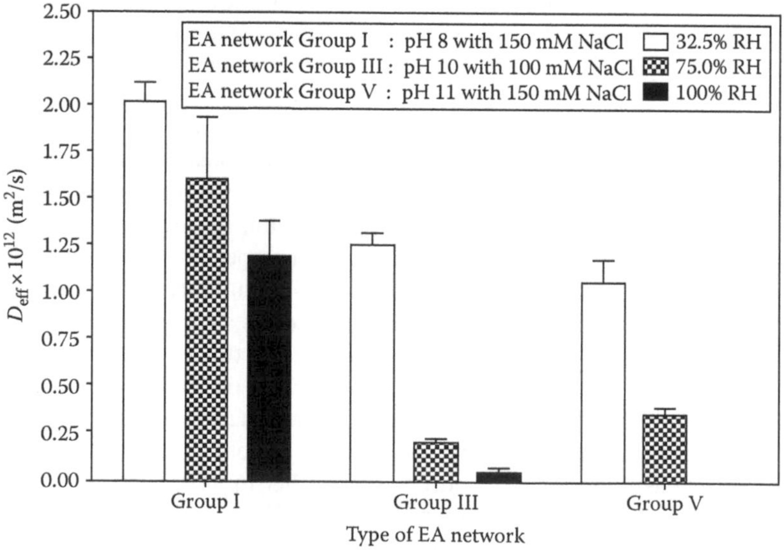

FIGURE 8.10 Effect of EA network type and RH during aging on D_{eff} of core material (carotenoid compound diffused from the composite film into the bulk oil phase). (From Wongsasulak, S. et al., *Food Res. Int.*, 40, 249, 2007. With permission.)

8.4 PHYSICOCHEMICAL CHANGES OF PROTEINS AND THEIR INFLUENCES ON FOOD STRUCTURE

8.4.1 Existence of Proteins as Colloidal Particles

8.4.1.1 Preferential Adsorption of Proteins at Oil–Water Interface

The preferential adsorption of caseinates over whey proteins was demonstrated by Britten and Giroux (1991b) and Hunt and Dalgleish (1994). α-la and β-lg showed similar affinity for adsorption at an interface (Dalgleish et al., 1991). Dalgleish et al. (1991) also compared surface concentration of ovalbumin with that of β-lg and reported that β-lg had a much greater affinity for the interface. The egg yolk protein phosvitin also exhibited lower affinity for the interface compared with β-casein and β-lg (Dickinson et al., 1991).

Preferential adsorption occurs in emulsions stabilized by mixed protein systems, i.e., caseinate (mixture of phosphorylated α_{s1}-, α_{s2}-, β-, and κ-caseins) and WPI (mixture of β-lg, α-la, bovine serum albumin [BSA], and immunoglobulin G [IgG]). Hunt and Dalgleish (1994) demonstrated that the adsorption behavior of proteins was affected by their concentration. These investigators reported that the maximum surface concentration of caseinate and WPI was $3.2\,mg/m^2$. When the protein concentration was the limiting factor (i.e., the concentration was less than the limiting surface concentration required to stabilize the emulsion), both WPI and caseinate adsorbed to the same extent. However, with an excess in protein concentration, caseinate adsorbed better than WPI.

Usually, emulsions stabilized by globular proteins are prone to flocculation and aggregation during thermal treatment (McClements, 2004). An o/w emulsion stabilized by soy proteins, for example, showed a high viscosity value of $400\,mPa \cdot s$ when the protein concentration was high enough to be present in the dispersed phase (Roesch and Corredig, 2002). Although SCN is extensively used as an excellent emulsifier in protein-stabilized emulsions, it is sensitive to ionic calcium (Agboola and Dalgleish, 1996; Dalgleish, 1997; Dickinson and Golding, 1998; Dickinson and Davies, 1999). The presence of phosphorylated α_{s1}-casein and β-casein, which are the major polypeptides in SCN, makes caseinate-stabilized emulsions prone to creaming and increased viscosity when the ionic calcium was added to above 10 mM. This is due to the presence of phosphoserine clusters in the casein molecules (Dickinson and Golding, 1998; Dickinson and Davies, 1999). Therefore, the structure of caseins at an oil–water interface may be altered in the presence of different ionic calcium concentrations.

The amphoteric characters of proteins and amino acids at an oil–water interface and in a bulk aqueous phase may further complex their behavior when the pH of an emulsion is altered; this is particularly true with an event occurring in the gastrointestinal tract. Figure 8.11 illustrates that a short digestion time of 15 min at acidic pH by pepsin resulted in an aggregation of protein-stabilized oil droplets. Prolonging the digestion time to 60 min resulted in larger oil droplet size, disintegration of aggregated oil droplets, and brighter signal of rhodamine B in light color at the interface, suggesting redepositing of peptic hydrolyzate at the oil–water interface during digestion. Over the course of digestion under stomach condition at pH 1–2 and intestinal condition at pH 7–7.5, the interface of protein-stabilized emulsion could undergo

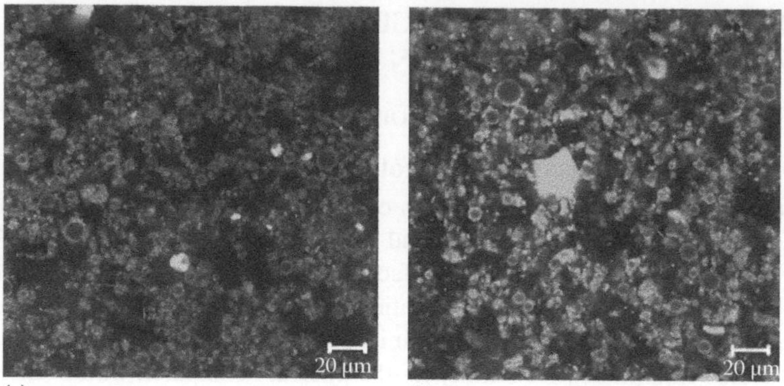

(a) (b)

FIGURE 8.11 Microstructure of heated mixed soy protein and SCN-stabilized emulsion during peptic digestion. (a) 15 min peptic digestion and (b) 60 min peptic digestion. Proteins were stained with rhodamine B, which appeared in light color.

several changes that alter the rates of oil release and peptide release, depending on the types of protein at the interface (Na Nakornpanom and Hongsprabhas, 2008, 2010).

8.4.1.2 Displacement of Proteins at Oil–Water Interface

After adsorption, adsorbed proteins can interact with each other since they have already denatured at an interface (Dickinson and Matsumura, 1991). This is evidenced by an increase in the interfacial viscosity with time after adsorption (Dickinson et al., 1990). Over time, the types and nature of proteins at the interface could further alter. This is crucial for storage stability. The ability of β-lg and α-la to displace one another previously adsorbed at the interface is slow and irreversible. In contrast, caseins show rapid and reversible displacement with α-la and β-lg (Dickinson et al., 1989). β-lg is more difficult to be displaced by β-casein than is α-la. Nevertheless, β-lg could not displace β-casein at the interface (Dalgleish et al., 1991). However, some β-lg could be found to adsorb at the interface of the emulsion droplets, which were previously stabilized by β-casein.

The extent of displacement depends on the duration before addition of the second protein or surfactant (aging time). Dalgleish et al. (1991) showed that once the layer of adsorbed ovalbumin was formed, it could not be displaced by β-casein; β-casein could not completely displace β-lg after aging. The displaced β-lg appeared to re-adsorb along with β-casein. Dickinson et al. (1990) showed that the exchange of casein was very rapid, whereas the processes involving serum proteins took several days to complete. The difficulty in the displacement might be due to the structured layer of the adsorbed proteins formed during aging. The increase in surface viscoelasticity of an adsorbed protein film as a function of time suggested that protein unfolding and the development of new protein–protein interactions at the interface took place over the period of aging.

The displacement of proteins at the oil–water interface is important in terms of the storage stability of protein-stabilized emulsions. Food emulsions requiring long-term stability need interfaces that could resist displacement to prevent coalescence during storage.

8.4.1.3 Influence of Small MW Emulsifiers

The presence of small MW emulsifiers may aid the creation of an interface since they promote higher reduction of interfacial tension than the action of proteins alone. This leads to smaller droplets during homogenization. However, the main purpose of adding emulsifiers is to perform functions not directly related to emulsification, e.g., to control fat polymorphism, to inhibit staling or promote fat globule clumping during whipping or freezing, or partial coalescence due to the displacement of interfacial proteins by a small MW emulsifier (Dickinson, 1992, 1993). This is important when partial coalescence of o/w emulsion is required to promote foam formation and stabilization in whipped cream and ice cream.

Small MW emulsifiers can play an important role in determining the properties of an emulsion. This is because these emulsifiers can compete with proteins for the interface. They can co-adsorb with the proteins, which results in a thicker and more viscoelastic adsorbed layer, or a more mobile adsorbed layer, depending on the strength and nature of the interactions (Dickinson, 1993, 1994).

Besides protein–protein interactions, adsorbed proteins can react with small MW emulsifiers. These interactions could lead to a reduction in the surface activity due to specific binding of the lipophilic tail of the emulsifier to a hydrophobic site on the proteins. Protein–emulsifier interactions may result in unfolding the tertiary structure, enhancing molecular flexibility and the rate of rearrangement at the interface. The binding of ionic emulsifiers can change the net charge on a stabilizing macromolecular layer. Interfacial protein–emulsifier complexing, an incorporation of proteins into emulsifier micelles, vesicles, and bilayers, or protein–emulsifier competitive adsorption, may change the viscoelasticity of the adsorbed layer (Dickinson, 1994).

The addition of an emulsifier before emulsification may alter the nature of surface interactions. For example, α-la could be adsorbed at an interface threefolds higher than β-lg when Tween 20, the water-soluble emulsifier, was added before homogenization (Courthaudon et al., 1991a). Courthaudon et al. (1991a,b,c) showed that the addition of Tween 20 immediately after emulsion formation led to a complete displacement of proteins from the interface at an emulsifier:protein molar ratio of 17:1. The presence of glycerol monostearate, the oil-soluble emulsifier, dissolved in the oil phase enhanced the displacement of proteins from the interface by Tween 20 (Dickinson and Tanai, 1992). It was also reported that the protein surface concentration in the fresh emulsion and the amount of water-soluble emulsifier required for complete displacement could be reduced by the oil-soluble emulsifier previously dissolved in the oil phase.

When an oil-soluble emulsifier is present in an oil phase, less proteins (e.g., β-lg and β-casein) are displaced by a water-soluble emulsifier if the emulsion has been aged before the addition of the water-soluble emulsifier (Chen and Dickinson, 1993; Chen et al., 1993). Although the aging time influences the competition of proteins at an interface, protein adsorption is likely to be noncompetitive once one protein has become established at the interface (Dickinson, 1992). The highest surface-active protein would be the major protein at the interface if it is present during emulsification. The protein likely to be predominant in an aged protein film would be the one that was first introduced to the interface, irrespective of whether or not it is the more surface-active of the two proteins.

8.4.2 Protein–Ingredient Interactions in Food Formulation

It is apparent that the structures of proteins, at both molecular and microstructural levels, such as those found at an interface, are dynamic and prone to changes during food processing and storage. The structural alterations may be accelerated or delayed upon further modifications of such parameters as pH and ionic strength, or upon addition of polysaccharides.

8.4.2.1 pH

Some foods require an adjustment of pH to enhance, control, or mask taste or flavor. pH can affect emulsification since protein solubility and charge repulsion can be reduced at isoelectric pH. However, the cohesiveness of protein films tends to be maximal near the isoelectric pH (Mangino, 1994). Near the isoelectric pH, proteins are able to form close-packed dense films, thus increasing the protein concentration at an interface. Waniska and Kinsella (1988) demonstrated that at isoelectric pH, electrostatic repulsion was minimized, allowing hydrophobic residues to stabilize a compact tertiary structure of β-lg. Klemaszewski et al. (1992) reported that the coalescence stability of emulsion stabilized by β-lg was dependent on pH. However, the coalescence stabilities of α-la- or SCN-stabilized emulsions were unaffected by pH.

8.4.2.2 Types of Salt and Ionic Strength

Ions can collapse a double layer and minimize charge interactions, and consequently destabilize an emulsion system. Increasing ion concentration changes the distance distribution of the repulsive potential, while the attractive potential is not affected, meaning that the net chemical potential drops faster with the distance (Friberg et al., 1990).

Dickinson et al. (1992) reported that when Ca^{2+} was added prior to homogenization, the extent of droplet flocculation was greater for β-casein-stabilized emulsion than for phosvitin-stabilized emulsion. It was suggested that β-casein had more tendency to precipitate by Ca^{2+} than phosvitin in egg yolk. However, when Ca^{2+} was added after homogenization, it was the phosvitin-stabilized emulsion that was more susceptible to flocculation. This was due to the greater binding affinity of phosvitin for Ca^{2+} than β-casein.

8.4.2.3 Polysaccharides

A polysaccharide can be added as a component in a protein system to produce a protein–polysaccharide composite structure. Tolstoguzov (2003) reviewed the main function of protein and polysaccharide in protein–polysaccharide food formulation. Generally, polysaccharides have less surface activity in comparison to proteins. This inferiority is related to their low flexibility and monotonic repetition of the monomer units in the backbone. The low surface activity of polysaccharides results in their inability to form a primary adsorbed layer in the system. The nature of interactions between polysaccharides and adsorbed proteins, as well as their influence on colloid stability, can either stabilize or destabilize the emulsions. Attractive protein–polysaccharide interactions can enhance the emulsion stability by forming a thicker and stronger steric-stabilizing layer. In contrast, the attractive interactions at higher protein and polysaccharide concentrations can destabilize an emulsion by bridging protein-coated droplets, a process so-called bridging flocculation. Repulsive protein–polysaccharide interactions, on the other hand, can stabilize the emulsions by immobilizing oil droplets in a weak or strong gel network.

Emulsions stabilized by a mixture of β-lg and dextran were noted to have poorer stability than those stabilized only by proteins (Dickinson and Galazka, 1991). The presence of unadsorbed polysaccharides gave rapid serum separation due to depletion flocculation. This problem could be overcome by using covalently linked protein–polysaccharide conjugates. Emulsions stabilized by β-lg–dextran conjugates (through Maillard reactions) had excellent stabilizing ability with respect to creaming, coalescence, and serum separation. The same principles are also applied in EA–polysaccharide conjugates; these protein–polysaccharide conjugates had superior emulsifying properties compared to native polysaccharides and EA (Kato et al., 1993).

The main advantage of the covalently linked protein–polysaccharide conjugates is their ability to maintain the solubility and molecular integrity over a wide range of conditions (pH, ionic strength, temperature, etc.). Another example of protein–polysaccharide conjugates, which is a natural stabilizer, is gum Arabic. Two percent protein associated with a high MW fraction of polysaccharide in gum Arabic is responsible for its surface activity. This fraction adsorbs strongly at an oil–water interface and is responsible for the emulsifying and stabilizing properties of gum Arabic in emulsion (Dickinson, 1992).

Carrageenans can be used as thickeners and stabilizers in milk products by binding to the casein micelles. The binding of a negatively charged carrageenan with caseins is performed through Ca^{2+}-crosslinking between α_s- and β-caseins and carrageenan. The interactions between κ-casein and carrageenan, which occur in the absence of Ca^{2+}, take place through specific positively charged regions of the κ-casein molecules (Dalgleish and Morris, 1988).

8.5 CONCLUDING REMARKS

Structural changes of proteins at the molecular level play an important role in the creation of a multiphase food structure during processing and storage. This chapter reviews the significance of proteins in determining the phase separation and their interactions with other ingredients in the creation of a multiphase food structure. Understanding how microstructural elements are assembled in each phase as well as how they physicologically respond during digestion and absorption in the gastrointestinal tract may shed some light in designing food structures appropriate for specific purposes or metabolic responses. Controlling the digestibility of macromolecules such as proteins, lipids, and starches can further influence post-prandial behavior, e.g., amino acid absorption, and blood plasma triglyceride and glucose concentrations, metabolically (Lundin et al., 2008). Subsequent bioavailability of several nutrients is thus influenced by the fate of these assembled microstructural elements after they have been ingested (Parada and Aguilera, 2007; Lentle and Janssen, 2008; Lundin et al., 2008; McClements et al., 2008). More work is needed to understand the fate of proteins in each phase during digestion. The surface characteristics and viscoelastic properties of digesta could influence the diffusivity of the nutrients and their absorption due to pH shift in the stomach and the small intestine as well as due to protein digestion and reassembly of peptides in the digesta.

REFERENCES

Agboola, S.O. and Dalgleish, D.G. 1996. Kinetics of the calcium-induced instability of oil-in-water emulsions: Studies under shearing and quiescent conditions. *LWT-Food Science and Technology* 29: 425–432.

Aguilera, J.M. 1992. Generation of engineered structures in gels. In: *Physical Chemistry of Foods*, eds. H.G. Schwartzberg and R.W. Hartel, pp. 387–422. New York: Marcel Dekker.

Aguilera, J.M. 2000. Microstructure and food product engineering. *Food Technology* 54: 56–65.

Aguilera, J.M. and Rojas, E. 1996. Rheological, thermal and microstructural properties of whey protein-cassava starch gels. *Journal of Food Science* 61: 962–966.

Aguilera, J.M. and Stanley, D.W. 1999. *Microstructural Principles of Food Processing and Engineering*, 2nd edn. Gaithersburg, MD: Aspen Publishers.

Antonov, Y.A., Grinberg, V.Y., and Tolstoguzov, V.B. 1977. Phase equilibria in water-protein-polysaccharide systems. *Colloid and Polymer Science* 255: 937–947.

Barbut, S. and Foegeding, E.A. 1993. Ca^{2+}-induced gelation of pre-heated whey protein isolate gels. *Journal of Food Science* 58: 867–871.

Bremer, L.G.B., Bijsterbosch, B.H., Schrijvers, R., van Vliet, T., and Walstra, P. 1990. On the fractal nature of the structure of acid casein gel. *Colloids and Surfaces* 51: 159–170.

Britten, M. and Giroux, H.J. 1991a. Emulsifying properties of whey protein and casein composite blends. *Journal of Dairy Science* 74: 3318–3325.

Britten, M. and Giroux, H.J. 1991b. Coalescence index of protein stabilized emulsions. *Journal of Food Science* 56: 792–795.

Britten, M., Giroux, H.J., Jean, Y., and Rodrigue, N. 1994. Composite blends from heat-denatured and undenatured whey protein: Emulsifying properties. *International Dairy Journal* 4: 25–36.

Chen, J. and Dickinson, E. 1993. Time-dependent competitive adsorption of milk proteins and surfactants in oil-in-water emulsion. *Journal of the Science of Food and Agriculture* 62: 283–289.

Chen, J., Dickinson, E., and Iveson, G. 1993. Interfacial interactions, competitive adsorption and emulsion stability. *Food Structure* 12: 135–146.

Cheng, M., Deng, J., Yang, F., Gong, Y., Zhao, N., and Zhang, X. 2003. Study on the properties and nerve cell affinity of composite films from chitosan and gelatin solutions. *Biomaterials* 24: 2871–2880.

Courthaudon, J.L., Dickinson, E., Matsumura, Y., and Clark, D.C. 1991a. Competitive adsorption of β-lactoglobulin + tween 20 at the oil-water interface. *Colloids and Surfaces* 56: 293–300.

Courthaudon, J.L., Dickinson, E., and Dalgleish, D.G. 1991b. Competitive adsorption of β-casein and nonionic surfactants in oil-in-water emulsions. *Journal of Colloid and Interface Science* 145: 390–395.

Courthaudon, J.L., Dickinson, E., Matsumura, Y., and Williams, A. 1991c. Influence of emulsifier on the competitive adsorption of whey protein in emulsions. *Food Structure* 10: 109–116.

Dalgleish, D.G. 1997. Structure-function relationships of caseins. In *Food Proteins and Their Applications*, eds. S. Damodaran and A. Paraf, pp. 199–223. New York: Marcel Dekker.

Dalgleish, D.G. and Leaver, J. 1991. Dimensions and possible structures of milk proteins at oil-water interfaces. In *Food Polymers, Gels and Colloids*, ed. E. Dickinson, pp. 113–122. Cambridge, U.K.: Royal Society of Chemistry.

Dalgleish, D.G. and Morris, E.R. 1988. Interactions between carrageenans and casein micelles: Electrophoretic and hydrodynamic properties of the particles. *Food Hydrocolloids* 2: 311–320.

Dalgleish, D.G., Euston, S.E., Hunt, J.A., and Dickinson, E. 1991. Competitive adsorption of β-lactoglobulin in mixed protein emulsions. In *Food Polymers, Gels and Colloids*, ed. E. Dickinson, pp. 485–489. Cambridge, U.K.: Royal Society of Chemistry.

Damodaran, S. 1996. Functional properties. In *Food Proteins: Properties and Characterization*, eds. S. Nakai and H.W. Modler, pp. 167–234. New York: Wiley-VCH.

Darling, D.F. and Birkett, R.J. 1987. Food colloids in practice. In *Food Emulsions and Foams*, ed. E. Dickinson, pp. 1–29. Cambridge, U.K.: Royal Society of Chemistry.

Dickinson, E. 1990. Particle gels. *Chemical Industry* 19: 595–599.

Dickinson, E. 1992. *An Introduction to Food Colloids*. New York: Oxford University Press.

Dickinson, E. 1993. Towards more natural emulsifier. *Trends in Food Science and Technology* 4: 330–334.

Dickinson, E. 1994. Protein-stabilized emulsions. *Journal of Food Engineering* 22: 59–74.

Dickinson, E. and Davies, E. 1999. Influence of ionic calcium on stability of sodium caseinate emulsions. *Colloids and Interfaces B* 12: 203–212.

Dickinson, E. and Galazka, V.B. 1991. Emulsion stabilization by ionic and covalent complexes of beta-lactoglobulin with polysaccharide. *Food Hydrocolloids* 5: 281–296.

Dickinson, E. and Golding, M. 1998. Influence of calcium ions on creaming and rheology of emulsions containing sodium caseinate. *Colloids and Surfaces A* 144: 167–177.

Dickinson, E. and Matsumura, Y. 1991. Time-dependent polymerization of β-lactoglobulin through disulphide bonds at the oil-water interface in emulsions. *International Journal of Biological Macromolecules* 13: 26–30.

Dickinson, E. and Tanai, S. 1992. Temperature dependence of the competitive displacement of protein from the emulsion droplet surface by surfactant. *Food Hydrocolloids* 6: 163–171.

Dickinson, E., Hunt, J.A., and Dalgleish, D.G. 1991. Competitive adsorption of phosvitin with milk proteins in oil-in-water emulsion. *Food Hydrocolloids* 4: 403–414.

Dickinson, E., Hunt, J.A., and Horne, D.S. 1992. Calcium induced flocculation of emulsions containing adsorbed β-casein or phosvitin. *Food Hydrocolloids* 6: 359–370.

Dickinson, E., Rolfe, S.E., and Dalgleish, D.G. 1988. Competitive adsorption of αs1-casein and β-casein in oil-in-water emulsions. *Food Hydrocolloids* 2: 397–405.

Dickinson, E., Rolfe, S.E., and Dalgleish, D.G. 1989. Competitive adsorption in oil-in-water emulsions containing α-lactalbumin and β-lactoglobulin. *Food Hydrocolloids* 3: 193–203.

Dickinson, E., Rolfe, S.E., and Dalgleish, D.G. 1990. Surface shear viscometry as a probe of protein-protein interactions in mixed milk protein films adsorbed at the oil-water interface. *International Journal of Biological Macromolecules* 12: 189–194.

Doi, E. 1993. Gels and gelling of globular proteins. *Trends in Food Science and Technology* 4: 1–5.

Fang, Y. and Dalgleish, D.G. 1993. Dimensions of the adsorbed layers in oil-in-water emulsion stabilized by caseins. *Journal of Colloid and Interface Science* 156: 329–334.

Friberg, S.E., Goubran, R.F., and Kayali, I.H. 1990. Emulsion stability. In *Food Emulsions*, 2nd edn., eds. K. Larrson and S.E. Friberg, pp. 1–40. New York: Marcel Dekker.

Hongsprabhas, P. and Barbut, S. 1997. Protein and salt effects on Ca^{2+}-induced gelation of whey protein isolate. *Journal of Food Science* 62: 382–385.

Hongsprabhas, P. and Dit-udom-po, S. 2006. Influence of flour type on mechanical properties of calcium-induced egg albumen-flour composite gel. *International Journal of Food Engineering* 2(1), article 2.

Hongsprabhas, P. and Israkarn, K. 2008. New insights on the characteristics of starch network. *Food Research International* 41: 998–1006.

Hongsprabhas, P., Barbut, S., and Marangoni, A.G. 1999. The structure of cold-set whey protein isolate gels prepared with Ca^{2+}. *LWT-Food Science and Technology* 32: 196–202.

Hongsprabhas, P., Israkarn, K., and Rattanawattanaprakit, C. 2007. Architectural changes of heated mungbean, rice and cassava starch granules: Effects of hydrocolloids and protein-containing envelope. *Carbohydrate Polymers* 67: 614–622.

Hunt, J.A. and Dalgleish, D.G. 1994. Adsorption behaviour of whey protein isolate and caseinate in soya oil-in-water emulsions. *Food Hydrocolloids* 8: 175–187.

Israkarn, K. and Hongsprabhas, P. 2008. Influence of waxy rice protein network on physical properties of waxy rice flour composites. *Kasetsart Journal (Natural Science)* 42: 376–386.

Kato, A., Minaki, K., and Kobayashi, K. 1993. Improvement of emulsifying properties of egg white protein by the attachment of polysaccharide through Maillard reaction in a dry state. *Journal of Agricultural and Food Chemistry* 41: 540–543.

Klemaszewski, J.L., Das, K.P., and Kinsella, J.E. 1992. Formation and coalescence stability of emulsion stabilized by different milk proteins. *Journal of Food Science* 57: 366–371.

Ledward, D.A. 1993. Creating textures from mixed biopolymer systems. *Trends in Food Science and Technology* 4: 39–42.

Lentle, R.G. and Janssen, P.W.M. 2008. Physical characteristics of digesta and their influence on flow and mixing in mammalian intestine; a review. *Journal of Comparative Physiology B* 178: 673–690.

Lundin, L., Golding, M., and Wooster, T.J. 2008. Understanding food structure and function in developing food for appetite control. *Nutrition and Dietetics* 65: S79–S85.

Mangino, M.E. 1994. Protein interactions in emulsions: Protein-lipid interactions. In *Protein Functionality in Food System*, eds. N.S. Hettiarachchy and G.R. Ziegler, pp. 147–180. New York: Marcel Dekker, Inc.

Marangoni, A.G., Barbut, S., McGauley, S.E., Marcone, M., and Narine, S.S. 2000. On the structure of particulate gels—the case of salt-induced cold gelation of heat-denatured whey protein isolate. *Food Hydrocolloids* 14: 61–74.

Maltais, A., Remondetto, G.E., and Subirade, M. 2008. Mechanisms involved in the formation and structure of soya protein cold-set gels: A molecular and supramolecular investigation. *Food Hydrocolloids* 22: 550–559.

Matveev, Y.I., Grinberb, V.Y., and Tolstoguzov, V.B. 2000. The plasticizing effect of water on proteins, polysaccharides and their mixtures. Glassy state of biopolymers, food and seeds. *Food Hydrocolloids* 14: 425–437.

McClements, D.J. 2004. Protein-stabilized emulsions. *Current Opinion in Colloid and Interface Sciences* 9: 305–313.

McClements, D.J., Decker, E.A., Park, Y., and Weiss, J. 2008. Designing food structure to control stability, digestion, release and absorption of lipophilic food components. *Food Biophysic* 3: 219–228.

Meakin, P. 1988. Fractal aggregates. *Advances in Colloid and Interface Science* 28: 249–331.

Mezzenga, R. 2007. Equilibrium and non-equilibrium structures in complex food systems. *Food Hydrocolloids* 21: 674–682.

Na Nakornpanom, N. and Hongsprabhas, P. 2008. Heating sequence and calcium lactate concentration effects on *in vitro* protein digestibility and oil release of emulsion. In *CORNUCOPIA and AGFD abstracts for the 236th American Chemical Society National Meeting*, vol. 29. Washington, DC: American Chemical Society.

Na Nakornpanom, N., Thongngam, M., and Hongsprabhas, P. 2009. Characteristics of heated mixed soy protein isolate and sodium caseinate. *Kasetsart Journal* (*Natural Science*) 43: 780–790.

Na Nakornpanom, N., Hongsprabhas, P., and Hongsprabhas, P. 2010. Effect of soy residue (okara) on in vitro protein digestibility and oil release in emulsion stabilized by heated mixed proteins. *Food Research International* 43: 26–32.

Oortwijn, H., Walstra, P., and Mulder, H. 1977. The membranes of recombined fat globules, 1: Electron microscopy. *Netherlands Milk and Dairy Journal (Netherlands)* 31: 134–147.

Parada, J. and Aguilera, J.M. 2007. Food microstructure affects the bioavailability of several nutrients. *Journal of Food Science* 72: R21–R32.

Renkema, J.M.S. and van Vliet, T. 2004. Concentration dependence of dynamic moduli of heat induced soy protein gels. *Food Hydrocolloids* 18: 483–487.

Richardson, R.K., Robinson, G., Ross-Murphy, S.B., and Todd, S. 1981. Mechanical spectroscopy of filled gelatin gels. *Polymer Bulletin* 4: 541–546.

Roesch, R.R. and Corredig, M. 2002. Characterization of oil-in-water emulsions prepared with commercial soy protein concentrate. *Journal of Food Science* 67: 2837–2842.

Ross-Murphy, S.B. and Todd, T. 1983. Ultimate tensile measurements of filled gelatin gels. *Polymer* 24: 481–486.

Schorsch, C., Jones, M.G., and Norton, I.T. 1999. Thermodynamic incompatibility and microstructure of milk protein/locuat bean gum/sucrose systems. *Food Hydrocolloids* 13: 89–99.

Shih, W.-H., Shih, W.Y., Kim, S.-I., Liu, J., and Aksay, I.A. 1990. Scaling behaviour of the elastic properties of colloidal gels. *Physical Review A* 42: 4772–4779.

Stading, M., Langton, M., and Hermansson, A.-M. 1993. Microstructure and rheological behaviour of particulate β-lactoglobulin gels. *Food Hydrocolloids* 7: 195–212.

Tanabe, T., Okitsu, N., Tachibana, A., and Yamauchi, K. 2002. Preparation and characterization of keratin-chitosan composite film. *Biomaterials* 23: 817–825.

Tolstoguzov, V.B. 1991. Functional properties of food proteins and role of protein-polysaccharide interaction. *Food Hydrocolloids* 4: 429.

Tolstoguzov, V. 2003. Some thermodynamic considerations in food formulation. *Food Hydrocolloids* 17: 1–23.

Tombs, M.P. 1974. Gelation of globular proteins. *Faraday Discussion Chemical Society* 57: 158–164.

van Vliet, T., Luyten, H., and Walstra, P. 1991. Fracture and yielding of gel. In *Food Polymers, Gels and Colloids*, ed. E. Dickinson, pp. 392–403. Cambridge, U.K.: The Royal Society of Chemistry.

Vreeker, R., Hoekstra, L.L., den Boer, D.C., and Agterof, W.G.M. 1992. Fractal aggregation of whey proteins. *Food Hydrocolloids* 6: 423–435.

Walstra, P., van Vliet, T., and Bremer, L.G.B. 1991. On the fractal nature of particle gels. In *Food Polymers, Gels and Colloids*, ed. E. Dickinson, pp. 369–382. Cambridge, U.K.: The Royal Society of Chemistry.

Waniska, R.D. and Kinsella, J.E. 1988. Foaming and emulsifying properties of glycosylated β-lactoglobulin. *Food Hydrocolloids* 2: 439–449.

Wongsasulak, S., Yoovidhya, T., Bhumiratana, S., Hongsprabhas, P., McClements, D.J., and Weiss, J. 2006. Thermo-mechanical properties of egg albumen-cassava starch composite films containing sunflower-oil droplets as influenced by moisture content. *Food Research International* 39: 277–284.

Wongsasulak, S., Yoovidhya, T., Bhumiratana, S., and Hongsprabhas, P. 2007. Physical properties of composite network of egg albumen and cassava starch formed by a salt-induced gelation method. *Food Research International* 40: 249–256.

9 Impacts of Freezing and Frozen Storage on Quality Changes of Seafoods

Soottawat Benjakul and Wonnop Visessanguan

CONTENTS

9.1 INTRODUCTION

Freezing is one of the most common preservation methods to maintain the quality of fish and shellfish. Although freezing or frozen storage is able to prevent microbial spoilage effectively, it cannot terminate chemical deteriorations, which mainly involve protein denaturation and lipid oxidation of the products.

The alteration of fish proteins during freezing and frozen storage is caused by different mechanisms: (1) partial dehydration of protein during freezing; (2) an increase in inorganic salts in an unfrozen phase; (3) interaction of lipids, free fatty acids, and/or lipid oxidation products with proteins; and (4) action of trimethylamine oxide demethylase (TMAOase) (Parkin and Hultin, 1982). These reactions are indeed associated with losses in quality and the consumer acceptability of the products, mostly related to losses in functionality such as gel-forming ability and water-holding capacity. The texture and muscle structure of fish and shellfish is also affected by freezing or frozen storage. The changes in texture are more pronounced after repeated freeze–thawing, leading to tough and rubbery texture together with the released water, known as drip loss. It is noted that, in general, muscle proteins in fish and shellfish are more susceptible to freeze denaturation compared with plant or land animal proteins. The reader is referred to Chapter 7 for information on freezing and thawing in general and to Chapter 10 for information on the effects of freezing and frozen storage on physicochemical changes of red meats.

9.2 ICE CRYSTAL FORMATION

Ice crystals formed from either intracellular or extracellular water during freezing or frozen storage can disrupt the muscle cells of fish and shellfish. The degree of muscle damage is dependent upon the size and shape of ice crystals, which are governed by the freezing rate or method, storage temperature, fluctuation of temperature, dehydration during extended storage, as well as additives used. During freezing, ice nuclei are generated between muscle cells (Goodband, 2002). As ice crystals grow, water is abstracted from myofibrils and cells become condensed. If the interaction between myofibrils occurs and the lattice is not able to fully recover, aggregation with coincidental reduced water-holding capacity can be observed (Goodband, 2002).

As ice crystallization takes place, extracellular salt becomes concentrated, leading to differences in osmotic pressure gradient across cell membrane (Xiong, 1997). Due to such differences, intracellular moisture flows outward. This causes the dehydration of muscle cell and an increase in ionic strength of the cell. The increased ionic strength in unfrozen phase is associated with the disruption of electrostatic interaction in protein molecules, leading to the denaturation or dissociation of protein molecules. The released water from cell further undergoes ice crystallization, resulting in an increase in the size of ice crystals.

As mentioned earlier, freezing can disrupt muscle cells, resulting in the release of mitochondrial and lysosomal enzymes into sarcoplasm (Hamm, 1979). α-glucosidase (AG) and β-*N*-acetyl-glucosidase (NAG) have been used as the markers of freezing and thawing process of fish muscle (Rehbein, 1979; Shimomura et al., 1987; Benjakul and Bauer, 2000). For example, both AG and NAG activities in croaker, lizardfish, threadfin bream, and bigeye snapper muscles were noted to increase during frozen storage at −18°C. The varying activities were found during the storage, depending on fish species (Benjakul et al., 2005a).

In general, drip is found in frozen stored fish and shellfish after thawing. Therefore, the losses of some nutrients and flavoring compounds can occur. Hsieh and Regenstein (1989) reported an increase in expressible drip in cod and ocean perch during frozen storage. Benjakul et al. (2003a) also found an increase in expressible drip of croaker, lizardfish, threadfin bream, and bigeye snapper during frozen storage at −18°C for up to 24 weeks.

9.3 PROTEIN DENATURATION

During freezing and frozen storage, fish muscle undergoes a number of changes including denaturation and aggregation of myofibrillar proteins. Fennema (1982) noted that all proteins would be expected to have an optimal stability at a temperature just above the freezing point of water. Fish myosin is very unstable in comparison with its mammal counterpart. Prolonged frozen storage at –20°C thus has a marked effect on the myosin transition in fish. Herrera et al. (2001) reported that the transition I of minced blue whiting muscle became less clearly defined and gradually flatted out over the storage period at –10°C. In contrast, transition II hardly changed during the whole storage period at this temperature, reflecting the high stability of actin. At lower temperature, intramolecular hydrogen bonding is enhanced (MacDonald and Lanier, 1991). Therefore, denaturation is accelerated as a result of destabilization of hydrophobic interaction. During frozen storage protein oxidation also occurs (Baron et al., 2007); this leads to a wide range of modifications, especially the formation of carbonyl groups (Kjersgard et al., 2006). These changes contribute to enhanced aggregation of proteins, which is most likely associated with the loss in protein extractability (Figure 9.1).

The extractability of fish myosin generally decreases during frozen storage. Huidobro et al. (1998) studied aggregation of myofibrillar proteins in hake, sardine, and mixed mince during frozen storage at –20°C for 1 year. During 8 months of storage the amount of protein linked by covalent bonds in hake increased with the storage time. For sardine and mixed sample a high percentage of protein linked by covalent bonds was detected earlier. Cod myosin and muscle also became less extractable in 0.6 M KCl (Tejada et al., 1996). Badii and Howell (2001) observed that the frozen storage of cod and haddock muscle at –10°C led to the denaturation of myosin, causing a reduction in solubility and precipitation of proteins. Srikar and Reddy (1991) also reported a decrease in solubility of pink perch protein during

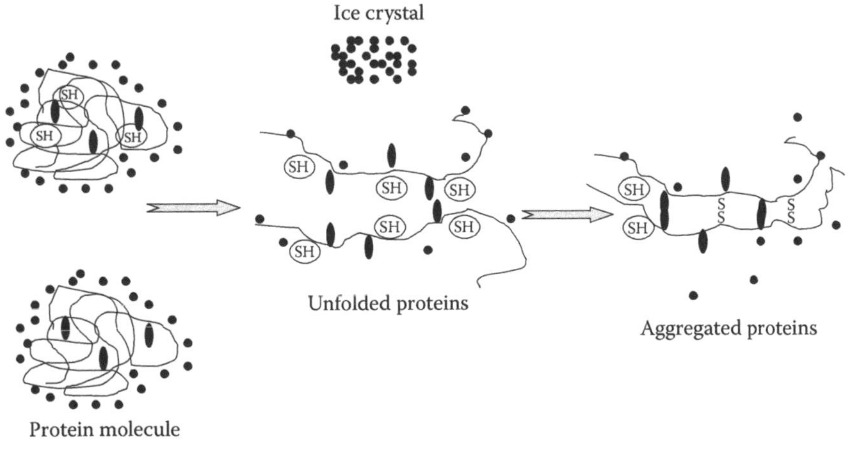

FIGURE 9.1 Scheme for denaturation and aggregation of muscle proteins during freezing and frozen storage.

frozen storage. Aggregate formed during frozen storage of mince cod were mediated by secondary interaction and disulfide bridges (Tejada et al., 1996).

During freezing and frozen storage the denaturation and aggregation of protein start from the formation of disulfide bonds, which is followed by a rearrangement of hydrophobic and hydrogen-bonded regions on an intra- and intermolecular basis (Buttkus, 1974). During storage various bonds stabilizing the secondary and tertiary structures of protein molecules are destroyed, causing changes in protein molecules. For instance, structural differences in protein were noted between fresh and frozen stored hake muscle after the 10 months of storage at −10°C and −30°C (Careche et al., 1999). Such structural changes mainly entailed an increase in β-sheets and the expense of α-helices. An increase in unordered protein structure was found only in sample stored at −10°C. However, the exposure of buried tryptophan residues was observed at both storage temperatures. A decrease in δCH_2 band upon storage suggested an increase in the hydrophobic interaction of aliphatic residues. Badii and Howell (2002a) found that the denaturation of proteins caused by ice crystal formation and/or resultant solute concentration, and also by lipid oxidation products, involved hydrophobic groups and the exposure of other polar groups due to unfolding of the molecules, accompanied by changes in the secondary structure. Structural changes in hake fillets were governed by freezing method and frozen storage temperature. Samples stored at −10°C showed greater alterations than at −30°C (Careche et al., 1999).

9.3.1 Changes in Ca^{2+}-ATPase Activity

Globular heads of myosin are responsible for ATPase activity, which is sensitive to alteration in the configuration of molecule (Sultanbawa and Li-Chan, 2001). ATPase is able to catalyze the decomposition of adenosine triphosphate (ATP) into adenosine diphosphate (ADP) and a free phosphate ion. During frozen storage myofibrillar proteins, especially myosin, are susceptible to denaturation as indicated by a decrease in Ca^{2+}-ATPase activity (Benjakul and Bauer, 2000). Jiang et al. (1985) reported a decrease in Ca^{2+}-ATPase activity of mackerel and amberfish actomyosin during frozen storage. Del Mazo et al. (1999) found that Ca^{2+}-ATPase activity of natural actomyosin (NAM) from hake fillet stored at −20°C and −30°C decreased during 15 days of frozen storage. Ca^{2+}-ATPase activity of fish myosin obtained from *Tilapia nilotica* decreased markedly during frozen storage, losing 89% of the initial value, suggesting that myosin head could be involved in fish myosin aggregation. Nambudiri and Gopakumar (1992) observed a decrease in ATPase activity of fresh water and brackish fish by 70%–90% after 6 months of frozen storage at −20°C. The 24% decrease in Ca^{2+}-ATPase was found in Alaska pollock after 226 days of frozen storage at −20°C.

The loss in ATPase activity is due to the tertiary structural changes caused by ice crystal and an increase in the ionic strength of the system (Benjakul and Bauer, 2000). Additionally, the rearrangement of protein via protein–protein interactions also contributes to the loss in ATPase activity (Benjaukl and Bauer, 2000). Ca^{2+}-ATPase activity in lizardfish muscle (*Saurida micropectoralis*) decreased continuously during frozen storage at −20°C (Leelapongwattana et al., 2005a).

9.3.2 Sulfhydryl Group and Disulfide Bonds

Sulfhydryl (SH) groups are considered the most reactive functional group in proteins, being oxidized to disulfide bond during the frozen storage of fish (Benjakul et al., 2003a). The formation of disulfide bonds via oxidation of SH groups or disulfide interchanges is coincidental with a decrease in total and surface SH contents (Hayakawa and Nakai, 1985). The oxidation of SH groups located in the head portion of myosin plays a role in the loss in Ca^{2+}-ATPase activity (Benjakul et al., 1997). Disulfide bridges occurring during frozen storage of muscle protein are associated with protein aggregation. Ramirez et al. (2000) suggested that the frozen storage of myosin in suspension resulted in aggregation involving the side-to-side interactions of the rod with a low formation of disulfide bonds. On the other hand, head-to-head interaction with a higher formation of disulfide bonds was involved in aggregation when myosin was solubilized prior to frozen storage. The protein denaturation of frozen mackerel and cod was mainly caused by the formation of disulfide, hydrogen, and hydrophobic bonds (Jiang et al., 1986). Higher disulfide bond formation was found in milkfish actomyosin stored at –20°C than at –35°C (Jiang et al., 1988). The addition of $NaNO_2$ and $NaBH_4$ to freeze-denatured actomyosin resulted in the recovered actomyosin solubility in 0.6 M KCl and an increase in SH group content. Therefore, the formation of disulfide bonds directly involve in denaturation of actomyosin during freezing and frozen storage (Jiang et al., 1988).

9.3.3 Surface Hydrophobicity and Hydrophobic Interaction

In general, the native protein structure is stabilized by many forces including hydrogen bond, dipole–dipole interaction, electrostatic interaction, and disulfide linkages. Protein can undergo unfolding, which increases with temperature abuse, such as repeated freeze–thawing (Srinivasan et al., 1997a,b). As a result, an exposure of hydrophobic portion localized inside molecules occurs, leading to an increase in surface hydrophobicity (Benjakul and Bauer, 2000). Moreover, ice crystallization may disrupt the water structure surrounding the area of hydrophobic interaction in proteins. Crystallization may also disrupt the water-mediated hydrophobic interactions, which contribute to protein stabilization (Sikorski and Kolakowski, 1990). Badii and Howell (2002b) reported an initial increase in the surfaced-protein hydrophobicity of cod muscle in the first month before decreasing during frozen storage at –10°C and –30°C. The increase in surface hydrophobicity could favor the association of hydrophobic groups via hydrophobic interaction (Benjakul and Bauer, 2000). The surface hydrophobicity of muscle proteins was found to be inversely related to solubility (Owusu-Ansah and Hultin, 1987). In addition, frozen storage of cod and haddock fillets resulted in an increase in hydrophobicity of their muscle proteins. This could be attributed to the unfolding of proteins and the exposure of hydrophobic aliphatic and aromatic amino acids. Hydrophobic interactions between the exposed groups resulted in a decrease in the solubility of proteins and the formation of aggregate. These changes were more pronounced in cod and haddock fillets stored at –10°C, compared with the fish stored at –30°C (Badii and Howell, 2001).

9.4 LIPID OXIDATION

Lipid oxidation in muscle foods is one of the major deteriorative reactions causing losses in quality during processing and storage. The oxidation of unsaturated fatty acids leads to formation of free radicals and hydroperoxide. These intermediary compounds are unstable and cause the oxidation of pigments, flavors, and vitamins. Oxidized unsaturated lipids bind to protein and form insoluble lipid–protein complexes. This accounts for toughened texture and poor flavor of frozen seafoods (Khayat and Schwell, 1983).

In the presence of oxidized lipids, protein oxidation is induced by free radical chain reaction (Kanner, 1994). Oxidation reactions can lead to the formation of protein radicals, polymers, and protein–lipid complexes. The sites of free radical attack on protein include both amino acid side chains and the peptide backbone. The attack of protein results in either protein polymerization or fragmentation (Xiong, 2000). In the presence of reactive oxygen species (ROS), chemical forces stabilizing protein are disrupted, resulting in the decreased thermal stability. An exposure to hydroxyl radicals causes reduction in thermal stability of myosin (Liu and Xiong, 2000). Additionally, hydroxyl radicals are capable of modifying side-chain groups of many amino acid residues.

Hamre et al. (2003) reported that during the storage of Norwegian spring-spawning herring for up to 9 weeks, there was an increase in lipid oxidation products as evidenced by an increase in thiobarbituric acid reactive substances (TBARS) and peroxide value (PV). The concomitant loss of red color and a substantial increase in yellow color indeed caused a visual change in appearance. Aubourg (1999) reported that blue whiting fillets frozen at −40°C, −30°C, and −10°C for up to 1 year underwent primary and secondary lipid oxidation, especially at −10°C; at −30°C, a significant increase in free fatty acid was obtained after 5 months of storage. TBARS of croaker, lizardfish, threadfin bream, and bigeye snapper increased with the frozen storage time at −18°C. Among all fish, croaker was more prone to lipid oxidation as indicated by the higher TBARS value, compared with other fish species (Benjakul et al., 2005a).

Lipid oxidation products are also involved in browning process by reacting with primary and secondary amino groups of protein- or amino-containing compounds such as phospholipids, e.g., phosphatidylethanolamine or phosphatidylserine (Pokorny and Kolakowska, 2002). Membrane phospholipids have a high content of highly polyunsaturated fatty acids. The membrane phospholipids exist primarily in the form of a bilayer, in which a much larger surface area is exposed per unit of lipid weight. They exist as inter- and intracellular fat droplet (Huang et al., 1993).

Pyrroles are common products in the reaction of different lipid oxidation products with primary amino groups of amines, amino acid, and protein (Zamora et al., 2000). For example, when squid microsomes were oxidized with iron and ascorbate, TBARS increased simultaneously with the b^*-value (yellowness) and pyrrole compounds, concomitantly with a decrease in free amines. Off-color formation in squid muscle could be due to the nonenzymatic browning reactions occurring between aldehydic lipid oxidation products and the amines on phospholipids head groups

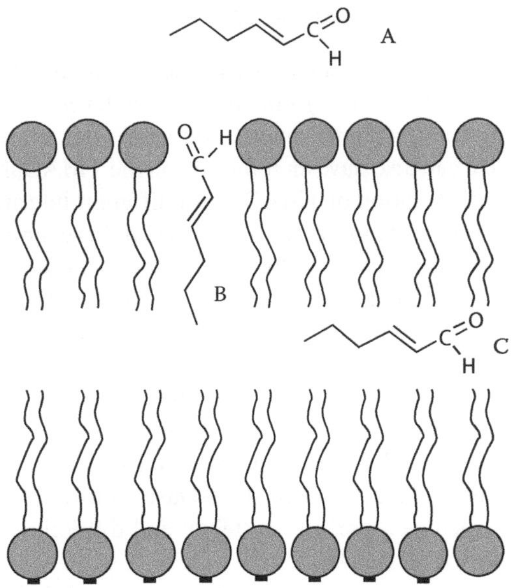

FIGURE 9.2 Proposed model for the impact of physical location of aldehydic lipid oxidation products on their reactivity with phospholipids amines. Physical locations include (A) aqueous, (B) interfacial, and (C) lipid phase. (From Thanonkaew, A. et al., *J. Agric. Food Chem.*, 54, 956, 2006a. With permission.)

(Thanonkaew et al., 2006a). The migration of aldehyde group to the cell membrane interface could place the carbonyl in close proximity to the phospholipids amine, thus facilitating nonenzymatic browning (Figure 9.2) (Thanonkaew et al., 2006a). Copper mainly caused the oxidation of protein, while iron induced lipid oxidation and the formation of yellow color in cuttlefish muscle, particularly with multiple freeze–thaw cycles (Thanonkaew et al., 2006b). As lipid oxidation and yellow pigment formation in the cuttlefish liposome proceeded, a loss of amine groups and pyrrolization was observed (Thanonkaew et al., 2007).

In general, the carbonyl groups of oxidized lipids participate in covalent bonding, leading to the formation of stable protein–lipid aggregates (Saeed and Howell, 2002). The reaction of proteins with lipid oxidation products results in the formation of protein-centered radicals (Saeed et al., 1999). Lipid degradation products are also able to cross-link polypeptides, thereby generating insoluble protein aggregate.

9.5 ROLE OF TRIMETHYLAMINE OXIDE DEMETHYLASE IN FROZEN SEAFOODS

Trimethylamine oxide demethylase (TMAOase) is one of the indigenous enzymes in fish exhibiting the adverse effect on fish quality. TMAOase converts trimethylamine oxide (TMAO) stoichiometrically to dimethylamine (DMA) and formaldehyde (FA) with 1:1 molar ratio (Benjakul et al., 2004).

$$TMAO \rightarrow DMA + FA$$

TMAOase is commonly found in some fish species. Among fish species Gadiform (families Gadidae and Merluccidae) contain a high level of TMAOase activity (Sotelo and Rehbein, 2000). Gender, maturation stage, the temperature of habitat, feeding status, and size can also have an influence on the TMAOase level (Sotelo and Rehbein, 2000). The distribution of TMAOase in different internal organs also varies. In the case of lizardfish (*Saurida micropectoralis*), for example, its kidney contains the highest TMAOase activity, followed by spleen, bile sac, intestine, and liver (Benjakul et al., 2004). Among several internal organs, hake, kidney, and spleen show the highest activity, while liver, heart, bile, and gall bladder have lower activities (Rey-Mansilla et al., 2001).

TMAOase from different fish species have different molecular characteristics. TMAOase from lizardfish kidney has the molecular weight of 128 kDa with the optimal pH and temperature of 7°C and 50°C, respectively. The activation energy is 30.5 kJ mol^{-1} K^{-1} (Benjakul et al., 2003b). TMAOase requires Fe^{2+} alone for activity and reducing agents such as ascorbate, cysteine, and dithiothreitol are required to maintain the active ferrous form (Fe^{2+}).

TMAOase is still active at frozen storage temperature. It has been reported that TMAOase is active down to −29°C; however, its activity decreases with decreasing frozen storage temperature (Rehbein, 1988; Sotelo et al., 1995). During extended storage, TMAOase activity in lizardfish muscle increased during frozen storage at −20°C for up to 12 weeks, followed by a continuous decrease up to 24 weeks (Leelapongwattana et al., 2005b).

FA has also been found to increase gradually in fish muscle during extended frozen storage. Benjakul et al. (2005a) reported an increase in formaldehyde in lizardfish during frozen storage at −18°C, while a small amount of FA was produced in other fish species (Figure 9.3). An increase in FA in hake fillet was also reported by Perez-Villarreal and Howgate (1991).

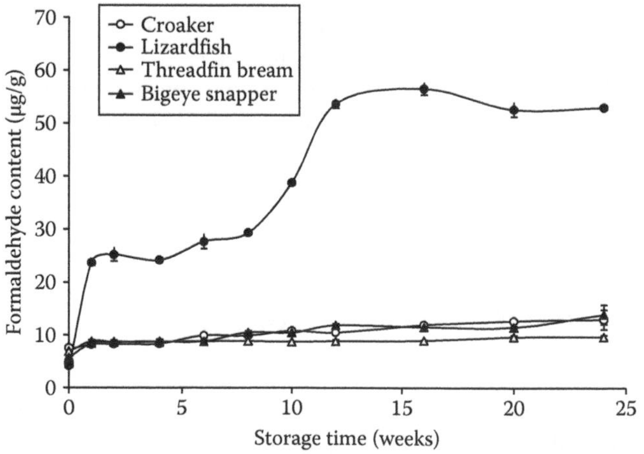

FIGURE 9.3 Changes in formaldehyde content in the muscle of different fish during storage at −18°C. (From Benjakul, S. et al., *Food Hydrocoll.*, 19, 197, 2005a. With permission.)

The susceptibility to conformational changes is also governed by the formation of FA in the fish muscle. Structural changes during frozen storage in cod myosin, for instance, were noted to be induced by the addition of FA (Careche and Li-Chan, 1997). Raman spectra for the simulated NAM system, either in the absence or presence of TMAOase, during chilled and frozen storage indicate changes in amide I band (1654 cm^{-1}), in which the maximum intensity of this band in the systems containing TMAOase shifts to the higher frequencies, probably associated with the decrease in α-helix content (the lower band intensity of Raman band at 940 cm^{-1}). The intensity of this band decreases as a function of storage time and TMAOase addition (Figure 9.4) (Leelapongwattana et al., 2008a). The C–H stretching region

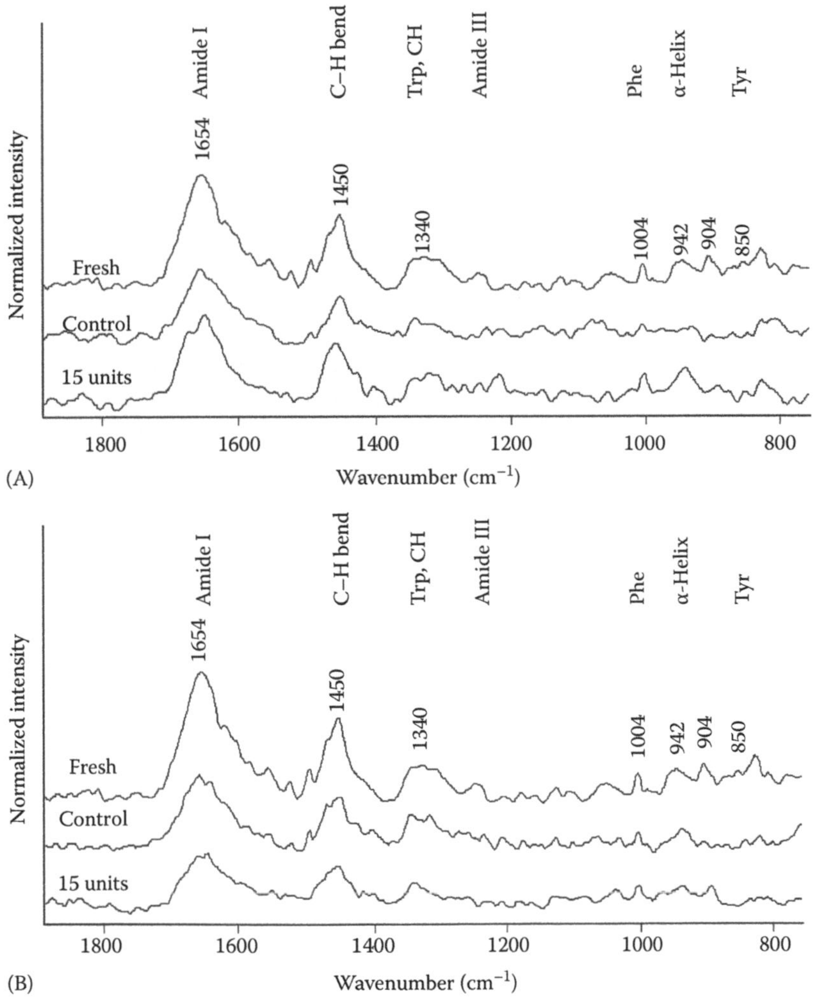

FIGURE 9.4 Raman spectra in the 800–1800 cm^{-1} wavenumber region of haddock NAM and simulated NAM system before and after storage (A) at 4°C for 15 days or (B) at −10°C for 8 weeks in the presence or absence of 15 units TMAOase per gram. (From Leelapongwatta, K. et al., *Food Chem.*, 106, 1253, 2008a. With permission.)

(2500–3400 cm^{-1}), corresponding to CH, CH_2, and CH_3 groups in the side chain of amino acids, is different between the systems without and with TMAOase at both 4°C and −10°C (Figure 9.5). In the presence of TMAOase, the structural changes of proteins are mostly caused by FA formation. Extensive cross-links induced by FA between methyl groups of hydrophobic amino side chains mainly contribute to

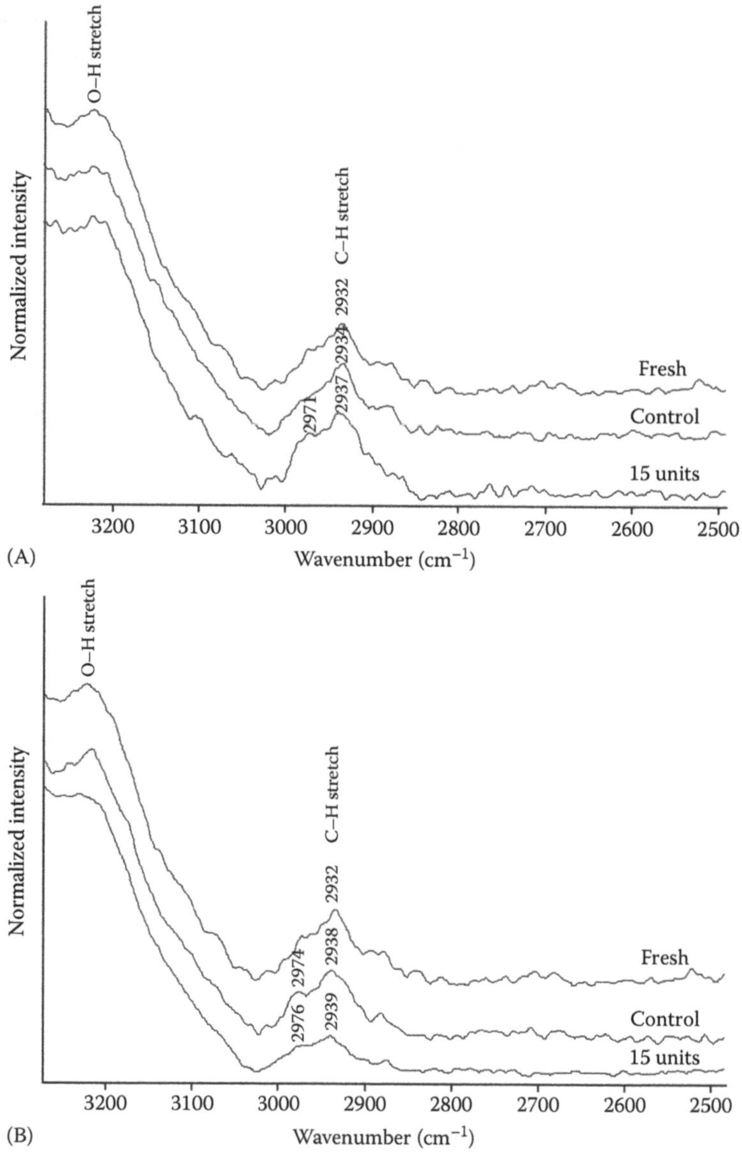

FIGURE 9.5 Raman spectra in the 2500–3400 cm^{-1} wavenumber region of haddock NAM before and after storage at (A) 4°C for 15 days or (B) −10°C for 8 weeks in the presence or absence of 15 units TMAOase per gram. (From Leelapongwatta, K. et al., *Food Chem.*, 106, 1253, 2008a. With permission.)

protein denaturation and aggregation (Leelapongwattana et al., 2008a). TMAOase was reconfirmed to exhibit the detrimental effect on haddock natural actomyosin, mainly caused by FA formation (Leelapongwattana et al., 2008b).

Dressing or processing fish or fish mince without the contamination of internal organs, particularly kidney, can retard the quality loss of frozen fish or mince associated with TMAOase activity during frozen storage. The removal of TMAOase-rich tissue prior to processing or storage could be another approach to extend the shelf life of frozen fish containing TMAOase. Evisceration before freezing or frozen storage is a recommended practice, thus preventing the contact of TMAOase-rich tissue with the substrate in muscle, TMAO (Sotero and Rehbein, 2000). The removal of red muscle, for example, was proved to improve the storage stability of gadoid fillets (Sotelo and Rehbein, 2000). The use of high-pressure treatments or heating of localized TMAOase-rich tissues could also be used to maximize the inhibition of TMAOase in fish (Sotelo and Rehbein, 2000). Lundstrom et al. (1982) reported that minimal DMA production rates were obtained by storing fresh red hake fillet or mince exposed to air or to 100% oxygen. On the other hand, maximal DMA production rates were obtained by storage under vacuum or 100% nitrogen.

9.6 ULTRASTRUCTURAL, TEXTURAL, AND RHEOLOGICAL CHANGES OF FROZEN SEAFOODS

The ultrastructure of fish muscle is affected by frozen storage. The changes take place during frozen storage include (1) the modification of the morphology of fibers due to ice-crystal formation, (2) the reduction of sarcoplasmic space between myofibrils, and (3) the reduction of interfilament distance (Herrero et al., 2005). However, the changes in ultrastructure are affected by species (Garcia et al., 1999). For example, muscle fibers of squid mantle were found to injure and aggregate when the frozen storage time increased, leading to the toughening of mantle (Ueng and Chow, 1998).

Textural changes in fish and shellfish muscle generally occur during extended frozen storage; the texture turns to be sponge-like structure with lower holding capacity. This makes the flesh dry and rubbery, especially for some fish species. During frozen storage muscle proteins undergo aggregation, in which the water-binding domain or hydrophilic portion of proteins is less available for water binding. This phenomenon is particularly found in fish containing TMAOase.

The rheological characteristics of fish muscle are considered to be governed by both myofibrillar and connective tissue proteins, while sarcoplasmic proteins contribute to texture at a lower degree (Dunajski, 1979). An increase in the storage modulus (G') of raw materials such as fish fillets is related to protein–protein and protein–lipid interactions, causing undesirable tough products (Badii and Howell, 2002a,b; Dileep et al., 2005). An increase in G' and loss modulus (G'') values, reflecting protein aggregation, was observed in Atlantic mackerel (*Scomber scombrus*) fillets stored at –20°C and –30°C for 2 years. The G' values were higher in mackerel fillets stored at –20°C than those stored at –30°C, indicating a greater degree of protein denaturation and cross-linking in fillets, which was correlated with activity loss and a decrease in protein solubility (Saeed and Howell, 2004).

9.7 CHANGES IN GEL-FORMING ABILITY

The properties of gel from fish mince or surimi are governed by the freshness of fish. Deterioration in gel properties is generally noted when poor-quality raw materials are used for surimi production. To prevent spoilage associated with long-distance handling before off-loading freezing can be used as a promising means to lower such deterioration. However, when frozen fish is used for surimi production, the resulting gel with decreased gel strength and darker color is obtained. The degree of decrease in gel-forming ability depends on fish species, freezing method, and frozen storage condition. With increasing frozen storage time, fish containing TMAOase mostly yields gel of poor quality with very low water-holding capacity. Benjakul et al. (2005a) found a continuous decrease in the breaking force and deformation of gel produced from four fish species, especially at extended storage time. Among all the species, surimi from lizardfish had a higher rate of decrease in gel-forming ability as the storage time increased (Figure 9.6). These changes were coincidental with the formation of FA in lizardfish muscle. FA could react with different functional groups of protein side chains and form intra- and intermolecular methylene bridges.

TMAOase localized in the internal organs of fish could be released into the muscle during processing, especially thawing. TMAOase in both muscle and from internal organs might work synergistically in formaldehyde formation. As a result, polymerized proteins could not undergo unfolding effectively, and the reactive groups could not interact with each other to form the gel network. In general, the regular aggregated structure with a well-organized three-dimensional network is observed with the gels prepared from fresh fish. On the other hand, gels from frozen fish have a coarser network with larger voids. Bead-type aggregate network was found in surimi gel from frozen lizardfish (Figure 9.7) (Benjakul et al., 2005a).

To improve the gel-forming ability of frozen fish, a renaturation process is required to dissociate cross-linked proteins prior to gelation. As a consequence, a more ordered network can be formed. Benjakul et al. (2005b) reported that the use of cysteine at levels of 0.05%, 0.1%, 0.05%, and 0.1% in surimi prepared from frozen croaker, lizardfish, threadfin bream, and bigeye snapper, respectively, could improve the breaking force and deformation of the gels. The addition of cysteine resulted in the reduction of disulfide bonds formed during extended frozen storage. A coincidental increase in Ca^{2+}-ATPase activity was noticeable, indicating a partial renaturation of myosin. An increase in transglutaminase activity in surimi sol was observed with the addition of cysteine (Table 9.1), causing an improved setting phenomenon of surimi. The increase in transglutaminase activity was possibly due to the renaturation of transglutaminase via the reduction of the disulfide bond in the active site (Benjakul et al., 2005b). Thus, appropriate reducing agents could partially recover the denatured muscle proteins as well as activate endogenous transglutaminase in frozen fish muscle. As a result, the gel quality of surimi produced from frozen fish can be improved and be equivalent to that of surimi prepared from fresh fish, particularly that from croaker (Benjakul et al., 2005b). However, the addition of cysteine above the optimal level resulted in the decreased breaking force and deformation of

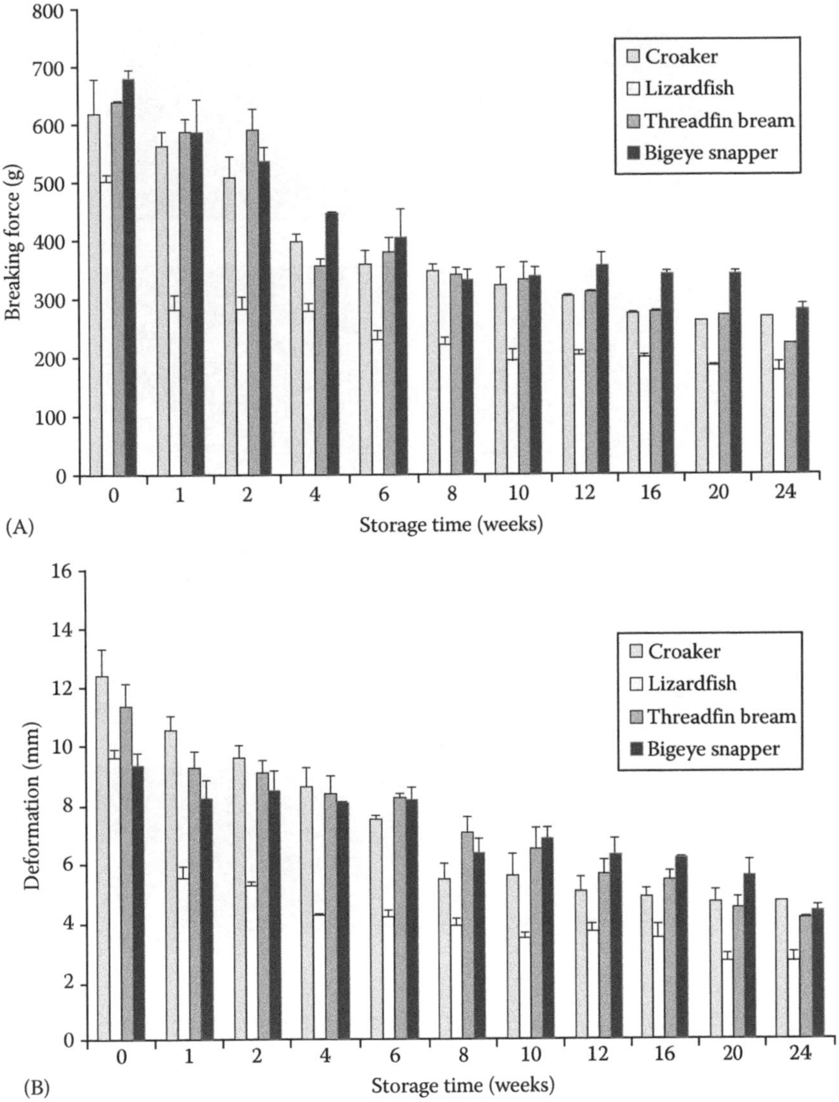

FIGURE 9.6 Changes in (A) breaking force and (B) the deformation of surimi gels prepared from four fish species during frozen storage at −18°C for 6 months. (From Benjakul, S. et al., *Food Hydrocoll.*, 19, 197, 2005a. With permission.)

the resulting gels (Benjakul et al., 2005b). The excessive amount of reducing agent might prevent the oxidation of the sulfhydryl group to disulfide bond during setting or heating. This leads to a weaker gel network with lower disulfide bond content. Note that disulfide bond is the covalent bond contributing to the stabilization of gel network (Benjakul et al., 2000).

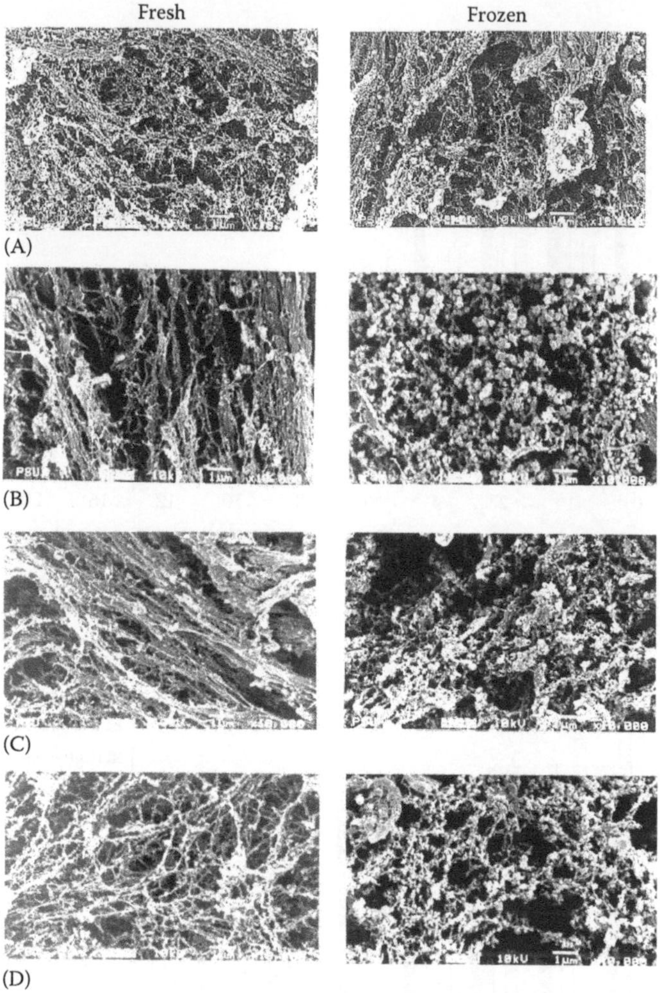

FIGURE 9.7 Microstructure of surimi gel from four fish species prepared from fresh and frozen fish (A) croaker, (B) lizardfish, (C) threadfin bream, and (D) bigeye snapper. (From Benjakul, S. et al., *Food Hydrocoll.*, 19, 197, 2005a. With permission.)

Some cross-linking enzyme, particularly microbial transglutaminase (MTGase) can induce a subsequent aggregation of renatured proteins. Furthermore, an addition of selected protein additives is also a means to improve the gel quality (Figure 9.8). Benjakul and Visessanguan (2004) found that the addition of 0.05%–0.1% cysteine in combination with 0.2% microbial transglutaminase and 0%–2.0% egg white or 1.0%–2.0% porcine plasma protein could improve the gels of surimi produced from frozen fish including croaker, lizardfish, threadfin bream, and bigeye snapper.

TABLE 9.1
Transglutaminase Activities of Surimi Sols from Four Frozen Fish with and without Addition of Cysteine

Fish Species	Amount of Cysteine (%)	Transglutaminase Activity (Milli Units/Gram)
Croaker	0	4.98 ± 0.27
	0.05	6.46 ± 0.89
Lizardfish	0	2.67 ± 0.42
	0.1	3.11 ± 0.57
Threadfin bream	0	2.80 ± 0.42
	0.05	3.75 ± 0.91
Bigeye snapper	0	3.21 ± 0.27
	0.1	3.81 ± 0.59

Source: Benjakul, S. et al., *Eur. Food Res. Technol.*, 220, 316, 2005b.

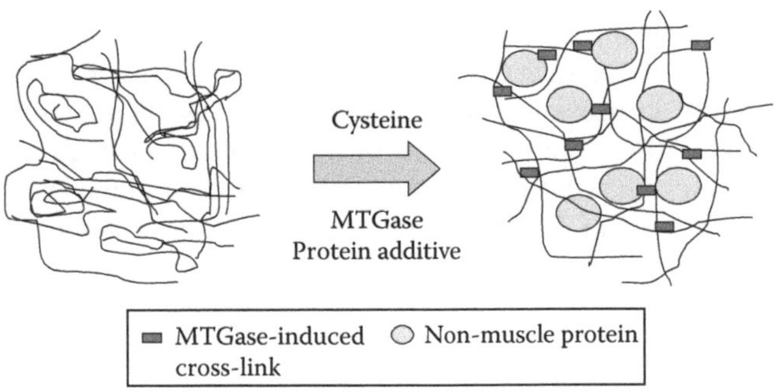

FIGURE 9.8 Scheme for improving the gel property of surimi from frozen fish.

9.8 USES OF CHEMICALS/ADDITIVES IN FROZEN SEAFOODS

9.8.1 Uses of TMAOase Inhibitors and Hydrocolloids

Various TMAOase inhibitors such as azide, cyanide, etc., have been shown to lower TMAOase activity; however, these inhibitors are toxic. Some food-grade inhibitors also exhibit inhibition toward TMAOase. Sodium citrate and H_2O_2 were found to slow down the rate of DMA and FA formation in frozen gadoid mince (Parkin and Hultin, 1982; Racicot et al., 1984). Treating Alaska pollack fillets with sodium citrate and sodium pyruvate prior to freezing resulted in a less tough texture (Krueger and Fennema, 1989). An addition of 0.5% alginate could lower DMA and FA formation in minced fillet of mackerel stored at –20°C for 2 months (Ayyad and Aboel-Niel, 1991). Leelapongwattana et al. (2008c) reported that sodium citrate and pyrophosphate could inhibit the activity of TMAOase from lizardfish (*Saurida tumbil*) muscle in a concentration-dependent manner, most likely due to their chelating property. Sodium alginate also exhibited the inhibitory activity toward TMAOase. During the storage of lizardfish mince at –20°C for 24 weeks, the addition of 0.5% sodium alginate and 0.3% pyrophosphate in combination with cryoprotectants (4% sucrose and 4% sorbitol) retarded an increase in TMAOase activity with the coincidental lowered formation of DMA and FA, compared with the control (without additives). The loss in solubility of muscle proteins was also impeded with the addition of those compounds, suggesting their role in the inhibition of TMAOase as well as in the retardation of protein denaturation induced by FA. The addition of TMAOase inhibitors (0.3% sodium pyrophosphate and 0.5% sodium alginate) in haddock mince containing 15 units of TMAOase per gram retarded FA formation throughout the storage at –10°C for 6 weeks (Leelapongwatta et al., 2008d). The addition of TMAOase inhibitor also lowered G' and G'' of mince containing TMAOase (Leelapongwattana et al., 2008d).

Hydrocolloids have been used to lower cross-linking of protein induced by FA produced during extended frozen storage. Herrera et al. (2000) reported that the addition of maltodextrins to minced blue whiting muscle inhibited FA production during storage at –10°C and –20°C. However, sucrose was effective only at –20°C. Herrera et al. (2002) also studied the effects of various cryostabilizers on the inhibition of FA production in frozen-stored minced blue whiting muscle. Several maltodextrins (DE, dextrose equivalent, of 9, 12, 18, and 28) or sucrose retarded a decrease in protein solubility during storage at –20°C and –10°C. DE 18 maltodextrin was the most effective at –20°C. The constraints of hydrocolloids on molecular diffusion reduce the exposure of reactive groups of proteins to be cross-linked by FA produced in the muscle. The uses of TMAOase inhibitor in combination with hydrocolloids therefore minimize the loss in quality of fish containing high level of TMAOase.

9.8.2 Use of Antioxidants

The prevention of oxidative deterioration of muscle foods during processing or storage can be achieved by the addition of antioxidants (Decker, 1998). Ascorbate and erythrobate are strong reducing agents and have been successfully used as antioxidants in many foods. Ascorbic acid efficiently scavenges hydrogen peroxide and free radicals. It may restore vitamin E by reducing the tocopheroxyl radical to its native state (Igoe and Hui, 2001). Chelating agents such as EDTA, citric acid, and phosphate, etc., can form complexes with undesirable trace metals, thus blocking the reactive sites of metal ions and rendering them inactive. However, EDTA and tripolyphosphate (TPP) were noted to have no antioxidative effect in frozen cuttlefish, while ascorbate and erythorbate showed a prooxidative effect (Thanonkaew et al., 2008).

9.8.3 Use of Phosphates

Phosphates have been widely accepted as potential additives in meat and seafood products to improve the functional properties of these products by increasing their water retention, especially in frozen–thawed products (Chang and Regenstein, 1997). Phosphates inactivate metal ions either by decreasing metal solubility or by complexing and maintaining them in a soluble redox inactive (Ellinger, 1975). Sodium tripolyphosphate has been used in many seafood products for prevention of drip loss and freezer burn. Additionally, phosphates can improve the shelf life and flavor of seafood products by retarding lipid oxidation. The treatment of Atlantic cod fillets in commercial tripolyphosphate or metaphosphate solution prior to frozen storage at −12°C or −30°C for up to 26 weeks decreased thaw drip and cooked drip and yielded a product with higher raw and cooked moisture (Woyewoda and Bligh, 2006). However, phosphate treatment had no impact on the formation of formaldehyde formation. Searobin fillet treated with 2% sodium tripolyphosphate or 2% blend (sodium tripolyphosphate + sodium tetra-pyrophosphate + NaCl) and pink cuskeel fillet, red shrimp, and mussel treated with 5% sodium tripolyphosphate or 5% blend for 120 min at 2°C improved weight gain after frozen storage for 15 days. In general, blends showed higher effectiveness in lowering weight loss (Woyewoda and Bligh, 2006). Polyphosphates (5.3% sodium tripolyphosphate and 4.7% sodium pyrophosphate) reduced the amount of expressible moisture of the striated adductor muscle of *Aulacomya ater ater* (Molina) stored at −30°C (Paredi et al., 1996). However, an excessive treatment of phosphate would result in undesirable flavor and sloppy texture.

9.8.4 Use of Cryoprotectants

Cryoprotectants are compounds that lower the protein denaturation of proteins during frozen storage. Frozen surimi requires cryoprotectant to maintain the gel-forming ability during extended storage. Sugar and polyhydric alcohols have been used widely to stabilize macromolecules including proteins by increasing the transition temperature of some proteins in aqueous solution (Arakawa and Timasheff, 1982).

Cryoprotectants are able to increase the surface tension of water (MacDonald and Lanier, 1991); sugar protects protein from freeze denaturation by increasing the surface tension of water as well as the amount of bound water. Therefore, the withdrawal of water molecules from the proteins is retarded, thus stabilizing the proteins. Water migration during freezing or frozen storage is associated with the disruption of hydrogen bonds between water–protein or protein–protein complex (Matsumoto and Noguchi, 1992). The stabilizing effect is also caused by enhancing the strength of intramolecular hydrophobic interaction of protein molecules.

Sucrose and sorbitol are commonly used in frozen surimi processing. However, sucrose imparts a sweet taste to surimi products, which is undesirable to the consumer (Sych et al., 1990; Auh et al., 1999; Sultanbawa and Li-Chan, 2001). Thus, the use of other cryoprotectants to reduce sweetness but exhibit the equivalent cryoprotective effect is required. Auh et al. (1999) used highly concentrated branched oligosaccharide mixture (HBOS) as cryoprotectant in fish protein. An addition of HBOS resulted in the remainder Ca^{2+}-ATPase activity of actomyosin extracted from Alaska pollock after freeze–thawing; the best stabilization effect of HBOS was observed at a concentration of 8%. Sych et al. (1990) studied the cryoprotective effects of lactitol dehydrate, polydextrose, and palitinit at 8% (w/w) in cod surimi in comparison with an industrial control (sucrose/sorbitol, 1:1). The addition of these cryoprotectants could maintain the solubility and myosin peak enthalpy for surimi as effectively as the commercial counterpart. Sultanbawa and Li-Chan (1998) studied cryoprotectant blends for their potential to stabilize ling cod surimi during frozen storage at −18°C for 4 months. Twenty-five blends containing lactitol, Litesse, sucrose, and sorbitol at a final total concentration of 4%–12% were used. All blends gave comparable surimi and cooked gels to those obtained with commercial mix (4% sucrose + 4% sorbitol) with the 4% blend containing 1% of each cryoprotectant being the most economical and having the lowest calorie content. Sultanbawa and Li-Chan (2001) reported that the commercial blend cryoprotectant (4% sucrose and 4% sorbitol) or the individual cryoprotectants (sorbitol, sucrose, lactitol, or Litesse) at a level of 8% were effective in maintaining the gel strength of ling cod NAM frozen at −10°C for 10 days and surimi after eight freeze–thaw cycles. Trehalose has also been used as a cryoprotectant in surimi. Zhou et al. (2006) reported that trehalose could decrease protein denaturation of tilapia surimi during frozen storage and an addition of 8% trehalose yielded the surimi gel with the highest breaking force and deformation.

9.9 GLAZING OF FROZEN SEAFOODS

Weight loss during frozen storage generally occurs, especially at the surface. This phenomenon leads to "freezer burn" with unacceptability of a product. Such dehydration can be solved by (1) covering the surface with packaging materials and (2) glazing. Glazing of small and irregularly shaped seafoods such as shrimp or mollusk is a means to lower dehydration when the products are stored without packagings (Jacobsen and Fossan, 2001; Goncalves and Gindri, 2009). After seafoods are frozen, glazing should be performed suddenly; the glazed product should then be transferred to low-temperature storage to rapidly refreeze as well as to minimize the thaw drip loss (Jacobsen and Pedersen, 1997; Jacobsen and Fossan, 2001). The amount of glaze depends on the following factors: (1) glazing time, (2) seafood temperature, (3) glazing water temperature, (4) product size, and (5) product shape (Johnston et al., 1994; Jacobsen and Pedersen, 1997; Jacobsen and Fossan, 2001; Goncalves and Gindri, 2009). For frozen shrimp glazing with the water uptake of 15% and 20% is recommended to guarantee the final quality.

9.10 CONCLUDING REMARKS

Freezing technology can be used as a potential preservative method for seafoods, in which microbial spoilage can be terminated. However, the quality of frozen seafoods can be altered to different degrees, depending on the freezing method, frozen storage conditions, etc. These changes can be minimized by introducing appropriate additives such as cryoprotectant, antioxidant, enzyme inhibitor, etc. Therefore, the further development of freezing technology to lower the denaturation of protein and other major constituents in seafoods as well as a search for safe additives together with the developed packaging technology to maintain the seafood quality are still needed. As a consequence, the prime quality equivalent to fresh counterpart can be gained for frozen seafoods.

REFERENCES

Arakawa, T. and Timasheffm, S.N. 1982. Stabilization of protein structure by sugars. *Biochemistry* 21: 6536–6544.

Aubourg, S.P. 1999. Lipid damage detection during the frozen storage of an underutilized fish species. *Food Research International* 32: 497–502.

Auh, J.H., Lee, H.G., Kim, J.W., Kim, J.C., Yoon, H.S., and Park, K.H. 1999. Highly concentrated branched oligosaccharides as cryoprotectant for surimi. *Journal of Food Science* 64: 418–422.

Ayyad, K. and Aboel-Niel, E. 1991. Effect of addition of some hydrocolloids on the stability of frozen minced fillet of mackerel. *Carbohydrate Polymers* 15: 143–149.

Badii, F. and Howell, N.K. 2001. A comparison of biochemical changes in cod (*Gadus morhua*) and haddock (*Melanogrammus aeglefinus*) during frozen storage. *Journal of the Science of Food and Agriculture* 82: 87–97.

Badii, F. and Howell, N.K. 2002a. Effect of antioxidants, citrate and cryoprotectants on protein denaturation and texture of frozen cod (*Gadus morhua*). *Journal of Agricultural and Food Chemistry* 50: 2053–2061.

Badii, F. and Howell, N.K. 2002b. Changes in the texture and structure of cod and haddock fillets during frozen storage. *Food Hydrocolloids* 16: 313–319.

Baron, C.P., Kiarsgard, I.V.H., Jessen, F., and Jacobsen, C. 2007. Protein and lipid oxidation during frozen storage of rainbow trout (*Oncorhynchus mykiss*). *Journal of Agricultural and Food Chemistry* 55: 8118–8125.

Benjakul, S. and Bauer, F. 2000. Physicochemical and enzymatic changes of cod muscle proteins subjected to different freeze-thaw cycles. *Journal of the Science of Food and Agriculture* 80: 1143–1150.

Benjakul, S. and Visessanguan, W. 2004. Improvement of gel surimi produced from frozen fish using additives. Thai Mini Patent. No. 1602.

Benjakul, S., Seymour, T.S., Morrissey, M.T., and An, H. 1997. Physicochemical changes in Pacific whiting muscle proteins during iced storage. *Journal of Food Science* 62: 729–733.

Benjakul, S., Visessanguan, W., Ishizaki, S., and Tanaka, M. 2000. Differences in gelation characteristics of natural actomyosin from two species of bigeye snapper, *Priacanthus tayenus* and *Priacanthus macracanthus*. *Journal of Food Science* 66: 1311–1318.

Benjakul, S., Visessanguan, W., Thongkaew, C., and Tanaka, M. 2003a. Comparative study on physicochemical changes of muscle proteins from some tropical fish during frozen storage. *Food Research International* 36: 787–795.

Benjakul, S., Visessanguan, W., and Tanaka, M. 2003b. Partial purification and characterization of trimethylamine-N-oxide demethylase from lizardfish kidney. *Comparative Biochemistry and Physiology* 135B: 359–371.

Benjakul, S., Visessanguan, W., and Tanaka, M. 2004. Induced formation of dimethylamine and formaldehyde by lizardfish kidney trimethylamine-N-oxide demethylase. *Food Chemistry* 84: 297–305.

Benjakul, S., Visessangaun, W., Thongkaew, C., and Tanaka, M. 2005a. Effect of frozen storage on chemical and gel-forming properties of fish commonly used for surimi production in Thailand. *Food Hydrocolloids* 19: 197–207.

Benjakul, S., Thongkaew, C., and Visessanguan, W. 2005b. Effect of reducing agents on physical properties and gel forming ability of surimi produced from frozen fish. *European Food Research and Technology* 220: 316–321.

Buttkus, H. 1974. On the nature of the chemical and physical bonds which contribute to some structural properties of protein foods: A hypothesis. *Journal of Food Science* 39: 484–489.

Careche, M. and Li-Chan, E.C.Y. 1997. Structural changes and Raman spectroscopic studies of cod myosin upon modification with formaldehyde or frozen storage. *Journal of Food Science* 62: 717–723.

Careche, M., Herrero, A.M., Rodriguez-Casado, M.L., Del Mazo, M.L., and Carmona, P. 1999. Structural changes of hake (*Merluccius merluccius* L.) fillets: Effects of freezing and frozen storage. *Journal of Agricultural and Food Chemistry* 47: 952–959.

Chang, C.C. and Regenstein, J.M. 1997. Water uptake, protein solubility and protein changes of cod mince storage on ice as affected by phosphates. *Journal of Food Science* 62: 305–309.

Decker, E.A.1998. Strategies for manipulating the prooxidative/antioxidative balance of foods to maximize oxidative stability. *Trends in Food Science and Technology* 9: 241–248.

Del Mazo, M.L., Torrejon, P., Careche, M., and Tejada, M. 1999. Characteristics of the salt-soluble fraction of hake (*Merluccius merluccius*) fillets stored at −20 and −30°C. *Journal of Agricultural and Food Chemistry* 47: 1372–1377.

Dileep, A.O., Shamasundar, B.A., Binsi, P.K., Badii, F., and Howell, N.K. 2005. Effect of ice storage on the physicochemical and dynamic viscoelastic properties of ribbonfish (*Trichiurus* spp.) meat. *Journal of Food Science* 70: 537–545.

Dunajski, E. 1979. Texture of fish muscle. *Journal of Texture Studies* 10: 301–318.

Ellinger, R.H. 1975. Phosphates in food processing. In: *Handbook of Food Additive* (Furia, T.E., ed.), p. 671. CRC Press: Boca Raton, FL.

Fennema, O. 1982. Behavior of proteins at low temperatures. In: *Food Protein Deterioration: Mechanisms and Functionality* (Cherry, J.P., ed.), p. 109. ACS Symposium: Washington, DC.

Garcia, M.L., Martin-Benito, J., Solas, M.T., and Fernandez, B. 1999. Ultrastructure of the myofibrillar component in cod (*Gadus morhua* L.) and hake (*Merluccius merluccius* L.) stored at −20°C as a function of time. *Journal of Agricultural and Food Chemistry* 47: 3809–3815.

Goncalves, A.A. and Gindri, C.S.G. Jr. 2009. The effect of glaze uptake on storage quality of frozen shrimp. The effect of glaze uptake on storage quality of frozen shrimp. *Journal of Food Engineering* 90: 285–290.

Goodband, R. 2002. Functional properties of fish proteins. In: *Seafoods-Quality, Technology and Nutraceutical Applications* (Alasalvar, C. and Taylor, T., eds.), pp. 73–82. Springer: Berlin, Germany.

Hamm, R. 1979. Delocalization of mitochondrial enzymes during freezing and thawing of skeletal muscle. In: *Protein at Low Temperature*, Vol. 180, Advances in Chemistry Series (Fennema, O.R., ed.), p. 191. American Chemical Society: Washington, DC.

Hamre, K., Lie, O., and Sandnes, K. 2003. Development of lipid oxidation on flesh color in frozen stored fillets of Norwegian spring-spawning herring (*Clupea harengus* L.): Effects of treatment with ascorbic acid. *Food Chemistry* 82: 445–453.

Hayakawa, J.J. and Nakai, S. 1985. Contribution of hydrophobicity, net charge and sulfhydryl groups to thermal properties of ovalumin. *Canadian Institute of Food Science and Technology Journal* 18: 290–295.

Herrera, J.J., Pastoriza, L., and Sampedro, G. 2000. Inhibition of formaldehyde production in frozen-stored minced blue whiting (*Micromesistius poutassou*) muscle by cryostabilizers: An approach from the glassy state theory. *Journal of Agricultural and Food Chemistry* 48: 5256–5262.

Herrera, J.J., Pastoriza, L., and Sampedro, G. 2001. A DSC study on the effects of various maltodextrins and sucrose on protein changes in frozen-stored minced blue whiting muscle. *Journal of the Science of Food and Agriculture* 81: 377–384.

Herrera, J.J., Pastroriza, L., and Sampedro, G. 2002. Effects of various cryostabilisers on protein functionality in frozen-stored minced blue whiting muscle: The importance of inhibiting formaldehyde production. *European Food Research and Technology* 214: 382–387.

Herroro, A.M., Carmona, P., Garcia, M.L., Solas, M.T., and Careche, M. 2005. Ultrastructural changes and structure and mobility of myowater in frozen-stored hake (*Merluccius merluccius* L.) muscle: Relationship with functionality and texture. *Journal of Agricultural and Food Chemistry* 53: 2558–2566.

Hsieh, Y.L. and Regenstein, J.M. 1989. Texture changes of frozen stored cod and ocean perch minces. *Journal of Food Science* 54: 824–826.

Huang, C.H., Hultin, H.O., and Jafar, S.S. 1993. Some aspects of Fe^{2+}-catalyzed oxidation of fish sarcoplasmic reticular lipid. *Journal of Agricultural and Food Chemistry* 41: 1886–1892.
Huidobro, A., Mohamed, G.F., and Tejada, M. 1998. Aggregation of myofibrillar proteins in hake, sardine and mixed minces during frozen storage. *Journal of Agricultural and Food Chemistry* 46: 2601–2608.
Igoe, R.S. and Hui, Y.H. 2001. *Dictionary of Food Ingredients*. Aspen: Gaithersburg, M.D.
Jacobsen, S. and Fossan, K.M. 2001. Temporal variations in the glaze uptake on individually quick frozen prawns as monitored by the CODEX standard and the enthalpy method. *Journal of Food Engineering* 48: 227–233.
Jacobsen, S. and Pedersen, W. 1997. Noncontact determination of cold-water prawn ice-glaze content using radiometry. *Lebensmittel-Wissenschaft und-Technologie* 30: 578–584.
Jiang, S.T., Ho, M.L., and Lee, T.C. 1985. Optimization of the freezing condition on mackerel and amberfish for manufacturing minced fish. *Journal of Food Science* 50: 727–732.
Jiang, S.T., Lan, C.C., and Tsao, C.Y. 1986. New approach to improve the quality of minced fish products from freeze-thawed cod and mackerel. *Journal of Food Science* 51: 310–312.
Jiang, S.T., Hwang, B.S., and Chen, C.S. 1988. Denaturation and change in SH group of actomyosin from milkfish (*Chanos chanos*) during storage at –20°C. *Journal of Agricultural and Food Chemistry* 36: 433–437.
Johnston, W.A., Nicholson, F.J., Roger, A., and Stroud, G.D. 1994. Freezing and refrigerated storage in fisheries. FAO Fisheries Technical Paper No. 340, 109 p.
Kanner, J. 1994. Oxidative process in meat and meat products: Quality implication. *Meat Science* 36: 169–189.
Khayat, A. and Schwall, D. 1983. Lipid oxidation in seafood. *Food Technology* 37: 130–140.
Kjersgard, I.V.H., Norrelykke, M.R., Baron, C.P., and Jessen, F. 2006. Identification of carbonylated protein in frozen rainbow trout (*Oncorhynchus mykiss*) fillets and development of protein oxidation during frozen storage. *Journal of Agricultural and Food Chemistry* 54: 9437–9446.
Krueger, D.J. and Fennema, O.R. 1989. Effect of chemical additives on toughening of fillets of frozen storage Alaska pollack (*Theragra chacogramma*). *Journal of Food Science* 54: 1101–1106.
Leelapongwattana, K., Benjakul, S., Visessanguan, W., and Howell, N.K. 2005a. Physicochemical and biochemical changes during frozen storage of minced flesh of lizardfish (*Saurida micropectoralis*). *Food Chemistry* 90: 141–150.
Leelapongwattana, K., Benjakul, S., Visessanguan, W., and Howell, N.K. 2005b. Physicochemical and biochemical changes in whole lizardfish (*Saurida micropectoralis*) muscles and fillets during frozen storage. *Journal of Food Biochemistry* 29: 547–569.
Leelapongwattana, K., Benjakul, S., Visessanguan, W., and Howell, N.K. 2008a. Raman spectroscopic analysis and rheological measurements on natural actomyosin from haddock (*Melanogrammus aeglefinus*) during refrigerated (4°C) and frozen (–10°C) storage in the presence of trimethylamine-N-oxide demethylase from kidney of lizardfish (*Saurida tumbil*). *Food Chemistry* 106: 1253–1263.
Leelapongwattana, K., Benjakul, S., Visessanguan, W., and Howell, N.K. 2008b. Effect of trimethylamine-N-oxide demethylase from lizardfish kidney on biochemical changes of haddock natural actomyosin stored at 4 and –10°C. *European Food Research and Technology* 226: 833–841.
Leelapongwattana, K., Benjakul, S., Visessanguan, W., and Howell, N.K. 2008c. Effect of some additives on the inhibition of lizardfish trimethylamine-N-oxide demethylase and frozen storage stability of minced flesh. *International Journal of Food Science and Technology* 43: 448–455.
Leelapongwattana, K., Benjakul, S., Visessanguan, W., and Howell, N.K. 2008d. Effect of inhibitors and antioxidants on physicochemical and biochemical changes of haddock muscle induced by lizardfish trimethylamine-N-oxide demethylase during frozen storage. *Journal of Food Biochemistry* (Submitted).

Liu, G. and Xiong, Y.L. 2000. Electrophoretic pattern, thermal denaturation on *in vitro* digestibility of oxidized myosin. *Journal of Agricultural and Food Chemistry* 48: 624–630.

Lundstrom, R.C., Correia, F.F., and Wilhelm, K.A. 1982. Enzymatic demethylation and formaldehyde production in minced American plaice and backbone flounder mixed with a red hake TMAOase fraction. *Journal of Food Science* 47: 1305–1310.

MacDonald, A.G. and Lanier, T.C. 1991. Carbohydrates as cryoprotectants for meats and surimi. *Food Technology* 45(4): 150–159.

Matsumoto, J.J. and Nouguchi, S.F. 1992. Cryostabilization of protein in surimi. In: *Surimi Technology* (Lanier, T.C. and Lee, C.M., eds.), pp. 357–388, Marcel Dekker: New York.

Nambudiri, D.D. and Gopakumar, K. 1992. ATPase and lactate dehydrogenase activities in frozen stored fish muscle as indices of cold storage deterioration. *Journal of Food Science* 57: 72–76.

Owusu-Ansah, Y.L. and Hultin, H.O. 1987. Effect of in situ formaldehyde production on solubility and cross-linking of proteins of minced red hake muscle during frozen storage. *Journal of Food Biochemistry* 11: 17–39.

Paredi, M.E., De Vido de Mattio, N.A., and Crupkin, M. 1996. Biochemical properties of actomyosin and expressible moisture of frozen stored striated adductor muscles of *Aulacomya ater ater* (Molina): Effects of polyphosphates. *Journal of Agricultural and Food Chemistry* 44: 3108–3122.

Parkin, K.L. and Hultin, H.O. 1982. Some factors influencing the production of dimethylamine and formaldehyde in minced and intact red hake muscle. *Journal of Food Processing and Preservation* 6: 73–97.

Perez-Villarreal, B. and Howgate, P. 1991. Deterioration of European hake (*Merluccius merluccius*) during frozen storage. *Journal of the Science of Food and Agriculture* 55: 455–469.

Pokorny, J. and Kolakowska, A. 2002. Lipid-protein and lipid-saccharide interactions. In: *Chemical and Functional Properties of Food Lipids* (Sikorski, Z. and Kolakowska, A., eds.), pp. 345–362. CRC Press: Boca Raton, FL.

Racicot, L.D., Lundstrom, R.C., Wilhelm, K.A., Ravesei, E.M., and Licciardello, J.J. 1984. Effect of oxidizing and reducing agents on trimethylamine oxide demethylase activity in red hake muscle. *Journal of Agricultural and Food Chemistry* 32: 459–464.

Ramirez, J.A., Martin-Polo, M.O., and Bandman, E. 2000. Fish myosin aggregation as affected by freezing and initial physical state. *Journal of Food Science* 65: 556–560.

Rehbein, H. 1979. Development of an enzymatic method to differentiate fresh and sea-frozen and thawed fish fillets I. Comparison of the applicability of some enzymes of fish muscle. *Zeitschrift Für Lebensmittel-Untersuchung und Forschung A* 169: 263–265.

Rehbein, H. 1988. Relevance of trimethylamine oxide demethylase activity and haemoglobin content to formaldehyde production and texture deterioration in frozen stored minced fish muscle. *Journal of the Science of Food and Agriculture* 43: 261–277.

Rey-Mansilla, M.M., Sotelo, C.G., Aubourg, S.P., Rehbein, H., Havemeister, W., Jorgensen, B., and Nielsen, M.K. 2001. Localization of formaldehyde production during frozen storage of European hake (*Merluccius merluccius*). *European Food Research and Technology* 213: 43–47.

Saeed, S. and Howell, N.K. 2002. Effect of lipid oxidation and frozen storage on muscle proteins of Atlantic mackerel (*Scomber scombrus*). *Journal of the Science of Food and Agriculture* 82: 579–586.

Saeed, S. and Howell, N.K. 2004. Rheological and differential scanning calorimetry studies on structural and textural changes in frozen Atlantic mackerel (*Scomber scombrus*). *Journal of the Science of Food and Agriculture* 84: 1216–1222.

Saeed, S., Fawthrop, S.A., and Howell, N.K. 1999. Electron spin resonance (ESR) study on free-radical transfer in fish lipid-protein interactions. *Journal of the Science of Food and Agriculture* 79: 1809–1816.

Shimomura, H., Takahashi, T., Morishita, T., and Ueno, R. 1987. Investigation of the differentiation of frozen-thawed fish from unfrozen fish by comparison of lysosomal enzyme activity. *Nippon Suisan Gakkaishi* 53: 1841–1845.

Sikorski, Z.E. and Kolakowska, A. 1990. Freezing of marine food. In: *Seafood: Resources, Nutritional Composition and Preservation* (Sikorski, Z.E., ed.), pp. 111–124. CRC Press, Inc.: Boca Raton, FL.

Sotelo, C.G. and Rehbein, H. 2000. TMAO-degrading enzymes. In: *Seafood Enzymes: Utilization and Influence on Post Harvest Seafood Quality* (Haard, N.F. and Simpson, B.K., eds.), pp. 167–190. Marcel Dekker: New York.

Sotelo, C.G., Gallardo, J.M., Pineiro, C., and Perez-Martin, R.I. 1995. Trimethylamine oxide and derived compounds, changes during frozen storage of hake (*Merluccius meluccius*). *Food Chemistry* 53: 61–65.

Srikar, N.N. and Reddy, G.V.S. 1991. Protein solubility and emulsifying capacity in frozen stored fish mince. *Journal of the Science of Food and Agriculture* 55: 447–453.

Srinivasan, S., Xiong, Y.L., and Blanchard, S.P. 1997a. Effect of freezing and thawing methods and storage time on thermal properties of freshwater prawns (*Macrobrachium rosenbergii*). *Journal of the Science of Food and Agriculture* 75: 37–44.

Srinivasan, S., Xiong, Y.L., Blanchard, S.P., and Tidwell, J.H. 1997b. Physicochemical changes in prawns (*Macrobrachium rosenbergii*) subjected to multiple freeze-thawed cycles. *Journal of Food Science* 67: 123–127.

Sultanbawa, Y. and Li-Chan, E.C.Y. 1998. Cryoprotective effects of sugar and polyol blends in ling cod surimi during frozen storage. *Food Research International* 31: 87–98.

Sultanbawa, Y. and Li-Chan, E.C.Y. 2001. Structural changes in natural actomyosin and surimi from ling cod (*Ophiodon elongates*) during frozen storage in the absence or presence of cryoprotectants. *Journal of Agricultural and Food Chemistry* 49: 4716–4725.

Sych, J., Lacroix, C., Adambounou, L.T., and Castaigne, F. 1990. Cryoprotective effects of lactitol, palatinit and polydextrose on cod surimi proteins during frozen storage. *Journal of Food Science* 55: 356–359.

Tejada, M., Careche, M., Torrejon, P., Del Mazo, M.L., Solas, M.T., Garcia, M.L., and Barba, C. 1996. Protein extracts and aggregates forming in minced cod (*Gados morhua*) during frozen storage. *Journal of Agricultural and Food Chemistry* 44: 3308–3314.

Thanonkaew, A., Benjakul, S., Visessanguan, V., and Decker, E.A. 2006a. Development of yellow pigmentation in squid (*Loligo peali*) as a result of lipid oxidation. *Journal of Agricultural and Food Chemistry* 54: 956–962.

Thanonkaew, A., Benjakul, S., Visessanguan, V., and Decker, E.A. 2006b. The effect of metal ions on lipid oxidation, color and physicochemical properties of cuttlefish (*Sepia pharaonis*) subjected to multiple freeze-thaw cycles. *Food Chemistry* 95: 591–599.

Thanonkaew, A., Benjakul, S., Visessanguan, V., and Decker, E.A. 2007. Yellow discoloration of the liposome system of cuttlefish (*Sepia pharaonis*) as influenced by lipid oxidation. *Food Chemistry* 102: 219–224.

Thanonkaew, A., Benjakul, S., Visessanguan, V., and Decker, E.A. 2008. The effect of antioxidants on the quality changes of cuttlefish (*Sepia pharaonis*) muscle during frozen storage. *LWT-Food Science and Technology* 41: 161–169.

Ueng, Y.E. and Chow, C.J. 1998. Textural and histological changes of different squid mantle muscle during frozen storage. *Journal of Agricultural and Food Chemistry* 46: 4728–4733.

Woyewoda, A.D. and Bligh, E.G. 2006. Effect of phosphate blends on stability of cod fillets in frozen storage. *Journal of Food Science* 51: 932–935.

Xiong, X.L. 1997. Protein denaturation and functionality losses. In: *Quality in Frozen Food* (Erickson, M.C. and Hung, Y.C., eds.), pp. 111–140. Chapman & Hall: New York.

Xiong, X.L. 2000. Protein oxidation and implication for muscle food quality. In: *Antioxidants in Muscle Food*, 1st edn. (Decker, E.A., Faustman, C., and Lopez-Bote, C.J., eds.), pp. 85–111. John Wiley & Sons, Inc.: New York.

Zamora, R., Alaiz, M., and Hidalgo, F.J. 2000. Contribution of pyrrole formation and polymerization to the nonenzymatic browning produced by amino-carbonyl reaction. *Journal of Agricultural and Food Chemistry* 48: 3152–3158.

Zhou, A., Benjakul, S., Pan, K., Gong, J., and Liu, X. 2006. Cryoprotective effects of trehalose and sodium lactate on tilapia (*Sarotherodon nilotica*) surimi during frozen storage. *Food Chemistry* 96: 96–103.

10 Effects of Some Common Processing Steps on Physicochemical Changes of Raw Red Meats

Bethany Uttaro

CONTENTS

10.1 INTRODUCTION

Some of the common processing steps used during the production and packaging of red meats include carcass handling, carcass chilling, freezing and thawing of meat, the application of differing degrees of pressure, packaging under different gases and atmospheric pressures, light exposure, etc. Meat responds differently to each of these processing steps and the effects of multiple processing steps can be cumulative, each one degrading the meat quality slightly more than the previous step. The purpose of this chapter is to provide the reader with a guide to the physicochemical characteristics and properties of raw red meats and how those change under typical processing steps.

The chapter is composed of two parts. The first part (Sections 10.2 through 10.4) briefly describes the basic structural and chemical components of meats, their interactions, and the processes that occur to convert muscles into meat. The second part (Section 10.5) describes the effects a number of processing steps have or can have on raw whole muscle red meats, using the understanding gained in the first part of the chapter. References have been chosen carefully to direct the reader to easily available and key sources, which would give more in-depth information (particularly fundamental diagrams and figures) on any particular aspect.

For more information on materials covered in Sections 10.2 through 10.4, the reader is referred to excellent references and tools listed at the end of the chapter, including those on bovine myology (Web-based atlas) and porcine myology (also a Web-based atlas). The reader is also encouraged to consult the *Encyclopaedia of Meat Sciences* (Jenses et al., 2004) as well as Judge et al. (1989), Lawrie (1998), and Warriss (2000).

10.2 MEAT STRUCTURE AND COMPONENTS

10.2.1 Muscle Structure and Contraction

A single muscle is covered by epimysial connective tissue, which binds together bundles of myofibers (fasciculi); each bundle is individually bound by perimysial connective tissue. Each myofiber is considered a cellular unit and is bound by sarcolemma, a double-walled transport system, within endomysial connective tissue. Myofibers are multinucleated cells containing a fluid phase (sarcoplasm) that contains organelles (mitochondria, lysosomes, peroxisomes, and lipid bodies), tubes leading from the sarcolemma, and a highly organized structure of contractile proteins (Lawrie, 1998). The principle contractile components are known as thick and thin myofilaments; each unit is called a sarcomere. Thin myofilaments are composed of anchored strands of actin, which are placed around strands of thick myofilaments (myosin) in a hexagonal arrangement. In combination, such as during contraction, they are known as actomyosin.

Actin is composed of globular proteins actin and troponin strung end-to-end, and a strand of tropomyosin, which are all wound into a single α-helical strand. Actin filaments are attached on opposite sides of a Z-disk. In a relaxed muscle, the ends of the actin strands between Z-disks do not meet. Myosin bridges this gap, overlaps the actin, and is equidistant from but does not reach the Z-disks. Myosin is composed of protein units and each unit contains a pair of heavy, globular meromyosins for the "head" with a light rod-like linear meromyosin "tail" attached to each head and then entwined with one another (Judge et al., 1989; Perry, 1996). These tails are grouped together along their lengths and arranged in a way that the heads protrude along the myosin filament in a spiral. Each head is offset from the next by 60° so that each one lines up with an actin filament. One half of the myosin filament heads and tails are oriented in one direction, while in the other half they are oriented in the opposite direction. Filaments are held in place by a number of structural proteins such as titan, nebulin, α-actinin, M-protein, and C-protein (Perry, 1996). The distance between the Z-disks is called a sarcomere and in red meat animals, the length of actin on one side of a Z-disk is approximately 1.0 μm; the full length of a myosin strand is 1.5 μm (Judge et al., 1989). Since actin filaments do not meet in relaxed muscle, there is a region at the middle of the sarcomere composed solely of myosin and regions adjacent to the Z-disks, which are solely actin. As muscle contracts, the sarcomere length decreases, increasing the overlap of actin and myosin, resulting in an increase in meat toughness. Conversely, stretched muscle with little myofilament overlap is tenderer. In some cases of extreme contraction, myosin may break through the Z-disk to disrupt the internal structure of the meat and produce meat that is tender, although with a mealy texture.

Muscle contraction is achieved through the release of an electric charge, which initiates a release of Ca^{2+} from the sarcolemma into the sarcoplasm, resulting in a cascade of chemical reactions culminating in the attachment of the myosin heads to receptor sites on the actin filament. The myosin heads rotate, release, and then reattach to the next site on the actin filament. This happens concurrently at both ends of the myosin strand, drawing the actin filaments together, shortening the sarcomere, and therefore the muscle. Muscle can be lengthened again by the combined release of the myosin heads from the actin filament as Ca^{2+} is removed from the cell and by contraction of an opposing muscle.

10.2.2 Muscle Fiber Types

Muscles are composed of differing proportions of red, white, and intermediate myofiber types, depending on the function of each muscle. The designation of myofiber color stems from the amount of myoglobin, which is a reflection of the amount of oxygen required by the muscle. Red fibers are found in greater numbers in muscles that are required for sustained contraction such as postural muscles. These myofibers are small in diameter; high in myoglobin, mitochondria, blood supply, and fat; and low in glycolytic enzymes. At the other end of the myofiber spectrum are white fibers, which are found in high proportions in muscles that are required for rapid, short-term contraction such as the *longissimus dorsi*. These myofibers are large in diameter; low in myoglobin, mitochondria, blood supply, and fat; and high in glycolytic enzymes (Judge et al., 1989).

10.2.3 Connective Tissues

Connective tissues knit the levels of a muscle together and contribute significantly to meat toughness (Lepetit, 2007). Connective tissues are composed of collagen, elastin, and reticulin. Collagen is a primary component of the structural connective tissues (Judge et al., 1989). Collagen fibers are composed of a series of strands, or fibrils, composed of tropocollagen molecules placed in parallel arrangement. Each molecule overlaps the adjacent one by one-quarter, producing a striated appearance under a microscope. These fibrils are held together by intermolecular permanent hydroxyproline cross-linkages, the number of which increases with maturity. Collagen is salt-soluble, although solubility decreases with animal age (Hill, 1966), and since it is inextensible, it is often found in lattice-like sheets around different levels of muscle (Rowe, 1974). The lattice construction allows for changes in muscle diameter during muscle extension and contraction. Certain collagens can be reduced to a gelatin-like consistency under moist heat-cooking conditions.

Elastin, as the name suggests, is quite elastic. It develops a large number of cross-links quickly, is insoluble in both water and salt solution, and is strongly resistant to most cooking methods. The *ligamentum nuchae* is almost pure elastin. This is the ligament that in grazing animals runs from the base of the skull, along the back of the neck, and then is anchored in progressively smaller sections along the vertebrae for most of the length of the back and serves to automatically lift the head when muscles on the underside of the neck are relaxed. When elastin is present in muscle, the amounts are usually very small with the exceptions of the *semitendinosus* and *latissimus dorsi* (Bendall, 1967), where it contributes significantly to toughness regardless of the pre- or post-rigor processing or cooking method used. Reticulin is composed of very fine, small fibers and is primarily associated with the fine connective tissue networks that hold blood vessels and nerves in place.

The amount of connective tissue and the degree of cross-linking varies from muscle to muscle within an animal (Burson and Hunt, 1986; Torrescano et al., 2003). Collagen with fewer cross-links contributes to meat being tenderer. The outermost layer of connective tissue on a muscle is known as the epimysium in scientific parlance and as "silverskin" in the trade. This layer is arranged in a combination of longitudinal and latticework structures. It may be thick on one side of the muscle and thin on the others; the thickness varies from muscle to muscle. The next layer of connective tissue groups myofibers together and is called the perimysium (Lepetit and Culioli, 1994). Once again the strands are in a precise lattice-like arrangement, which changes shape with contraction and relaxation. The innermost connective tissue layer (endomysium) groups the myofibrils together. The endomysium is composed of a mesh of randomly placed strands of reticulin.

10.2.4 Myoglobin

Myoglobin (Mb) is a compound that gives meat its color and almost anything that is done to either animal or meat affects Mb to some degree. The function of Mb is to transport oxygen (O_2) within muscle cells. Mb is composed of an iron molecule, or heme group, encased in a protein or globin ring with one docking site for an

O_2 molecule. It is one-quarter the size of hemoglobin (Hb), which transports O_2 in blood and responds to gases in much the same manner. However, Mb has a higher affinity for O_2 than Hb, and that affinity is not pH dependent (Stryer, 1981). One Hb unit can loosely be thought of as 4 Mb units in size. As such Hb is far too large to pass through the capillaries itself; therefore the O_2 it carries from the lungs must be passed through the vascular barrier to Mb for delivery within the muscle cells. Unoxygenated Mb is a dark purple-red and known as deoxymyoglobin (DMb). This is the color one sees immediately on slicing open an intact piece of raw meat. When Mb becomes oxygenated, it turns red, producing the familiar color of fresh meat, and is called oxymyoglobin (OMb). Although the formation of OMb on exposure to air is instantaneous, the color of fresh meat takes 20–30 min to develop fully after the initial cut (bloom time) because under normal atmospheric pressure this reaction occurs not only on the surface of the meat, but also to a depth of several millimeters below the surface. It is this accumulated depth of OMb that consumers are accustomed to seeing in the retail case. As OMb becomes oxidized, brown metmyoglobin (MMb) is formed and once the proportion of this becomes high enough the meat appears brown. The oxidation of MMb, irradiation, or microbial spoilage can lead to the development of various green colors such as sulfmyoglobin (Lin et al., 1977; Faustman and Cassens, 1990), cholemyoglobin (George and Irvine, 1952; Fox, 1966), and nitrimetmyoglobin in cured meats (Fox, 1966). The characteristic pink of cured meats occurs when Mb is reduced with nitrate, nitrite salts, or nitrogen dioxide. This may also occur unintentionally in uncured meats from exposure to incompletely combusted gas or from nitrite or nitrate contamination from water, processing equipment, or spices. Denaturation of Mb, such as during cooking, causes the globin to unfold, exposing the heme group, which undergoes rapid oxidation. These structural and chemical changes contribute to the grey-brown color of cooked meat.

Mb has been shown to accept a number of different gases in addition to O_2. These include carbon monoxide (CO) used in packaging, nitric oxide, and, experimentally, xenon and cyanide (Cohen et al., 2006). Carboxymyoglobin (COMb) forms because of the greater affinity of Mb for CO than for O_2 and is discussed in more detail below. OMb is photosensitive, forming MMb on exposure to light, particularly at 254 nm, which is approaching the deep ultraviolet region of the spectrum. This rate of oxidation is 4700 times greater than the rate due to 546 nm (green) light in the visible region of the spectrum (Bertelsen and Skibsted, 1987).

With exposure to heat Mb in the center of an intact piece of meat changes from blue-red to red to pinkish to brown due to thermal denaturation of Mb and proteins. Thus, in meat cooked to rare (approximately 50°C) and then sliced open, it is often possible to see the whole range of color changes present, with fully denatured Mb at the outside edges and untouched Mb at the center.

10.2.5 Fat

Fat is deposited in a number of key areas in an animal body. These are under the skin (subcutaneous), in the body cavity and around the internal organs (known variously as leaf lard, kidney fat, or body cavity fat), between the muscles (intermuscular or seam fat), and within the muscles (intramuscular or marbling fat). The fatty acid composition, and therefore saturation and firmness of fat, varies from one depot to the next as well as within different layers of the same depot. The latter is very noticeable in the subcutaneous fat in pigs where there may be up to three layers separated by membranes. The innermost layer has the lowest degree of saturation and so is the softest. Meat with lower fat saturation levels undergoes more rapid oxidative changes during storage than meat with more highly saturated fat because of the greater propensity of unsaturated fats to undergo more rapid oxidation.

10.3 PROCESSING MUSCLES INTO MEATS

In an ideal slaughter situation, calm animals are stunned and exsanguinated. With the removal of blood, body temperature rises and cell respiration becomes anaerobic, causing adenosine triphosphate (ATP) production to fall drastically. Without the blood to remove metabolic by-products, the cell environment starts to become acidic due to accumulation of lactic acid; the pH begins to drop from the antemortem pH of 7–7.3. The ultimate or final pH of a cooled carcass is strongly influenced by antemortem handling and/or carcass cooling rate. The overexcitement of animals prior to slaughter results in a more rapid postmortem pH drop, in part because of a buildup of lactic acid in the cells. A very slow postmortem pH drop is seen when animals have been exhausted prior to slaughter and have depleted the muscle glycogen reserves, thus removing the energy source for anaerobic metabolism and preventing the accumulation of lactic acid. The rate at which body temperature rises postmortem is somewhat slowed when the body cavity is exposed to the relatively cooler ambient air during evisceration, and then is decreased further by moving carcasses through cool-water showers and into chilling coolers. Rigor mortis, or the "stiffness of death," commences as muscle glycogen, and therefore ATP, is depleted. The rate of depletion is affected by the amount of available O_2, muscle pH, and temperature (Lawrie, 1998). Under normal cooling conditions, the contraction state of the muscle at the time of rigor is retained in fully cooled carcasses.

10.4 RELATIONSHIP BETWEEN pH AND WATER-HOLDING CAPACITY

The relationship between pH and water-holding capacity is very important to the understanding of keeping color, eating quality, physical appearance, response to heating, and the ability of fresh meat to bloom. Normally, the 24 h pH falls to between approximately 5.5 and 5.8 in muscles with a high proportion of white fibers, but may be higher in muscles with a high proportion of red fibers. A high pH (>5.8) at 24 h postmortem is associated with high water-holding capacity, dark color, and short keeping time due to the improved conditions for bacterial growth. Should the pH drop to the isoelectric point of meat (approximately pH 5.3), the ability to hold water decreases and the color becomes lighter (Hamm, 1960). If the pH can be induced to drop further, such as when meat is treated with an acid marinade, water retention increases once again (Gault, 1991), but may also be accompanied by acid-induced protein denaturation.

Water is held by capillary force in the interfilament space of meat; this spacing is affected by electrostatic forces (Offer and Trinick, 1983; Lawrie, 1998). Water molecules closest to the filaments are held tightly, those further away are held loosely, and those furthest away are free and eventually drip out of the meat. As the muscle pH drops, more water becomes loosely bound and free, increasing the amount of drip. Since loosely bound and free water is highly reflective, this meat appears light-colored. At a higher pH, there is less loosely bound or free water, reducing drip and absorbing more light, making the meat appear darker.

10.5 EFFECTS OF DIFFERENT COMMON PROCESSING STEPS ON RAW MEATS

10.5.1 Cold

A primary purpose of chilling is to reduce the carcass temperature to low enough levels to avoid pathogen growth. Freezing is used for extended storage purposes and, in some cases, to prepare some meats for slicing or grinding steps in a production line/process.

10.5.1.1 Chilling Rate

The rate at which a carcass is cooled can have a marked effect on the quality of meat. Rapid chilling produces dark and dry meat, which may be tougher, whereas very slow chilling produces more tender meat, but with poorer keeping qualities because of bacterial growth.

In conventional or classic chilling, the temperature of a carcass is reduced by showering the carcass with cold water, placing it in a prechilled cooler for a number of hours, exposing it to circulating cold air, or a combination of these sequentially or in concert. Fatter carcasses cool more slowly than leaner ones and heavier carcasses more slowly than lighter ones. Conventional cooling produces beef of the familiar bright red color, or light pink moderately firm pork. Conventional chilling can induce some cold shortening, although this occurs more often in beef than in pork because the hide and part of the subcutaneous fat is removed from the former during the dressing procedure, leaving it with less insulation from the cold; pork, on the other hand, is chilled with the skin on. Cold shortening is the pre-rigor contraction of muscle, which occurs when the temperature drops too quickly, and is associated with increased toughness because of shorter sarcomere lengths.

Following dressing, blast chilled carcasses are exposed for 30–60 min or more to −30°C to −40°C temperatures, which can be accompanied by circulated air for additional temperature drop. This reduces the carcass temperature very quickly, halting the postmortem pH drop, which results in darker meat with reduced drip and contributes to uniform eating quality. It also freezes the carcass surface. Carcasses are then moved to a holding cooler at 1°C–4°C for the remainder of the chilling period. For hog carcasses, this process has been very successful in reducing the incidence of pork meat that is pale, soft, and exudative. Very rapid pre-rigor chilling or even freezing can result in unrestrained muscle undergoing extreme contraction immediately (cold shortening) or upon thawing ("thaw-rigor" or "thaw-shortening") due to the release of Ca^{2+} from ice-damaged sarcolemma (Joseph, 1996), making meat very tough. However, when the surface of a hanging carcass is frozen and later thawed, the meat from that carcass has been found to be tenderer than it would have been under conventional chilling conditions. This may be due to the frozen surface, providing a rigid scaffold for inner, warmer, fibers to pull against, preventing the interior of the muscle from shortening as it goes into rigor (Davey and Garnett, 1980; Joseph, 1996). By the time the exterior crust thaws, the interior in turn serves as a relatively rigid scaffold, preventing shortening of the exterior fibers.

10.5.1.2 Freezing and Thawing

The purpose of freezing is to extend the storage time. Since sarcoplasm is not pure water, meat starts to freeze at approximately –2°C. Partial or full thawing is often used in industry for further processing of held meat. Slicing and blade tenderization, in particular, proceed more smoothly on meat that is still partially frozen. The rates of both freezing and thawing influence the manner and size of ice-crystal growth, which affects cell integrity and hence the pH, purge/drip, and color of meat. Freezing and thawing is conventionally achieved through changes in ambient temperature. For freezing and thawing through pressure modifications, see Section 10.5.3.

Bevilacqua et al. (1979) viewed meat as a multicomponent system made up of insoluble and soluble components in free and bound water with the additional complication that the solutions are both intercellular and intracellular. Thus, ice crystals could form both within and between muscle fibers. Since the salt concentration of the intercellular liquid is lower than that of the intracellular liquid, under slow freezing conditions large ice crystals grow from a small number of crystal nuclei, which form first between fibers (Grujić et al., 1993). Slow freezing (≥20°C) allows time for the movement of water from inside the fiber to contribute to the growth of large crystals outside the fibers, resulting in fiber dehydration and deformation (Grujić et al., 1993; Devine et al., 1996). During the growth of large crystals, fibers and sarcomeres are torn apart. In addition, a larger volume of liquid remains unfrozen due to an increase in the solute concentration as water is drawn away and frozen (Pincock and Kiovsky, 1966). Under rapid freezing conditions, numerous crystal nuclei form within fibers and crystal growth is small and fine. There is little to no migration of water to the inside of the cell to contribute to this growth, which crowds and compresses the cell contents but does relatively little damage to the ultrastructure (Grujić et al., 1993).

Industrially, very large pieces of meat are processed and the impact of temperature gradients on meat ultrastructure during freezing and thawing is of great importance. These sorts of problems are encountered in such situations as plate freezing and freezing full boxes of meat (de Michelis and Calvelo, 1982). Ice-crystal growth occurs in the direction of heat flow (Bevilacqua et al., 1979). If freezing is slow and the direction of heat flow is perpendicular to meat fibers, more ultrastructural damage from the formation of large ice crystals may be encountered than if the heat flow is parallel to fibers. During thawing, a heat absorption gradient develops. Meat quickly reaches the freezing point and then remains there until thawing is complete before it warms further (Judge et al., 1989). Under these slow thawing conditions there is opportunity for the formation of large ice crystals, resulting in ultrastructural damage. However, Gonzalez-Sanguinetti et al. (1985), who froze meat slowly (–20°C) to create only intercellular ice crystals and then measured the amount of exudate produced, found that rapid thawing produced greater amounts of exudate than slow thawing despite the greater potential for large ice crystal formation. This is because the extended thawing time was found to permit the reabsorption of extracellular water into intracellular spaces. On the other hand, Ambrosiadis et al. (1994) noted that slow thawing rates (e.g., in air at 4°C for 28 h) always caused more damage than fast thawing rates (e.g., in water at 16°C–18°C for 1.5 h). The study of morphological changes led these investigators to conclude that the damage was caused by the formation of large extracellular ice crystals during the thawing process. A similar trend was observed by Ngapo et al. (1999). If slow freezing is followed by slow thawing, more ultrastructural damage but less exudate results than if it is followed by rapid thawing.

The characteristic freezing time (t_{CF}) of meat, which is the time required for a drop of temperature at the center of a sample from –1°C to –7°C, can be used to identify the rate of freezing (Bevilacqua et al., 1979). In experiments with 2 cm thick beef steaks frozen at temperatures ranging from –14°C to –78°C, and time of 257–10.6 min (t_{CF} of 216–8.2 min), Grujić et al. (1993) reported that when thawed at the same rate the greatest moisture losses occurred at freezing temperatures of ≥40°C. Slower freezing rates produced meat with lower pH and higher protein solubility. Below –40°C the eating quality noticeably changed and meat became juicier but less tender. Extremely low freezing temperatures (–78°C) caused more damage to the ultrastructure than the slower freezing rates. The initial conclusion was that portioned beef was best frozen at between –40°C and –50°C, or a rate of between 3 and 4 cm/min. Later, this conclusion was revised and the range broadened to a rate of between 2 and 5 cm/min (Petrović et al., 1993).

Crystal-mediated cellular damage can take place during frozen storage due to temperature oscillations, resulting in a net increase in the crystal size (Martino and Zaritzky, 1988); this type of damage occurs more rapidly at the surface of meats (Devine et al., 1996).

Freezing detrimentally affects the color stability of thawed meat. Jeong et al. (2004) hypothesized that the rapid development of a brown color (MMb) might be due to the autooxidation of Mb by mitochondrial enzymes that were released due to ice-crystal damage. Greer and Murray (1991) noted that although the shelf life of frozen and thawed meat decreased because of color change, it did not concurrently decrease due to spoilage.

10.5.2 Heat

10.5.2.1 Effect on Color

The application of heat denatures proteins and causes changes to the connective tissues of raw meats. These events take place at different temperatures as the meat heats; each of these thermally induced physical changes in turn affects appearance and eating quality of the meats. Mb is relatively heat-stable, so meat cooked to an internal temperature of 60°C is bright red, to 60°C–70°C is pink, to 70°C–80°C is grayish brown (Renerre, 1990; Lawrie, 1998); complete denaturation takes place around 85°C (Renerre, 1990). External browning also occurs from carbohydrates caramelizing and from Maillard-type reactions between sugars and proteins (Renerre, 1990).

The denaturation of other meat components take place at the following temperatures although there are some variations from these temperatures among breeds and muscles (Voutila et al., 2007): myosin tails at 40°C–54°C, myosin heads at 53°C–60°C, collagen at 56°C–62°C, and actin at 66°C–73°C (Martens et al., 1982).

Between 60°C and 75°C, both the pH and respiration activity drop, and therefore so does the depth of the surface OMb layer (Gašperlin et al., 2001). Trout (1989) observed that when meat with a starting pH > 6 was cooked to an end temperature of 90°C, the interior still appeared pink although all the Mb had been denatured. Ghorpade and Cornforth (1993) noticed that the interior color of pork cooked to an internal temperature of 82°C was initially brown, but on refrigerated storage became pink; the pink color faded after it had been exposed to the air for a few minutes. These investigators hypothesized that although the Mb was certainly denatured, the pink color might have been from denatured globin hemochromes or from non-nitrosyl hemochromes. As observed by Tuomy et al. (1963), a meat of normal pH could be held for an extended period of time between the denaturation temperatures of actin and myosin and on the verge of Mb denaturation and still appear done to rareness.

10.5.2.2 Effect on Eating Quality

The eating quality of meat cooked to end temperatures corresponding to specific denaturation temperatures differs markedly. Heating at 40°C, which is just above the body temperature of cattle and swine, generally causes little change in the structural appearance (Hearne et al., 1978). By 50°C, sarcomeres show some swelling (Jones et al., 1977; Palka and Daun, 1999); some small cracks and breaks appear in the fibers (Hearne et al., 1978) and tenderness decreases due to a combination of changes in the mechanical properties of perimysium (Christensen et al., 2000) and the denaturation of myosin. By 60°C, microscopy shows rigid blocks of actomyosin and some dehydration (Hearne et al., 1978; Palka and Daun, 1999) along with many cracks and breaks in the fibers (Hearne et al., 1978); tenderness is improved (Christensen et al., 2000) and 10%–12% of total collagen is soluble (Palka and Daun, 1999). By 70°C, the fiber structure would disintegrate, showing complete breaks in the fibers (Hearne et al., 1978); almost 20% of the collagen present is soluble (Palka and Daun, 1999) and some thermal contraction of collagen is evident (Bailey and Sims, 1977), while tenderness decreases (Christensen et al., 2000).

From a sensory point of view, Martens et al. (1982) made the following observations: firmness increased as myosin denatured, decreased when collagen denatured, and then increased again when actin denatured. There was a small drop in juiciness when myosin and collagen denatured; a further larger drop was observed on actin denaturation. Fiber cohesivity became quite poor after collagen denatured. This was when texture preferences were highest, and they dropped off again as actin denatured.

As the cooking temperature rises to that of overdone meat (80°C) and further, to that of retorting or canning (>100°C), collagen solubility drops to <6% of the total collagen (Palka and Daun, 1999); sarcomere lengths decrease to the length of myosin (1.5 μm) and cooking losses increase (Tischer et al., 1953; Jones et al., 1977; Palka and Daun, 1999). Undesirable eating characteristics such as chewiness and hardness generally peak at 80°C (Palka and Daun, 1999). Tenderness first decreases and then slowly increases although the rate at which this occurs is dependent on the processing temperature (Tischer et al., 1953).

Under dry-heat cooking conditions, e.g., in the case of an uncovered roast in a home oven, slow heating results in a large amount of myofiber granulation, whereas rapid heating shows less granulation but more cracks and breaks (Hearne et al., 1978). From time and temperature experimentation, Bard and Tischer (1951) reported that the rates of change of cooking losses and tenderness were directly proportional to the cooking temperature. Tuomy et al. (1963) later found that under cooking temperatures of 60°C–99°C and heating time of 0–7 h, initial toughening occurred rapidly and independently of the cooking temperature. At heating temperatures of <82°C, subsequent tenderness changes were temperature-dependent with little or no change even after 7 h of heating. On the other hand, at heating temperatures of ≥82°C, tenderness changes were governed by both temperature and heating time, so that by 7 h cooking at 99°C meat fell apart and could not undergo testing.

10.5.3 Pressure

High-pressure processing normally involves immersion of packaged meat in a liquid-filled chamber and can be either static with a steady increase in pressure or dynamic with a sudden increase in pressure such as from a shock wave. For additional information on the effects of high-pressure processing on the quality of various foods, including animal products, the reader is referred to Chapter 5.

10.5.3.1 Static Pressure

There are many factors that affect the outcome of high-hydrostatic pressure (HSP) processing, including the rate of compression and decompression, starting temperature, holding temperature, and holding time. As a result, the discussion below addresses the general response of meat to differing HSP. More detailed discussion may be found in Montero and Gómez-Guillén (2005) and Suzuki et al. (2006).

The primary reason for the application of HSP to foods has been to improve food safety through inactivation of microbial flora. However, at the pressures required for HSP, meat quality is severely compromised. Jung et al. (2003) reported that a pressure of 130 MPa had little effect on total flora of beef samples and was minimally detrimental to meat quality, while a pressure of 520 MPa resulted in a 2.5-log reduction of bacteria and a 1 week delay in growth, but also immediately caused a meat color change from red to brown along with a complete modification of the ultrastructure (Jung et al., 2000a). Rodríguez et al. (2003) noted that most microbial inactivation took place between pressures of 300 and 600 MPa, while Lee et al. (2007) observed that at pressures greater than 200 MPa, meat proteins denatured. Other reasons to apply HSP commercially are to modify enzyme activity and, for non-intact meat systems, to change the texture and gelation properties in the development of new products (Jiménez-Colmenero and Borderias, 2003).

For an application of HSP, meat is usually vacuum packed, then place in a chamber containing water or oil. A steady increase in pressure is then applied, which affects all parts of the sample at one time. Numerous researchers have determined that HSP has the largest effect on the protein portions of meat and little effect on the collagen (Suzuki et al., 1993; Jung et al., 2000b; Ma and Ledward, 2004), possibly because the protein portion of meat is 70%–75% water. The protein structure is forcibly compressed, and meat exposed to pressures above 200 MPa invariably shows very short sarcomere lengths, decreased tenderness, increased fiber diameter, protein denaturation, increased cooking losses, and increased formation of metmyoglobin. Jung et al. (2000a) reported an increase in protein solubility with the application of as little pressure as 100 MPa. By 325 MPa cellular fragments were apparent in interfibrillar spaces and by 520 MPa the ultrastructure was completely modified. At 520 MPa, sarcomeres were shortened to less than the normal length of myosin (Jung et al., 2000b). When sarcomeres are severely shortened, fiber diameter increases and meat can become very tough due to the overlap of actin filaments, which is in addition to the normal overlap of actin and myosin. However, if the myosin filaments break through the ends of the sarcomere and disrupt the Z-line causing intrafibrillar tears, the meat can become tender although the texture of the cooked product is often mealy.

HSP has little effect on the collagen ultrastructure of meat (Suzuki et al., 1993; Cheftel and Culioli, 1997; Ananth et al., 1998; Jung et al., 2000b; Ma and Ledward, 2004), although recently Ichinoseki et al. (2006) showed that at sufficiently high pressures (>200 MPa), the intramuscular collagen structure could be affected enough to increase its solubility on the application of heat.

The application of HSP to raw meats has resulted in a marked alteration of color. For beef the color was reported to change from the familiar medium bright red to a light brown, which is somewhat similar to cooked meat (Jung et al., 2000b, 2003). For pork the change was from the familiar medium light pink to light brown (Ananth et al., 1998). Study in this area has shown that although the color change is due to protein denaturation through the application of pressures above 200 MPa, the mechanism for HSP-induced protein denaturation is different from that of heat-induced denaturation (Lee et al., 2007).

10.5.3.2 Dynamic Pressure

Hydrodynamic pressure (HDP) processing involves a sudden pressure increase in the form of a shock wave, which passes through the meat, causing myofibrillar fragmentation and sarcomere disruption, immediately tenderizing the meat (Zuckerman and Solomon, 1998; Solomon et al., 2006). Connective tissue is more resistant to this treatment and is minimally impacted with small disruptions of the endomysium (Solomon et al., 2006) and slight increases in both soluble and insoluble collagen (Bowker et al., 2007); therefore meat treated in this manner can become only as tender as the type, amount, and structure of the connective tissue present. Although there has been little evidence that protein solubility is affected (Schilling et al., 2002), delayed tenderization similar to the tenderization obtained with aging meat has been observed a few days following the treatment. Solomon et al. (2006) and Bowker et al. (2007) hypothesized that this might be due to direct changes in muscle proteins. An application of this treatment to frozen and fresh cuts of meat resulted in increased tenderness of both types of meat, but with a greater response from fresh meat (Solomon et al., 1997). When compared to meat tenderized through aging for 6 days or through conventional freezing and thawing, the application of HDP was found to have the greatest effect (Solomon et al., 2008). When meat quality parameters were examined on previously frozen and thawed pork shortly following HDP treatment (Moeller et al., 1999), no differences were found for subjectively evaluated color, or firmness/wetness, although the objective redness (a^*) was higher, indicating rosier meat. Treatment effects on appearance, eating qualities, and keeping quality in longer-term storage have not yet been reported, however.

The advantage of hydrodynamic tenderizing over the common industry practice of blade tenderization is that the former is noninvasive, relieving the threat of bacterial contamination of the interior of the meat, which, until it is cut into, is considered sterile.

10.5.3.3 Freezing and Thawing

The freezing and thawing discussion above revolved around the conventional methods in which freezing or thawing rate is directly related to the ambient temperature. However, the rate of freezing and thawing can be markedly increased by the application of multi-atmosphere pressure changes. Freezing with the aid of pressure has the potential in the food industry for the formation of smaller ice crystals, reduced freezing time, and reduced thawing time. Cheftel et al. (2000) and Sanz (2005) have indeed presented excellent reviews of freezing and thawing foods under pressure.

Sanz (2005) distinguished between the two basic types of high-pressure freezing. In the first type of freezing (pressure-assisted freezing), the temperature is lowered as a constant pressure greater than atmospheric pressure is maintained. In the second type of freezing (high-pressure shift freezing), temperature drop is affected by either slowly or rapidly reducing the pressure. This distinction is important because there is a difference in the way freezing occurs and therefore affects the product structure. In pressure-assisted freezing, the sample cools from the outside to the inside, as it does with conventional freezing, and a thermal gradient forms. However, unlike conventional freezing, ice nuclei only form on the outside edge of the sample, while long, needle-like ice crystals grow to the center. In high-pressure shift freezing, each change in pressure results in the formation of ice nuclei specific for that pressure. Since pressure changes affect the whole sample at one time, ice nuclei form throughout the sample and no temperature gradient forms. Martino et al. (1998) found that this scheme worked very well for large pieces of pork *longissimus dorsi* (5 × 7.5 × 11 cm). Fernández et al. (2008) explored the possibility that extended frozen storage following high-pressure shift freezing might result in the recrystallization and growth of ice crystals. This was found to be true over a matter of hours for a pure salt solution with no cell wall restrictions, but it remains to be seen if it holds for intact systems over extended storage time of days and months.

Thawing may also be accomplished with pressure changes. Knorr et al. (1998) identified "pressure-assisted" thawing, in which the pressure was kept constant while the temperature increased; and "pressure-induced" thawing, in which the pressure was changed. For both techniques, melting heat must be removed as it forms, resulting in thermal gradients. Following thawing, the sample temperature must be increased enough so that the final release of pressure does not cause recrystallization (Cheftel et al., 2000; Sanz, 2005). Park et al. (2006) explored thaw-assisted pressures up to 200 MPa and found this a favorable way to thaw large cylinders of pork *longissimus dorsi* (5 cm in diameter × 10 cm in length). Up to 100 MPa, meat quality characteristics were comparable to conventional thawing. Above that pressure threshold, although water-holding capacity and thawing losses were improved, cooking losses and shear forces increased, and color was compromised.

10.5.4 Gases

A number of different gases are used in the processing and packaging of fresh meats. The most common are oxygen (O_2), nitrogen (N_2), carbon dioxide (CO_2), and, more recently, carbon monoxide (CO). The use of other gases has been investigated and includes sulfur dioxide (SO_2), argon (Ar) (Walsh and Kerry, 2002), helium (He), hydrogen (H_2), neon (Ne), nitrous oxide (N_2O), nitric oxide (NO), ozone (O_3), chlorine (Cl_2), ethylene, and propylene oxide (Church, 1994). Air is composed of approximately 21% O_2, 78% N_2, and 0.3% CO_2 (Sebranek and Houser, 2006). Walsh and Kerry (2002) and Sebranek and Houser (2006) covered the details of gas proportions used in modified atmosphere packagings; however, for the purposes of this discussion, only the response of meats to single gases will be discussed.

The characteristics of the commonly used pure gases, according to the Merck Index (Budavari et al., 1989), are as follows: O_2 is a colorless, odorless, and tasteless neutral gas, which supports combustion and is soluble in water in a ratio of 1 volume O_2 to 32 volumes H_2O at 20°C. Nitrogen is also an odorless gas but is only sparingly soluble in water (0°C: 2.4 volumes N_2 to 100 volumes H_2O; 20°C: 1.6 volumes N_2 to 100 volumes H_2O). Carbon dioxide is colorless, odorless, and noncombustible with a faint acid taste. At one atmosphere pressure and 0°C, 171 mL of CO_2 is soluble in 100 mL H_2O, whereas at 20°C, only 88 mL is soluble in the same volume. Carbon monoxide is highly poisonous, odorless, colorless, tasteless, and flammable. It is sparingly soluble in water (0°C: 3.3 mL CO in 100 mL H_2O; 20°C: 2.3 mL CO in 100 mL H_2O) and is approximately seven times more soluble in methanol or ethanol than in water. The toxicity of CO is due to the much greater affinity of Hb and Mb to it compared to O_2, effectively causing suffocation if inhaled in large enough quantities for a long enough period.

10.5.4.1 Nitrogen

Since N_2 has a low solubility in water and in fat, there is little reaction with meats (which are 70%–75% water). Therefore, it is useful for displacing O_2 in packagings (Church, 1994; Sebranek and Houser, 2006). With little or no O_2 present, meat color is the brown of MMb and aerobic bacteria do not thrive.

10.5.4.2 Oxygen

There are three sources for color changes of meat in response to O_2. Color changes associated with binding to Mb were discussed above, but O_2 also binds to fat and is a respiration source for aerobic bacteria on the surface of the meat. Following a few days of cooler storage after initial exposure to O_2, the OMb formed oxidizes to MMb; fat starts to become rancid and may yellow and there might be additional color and odor changes due to bacterial growth.

The concentration of O_2 is often increased in packaging to extend the amount of time that meat appears red (Zakrys et al., 2008). This is because higher O_2 levels form a thicker layer of OMb in the meat, masking the formation of MMb for a longer period (Taylor and MacDougall, 1973). However, as O_2 concentration increases above 50%, there is a decrease in both color and lipid stability (Zakrys et al., 2008). The blooming response of muscles to O_2 varies, with the *longissimus dorsi* the most stable, and the *psoas major* or tenderloin the least stable (MacDougall and Taylor, 1975; Beggan et al., 2005). Modern packaging methods aim to reduce the amount of O_2 in the package to extend shelf life, but this must be done carefully. Ledward (1970) indeed reported that the maximum rate of MMb formation occurs at 0°C under 6mm Hg, such as may occur in a vacuum package. This investigator recommended that to avoid irreversible browning from low levels of O_2, the O_2 partial pressure should be above approximately 5%.

10.5.4.3 Carbon Monoxide

The previously mentioned preference of Mb for CO over O_2 results in the formation of the highly stable carboxymyoglobin (COMb), which is virtually indistinguishable by eye from OMb. COMb can form from any one of DMb, OMb, or MMb (Suman et al., 2006) and is much longer-lived than OMb. CO is usually used at levels of less than 1% to give a fresh color to meats packaged in anoxic conditions, which would otherwise appear brown (Walsh and Kerry, 2002).

10.5.4.4 Carbon Dioxide

Since CO_2 is almost 60 times more soluble in water than O_2, it is readily absorbed by the lean portion of meat, although it is not bound by Mb. In addition, the linear structure of this gas molecule is not strongly polar, making it readily soluble in the fat portion of the meat (Sebranek and Houser, 2006). Should enough CO_2 be absorbed, small holes and fissures may be seen in the cooked meat due to expansion during cooking of the absorbed gas (Bruce et al., 1996). On reacting with water, this gas forms carbonic acid (H_2CO_3), which can decrease meat pH (Sebranek and Houser, 2006) and therefore increase drip (Church, 1994). Since CO_2 is used in meat packaging primarily as an antioxidant and antimicrobial agent (Church, 1994; Walsh and Kerry, 2002), the corresponding lack of O_2 in the environment results in the meat appearing brown instead of red (Church, 1994).

10.5.4.5 Nitrogen Dioxide

An unintentional gas-induced color change to meat occurs when it is exposed to sufficiently high levels of nitrogen dioxide (NO_2) such as during incomplete combustion of gases in gas ovens. The surface of the meat becomes pink, much like the characteristic pink of cured meat, due to the formation of nitrosylhemochrome (Cornforth et al., 1998).

10.6 CONCLUDING REMARKS

Each of the processing steps has a detrimental effect on the final meat quality; therefore the number of times each process is used, or the sequence in which they are used during the production of a particular product must be carefully considered.

REFERENCES

Ambrosiadis, I., N. Theodorakakos, S. Georgakis, and S. Lekas. 1994. Influence of thawing methods on the quality of frozen meat and the drip loss. *Fleischwirtschaft/Fleischwirtschaft International* 72: 284–287/252–255.

Ananth, V., J. S. Dickson, D. G. Olson, and E. A. Murano. 1998. Shelf life extension, safety, and quality of fresh pork loin treated with high hydrostatic pressure. *Journal of Food Protection* 61: 1649–1656.

Bailey, A. J. and T. J. Sims. 1977. Meat tenderness: Distribution of molecular species of collagen in bovine muscle. *Journal of the Science of Food and Agriculture* 28: 565–570.

Bard, J. C. and R. G. Tischer. 1951. Objective measurement of changes in beef during heat processing. *Food Technology* 5: 296–300.

Beggan, M., P. Allen, and F. Butler. 2005. The use of micro-perforated lidding film in low-oxygen storage of beef steaks. *Journal of Muscle Foods* 16: 103–116.

Bendall, J. R. 1967. The elastin content of various muscles of beef animals. *Journal of the Science of Food and Agriculture* 18: 553–558.

Bertelsen, G. and L. H. Skibsted. 1987. Photooxidation of oxymyoglobin. Wavelength dependence of quantum yields in relation to light discoloration of meat. *Meat Science* 19: 243–251.

Bevilacqua, A., N. E. Zaritzky, and A. Calvelo. 1979. Histological measurements of ice in frozen beef. *Journal of Food Technology* 14: 237–251.

Bovine Myology. http://bovine.unl.edu/ (accessed April 7, 2010).

Bowker, B. C., M. N. Liu, M. B. Solomon, J. S. Eastridge, T. M. Fahrenholz, and B. Vinyard. 2007. Effects of hydrodynamic pressure processing and blade tenderization on intramuscular collagen and tenderness-related protein characteristics of top rounds from Brahman cattle. *Journal of Muscle Foods* 18: 35–55.

Bruce, H. L., F. H. Wolfe, S. D. M. Jones, and M. A. Price. 1996. Porosity in cooked beef from controlled atmosphere packaging is caused by rapid CO_2 gas evolution. *Food Research International* 29: 189–193.

Budavari, S., M. J. O'Neil, A. Smith, and P. E. Heckelman (Eds.). 1989. *The Merck Index: An Encyclopedia of Chemicals, Drugs, and Biologicals*, 11th edn. Rahway, NJ: Merck & Co. Inc.

Burson, D. E. and M. C. Hunt. 1986. Proportion of collagen types I and III in four bovine muscles differing in tenderness. *Journal of Food Science* 51: 51–53.

Cheftel, J. C. and J. Culioli. 1997. Effects of high pressure on meat: A review. *Meat Science* 46: 211–236.

Cheftel, J. C., J. Lévy, and E. Dumay. 2000. Pressure-assisted freezing and thawing: Principles and potential applications. *Food Reviews International* 16: 453–483.

Christensen, M., P. P. Purslow, and L. M. Larsen. 2000. The effect of cooking temperature on mechanical properties of whole meat, single muscle fibres and perimysial connective tissue. *Meat Science* 55: 301–307.

Church, N. 1994. Developments in modified-atmosphere packaging and related technologies. *Trends in Food Science and Technology* 5: 345–352.

Cohen, J., A. Arkhipov, R. Braun, and K. Schulten. 2006. Imaging the migration pathways for O_2, CO, NO, and Xe inside myoglobin. *Biophysical Journal* 91: 1844–1857.

Cornforth, D. P., J. K. Rabovitser, S. Ahuja et al. 1998. Carbon monoxide, nitric oxide, and nitrogen dioxide levels in gas ovens related to surface pinking of cooked beef and turkey. *Journal of Agricultural and Food Chemistry* 46: 255–261.

Davey, C. L. and K. J. Garnett. 1980. Rapid freezing, frozen storage and the tenderness of lamb. *Meat Science* 4: 319–322.

de Michelis, A. and A. Calvelo. 1982. Mathematical models for nonsymmetric freezing of beef. *Journal of Food Science* 47: 1211–1217.

Devine, C. E., R. G. Bell, S. Lovatt, B. B. Chrystall, and L. E. Jeremiah. 1996. Red meats. In *Freezing Effects on Food Quality*, L. E. Jeremiah (ed.), pp. 51–84. New York: Marcel Dekker, Inc.

Faustman, C. and R. G. Cassens. 1990. The biochemical basis for discoloration in fresh meat: A review. *Journal of Muscle Foods* 1: 217–243.

Fernández, P. P., L. Otero, M. M. Martino, A. D. Molina-García, and P. D. Sanz. 2008. High-pressure shift freezing: Recrystallization during storage. *European Food Research and Technology* 227: 1367–1377.

Fox, J. B. Jr. 1966. The chemistry of meat pigments. *Journal of Agricultural and Food Chemistry* 14: 207–210.

Gašperlin, L., B. Žlender, and V. Abram. 2001. Colour of beef heated to different temperatures as related to meat ageing. *Meat Science* 59: 23–30.

Gault, N. F. S. 1991. Marinaded meat. In *Developments in Meat Science*, 5th edn., pp. 191–246. London: U.K.: Elsevier Applied Science.

George, P. and D. H. Irvine. 1952. The reaction between metmyoglobin and hydrogen peroxide. *Biochemical Journal* 52: 511–517.

Ghorpade, V. M. and D. P. Cornforth. 1993. Spectra of pigments responsible for pink color in pork roasts cooked to 65 or 82°C. *Journal of Food Science* 58: 51–52, 89.

Gonzalez-Sanguinetti, S., M. C. Añon, and A. Calvelo. 1985. Effect of thawing rate on the exudate production of frozen beef. *Journal of Food Science* 50: 697–700, 706.

Greer, G. G. and A. C. Murray. 1991. Freezing effects on quality, bacteriology and retail-case life of pork. *Journal of Food Science* 56: 891–894, 912.

Grujić, R., L. Petrović, B. Pikula, and L. Amidžić. 1993. Definition of the optimum freezing rate-1. Investigation of structure and ultrastructure of beef *m. longissimus dorsi* frozen at different freezing rates. *Meat Science* 33: 301–318.

Hamm, R. 1960. Biochemistry of meat hydration. *Advances in Food Research* 10: 355–463.

Hearne, L. E., M. P. Penfield, and G. E. Goertz. 1978. Heating effects of bovine semitendinosus: Phase contrast microscopy and scanning electron microscopy. *Journal of Food Science* 43: 13–16.

Hill, F. 1966. The solubility of intramuscular collagen in meat animals of various ages. *Journal of Food Science* 31: 161–166.

Ichinoseki, S., T. Nishiumi, and A. Suzuki. 2006. Tenderizing effect of high hydrostatic pressure on bovine intramuscular connective tissue. *Journal of Food Science* 71: E276–E281.

Jenses, W. K., C. Devine, and M. Dikeman (Eds.). 2004. *Encyclopedia of Meat Sciences*, 3 Vols. Oxford, U.K.: Elsevier Academic Press.

Jeong, J.-Y., H.-S. Yang, S.-H. Moon, G.-B. Park, and S.-T. Joo. 2004. Effect of freeze-thaw process on myoglobin oxidation of pork loin during cold storage. *Proceedings 50th International Congress of Meat Science and Technology*, Helsinki, Finland.

Jiménez-Colmenero, F. and A. J. Borderias. 2003. High-pressure processing of myosystems. Uncertainties in methodology and their consequences for evaluation of results. *European Food Research and Technology* 217: 461–465.

Jones, S. B., R. J. Carroll, and J. R. Cavanaugh. 1977. Structural changes in heated bovine muscle: A scanning electron microscope study. *Journal of Food Science* 42: 125–131.

Joseph, R. L. 1996. Very fast chilling of beef and tenderness—A report from an EU concerted action. *Meat Science* 43: S217–S227.

Judge, M., E. Aberle, J. Forrest, H. Hedrick, and R. Merril. 1989. *Principles of Meat Science*, 2nd edn. Dubuque, IA: Kendall/Hunt Publishing Company (*Note*: Aberle is the first author on more recent editions of this text).

Jung, S., M. de Lamballerie-Anton, and M. Ghoul. 2000a. Modifications of ultrastructure and myofibrillar proteins of post-rigor beef treated by high pressure. *Lebensmittel-Wissenschaft Und-Technologie-Food Science* 33: 313–319.

Jung, S., M. Ghoul, and M. de Lamballerie-Anton. 2000b. Changes in lysosomal enzyme activities and shear values of high pressure treated meat during ageing. *Meat Science* 56: 239–246.

Jung, S., M. Ghoul, and M. de Lamballerie-Anton. 2003. Influence of high pressure on the color and microbial quality of beef meat. *Lebensmittel-Wissenschaft Und-Technologie-Food Science* 36: 625–631.

Knorr, D., V. Heinz, O. Schlüter, and M. Zenker. 1998. The potential and impact of high pressure as unit operation for food processing. In *High Pressure Food Science, Bioscience and Chemistry*, N. S. Isaacs (ed.), pp. 227–235. Cambridge, U.K.: The Royal Society of Chemistry.

Lawrie, R. A. 1998. *Lawrie's Meat Science*, 6th edn. Cambridge, England: Woodhead Publishing Limited.

Ledward, D. A. 1970. Metmyoglobin formation in beef stored in carbon dioxide enriched and oxygen depleted atmospheres. *Journal of Food Science* 35: 33–37.

Lee, E.-J., Y.-J. Kim, N.-H. Lee, S.-I. Hong, and K. Yamamoto. 2007. Differences in properties of myofibrillar proteins from bovine *semitendinosus* muscle after hydrostatic pressure or heat treatment. *Journal of the Science of Food and Agriculture* 87: 40–46.

Lepetit, J. 2007. A theoretical approach of the relationships between collagen content, collagen cross-links and meat tenderness. *Meat Science* 76: 147–159.

Lepetit, J. and J. Culioli. 1994. Mechanical properties of meat. *Meat Science* 36: 203–237.

Lin, T.-S., R. E. Levin, and H. O. Hultin. 1977. Myoglobin oxidation in ground beef: microorganisms and food additives. *Journal of Food Science* 42: 151–154.

Ma, H.-J. and D. A. Ledward. 2004. High pressure/thermal treatment effects on the texture of beef muscle. *Meat Science* 68: 347–355.

MacDougall, D. B. and A. A. Taylor. 1975. Colour retention in fresh meat stored in oxygen—A commercial scale trial. *Journal of Food Technology* 10: 339–347.

Martens, H., E. Stabursvik, and M. Martens. 1982. Texture and colour changes in meat during cooking related to thermal denaturation of muscle proteins. *Journal of Texture Studies* 13: 291–309.

Martino, M. N., L. Otero, P. D. Sanz, and N. E. Zaritzky. 1998. Size and location of ice crystals in pork frozen by high-pressure-assisted freezing as compared to classical methods. *Meat Science* 50: 303–313.

Martino, M. N. and N. E. Zaritzky. 1988. Ice crystal size modifications during frozen beef storage. *Journal of Food Science* 53: 1631–1637, 1649.

Moeller, S., D. Wulf, D. Meeker, M. Ndife, N. Sundararajan, and M. B. Solomon. 1999. Impact of the hydrodyne process on tenderness, microbial load, and sensory characteristics of pork longissimus muscle. *Journal of Animal Science* 77: 2119–2123.

Montero, P. and M. C. Gómez-Guillén. 2005. High-pressure applications on myosystems. In *Novel Food Processing Technologies*, G. V. Barbosa-Cánovas, M. S. Tapia, M. P. Cano, O. Martín-Belloso, and A. Martínez (eds.), pp. 311–342. Boca Raton, FL: CRC Press, Marcel Dekker.

Ngapo, T. M., I. H. Babare, J. Reynolds, and R. F. Mawson. 1999. Freezing and thawing rate effects on drip loss from samples of pork. *Meat Science* 53: 149–158.

Offer, G. and J. Trinick. 1983. On the mechanism of water holding in meat: The swelling and shrinking of myofibrils. *Meat Science* 8: 245–281.

Palka, K. and H. Daun. 1999. Changes in texture, cooking losses, and myofibrillar structure of bovine *M. semitendinosus* during heating. *Meat Science* 51: 237–243.

Park, S. H., H. S. Ryu, G. P. Hong, and S. G. Min. 2006. Physical properties of frozen pork thawed by high pressure assisted thawing process. *Food Science and Technology International* 12: 347–352.

Perry, S. V. 1996. *Molecular Mechanisms in Striated Muscle*. Cambridge, U.K.: Cambridge University Press.

Petrović, L., R. Grujić, and M. Petrović. 1993. Definition of the optimal freezing rate-2. Investigation of the physico-chemical properties of beef *m. longissimus dorsi* frozen at different freezing rates. *Meat Science* 33: 319–331.

Pincock, R. E. and T. E. Kiovsky. 1966. Kinetics of reactions in frozen solutions. *Journal of Chemical Education* 43: 358–360.

Porcine Myology. http://porcine.unl.edu/porcine2005/pages/index.jsp (accessed April 7, 2010).

Renerre, M. 1990. Review: Factors involved in the discoloration of beef meat. *International Journal of Food Science and Technology* 25: 613–630.

Rodríguez, J. J., G. V. Barbosa-Cánovas, G. F. Gutiérrez-López, L. Dorantes-Alvárez, H. W. Yeom, and Q. H. Zhang. 2003. An update on some key alternative food processing technologies: Microwave, pulsed electric field, high hydrostatic pressure, irradiation, and ultrasound. In *Food Science and Food Biotechnology*, G. F. Gutiérrez-López and G. V. Barbosa-Cánovas (eds.), pp. 279–312. Boca Raton, FL: CRC Press.

Rowe, R. W. D. 1974. Collagen fibre arrangement in intramuscular connective tissue. Changes associated with muscle shortening and their possible relevance to raw meat toughness measurements. *Journal of Food Technology* 9: 501–508.

Sanz, P. D. 2005. Freezing and thawing of foods under pressure. In *Novel Food Processing Technologies*, G. V. Barbosa-Cánovas, M. S. Tapia, M. P. Cano, O. Martín-Belloso, and A. Martínez (eds.), pp. 233–260. New York/Boca Raton, FL: Marcel Dekker/CRC Press.

Schilling, M. W., J. R. Claus, N. G. Marriott, M. B. Solomon, W. N. Eigel, and H. Wang. 2002. No effect of hydrodynamic shock wave on protein functionality of beef muscle. *Journal of Food Science* 67: 335–340.

Sebranek, J. G. and T. A. Houser. 2006. Modified atmosphere packaging. In *Advanced Technologies for Meat Processing*, L. M. L. Nollet and F. Toldrá (eds.), pp. 419–447. Boca Raton, FL: CRC Press/Taylor & Francis.

Solomon, M. B., M. N. Liu, J. R. Patel, B. C. Bowker, and M. Sharma. 2006. Hydrodynamic pressure processing to improve meat quality and safety. In *Advanced Technologies for Meat Processing*, L. M. L. Nollet and F. Toldrá (eds.), pp. 219–244. Boca Raton, FL: CRC Press/ Taylor & Francis.

Solomon, M. B., M. N. Liu, J. Patel, E. Paroczay, J. Eastridge, and S. W. Coleman. 2008. Tenderness improvement in fresh or frozen/thawed beef steaks treated with hydrodynamic pressure processing. *Journal of Muscle Foods* 19: 98–109.

Solomon, M. B., J. B. Long, and J. S. Eastridge. 1997. The hydrodyne: A new process to improve beef tenderness. *Journal of Animal Science* 75: 1534–1537.

Stryer, L. 1981. *Biochemistry*, 2nd edn. San Francisco, CA: W. H. Freeman and Company.

Suman, S. P., R. A. Mancini, and C. Faustman. 2006. Lipid-oxidation-induced carboxymyoglobin oxidation. *Journal of Agricultural and Food Chemistry* 54: 9248–9253.

Suzuki, A., K. Kim, H. Tanji, T. Nishiumi, and Y. Ikeuchi. 2006. Application of high hydrostatic pressure to meat and meat processing. In *Advanced Technologies for Meat Processing*, L. M. L. Nollet and F. Toldrá (eds.), pp. 193–217. Boca Raton, FL: CRC Press/Taylor & Francis Group.

Suzuki, A., M. Watanabe, and Y. Ikeuchi. 1993. Effects of high-pressure treatment on the ultrastructure and thermal behaviour of beef intramuscular collagen. *Meat Science* 35: 17–25.

Taylor, A. A. and D. B. MacDougall. 1973. Fresh beef packed in mixtures of oxygen and carbon dioxide. *Journal of Food Technology* 8: 453–461.

Tischer, R. G., H. Hurwicz, and J. A. Zoellner. 1953. Heat processing of beef. III. Objective measurement of changes in tenderness, drained juice, and sterilizing value during heat processing. *Journal of Food Science* 18: 539–554.

Torrescano, G., A. Sánchez-Escalante, B. Giménez, P. Roncalés, and J. A. Beltrán. 2003. Shear values of raw samples of 14 bovine muscles and their relation to muscle collagen characteristics. *Meat Science* 64: 85–91.

Trout, G. R. 1989. Variation in myoglobin denaturation and color of cooked beef, pork, and turkey meat as influenced by pH, sodium chloride, sodium tripolyphosphate, and cooking temperature. *Journal of Food Science* 54: 536–540, 544.

Tuomy, J. M., R. J. Lechnir, and T. Miller. 1963. Effect of cooking temperature and time on the tenderness of beef. *Food Technology* 17: 1457–1460.

Voutila, L., A. M. Mullen, M. Ruusunen, D. Troy, and E. Poulanne. 2007. Thermal stability of connective tissue from porcine muscles. *Meat Science* 76: 474–480.

Walsh, H. M. and J. P. Kerry. 2002. Meat packaging. In *Meat Processing. Improving Quality*, J. Kerry, J. Kerry, and D. Ledward (eds.), pp. 417–451. Boca Raton, FL/Cambridge, England: CRC Press/Woodhead Publishing Limited.

Warriss, P. D. 2000. *Meat Science: An Introductory Text*. Wallingford, U.K.: CABI Pub.

Zakrys, P. I., S. A. Hogan, M. G. O'Sullivan, P. Allen, and J. P. Kerry. 2008. Effects of oxygen concentration on the sensory evaluation and quality indicators of beef muscle packed under modified atmosphere. *Meat Science* 79: 648–655.

Zuckerman, H. and M. B. Solomon. 1998. Ultrastructural changes in bovine longissimus muscle caused by the hydrodyne process. *Journal of Muscle Foods* 9: 419–426.

11 Pet Foods and Their Physicochemical Properties as Affected by Processing

Chalida Niamnuy and Sakamon Devahastin

CONTENTS

11.1 INTRODUCTION

Pet foods are special foods that are formulated to meet nutritional needs of animals. Pet foods generally consist of meats, meat by-products, cereal grains, vitamins, and minerals. Pet foods are mainly available in three basic forms, namely, dry pet foods, semimoist pet foods, and moist pet foods. The first commercial pet food was dry food for dogs, which was developed in England in 1860 (Corbin, 2003). Moist pet foods made their appearance in canned form in 1922, while semimoist pet foods were later developed in the United States (Corbin, 2003). Demands for dry pet foods are nevertheless higher than those for moist and semimoist pet foods.

The conventional manufacturing processes of pet foods are similar to those of human foods. The exact ingredients and processes depend on the types of pet foods, however. Physicochemical properties of pet foods, which are related to the quality and acceptability of the products by both pets and pet owners, are largely affected by ingredients, formulas, and processes employed to manufacture the products. Understanding the effects of various factors on the physicochemical properties of pet foods can thus be a way to enhance the quality and value of these products.

11.2 DRY PET FOODS

The Association of American Feed Control Officials (AAFCO) defines the category of pet foods based on their moisture content. Conventional dry pet foods are characterized as having a moisture content of 15% by weight or less, with a typical value of around 8%–10% by weight (Dzanis, 2003). These products are the most microbiologically stable and require no special storage conditions or packages during distribution. Dry pet foods are generally high in starch (e.g., corn, wheat, and soybean) compared with other types of pet foods. Dry pet foods have low energy content, have relatively poor palatability, are characterized as having hard and brittle structures, and have only a nominal resemblance to meats.

11.2.1 Production of Dry Pet Foods

Typical quantitative ranges of dry pet food ingredients are 24%–50% for amylaceous (starchy) products, 5%–25% for animal proteins, 10%–30% for vegetable proteins, 5%–15% for fats, 1%–5% for supplements (e.g., vitamins, minerals, and flavors), and 5%–15% for water by weight. Balaz et al. (1977), for example, proposed dry pet foods that were composed of amylaceous ingredients such as cereal grains, flours, and starches; fats; sugar; proteinaceous adhesives; and plasticizing agents. Since amylaceous ingredients (including corn, oat, wheat, barley, rice, by-products of these cereal grains, and other edible grains or tuberous starches and modified starches) are the main ingredients of dry pet foods, many research studies have been conducted on the preparation and/or processing of amylaceous ingredients for pet foods; some of the works are summarized in Table 11.1.

Various protein ingredients are utilized to satisfy nutritional requirements for protein quantity and quality in pet foods. Typical sources of protein ingredients are soybean oil meal, soybean flour, soy protein concentrate, soy protein isolate, meat meal, meat and bone meal, meat by-products, fish meal, blood meal, dried blood plasma, yeast, and milk protein (Balaz et al., 1977). In addition, vitamins, minerals, colorants, and other supplements such as choline chloride; MgO; vitamins A, B_{12},

TABLE 11.1
Effects of Process Parameters on Selected Quality Attributes of Amylaceous Ingredients

Source of Ingredient	Processing Step	Quality Attribute	References
Pearl millet flour	Fermentation	Phytate and polyphenolic content, digestibility of starch and proteins	Kheterpaul and Chauhan (1991)
Corn, wheat, sorghum	Grinding	Texture interactions	Nir et al. (1994)
Sorghum and pearl millet	Parboiling, decortication	Nutritional value	Serna-Saldivar et al. (1994)
Soybean	Heating	Bioavailability and digestibility of lysine	Fernandez and Parsons (1996)
Peanut	Autoclaving	Digestibility of amino acids, protein solubility	Zhang and Parson (1996)
Dry corn fractions	Cooking	Distribution of fumonisin B_1, distribution of aflatoxins and zearalenone	Brera et al. (2004, 2006)
Soy residue (okara)	Drying	Color, oxidation level, rehydration ability, urease activity, protein solubility	Wachiraphansakul and Devahastin (2007)

TABLE 11.2
Sample Formula for a Dry Pet Food

Ingredient	Percent by Weight
Modified soy protein isolate	19.96
Defatted soy flour	4.96
Meat and bone meal	4.96
Sodium caseinate	4.96
Sucrose	28.80
Beef trimmings	4.96
Sorbitol	9.93
Prime steam lard	1.99
Dicalcium phosphate dehydrate	4.86
Potassium chloride	0.10
Potassium sorbate	0.09
Trace minerals/vitamins	0.60
Coloring agent	
Red No. 40	0.02
Violet aluminum lake No. 1	0.0022
Water	12.91

Source: Balaz, A. et al., Method of making a dry-type pet food, U.S. Patent 4,055,681, October 25, 1977.

D_3, and E; riboflavin; niacin; folic acid; and pyridoxine HCl may also be added to enhance the product properties. Such supplements are commonly fortified to about 5% by weight of the final product. A typical example of dry pet food formula is given in Table 11.2. There are several means by which dry pet foods can be prepared; these include baking, pelletizing, and extrusion, which is the most popular. However, prior to these final steps, several other upstream processing steps are necessary and need to be performed with care as well.

The first step for the production of dry pet foods is raw materials preparation. This starts with obtaining dry and liquid ingredients and is followed by grinding, batching, and mixing of the dry ingredients (see Figure 11.1). Grinding is necessary in order to increase the availability of nutrients as well as to improve the ease in which the dry ingredients are processed; most dry mixes are ground to coarse flour prior to further processing. A hammer mill is often used for this purpose.

Further processing steps for dry pet foods' production are shown in Figure 11.2. The ground dry ingredients are mixed with liquid ingredients, which are typically premixed with hot water for partial melting of the present fat. The liquid mixes are poured onto the dry components in a mixer, which is generally called a preconditioner or conditioner, where the ingredients are hydrated and cooked. During preconditioning, starch in the ingredients may be cooked by about 40%–50% (Mair, 2003a).

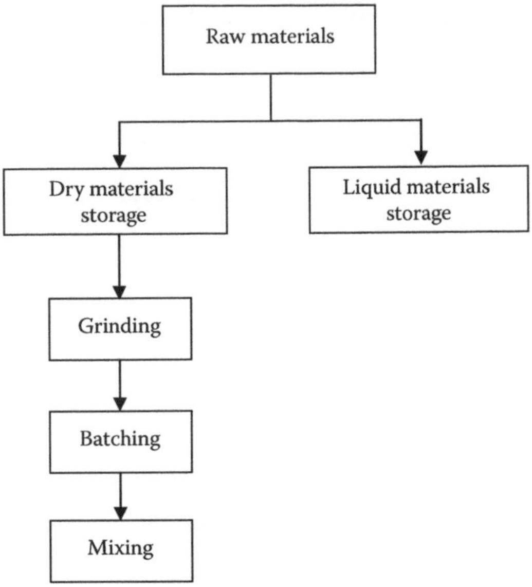

FIGURE 11.1 Raw materials preparation process.

The preconditioned mix is then fed into an extruder. Extrusion is defined as a process of forcing a feed material to flow through a die, which is designed to form and/or puff a dry ingredient (Rossen and Miller, 1973). Extrusion combines several unit operations including mixing, cooking, kneading, shearing, shaping, and forming. A typical extruder consists of a cylindrical multi-segmented barrel with a screw that propels, mixes, and further cooks the material, and then forces it through a die where it is cut to the desired length by a knife. During product movement through an extruder, heat is produced by friction, which then cooks the product. The speed and friction levels must be properly set to ensure that the product is cooked at the right temperature for the right period of time. The temperature within an extruder ranges from 112°C to 132°C, while the temperature of the extrudate leaving the die is in the range of 120°C–127°C. The extrudate is immediately cooled and cut into small-sized pieces to prevent the formed fibers from coalescing. Dry pet foods typically leave an extruder with the moisture content in the range of 20%–30% (w.b.). Many types of extruders are available in the marketplace of pet foods' production. Most commonly, extruders used for pet foods production are single-screw and twin-screw extruders.

To prevent quality deterioration and microbial growth, extruded pet foods need to be dried. Most pet foods are dried in forced-air convection conveyor dryers; however, some pet food producers may use vertical dryers or other dryer configurations. Extruded pet foods are spread onto a conveyor belt, and hot air is forced through the conveyor. Pet food dryers commonly have at least two or three levels of conveyors. The multiple-pass conveyors can increase the product residence time in the dryers and, at the same time, reduce the floor space required for drying. Pet food dryers typically operate at an air temperature in the range of 100°C–180°C (Poirier, 2003); if pet foods are dried too quickly or at a very high temperature, they would become

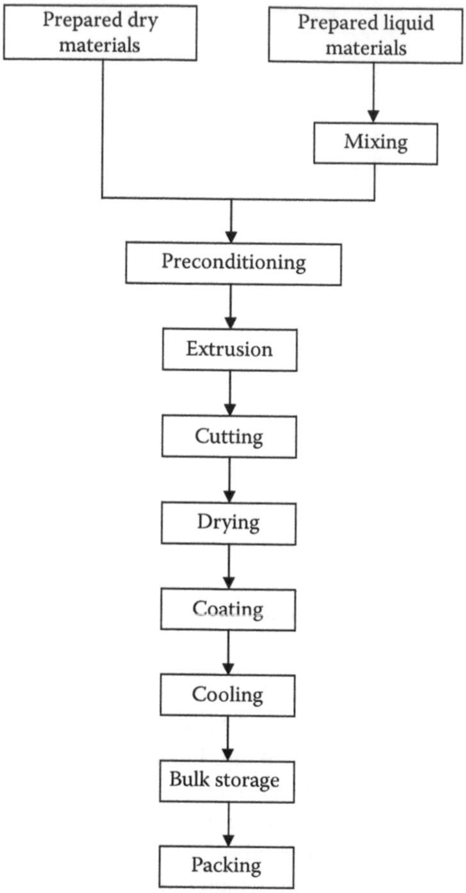

FIGURE 11.2 Dry pet food production process.

fragile and easily break during handling. Possible heat sources of dryers are gas burners, steam coils, and electric heaters, but the most common one is burners using a gaseous fuel, such as natural gas.

After drying, pet foods are cooled down to within 5°C–10°C of ambient temperature before being sent for storage or packaging. If pet foods are not properly cooled, condensation can occur inside the packages of finished products, causing growth of mold and bacteria. Generally, cooling is performed in the cooling zone, which is located toward the end of a dryer. However, cooling is sometimes separated from the dryer, since coating and/or enrobing of the products may need to be done at higher temperatures to improve penetration of the coatings into the products. The coating step can be done by adding various liquids or powders, including liquid animal digests, fats, and tallow; and dried coatings, such as dried yeast, onto the outer surface of the products (Ernst et al., 2002; Fritz-Jung et al., 2001). Coatings can improve shape, appearance, and palatability of dry pet foods. When the cooling operation is separated from a drying unit, a vertical cooler or a conveyor cooler is commonly used (Poirier, 2003).

Conventional dry pet foods are typically hard and brittle and have poor palatability to the point where it is unappetizing to both pets and pet owners. Hence, new types of dry pet foods are constantly being developed and marketed.

11.2.1.1 Soft Dry Pet Foods

Soft dry pet foods are those pet foods that are softer than conventional dry pet foods. Bone (1977) formed a soft dry pet food from proteinaceous adhesive; a polyhydric plasticizing agent, which had both plasticizing and humectant functions; a starch-derived polysaccharide; various additives such as an emulsifying agent, which was used to provide an emulsion form of material, and a texturizing agent, which was used for developing a soft texture; and a microorganism inhibitor. A starch-derived polysaccharide is an important ingredient because it provides nutritional calories and functionally aids the formation of a desired soft texture. This ingredient is generally in soft dry pet foods in the range of 20%–40%.

The production process of soft dry pet foods starts with the preparation of a mixture of a proteinaceous adhesive and water, which is then heated to 100°C. After this, the mixture is blended with various ingredients at room temperature, during which time the dough develops into a cohesive, elastic, meat-like-textured mass. The completely developed dough is then passed through an extruder. In this case, the friction-induced heat is controlled (via a control of throughput or by cooling), so that the temperature is kept below 55°C. The extrudate is then cut into bite-sized pieces, placed on perforated trays, and dried to less than 15% moisture content by forced air at a temperature below 55°C, and then coated with 1% by weight corn oil. Generally, the water activity of a soft dry pet food is in the range of 0.60–0.75. More preferably, the water activity is 0.65–0.75. The relative hardness of this pet food is in the range of 1.5–3.5.

11.2.1.2 Marbled Meat-Like-Textured Dry Pet Foods

A type of dry pet food that has a soft, elastic, meat-like texture and is high in palatability is marbled meat-like-textured dry pet food. Bone and Shannon (1977b) formulated a marbled meat-like-textured dry pet food that contained various ingredients, namely, amylaceous ingredients, coloring additives, proteinaceous adhesives, and plasticizing agents. The use of amylaceous ingredients was limited to 25% by weight or less. Generally, the product can be produced by blending a mixture of the above ingredients and a sufficient amount of water to form moist dough. The dough is then passed into an extruder and heated to a temperature in the range of 70°C–150°C, which is a sufficient temperature to develop a meat-like texture and appearance. The extrudate is cooled on an air- or water-cooled conveyor. The product is then cut into bite-sized pieces. This type of pet food has a soft, marbled meat appearance, which is firm, non-sticky, and temperature stable.

11.2.1.3 Mixed Soft and Hard Dry Pet Foods

Although soft dry pet foods have a highly palatable soft texture, they do not provide the desirable teeth-cleaning characteristics of a conventional dry pet food. Bone and Shannon (1977a) thus developed a dry pet food, which contained both a soft dry pet food component and a hard dry pet food component; the product had both a soft meat-like texture and teeth-cleaning characteristics. The hard component consisted of vegetable protein sources; amylaceous ingredients such as cereal grains or starch; fats; and animal protein sources. All ingredients were blended and mixed with water in a continuous extrusion mixer, during which time the dough developed. The extrudate was passed through sheeting rolls to form a sheet. The sheet of dough was then transported on a belt through a baking oven containing heating zones at a temperature in the range of 230°C–290°C. The baked product was then passed through kibbling rolls. The moisture content of this hard component was set about 12% by weight.

On the other hand, the soft component consisted of sugar, a proteinaceous adhesive, animal protein sources, vegetable protein sources, fat, and a plasticizing agent. The various ingredients were mixed and blended with water. Then, the mixture was fed into a screw extruder. After leaving the extruder, the cooked extrudate was discharged into an air-cooled conveyor. The moisture content of the soft component was about 11%–14% by weight. The soft component was tumbled with a hard component at a ratio of 40–60 before being packaged.

11.2.1.4 Yieldable Elastic Dry Pet Foods

One of the special dry pet foods is yieldable elastic dry pet food. Brown and Karrasch (1979) discovered that the utilization of monosaccharides, such as sucrose, in a dry pet food would result in a product having a unique, soft, yieldingly elastic structure, which would continue despite plastic deformation. This type of product includes at least 50% of cereal grains, 15%–35% of protein sources, 10%–15% moisture content, and 10%–15% of a monosaccharide. The various ingredients are mixed with water at a temperature below the boiling point; the mixture is then introduced into an extruder with enough pressure, leading to a solution of sugar and water to permeate the meal ingredients at high temperature. The product is then emitted from the extruder at a temperature higher than 100°C; a portion of the moisture flashes off and a puffed product, whose moisture content is about 20%, is obtained. The product is cut into small-sized pieces and then dried at lower temperature and under controlled humidity to prevent case hardening. The moisture content of the product is about 15% or less, but not below 10%.

11.2.2 Some Physicochemical Properties of Dry Pet Foods as Affected by Processing

11.2.2.1 Gelatinization Behavior

Various properties of dry pet foods, such as texture and digestibility, are largely affected by gelatinization of the dough (Holm et al., 1985). Extrusion cooking is an important process for the gelatinization of starches. The degree of starch gelatinization of extrudate mainly depends on extrusion conditions, such as the initial moisture content of materials, screw speed, and heating temperature. Water is generally added to the ingredients to provide enough moisture content for gelatinization (Henderson et al., 1981). However, excessive initial moisture content may reduce the degree of starch gelatinization of the extrudate. The heating level and retention time in an extruder are also important factors affecting the gelatinization of starches.

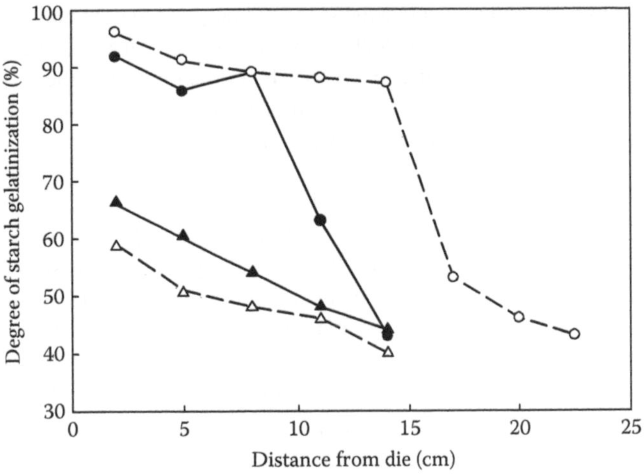

FIGURE 11.3 Degree of starch gelatinization as a function of distance from the extruder die. (●) 25 g/kg fat, 200 rpm; (○) 25 g/kg fat, 400 rpm; (▲) 75 g/kg fat, 200 rpm; (△) 75 g/kg, 400 rpm. (From Lin, S. et al., *LWT-Food Sci. Technol.*, 30, 754, 1997. With permission.)

The temperature has to be above the gelatinization temperature of materials and must be maintained for a suitable retention time. Heat is introduced into the feed dough during its passage through the extruder by one or more of the following three mechanisms: (1) viscous dissipation of mechanical energy being added to the shaft of the screw, (2) heat transfer from steam or electric heaters surrounding the barrel, and (3) direct injection of steam, which is then mixed with the dough in the screw. A temperature higher than 120°C and a moisture content of dough of 20%–30% (or even lower, namely, 10%–20%) generally provide high enough shear and heat that lead to complete gelatinization of the dough (Harper, 1979; Cheftel, 1986).

Lin et al. (1997) studied the effect of fat content and processing conditions on the degree of starch gelatinization of extruded matter. Figure 11.3 shows the degree of starch gelatinization as a function of distance from the die of an extruder at various levels of fat addition and screw speed. The degree of starch gelatinization increased gradually along the cooking zone. In a low-fat-content treatment, increasing the screw speed increased the length of the cooking zone, while an opposite result was observed when using a higher fat content. The lubricating effect of lipid at a higher fat content reduced the friction generated at a higher screw speed, and therefore resulted in a lower degree of starch gelatinization.

11.2.2.2 Texture

Texture is an important quality influencing the acceptability of a pet food product. The texture of dry pet foods is largely affected by ingredient compositions, especially amylaceous and/or proteinaceous ingredients, which are commonly used to prepare conventional dry pet foods (Thomas et al., 1998). The starch and/or proteinaceous continuous phase, being in a crystalline and/or a glassy state of matter, is produced by the action of hydration–dehydration during thermal processing at temperatures in excess of the controlled temperature required to hydrate starch and/or proteinaceous sources, and as a consequence of subsequent drying and cooling as well (Bone, 1977). In addition to the effect of hydration–dehydration, the shear, which acts on the ingredients during extrusion, also affects the texture of the product. A high shear rate tends to produce long-molecule components, giving rise to cross-linking or restructuring of the extruded product (Harper, 1979).

11.2.2.3 Density

Bulk density is an important property for the commercial implication of dry pet foods, which are typically sold by weight. High-bulk-density pet foods can be stored in small packages; hence, packaging and storage costs could be reduced. Thus, there is a need for relatively high-density pet food products that are accepted by animals. However, high-bulk-density products would become too hard if they are produced without the addition of fat, emulsifier, and other additives such as gums and hydrocolloids.

Enzymes, namely, amylase, are used to produce dry pet foods, which possess high bulk density and a soft texture. The function of amylase is breaking down starches into sugars, leading to materials that do not expand upon extrusion. As a result, high-bulk-density dry pet foods are produced. The cooking temperature and pH of the materials during processing must be controlled, because these parameters may affect the activity of amylase; the cooking temperature should normally be in the range of 93°C–110°C to produce high-bulk-density dry pet foods (Parthasarathy, 2004). A consistent bulk density also indicates that operations of the extruder, the dryer, and other units in the process are well controlled.

11.2.2.4 Palatability

Palatability is a combination of many qualities, such as taste, aroma, mouth feel, and ingestive behavior of pet foods. Raw materials and the formula have a significant impact on palatability; hence, producers should be careful to select ingredients and recipe in order to produce high-quality and high-yield dry pet foods. Fats, which are one of the main ingredients, should be low in free fatty acids and peroxide contents to prevent or reduce rancidity, which would lead to a dramatic loss in the palatability of pet foods. Various antioxidants could be added to prevent oxidation reactions in pet foods.

A suitable preconditioning process can improve the palatability of dry pet foods. Heating a feed mixture to 95°C prior to feeding it into an extruder can improve the palatability of a product (Trivedi and Benning, 2003). High-shear extrusion and a suitable moisture content of dough give perfectly formed pieces and constant density, leading to a high-palatability product. Drying also significantly affects the palatability of dry pet foods. An appropriate drying temperature and drying time can prevent case hardening on the outer surface of a product without excessive drying of the interior of the product; generally, the final moisture content of a product should be less than 9% by weight (Trivedi and Benning, 2003). In addition, palatability of dry pet foods can be increased by coating the surface of a product by liquid or dry palatability enhancers, such as phosphoric, citric, and hexamic acids.

Since dry pet foods are the most popular pet foods for various animals, several researchers have reported studies on the effects of various ingredients and process parameters on the qualities of dry pet foods, some of which are summarized in Table 11.3.

TABLE 11.3
Effects of Process Parameters on Various Qualities of Dry Pet Foods

Unit Operation	Processing Parameter	Quality Attribute	References
Extruder	Extrusion temperature, retention time	Bioavailability, level of chemical modifications of proteins and starches	Cheftel (1986)
Extruder	Screw speed, diameter of dies, lipid content	Puff ratio, bulk density, shear strength, water-holding capacity, texture	Bhattacharya and Hanna (1988), Ilo et al. (2000)
Extruder	Extrusion temperature	Oil stability	Rao and Artz (1989)
Extruder	Retention time, pressure, thermal and mechanical energy input	Expansion ability, texture	Kokini (1991)
Extruder	Fat types and contents, moisture content, screw speed	Degree of gelatinization	Lin et al. (1997)
Conditioner	Steam pressure, paddle configuration, retention time	Pellet durability, hardness	Thomas et al. (1998), Briggs et al. (1999)
Grinder, extruder	Grinder type, particle size	Volumetric expansion index	Mathew et al. (1999a)
Conditioner, extruder	Corn moisture, preconditioning retention time	Expansion behavior	Mathew et al. (1999b)
Pellet machine	Temperature, humidity, pellet size	Lysine bioavailability	Mavromichalis and Baker (2000)
Conditioner	Moisture content, retention time, steam quality	Pellet durability	Gilpin et al. (2002)
Mixer	Moisture content, ingredient compositions	Pellet durability, pellet mill energy usage	Moritz et al. (2002), Cavalcanti and Behnke (2005)
Extruder	Extrusion conditions	Chemical compositions and digestibility of grains	Dust et al. (2004)
Grinder, pellet machine	Particle size	Degree of gelatinization, pellet strength	Svihus et al. (2004)
Extruder	Sources of proteins	Expansion ability, density, sinking rate, fat leakage, durability	Overland et al. (2007)
Pellet machine	Particle size, ingredient composition, temperature, retention time	Water adsorption ability, rehydration ability	Hemmingsen et al. (2008)
Extruder	Extrusion temperature	Palatability, digestibility, destruction level of undesirable, nutritionally active factors	Tran et al. (2008)

11.3 SEMIMOIST PET FOODS

Semimoist pet foods are those pet foods that have moisture content in the range of 15%–50% by weight (Priegnitz, 1980; Parthasarathy, 2004). Semimoist pet foods represent more expensive products with more meat-like characteristics and better appearances as well as improved palatability as compared to dry pet foods. Semimoist pet foods tend to maximize the advantages of both dry and moist pet foods, while minimizing the disadvantages of both products.

The main ingredients of semimoist pet foods are meat or meat by-products, or a combination of meat and vegetable proteins. Typical vegetables used for the production of semimoist pet foods include peanut grits, peanut flour, cottonseed protein, and other proteins. Soy protein is one of the preferred vegetable proteins for the production of semimoist pet foods because it has a high protein content and can improve the quality of semimoist pet foods. Soy flakes provide coarseness, which allows a firm product, while soy flour provides a water-binding function. Since semimoist pet foods are high in moisture content than dry pet foods, special technology is required to ensure their microbiological stability without refrigeration. Humectants and bacteriological stabilizers are recommended as major components that contribute to the stability of semimoist pet foods (Foulkes, 1977).

11.3.1 Production of Semimoist Pet Foods

Generally, the ingredients of semimoist pet foods include meat and meat by-products, vegetable proteins, stabilizers, salt, water binder, fat, and some nutritional additives. The meat content typically ranges from 16% to about 45% due to palatability maximization and cost minimization at this meat content (Foulkes, 1977). Meat in this case could mean the flesh of cattle, swine, sheep, goat, horse, poultry, fish, or by-products of these animal meats. The vegetable proteins' content is generally in the range of 15%–45% by weight. These ingredients may include peanut grits, peanut flour, cottonseed protein, and soy protein.

Stabilizers consist of at least three components, namely, a polyhydric component, food grade acids, and an antimycotic agent. Of the polyhydric components, propylene glycol is preferred at a concentration of about 2% to about 8% of a product. Food grade acids, such as phosphoric, acetic, and citric acids, are added up to about 1% of a product to reduce its pH to 4.0–5.0. Up to about 0.5% of an antimycotic agent is included. Among the most effective antimycotic agents is potassium sorbate; this antimycotic agent also provides the potassium nutrient value.

Various salts such as dicalcium phosphate, sodium chloride, or potassium chloride could also be added up to 5% of a product to act as both a water binder and a buffered stabilizer. A suitable fat, such as prime steam lard, tallow, or white grease, is also generally included up to 10% of a product.

Low-molecular-weight sugars, such as dextrose, sucrose, glucose, and corn syrup, are also an important ingredient that acts as a humectant, which controls the water activity of semimoist pet foods. The total sugar content of semimoist pet foods is generally in the range of 6% to about 30% by weight. Corn molasses may be

TABLE 11.4
Sample Formula for a Semimoist Pet Food

Ingredient	Percent by Weight
Beef tripe	30
Soybean flour	26
Wheat flour	13.85
Corn syrup	10
Propylene glycol	6
Phosphoric acid	1.5
Salt, vitamins, flavor	6
Mono glyceride emulsifier	0.15
Water	6.5
Total	100
Moisture content after extrusion	35

Source: Priegnitz, R.D., Semi-moist pet food product and process, U.S. Patent 4,228,195. October 14, 1980.

used as a partial replacement for sugar; these ingredients allow improvement of the product palatability upon storage. Various vitamins and minerals are also included up to 0.5% (Foulkes, 1977). A sample formula for a semimoist pet food is given in Table 11.4.

The production process of semimoist pet foods is similar to that of dry pet foods. However, there are some major differences, as shown in Figure 11.4. Semimoist pet foods require less energy inputs for processing because of their unexpanded nature (bulk density of greater than 480 g/L). Some semimoist pet foods are low-temperature formed with little or no cooking steps involved; ingredients such as humectants are selected to control the water activity.

Small-sized ingredients such as cereal grain flours (sizes smaller than 60 mesh) are required for the extrusion-forming process. Dry ingredients, such as cereal flours, and liquid ingredients are put into a preconditioner. They are then mixed, heated, and hydrated via hot water (60°C) or steam injection. A rigorous mixing between ingredients and hot water helps extend the retention time in the preconditioner to allow the moisture content to be at an optimum level, hence a reduction in the power requirement for the subsequent extrusion step. Semimoist pet food recipes contain large amounts of liquid additives, so good mixing in the preconditioner is critical for product uniformity.

After mixing, preconditioned materials are discharged into an extruder. The final product characteristics are affected by screw and barrel profiles, screw speed, processing conditions (temperature, moisture content, etc.), raw material properties, and die/knife type. A single-screw extruder is used in 85%–90% of all semimoist pet food production. All equipment requires stainless steel construction due to low pH (4.0–4.5) of most formulas. The extrusion temperature is maintained

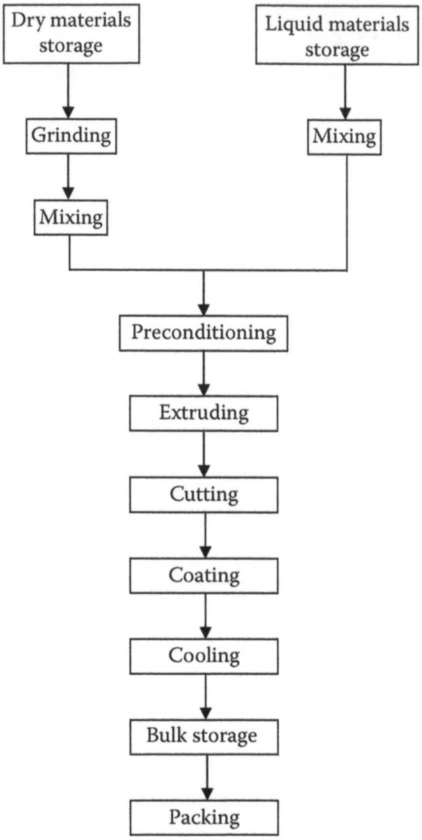

FIGURE 11.4 Semimoist pet food production process.

considerably below 100°C (Priegnitz, 1980). A normal extruder is usually designed for cooling the final product in the last portion of the extruder barrel (prior to the final die) to form minimum expanded products as semimoist pet foods before cutting.

Drying is not required after extrusion for semimoist pet foods. Instead, extruded semimoist pet foods are introduced to low-agitation coating drums, where water, humectants, and acids are added. After leaving the coating drums, the products are conveyed to a cooler by belts or pneumatic systems to set the structure, so the products would maintain their high moisture content and porous structure; semimoist products are transported to the cooler at a moisture content of 20%–28% by weight and a temperature of 85°C–95°C. The cooling process must cool down the products to within 10°C of the ambient temperature without excessive moisture loss; excessive moisture loss during cooling would lead to rigid layer formation on the surface of the products, which could in turn affect the product appearance and texture (Rokey, 2003). Most semimoist pet foods do not require a post-extrusion enrobing step. However, if an enrobing step is required, the same unit employed for coating dry pet foods could be used.

11.3.1.1 Expanded Semimoist Pet Foods

Expanded semimoist pet foods are alternative semimoist pet foods with enhanced appearance and palatability. Normally, the expansion of semimoist pet foods is difficult due to the high moisture content and predominance of non-expandable materials, such as meat, sugar, and polyhydric alcohols. On the other hand, sugar is needed to provide stability to semimoist pet foods. Thus, it is generally difficult to obtain the desired expansion, while achieving the stability of semimoist pet foods.

Bartsch (1977) formulated sugarless, expanded semimoist pet foods that were microbiologically stable through the adjustment of pH and the use of an antimycotic agent in combination with various ingredients, which helped expand and maintain the expanded structure. The preferred pH was in the range of 4.6–5.6.

Generally, expandable semimoist pet foods are composed of protein sources; amylaceous ingredients, such as cereals; polyhydric components; animal or vegetable fats; food grade acids; antimycotic agents; and other supplements, such as vitamins and minerals. Water is added to maintain the desired moisture content. The various ingredients are mixed to form a dough, which is then transported into an extruder and heated to about 93°C–127°C (Bartsch, 1977). The degree of expansion is commonly associated with the extrusion temperature. As the product exits from the extruder, it is passed through an appropriate die; some steam flashes off as the product expands. It is then cut into the desired size. The product is then cooled and packaged.

11.3.1.2 Meatless Marbled Semimoist Pet Foods

Due to insufficiency of meat throughout the world, it is sometimes desirable to avoid the use of meat as an ingredient of pet foods. However, the elimination of meat in pet food formula has led to a product that lacks palatability and is often hard and brittle. Burkwall et al. (1976) developed a meat-like semimoist pet food with a high level of palatability and acceptability even though meat was not used. The product was produced from two types of doughs. The first dough contained the first coloring agent, especially red and red-brown color, so that a meat-like appearance could be obtained. The second dough contained the second coloring agent, preferably in white as a fat color. Each dough consisted of 1%–30% of vegetable protein; 0.5%–20% of amylaceous ingredients; 10%–45% of a water-soluble solute that was capable of raising the osmotic pressure of water, such as sugar and polyhydric alcohols; and 5%–30% of casein salt. An antimycotic agent was also added to prevent the possible activity of yeast and mold. Each dough ingredient was mixed in a blender and was then introduced into an extruder. The temperature in the extruder was in the range of 85°C–150°C. A suitable pressure was used to prevent substantial expansion of the dough within the extruder. The final product contained 70%–90% of the red dough and 10%–30% of the white dough. Therefore, the extrusion rates of both doughs must be adjusted accordingly. Upon cooling, a marbled, meat-like semimoist pet food was obtained; the product was noted to be highly palatable.

11.3.2 Some Physicochemical Properties of Semimoist Pet Foods as Affected by Processing

11.3.2.1 Water Activity

The water activity of semimoist pet foods should preferably be 0.65 or lower to prevent mold and microbiological growth. An important ingredient, namely, humectant, is typically used to lower the product water activity to increase the shelf life of the products. Some examples of humectants commonly used in the production of semimoist pet foods are glycerin, corn syrups, maltodextrins, sugars, salts, propylene glycol, and honey. In addition, some humectants also exhibit other desirable actions such as antimicrobial properties, and texturizing and sweetening capacities. Humectants are conventionally added at a level of 10%–18% to control the water activity below 0.65 (Rokey, 2003). Humectants are usually added in the preconditioning stage before extrusion. If humectants are needed at a higher level (>12% by weight), a portion of them might need to be added in an extruder barrel to avoid the formation of a sticky dough.

Some alternative ingredients, such as low-molecular-weight polypeptides, could also help reduce the water activity of semimoist pet foods. Preferred sources of low-molecular-weight polypeptides are autolysates of protein sources such as fish and meat. The materials are acidified with lactic acid to a pH below 4.5 prior to autolysis by storing at 20°C for 5 days and concentrating the filtered autolyzed materials to about 50% by weight solid. Adding low-molecular-weight polypeptides can also help improve the performance of conventional humectants (Barker et al., 1977).

11.3.2.2 Color

Color is an important property for pet foods. Semimoist pet foods offer a meat-like appearance, to which the red color contributes greatly. The stabilization of product color is sometimes problematic in semimoist pet food production. A reaction of animal protein sources with carbon monoxide, which causes the formation of color-stabilized protein sources, is used to generate color-stabilized semimoist pet foods with a heat-stable bright red color (Hood, 1978).

The key variables influencing the color of semimoist pet foods are (1) the extent of saturation of carbon-monoxide-bonding sites, (2) the oxidation of iron in the meat protein, (3) the final concentration of meat or blood protein, and (4) the type of heat (dry or moist) and its process, which maintains the color of the carbon-monoxide-treated materials. The carbon monoxide treatment is indeed necessary to stabilize the color of the products (Hood, 1977).

The carbon monoxide treatment is provided by grounding and emulsifying ingredients to form a slurry, which is then placed in a sealable vessel. The vessel has two gas valves, one for the entrance and another for the exit of air and reaction gases. It must be ensured that the space above the slurry in the vessel is totally occupied with carbon monoxide and agitated at high speed. The agitation time of 30 min is adequate for the complete reaction of carbon monoxide with the myoglobin binding site. The slurry is then used as the meat ingredients for semimoist pet food production (Hood, 1978).

11.3.2.3 Coherence

To enhance the coherent attribute of semimoist pet foods, it is preferred to use coagulable binders, such as gluten, soluble meat protein, protein extracts, egg albumin, and starches. These binders may be coagulated by conventional heating techniques, such as baking, drying, and frying, after extrusion. The coagulable binder solids are included at 0%–30% of all ingredients.

In a typical production process, dried protein sources such as fish meal; carbohydrate materials such as cereal flour; heat coagulable binders such as gluten; and an antimycotic agent are mixed as dry ingredients. A slurry, which may be made of fish soluble and emulsion of muscle meat, is added to convert the mix into a stiff dough, which is then extruded and cut into pieces and dried to coagulate the proteins and to lower the moisture content (Priegnitz, 1980).

11.3.2.4 Palatability

Flavor greatly affects the palatability of semimoist pet foods. Some ingredients that are used to lower the water activity of the products also affect the flavor and palatability of semimoist pet foods. Such ingredients include sugars, soluble salts, acids, and humectants. Of these, sugars positively affect the palatability of pet foods, since sweetness is a positive factor for various kinds of pets, including dogs (Foulkes, 1977). However, sugars are not used at all or in limited amounts in foods for cats, because cats seldom respond to sweetness; acids that are used to lower pH in semimoist pet foods are a positive factor for palatability of cat foods. Foulkes (1977) advised that using corn molasses as humectants could improve the palatability of semimoist pet foods upon storage. Addition of corn molasses at 5%–15% of all ingredients could enhance the palatability of semimoist pet foods.

11.4 MOIST PET FOODS

Moist pet foods are usually defined as pet foods having a meat-like character and a moisture content of 50% by weight or higher. The maximum allowable moisture content for moist pet foods is 78% by weight (AAFCO, 2003); the moisture content in most moist pet foods is two to three times of that in semimoist pet foods and up to many times of that in dry pet foods.

Moist pet foods contain separable meat particles with a moisturized appearance; hence, these types of pet foods are received most favorably by pets. Major components of moist pet foods are meats and grains, while minor components are vitamins, vitamin-like materials, colorants, oils, starches, flavoring agents, emulsifiers, and gelling and thickening agents. Compared with other types of pet foods, the recipes of moist pet foods have higher adaptability. Some may contain a mixture of grains and fats from animal sources in contents similar to dry or semimoist pet foods, but more often moist pet foods contain much larger contents of meat, poultry, or fish. As a result, the carbohydrate and fat portions of moist pet foods are typically lower, while the protein concentrations are much higher on a dry matter basis. The higher moisture content of moist pet foods requires thermal processing in sealed containers to obtain commercially sterile products with an extended shelf life. Traditionally, moist pet foods can be divided into two categories: chunks in gravy/jelly (CIG/J) products, wherein meat or meat-like chunks can be individually separated in a liquid; and chunks in loaf (CIL) product, which consists of a mixture of meat meals and cereals (Edley et al., 2003).

11.4.1 Production of Moist Pet Foods

Conventional moist pet foods have approximately 10% protein, 4% fat, 78% moisture, and 8% carbohydrate, with other nutritional supplements making up the rest of the whole proportion (Plant and Aldrich, 2003). A majority of moist pet foods is based on meat; in general, the animal materials used are by-products, i.e., materials in excess of human food requirements, namely, lung, liver, spleen, stomach, ground bone, meat trimmings, chicken trimmings, and fish trimmings. These by-products may be in fresh, frozen, or dried form. Liver is an important ingredient in terms of texture due to its cohesive forming role, which alleviates leaching of fats and gellable proteins from chunks into gravy during retorting. Blood may be required for a heat-coagulating function, and binding of fats for chunk preparation. Care must be exercised that blood has not been denatured to a point that the coagulating properties have been lost. The other ingredients of moist pet foods are defatted soybean flour; fat; and other heat-stable ingredients, which act as binders, such as wheat gluten (WG), egg albumen, starches, cellulose, cottonseed protein isolates, and gelatin. Various minor ingredients such as nutritional supplements, coloring agents, and antioxidants are also commonly added. The overall combinations affect product texture, integrity, and flavor of the products. A sample formula for a moist pet food is given in Table 11.5.

There are two major product formats for moist pet foods: CIG/J and CIL. CIG/J is recognized as a premium product, while CIL is a less expensive product type. Both types of products are packaged in three main container types: cans, trays, and pouches. Cans still remain the main format; trays and pouches, however, are increasingly favored by pet owners today. Different equipment and processes are required for these different products and container formats.

11.4.1.1 Meat Preparation

Meats used in the production of moist pet foods are generally in the form of fresh or frozen meats. Meats are a combination of high-quality offals, such as liver, kidney, and heart, and those undesired for human consumption, such as lung, stomach, intestines, skin, bone, and blood. Meat preparation involves several stages, depending on the expected use. A preprocessing step (only for frozen meats) involves pre-breaking meat into suitable-sized chunks with a typical piece size of 50–100 mm. These chunks are then coarse-ground to a piece size of 10–25 mm. The pre-breaking unit and coarse-grinding unit may be combined into one unit. Since meat emulsion is required in case of CIL, further grinding is required. The general piece size of meat is in the range of 0.5–2 mm (Edley et al., 2003).

TABLE 11.5
Sample Formula for a Moist Pet Food

Ingredient	Beef Chunks (52%) Percent by Weight	Gravy (48%) Percent by Weight
Beef chunks in gravy		
Beef	30	—
Beef lungs	20	—
Pork liver	20	—
Beef spleen	13	—
Defatted soybean flour	7	—
Sugar	—	5
Powdered blood plasma	4.5	—
Modified waxy maize starch	—	4
Animal fat	2	—
Salt	1	0.5
Dicalcium phosphate	1	—
Coloring agent	—	0.5
Vitamins, minerals, and antioxidants	0.5	—
Water	1	90

Source: Baker, G.J. et al., Canned meat and gravy pet food and process, U.S. Patent 4,895,731, January 23, 1990.

11.4.1.2 Gravy Preparation

The gravy used for pet food production is assessed in terms of various properties, such as viscosity, product consistency, pH, flavor, color, and nutrition (such as vitamins and minerals). Gels and thickeners are generally mixed in a vessel with high-speed agitation to ensure that each solid particle is surrounded by water before hydration to avoid lump formation; the formation of lump is indeed related to the viscosity and consistency of gravy. Heat may be required for correct functionality of the gels and thickeners; heat requirement can be satisfied by adding hot water or injecting steam into the vessel. Hot water can be used for makeup, but this can exacerbate powder clumping, so steam injection is sometimes preferred.

11.4.1.3 Preparation of Chunks for CIG/J

Manufactured chunks are commonly used for CIG/J, except for some premium-grade products, which would include diced natural chunks. The technologies used to produce chunks are heat setting for meat proteins and gel setting. Chunk strength is an important property and is directly related to the acceptability of CIG/J. Heat set chunks occur by creation of strong matrix of soluble meat proteins when heated above 80°C with addition of some additives such as blood plasma and wheat flour (Edley et al., 2003).

The first step in preparing chunks is to mix fine meats, powder (wheat flour, sugar etc.) and liquid ingredients (water, gravy, blood, etc.) in a mixer. Then, a fine grinder is used to decrease the sizes of all ingredients to produce an emulsion. The grinding process needs no overheating to avoid pre-coagulation of meat proteins. However, controlling the temperature at least to 85°C after cooking is necessary to ensure that all meat proteins are coagulated and starch from wheat flour is gelatinized (Edley et al., 2003). The cooked products exiting a steam oven may be very elastic to generate soft texture. The products should be air-cooled to at least 50°C–55°C to avoid too soft texture after exiting the steam oven. After cutting the chunks should be cooled again to less than 20°C if they are to be stored for any length of time before being used as an ingredient of the CIG/J products.

11.4.1.4 Preparation of Chunks for CIL

Coarse-ground meats are generally used for this chunk format. Chunks are usually filled into cans or trays. Meat and liquid ingredients are passed into a mixer, which is used to generate a homogenous mix in a short time without damaging the ingredients. Too much energy input may cause unexpected emulsification of fat, which would lead to an undesirable product. The best mixer design is twin intermeshing paddles that give rapid and intimate mixing through a gentle action of slow rotors. Hot water/steam jackets on the mixer can be used to control the mix temperature. The mixed material has to be filled and entered into a sterilizer within 1 h after mixing because it has short shelf life. Before filling and entering the sterilizer, the mix must be screened by a metal detector. For the filling step, a piston is used to suck the product from a reservoir and discharge it into a container. Steam is blown over the top of the filled container to provide heat to the product; this is done so that when the cans cool, they could be vacuum-sealed to prevent microbiological growth (Ranganna, 2000).

11.4.1.5 Sterilization

This is the most critical operation in the manufacture of moist pet foods. The most important objective of thermal sterilization is to achieve commercial sterility. After sealing, moist pet foods in containers are sent to a sterilization unit. A conventional method for sterilization is retorting. To achieve sterility, the products must receive sufficient heat treatment to destroy the heat-resistant spores of *Clostridium botulinum* and heat-resistant spoilage organisms such as *Clostridium stearothermophilus* and other microorganisms, which could lead to spoilage under conditions of normal, non-refrigerated storage in sealed containers. Retorting is normally performed using steam or water heating for a holding time of at least 15 min at 121°C or 3 min at 134°C (Hersom and Hulland, 1980; Ranganna, 2000). The most concerned bacterium is *Clostridium botulinum*, which is destroyed at temperatures above 116°C. The retorting time, temperature, and pressure must be carefully evaluated for each product type, package type, and pack size. After thermal treatment, the products are rapidly cooled to avoid too soft a texture or color change (Edley et al., 2003).

In order to improve the quality of moist pet foods, several interesting alternatives of moist pet foods have been developed; some of them are discussed below.

11.4.1.6 Moist Pet Foods Containing Filamentous Fungal Biomass

To improve the palatability of moist pet foods, alternative moist pet foods containing a filamentous fungal biomass as a source of protein have been developed. The use of the filamentous fungal biomass as a replacement for fresh meat allows the production of a product that has lower price and, at the same time, higher palatability compared with conventional moist pet foods that contain only meat. The class of fungi that is identified as filamentous generally refers to a group of fungi having hyphae or mycelia. Hogan and Gierhart (1989) introduced a moist pet food containing filamentous fungal biomass. The ingredients of this product consist of a filamentous fungal biomass, meat, and several nutritional balancing ingredients such as fat, vitamins, and coloring materials, which are used to improve the appearance and properties of the product. All ingredients are mixed and then heated at a temperature in the range of 48°C–60°C for 45–75 min. This heating step can also be used to pasteurize the ingredients. The mixture is then poured into a container and carried to a retort for sterilization. The retorting temperature is in the range of 118°C–121°C at 15 psi and the retorting time is approximately 65 min. The retorted product is then cooled, rinsed, and allowed to dry.

11.4.1.7 Moist Pet Foods with Sliced Meat Analogue

Premium and "superpremium" grades of products are needed to satisfy the demands of some groups of pet owners. Gifford (2003) patented a premium moist pet food with a sliced meat analogue. To produce these types of moist pet foods, extruded fibrous proteinaceous ingredients are first shredded to proximate-sized chunks. Then, liquid ingredients consisting of water, binders, and red iron oxides are mixed. The binders are used for holding the structure of the analogue together, while the iron oxides serve as an authentic color to the analogue. Subsequently, the mixed liquids, ground meats, and extruded materials are mixed until homogeneity is achieved. The mixture is then filled into cases via a vacuum filler. The filled and sealed cases are retorted at a temperature of 95°C for 60 min. The thermal processing leads to a denatured protein matrix, thereby binding of the extruded materials. After thermal processing and chilled storage, the cases are removed and the analogue material is cut as slices and collected into containers. Vegetable materials may be added in order to provide more nutrition.

Separately, gravy is prepared by blending starches, gum, and water; all the ingredients are then added into the containers. The containers are then sealed and retorted to achieve commercial sterilization.

11.4.2 Some Physicochemical Properties of Moist Pet Foods as Affected by Processing

11.4.2.1 Emulsion Formation

Generally, chunks are produced by forming emulsions of meats and binders through the application of heat and pressure. Binders that are generally used in pet food production include starches, carrageenan, WG, egg albumen, konjac powder, etc. A more heat-stable binder can be applied in a formula to reduce syneresis, which is the separation of water from a product after retorting (Shi and Tang, 2003).

Blending is an important step that affects the formation of emulsions in moist pet foods. A twin-shaft ribbon blender is recommended for this task. A uniform fine size of particles of all raw materials is desired, as this would increase the degree of binding and syneresis of emulsions during subsequent cooking, which is normally performed until the product core temperature reaches 85°C–90°C. The cooking time must be carefully controlled (by controlling the speed of a conveyor belt that transports the product through a cooker, for example) to ensure that the chunks have the expected temperature at the end of the cooking process. The temperature in the cooker should be maintained at about 100°C and the steam pressure should be at about 3 bars to guarantee a good cooking process of emulsions (Dubbelink, 2003).

11.4.2.2 Texture

Texture is an important quality and has a close relationship with the palatability of moist pet foods. Textural properties, including chewiness, hardness, and elasticity of loaf and chunk, significantly affect the mouth feel of moist pet foods. A proper texture measurement can thus help indicate a suitable portion of ingredients to create high-quality moist pet foods. Binders such as guar gums, carrageenan, starches, and caseinate are the important ingredients for developing a desirable texture of a product. However, different binders in a particular formula can produce a significant difference in the textural properties of that product, which in turn influence the acceptability of the product.

Mixing also affects the texture of a product. Excessive mixing may lead to a product that is too sticky and pasty, which is unacceptable for pet foods. In addition, a short holding time between mixing and filling is found to improve the texture of moist pet foods. The speed of a filling machine should be corresponded to the size of a mixer to allow enough time for mixing and complete heating up of the ingredients before the filling step (Heinicke, 2003). Sterilization is also an important step that largely affects the texture of moist pet foods. An excessive heating temperature and time may overcook and produce undesirable textural characteristics of a product.

The effects of added poultry fat and type of binder on the texture of cooked chunks were investigated by Polo et al. (2007). Figure 11.5 illustrates the relationship between the hardness of the chunk and the poultry fat content for a recipe containing spray-dried animal plasma (SDAP) and WG as the binders. The results showed that the hardness of the chunk decreased with an increase in the level of fat addition, indicating that the fat reduced the viscosity of the emulsion. Nevertheless, the hardness of the chunk containing SDAP was greater than that of the chunk containing WG.

11.4.2.3 Flavor

Ingredients and heating condition are important factors influencing the flavor of moist pet foods. The main ingredient affecting the flavor of moist pet foods is fat, because crude fat acts as an oil-soluble flavor carrier. Bacon and chicken fats, for example, have strong aroma characteristics and tastes (Hanna, 1976). The fat flavor is influenced by relative protein tissues and the nature of the fat itself. A suitable amount of fats in moist pet foods is around 3%–6% of a final product. Excess fats in moist pet foods may lead to a problem of nutrition balance and inversely affect the

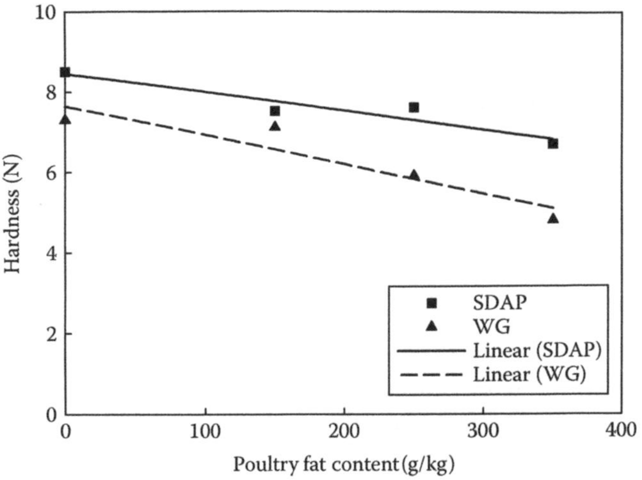

FIGURE 11.5 Relationship between hardness of cooked chunk and poultry fat content, and use of SDAP and WG. (Modified from Polo, J. et al., *Anim. Feed Sci. Technol.*, 133, 309, 2007.)

acceptability of the product. It is noted that the characteristic aromas of fats are more intensely perceived by pets than by humans.

Thermal processing can generate many volatile compounds, some of which are desirable while some are not. Maillard reactions can occur during retorting, especially in the case of pet foods containing high protein and carbohydrate contents; these reactions have a very strong effect on the flavor of moist pet foods. Under high temperature and pressure of retorting, desirable flavor compounds, such as sulfur-containing thiazole and thiophene compounds, can be developed. Heterocyclic aroma-intensive compounds, such as pyrrole, pyridine, and pyrazine, can also be produced. Maillard reactions, however, can give undesirable flavors as "burnt" flavors during retorting; these types of off-flavors can be inhibited by *N*-acetyl-l-cysteine (Shi and Tang, 2003). Some flavor enhancers such as monosodium glutamate can be destroyed during retorting and reduce the flavor enhancement performance (Shi and Tang, 2003).

11.4.2.4 Palatability

In general, the main ingredients of moist pet foods are meat and meat by-products, which largely and positively influence the palatability of the products. The format of raw materials also affects the palatability of moist pet foods. Frozen and dried meat may give lower-palatability pet foods than fresh meat, because some deterioration can occur during freezing or drying. The type of added cereals can also affect the palatability of a product. For example, yellow and white milos are more suitable to be used as ingredients of moist pet foods than red milo because red milo contains higher tannin, which gives a bitter flavor (Heinicke, 2003).

Low moisture content also has a negative effect on the palatability of moist pet foods. Several producers thus try to increase the moisture content of the products to achieve higher palatability. Another important parameter affecting the palatability of moist pet foods is heating. A long heating time generally leads to poor palatability. Heating does not only negatively impact palatability but also reduces the nutritive values of ingredients, especially various vitamins, which are heat-sensitive in nature. An appropriate heating time should thus be evaluated through a heat penetration calculation (Heinicke, 2003).

11.5 OTHER TYPES OF PET FOODS

11.5.1 Treats

Treats are one of the popular manufactured pet foods. Most treats are commercially available as soft-moist products, which have moisture content in the range of 18%–30% by weight. The main ingredients of treats are 45%–70% of grain flour, 5%–15% of meat, and 2%–8% of fat (Mair, 2003b). These pet foods are stabilized against microbial deterioration, especially mold growth, by the inclusion of specific ingredients that control the availability of water in the products; a combination of salts and acids is also used to limit mold growth. Additives such as flavors, coloring agents, and antioxidants are also generally added to increase the palatability of the products.

The preparation of soft-moist treats starts with cooking the ingredients in a thermal cooking system. Examples of a thermal cooking system are a steam-jacketed mixer and an extruder with low shear; the extruder exit temperature should be kept below 100°C to induce a dense cell-free structure of the products. Then, the products are transported to a forming machine, which pumps the cooked ingredients through a die for the final shape creation. The moving rate of products for cutting must be controlled, because moving the products too fast may lead to deformation due to an impact between the products and guards or conveying surfaces (Mair, 2003b). It is indeed recommended that the products be kept for a short time prior to cutting to allow a starch structure to set into a firm shape to prevent deformation and further breakage. Fiber-bearing ingredients can be included at 2%–10% by weight of extruded treats to decrease the breakage level of the products (Collings et al., 1996; Mair, 2003b).

Baked treats are one of the most popular treats, which are available as hard biscuit products. These products are produced by mixing grain flour and water to produce a dough, which is then shaped by rotary molding and stamping. The molded and stamped dough is baked in a relatively cool oven to make hard and crunchy products. Some producers may produce glaze-coated treats to increase the product values (Mair, 2003b).

11.5.2 Frozen Pet Foods

Frozen pet foods are products in the form of raw-ingredient diets. Since the raw materials are not subject to thermal processing, freezing is required to extend the product shelf life. Air-blast freezing and flash freezing are common methods used to freeze pet food items. The freezing rate clearly affects the structure of a product. Slow freezing tends to produce large ice crystals, while rapid freezing yields smaller ice crystals. The size of ice crystals in pet foods may affect the texture and palatability of thawed pet foods. Generally, pet foods can be preserved for several months by freezing; long-term freezing requires a constant temperature of −18°C or less.

11.5.3 Dog Chews

Most dogs enjoy chewing on various things such as cow bones, wood, nylon, and rubber. Edible dog chews have thus been developed to serve the requirements of dogs and dog owners. One of the conventional edible dog chews is the rawhide dog chew. A typical rawhide production process is similar to a rendering leather process, which starts with rendering or tanning animal hides. The hides are then cut and trimmed into desired sizes; a final product shape is formed with multiple layers of hides. After this, it is loaded into a drying system where it is dried and cooled (Barber, 2003).

A critical factor that must be considered during the production of rawhide dog chew is pathogen contamination. To ensure that pathogens such as *Salmonella* are killed, a product is normally heated up to a temperature of 70°C and is held for a period of time. Chiewchan et al. (2007) reported that the initial water activity and drying temperature significantly affected the heat resistance of *Salmonella* Krefeld attached to the rawhide surface. The different initial loads of *Salmonella* on the surface and inside the rawhide required different drying conditions. A good manufacturing practice (GMP) is recommended to control *Salmonella* contamination on rawhide.

Since many indigestible objects given to dogs as a chew could affect dogs' digestive systems, edible dog chews, which are completely digestible and nutritious, have been developed (Axelord, 2000). A mixture of potato starch, water, calcium carbonate, natural vegetable additives, and meat powder is molded under heat at a temperature in the range of 120°C–205°C and a pressure of 1000–2500psi into a desired form such as animal bone. The texture of molded dog chew can be improved by microwave heating; the expansion of moisture within the chew during microwave heating causes the expansion of the product structure (Axelrod, 1993).

11.6 CONCLUDING REMARKS

The manufacture of commercial pet foods is actually very similar to that of human foods. Understanding the effect of the manufacturing process on physicochemical properties of pet foods can help the producers develop and/or improve pet food production processes more effectively. A well-developed process allows the production of a pet food that can satisfy both nutritional needs of pets and emotional needs of pet owners.

REFERENCES

Association of American Feed Control Officials (AAFCO). 2003. *Official Publication*. Oxford, IN: AAFCO.

Axelrod, G.S. Edible dog chew. U.S. Patent 6,126,978. October 3, 2000.

Axelrod, H.R. Dog chew with modifiable texture. U.S. Patent 5,200,212. April 6, 1993.

Baker, G.J., Bansal, A.K., Konieczka, J.L., Kuntz, D.A. Canned meat and gravy pet food and process. U.S. Patent 4,895,731. January 23, 1990.

Balaz, A., Bone, D.P., Shannon, E.L. Method of making a dry-type pet food. U.S. Patent 4,055,681. October 25, 1977.

Barber, T. 2003. Manufacturing rawhides and animal part treats. In *Petfood Technology*, eds. J.L. Kvamme and T.D. Phillips, pp. 400–402. Mount Morris, IL: Watt Publishing.

Barker, D., Burrows, I.E., Buckley, K. Semi moist animal food. U.S. Patent 4,054,674. October 18, 1977.

Bartsch, A.G. Expanded semi-moist pet food. U.S. Patent 4,001,345. March 8, 1977.

Bhattacharya, M., Hanna, M.A. 1988. Effect of lipids on the properties of extruded products. *Journal of Food Science* 53: 1230–1231.

Bone, D.P. Soft dry pet food product and process. U.S. Patent 4,039,689. August 2, 1977.

Bone, D.P., Shannon, E.L. Process for making a dry pet food having a hard component and a soft component. U.S. Patent 4,006,266. February 1, 1977a.

Bone, D.P., Shannon, E.L. Method of making a dry pet food having a marbled meat-like texture. U.S. Patent 4,029,823. June 14, 1977b.

Brera, C., Debegnach, F., Grossi, S., Miraglia, M. 2004. Effect of industrial processing on the distribution of fumonisin B_1 in dry milling corn. *Journal of Food Protection* 67: 1261–1266.

Brera, C., Catano, C., De Santis, B., Debegnach, F., De Giacomo, M., Pannunzi, E., Miraglia, M. 2006. Effect of industrial processing on the distribution of aflatoxins and zearalenone in corn-milling fractions. *Journal of Agricultural and Food Chemistry* 54: 5014–5019.

Briggs, J.L., Maier, D.E., Watkins, B.A., Behnke, K.C. 1999. Effect of ingredients and processing parameters on pellet quality. *Poultry Science* 78: 1464–1471.

Brown, A.V., Karrasch, R.J. Monosaccharide-containing dry pet food having yieldable elastic structure. U.S. Patent 4,162,336. July 24, 1979.

Burkwall, M.P., Leyh, J.C. Jr., Reagan, J.G. Meatless marbled, semi-moist pet food. U.S. Patent 3,984,576. October 5, 1976.

Cavalcanti, W.B., Behnke, K.C. 2005. Effect of composition of feed model systems on pellet quality: A mixture experimental approach. I. *Cereal Chemistry* 82: 455–461.

Cheftel, J.C. 1986. Nutritional effects of extrusion-cooking. *Food Chemistry* 20: 263–283.

Chiewchan, N., Pakdee, W., Devahastin, S. 2007. Effect of water activity on thermal resistance of *Salmonella* Krefeld in liquid medium and on rawhide surface. *International Journal of Food Microbiology* 144: 43–49.

Collings, G.F., Stout, N.P., Cowell, C.S., Plas, S.J. Extended dog treat food product having improved resistance to breakage. U.S. Patent 5,501,868. March 26, 1996.

Corbin, J. 2003. The history of petfood. In *Petfood Technology*, eds. J.L. Kvamme and T.D. Phillips, pp. 514–516. Mount Morris, IL: Watt Publishing.

Dubbelink, O. 2003. Steam tunnel cooking. In *Petfood Technology*, eds. J.L. Kvamme and T.D. Phillips, pp. 389–391. Mount Morris, IL: Watt Publishing.

Dust, J.M., Gama, A.M., Flickinger, E.A., Burkhalter, T.M., Merchen, N.R., Fahey, G.C. Jr. 2004. Extrusion conditions affect chemical composition and in vitro digestion of select food ingredients. *Journal of Agricultural and Food Chemistry* 52: 2989–2996.

Dzanis, D.A. 2003. Petfood types, quality assessment and feeding management. In *Petfood Technology*, eds. J.L. Kvamme and T.D. Phillips, pp. 68–73. Mount Morris, IL: Watt Publishing.

Edley, D., Moss, J., Plant, T.A. 2003. Wet petfood manufacture. In *Petfood Technology*, eds. J.L. Kvamme and T.D. Phillips, pp. 382–388. Mount Morris, IL: Watt Publishing.

Ernst, T.J., Lepp, R.S., Jackson, J.R. Inhibition of *Tyrophagus putrescentiae* in pet food products. U.S. Patent 2002/0172740 A1. November 21, 2002.

Fernandez, S.R., Parsons, C.M. 1996. Bioavailability of digestible lysine in heat-damaged soybean meal for chick growth. *Poultry Science* 75: 224–231.

Foulkes, P.H. Semi-moist pet food containing corn molasses. U.S. Patent 4,018,909. April 19, 1977.

Fritz-Jung, C., Singh, B., Bhatnagar, S., Woodward, G.J., Kettinger, K.L., Speck, D.R., Stoll, J.A., Woerz, S.E., Zhang, P. Process for dry stable intermediate pet food composition. U.S. Patent 6,270,820 B1. August 7, 2001.

Gifford, C. Canned pet food with sliced meat analogue. U.S. Patent 2006/0210675 A1. November 18, 2003.

Gilpin, A.S., Herrman, T.J., Behnke, K.C., Fairchild, F.J. 2002. Feed moisture, retention time, and steam as quality and energy utilization determinants in the pelleting process. *Applied Engineering in Agriculture* 18: 331–338.

Hanna, K.L. Canned pet food. U.S. Patent 3,946,123. March 23, 1976.

Harper, J.M. 1979. Food extrusion. *Critical Reviews in Food Science and Nutrition* 11: 155–215.

Heinicke, H.R. 2003. Factors affecting the palatability of canned and semi-moist petfoods. In *Petfood Technology*, eds. J.L. Kvamme and T.D. Phillips, pp. 183–186. Mount Morris, IL: Watt Publishing.

Hemmingsen, A.K.T., Stevik, A.M., Claussen, I.C., Lundblad, K.K., Prestløkken, E., Sorensen, M., Eikevik, T.M. 2008. Water adsorption in feed ingredients for animal pellets at different temperatures, particle size, and ingredient combinations. *Drying Technology* 26: 738–748.

Henderson, G.A., Thompson, W.J.C., Thatcher, J.T. Continuous gelatinization process. U.S. Patent 4,256,771. March 17, 1981.

Hersom, A.C., Hulland, E.D. 1980. *Canned Food: Thermal Processing and Microbiology.* Edinburgh, U.K.: Churchill Livingstone.

Hogan, W.C., Gierhart, D.O. High moisture animal food product containing a filamentous fungal biomass. U.S. Patent 4,800,093. January 24, 1989.

Holm, J., Bjorck, I., Asp, N.G., Sjoberg, L.B., Lundquist, I. 1985. Starch availability in vitro and in vivo after flaking, steam-cooking and popping of wheat. *Journal of Cereal Science* 3: 193–206.

Hood, L.L. Color stabilized product process. U.S. Patent 4,001,446. January 4, 1977.

Hood, L.L. Color-stabilized semi-moist food and process. U.S. Patent 4,089,983. May 16, 1978.

Ilo, S., Schoenlechner, R., Berghofe, E. 2000. Role of lipids in the extrusion cooking processes. *Grasas y Aceites* 51: 97–110.

Kheterpaul, N., Chauhan, B.M. 1991. Effect of natural fermentation on phytate and polyphenolic content and in-vitro digestibility of starch and protein of pearl millet (*Pennisetum typhoideum*). *Journal of the Science of Food and Agriculture* 55: 189–195.

Kokini, J.L. 1991. Physicochemical changes and rheological properties of starch during extrusion (a review). *Biotechnology Progress* 7: 251–266.

Lin, S., Hsieh, F., Huff, H.E. 1997. Effects of lipids and processing conditions on degree of starch gelatinization of extruded dry pet food. *LWT-Food Science and Technology* 30: 754–761.

Mair, C. 2003a. Conditioning. In *Petfood Technology*, eds. J.L. Kvamme and T.D. Phillips, pp. 342–346. Mount Morris, IL: Watt Publishing.

Mair, C. 2003b. Petfood treat production: General trends. In *Petfood Technology*, eds. J.L. Kvamme and T.D. Phillips, pp. 392–395. Mount Morris, IL: Watt Publishing.

Mathew, J.M., Hoseney, R.C., Faubion, J.M. 1999a. Effects of corn sample, mill type, and particle size on corn curl and pet food extrudates. *Cereal Chemistry* 76: 621–624.

Mathew, J.M., Hoseney, R.C., Faubion, J.M. 1999b. Effect of corn moisture on the properties of pet food extrudates. *Cereal Chemistry* 76: 953–956.

Mavromichalis, I., Baker, D.H. 2000. Effects of pelleting and storage of a complex nursery pig diet on lysine bioavailability. *Journal of Animal Science* 78: 341–347.

Moritz, J.S., Wilson, K.J., Cramer, K.R., Beyer, R.S., McKinney, L.J., Cavalcanti, W.B., Mo, X. 2002. Effect of formulation density, moisture, and surfactant on feed manufacturing, pellet quality, and broiler performance. *Journal of Applied Poultry Research* 11: 155–163.

Nir, I., Hillel, R., Shefet, G., Nitsan, Z. 1994. Effect of grain particle size on performance. 2. Grain texture interactions. *Poultry Science* 73: 781–791.

Overland, M., Romarheim, O.H., Ahlstrom, O., Storebakken, T., Skrede, A. 2007. Technical quality of dog food and salmon feed containing different bacterial protein sources and processed by different extrusion conditions. *Animal Feed Science and Technology* 134: 124–139.

Parthasarathy, M. Increased density pet food product and method of production. U.S. Patent 2004/0076715 A1. April 22, 2004.

Plant, T.A., Aldrich, G. 2003. Wet petfood formulation. In *Petfood Technology*, eds. J.L. Kvamme and T.D. Phillips, pp. 144–147. Mount Morris, IL: Watt Publishing.

Poirier, D. 2003. Drying and cooling properties. In *Petfood Technology*, eds. J.L. Kvamme and T.D. Phillips, pp. 366–369. Mount Morris, IL: Watt Publishing.

Polo, J., Rodriguez, C., Rodenas, J., Morera, S., Saborido, N. 2007. Use of spray-dried animal plasma in canned chunk recipes containing excess of added water or poultry fat. *Animal Feed Science and Technology* 133: 309–319.

Priegnitz, R.D. Semi-moist pet food product and process. U.S. Patent 4,228,195. October 14, 1980.

Ranganna, S. 2000. *Handbook of Canning and Aseptic Packaging*. New Delhi, India: Tata McGraw-Hill.

Rao, S.K., Artz, W.E. 1989. Effect of extrusion on lipid oxidation. *Journal of Food Science* 54: 1580–1583.

Rokey, G. 2003. Semi-moist/semi-expanded petfoods. In *Petfood Technology*, eds. J.L. Kvamme and T.D. Phillips, pp. 376–379. Mount Morris, IL: Watt Publishing.

Rossen, J.L., Miller, R.C. 1973. Food extrusion. *Food Technology* 27: 46–53.

Serna-Saldivar, S.O., Clegg, C., Rooney, L.W. 1994. Effects of parboiling and decortication on the nutritional value of sorghum (*Sorghum bicolor* L. Moench) and pearl millet (*Pennisetum glaucum* L.). *Journal of Cereal Science* 19: 83–89.

Shi, Z., Tang, G. 2003. Development of palatants for canned petfoods. In *Petfood Technology*, eds. J.L. Kvamme and T.D. Phillips, pp. 180–182. Mount Morris, IL: Watt Publishing.

Svihus, B., Klovstad, K.H., Perez, V., Zimonja, O., Sahlström, S., Schüller, R.B., Jeksrud, W.K., Prestlokken, E. 2004. Physical and nutritional effects of pelleting of broiler chicken diets made from wheat ground to different coarsenesses by the use of roller mill and hammer mill. *Animal Feed Science and Technology* 117: 281–293.

Thomas, M., van Vliet, T., van Der Poel, A.F.B. 1998. Physical quality of pelleted animal feed 3. Contribution of feedstuff components. *Animal Feed Science and Technology* 70: 59–78.

Tran, Q.D., Hendriks, W.H., van Der Poel, A.F.B. 2008. Effects of extrusion processing on nutrients in dry pet food. *Journal of the Science of Food and Agriculture* 88: 1487–1493.

Trivedi, N., Benning, J. 2003. Palatability keys. In *Petfood Technology*, eds. J.L. Kvamme and T.D. Phillips, pp. 178–179. Mount Morris, IL: Watt Publishing.

Wachiraphansakul, S., Devahastin, S. 2007. Drying kinetics and quality of okara dried in a jet spouted bed of sorbent particles. *LWT-Food Science and Technology* 40: 207–219.

Zhang, Y., Parsons, C.M. 1996. Effects of overprocessing on the nutritional quality of peanut meal. *Poultry Science* 75: 514–518.

Index

D

E

F

G

H

I

L

M

R

S

T

U

V

W

X

Y

Milton Keynes UK
Ingram Content Group UK Ltd.
UKHW020317111024
449327UK00040B/1360

9 780367 383732